Air Conditioning

and Refrigeration

WILLIAM H. SEVERNS, M.S.

*Late Member American Society of Heating
and Air-Conditioning Engineers
Professor of Mechanical Engineering
University of Illinois*

and

JULIAN R. FELLOWS, M.S.

*Member American Society of Heating
and Air-Conditioning Engineers
Member American Society of Mechanical Engineers
Member American Society of Refrigerating Engineers
Member American Society for Engineering Education
Professor of Mechanical Engineering
University of Illinois*

New York

Air Conditioning and Refrigeration

John Wiley & Sons, Inc.

London

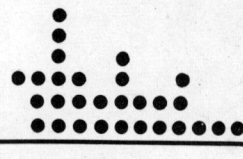

Authorized reprint of the edition published by John Wiley & Sons, Inc., New York and London.

Copyright © 1958 John Wiley & Sons, Inc.

All Rights Reserved.

No part of this book may be reproduced in any form without the written permission of John Wiley & Sons, Inc.

Library of Congress Catalog Card Number : 58-7908

Printed in Japan
By TOPPAN PRINTING COMPANY, LTD.

Preface

Air Conditioning and Refrigeration is written principally as a textbook for undergraduate courses in mechanical engineering and architecture. It will, however, be of considerable interest as a reference volume to those engineers practicing in this field.

The coverage is broad to the extent that instructors will find more than adequate treatment for the general three-semester hour courses. Parts of the text can be omitted. Portions of Chapters 1 and 6 are not necessary for students who have already covered the subjects of thermodynamics and fluid mechanics. Chapter 5 has been written in such a way that duplication of course work in combustion is avoided, with the basic information needed for fuel selection included. If the curriculum provides for a separate course in refrigera h, Chapter 17 may be omitted. Chapter 21, which describes devices employed in the control of air-conditioning apparatus, may be excluded if there is not sufficient time available. Problems dealing with the control of specific types of plants are discussed in various chapters. Some of the chapters can be covered in a single assignment if the instructor selects the type of equipment he wishes to emphasize and briefs the student on the material and problems he wants read and solved.

Examples of typical calculations have been included throughout the book wherever it was felt that they would add to the clarity of the presentation. Pictures of commercial equipment have been used only to illustrate general types. A limited amount of tabular data pertaining to specific models has been included to make it possible for the instructor to assign problems involving the selection of units of the proper size. An extensive index, designed to increase the usefulness of the volume both as a textbook and a reference, is included at the end.

The book is based on the second edition of *Heating, Ventilating, and Air Conditioning Fundamentals* by the late Professor Severns and the undersigned. We began writing this book as a third edition of the above title, but as the work approached the final manuscript stage it became apparent that we had really created a new book, serving a slightly different purpose. It will be apparent to readers that the new book emphasizes air conditioning and refrigeration, inasmuch as heating is now regarded as winter air conditioning and the coverage of refrigeration in the new book is quite thorough. Professor Severns worked on the book until his untimely death in 1956, and is responsible for much of its material.

JULIAN R. FELLOWS

Urbana, Illinois
January 1958

Contents

ABBREVIATIONS, SYMBOLS, AND CONVERSION EQUATIONS

Chapter

1. DEFINITIONS, LAWS, AND PROPERTIES OF WATER AND STEAM   *1*
2. PSYCHROMETRIC PROPERTIES OF AIR—HUMIDIFICATION AND DEHUMIDIFICATION   *16*
3. FACTORS AFFECTING HUMAN COMFORT   *44*
4. HEAT TRANSMISSION, VAPOR TRANSMISSION, AND HEAT LOSSES FROM BUILDINGS   *57*
5. FUEL SELECTION AND UTILIZATION   *95*
6. FLUID FLOW AND PRESSURE LOSSES IN DUCTS AND PIPES   *114*
7. HEATING WITH WARM-AIR FURNACES   *135*
8. RADIATORS, CONVECTORS, AND BASEBOARD HEATING UNITS   *154*
9. HEATING BOILERS AND APPURTENANCES   *169*

10. Pipe, Tubing, Fittings, Coverings, and Piping Details   188
11. Heating with Steam   211
12. Heating with Hot Water   234
13. Panel Heating, Snow Melting, and Panel Cooling   262
14. Air Conveying and Distribution, Fans, Duct Design, and Diffusion   279
15. Ventilation and Air Purification   324
16. Heating with Central Fan-Coil Systems and Unit Heaters   349
17. Mechanical Refrigeration—Applications to Cooling and Heating   365
18. Estimation of Cooling Loads   427
19. Apparatus for Producing Comfort in Summer   460
20. All-Year Air-Conditioning Methods and Equipment   490
21. Automatic Controls for Air-Conditioning Systems   517

Appendix

Table A.1.   Areas and Circumferences of Circles   537
Table A.2.   Graphical Symbols for Drawings   538

Index   545

Abbreviations, Symbols, and Conversion Equations

Abbreviations. The abbreviations which will be used in this book are taken from the Heating Ventilating Air-Conditioning Guide published annually by the American Society of Heating and Air-Conditioning Engineers (ASHAE).

Absolute	abs
Air horsepower	air hp
Average	avg
Barometer	bar.
Brake horsepower	bhp
Brake horsepower-hour	bhp-hr
British thermal unit	Btu
British thermal units per hour	Btuh
Cubic feet per minute	cfm
Cubic feet per second	cfs
Cubic foot	cu ft
Decibel	db
Degree	deg
Degree, centigrade	C
Degree, Fahrenheit	F
Feet per minute	fpm
Feet per second	fps
Foot	ft
Foot-pound	ft-lb
Gallon	gal

ix

ABBREVIATIONS, SYMBOLS, AND CONVERSION EQUATIONS

Gallons per minute..gpm
Horsepower..hp
Horsepower-hour...hp-hr
Hour..hr
Inch..in.
Inch-pound..in.-lb
Kilowatt..kw
Miles per hour..mph
Minute..min
Ounce...oz
Pound...lb
Pounds per square inch..psi
Pounds per square inch, absolute....................................psia
Pounds per square inch, gage..psig
Revolutions per minute..rpm
Revolutions per second..rps
Second..sec
Specific gravity..sp gr
Specific heat...sp ht
Square foot...sq ft
Square inch...sq in.

Symbols. Symbols which are employed repeatedly are defined here. Symbols used in one discussion only will be defined in text.

A, area, sq ft.

C, conductance of any material in the thickness stated, Btuh per sq ft per deg F. Also represents a constant.

C_a, conductance of an air space.

C_e, coefficient of linear expansion (Table 1.4).

c_p, specific heat of a gas when heated or cooled at constant pressure, Btu per lb per deg F. [c_{pa} when applied to air; c_{pw}, water; c_{ps}, superheated steam (used as 0.45 in air-conditioning work).]

c_v, specific heat of a gas when heated or cooled at constant volume.

COP, coefficient of performance (Art. 17.4).

D, internal diameter of a pipe or duct, ft.

d, internal diameter of a pipe or duct, in.

d_a, density of air (not dry air), lb per cu ft.   (d_{da} when applied to dry air.)

d_c, circular equivalent (Art. 14.19).

D_e, equivalent diameter (Art. 6.7).

d_s, density of steam, lb per cu ft.

d_w, density of water, lb per cu ft.

dbt, dry-bulb temperature.

E, efficiency, per cent. (E_b, boiler efficiency; E_m, mechanical efficiency; E_h, humidifying efficiency; E_s, static efficiency.)

ABBREVIATIONS, SYMBOLS, AND CONVERSION EQUATIONS

e, emissivity of a surface (Art. 4.7). Also, absolute roughness (Art. 6.6).

e_c, combined emissivity of two surfaces (Art. 4.7).

E_e, the number of standard elbows equivalent to a fitting as regards friction loss.

E_x, expansion, or change of linear length, in.

f, friction factor (Art. 6.6).

f_i, conductance of an inside air film, Btuh per sq ft per deg F.

f_o, conductance of an outside air film.

g, acceleration of gravity, ft per sec^2.

I_D, solar radiation received on a surface (Art. 18.4).

H, heat transmitted (or transferred), Btuh. (H_c by conductance; H_r by radiation; H_i by infiltration.)

H_t, total heat gain from ventilation air, Btuh.

h, enthalpy of 1 lb of dry air and its contained water vapor when it is not saturated, Btu per lb of dry air. (h_i, inside; h_o outside; h_1, initial; h_2, final; h_a, dry air; h_s, saturated air.)

h_d, enthalpy of refrigerant vapor as discharged from a compressor, Btu per lb.

h_f, enthalpy of liquid water, Btu per lb. (h_{fc}, condensate; h_{ff}, feedwater.)

h_{fg}, enthalpy of vaporization at any specific vapor pressure, Btu per lb.

h_{fg}', enthalpy of vaporization of saturated vapor at the wet-bulb temperature, Btu per lb.

h_g, enthalpy of saturated steam, Btu per lb.

h_{su}, enthalpy of superheated steam, Btu per lb.

K, kinematic viscosity (Art. 6.6). Also a constant in equation 1.2.

k, thermal conductivity, Btuh per sq ft per deg F per in. of thickness.

L, length, ft.

L_e, equivalent length, ft.

M, mass handled, lb per hr. (M_a, air; M_w, water.) Also, mass in lb in $PV = MRT$.

M_m, mass of 1 mole, lb.

m, mass handled, lb per min.

P, absolute total pressure, lb per sq ft ($144 \times p$).

p, absolute total pressure, psia.

p', saturation water vapor pressure at the wet-bulb temperature, psia.

p_a, partial pressure of the dry air in a mixture of dry air and water vapor, psia.

P_e, permeance (Art. 4.27).

PF, performance factor (Art. 19.14).

P_f, pressure lost due to friction, ft of fluid. (p_f when in in. of water.)

p_s, saturation water vapor pressure at the dry-bulb temperature, psia.

P_{sh}, pressure lost due to shock (Art. 6.9), ft of fluid. (p_{sh} when in in. of water.)

P_t, total head developed by a fan or pump, ft of fluid. (p_t when in in. of water.)

p_{st}, static pressure, in. of water.

xii ABBREVIATIONS, SYMBOLS, AND CONVERSION EQUATIONS

P_v, velocity pressure, ft of fluid. (p_v when in in. of water.)

p_w, partial pressure of water vapor (Art. 2.3), psia.

Q, volume flowing, cu ft per hr. (q_i, infiltration volume, cu ft per ft of crack per hr.)

q, volume flowing, cfm.

R, gas constant (Art. 1.12). Also, resistance in heat transmission and the ratio of the center-line radius of an elbow to the pipe (or duct) diameter.

R_a, radiation factor (Art. 8.4).

R_c, compression ratio.

R_h, hydraulic radius (Art. 6.6) (equation 6.2).

R_N, Reynolds number (Art. 6.6).

T, absolute temperature, deg F ($t + 459.6$). (T_c, cold; T_h, hot.)

t, dry-bulb temperature, deg F. Also, room temperature at the 5-ft level, deg F. (t_a, attic; t_c, cold; t_g, ground; t_h, hot; t_i, inlet; t_o, outlet or outside; t_p, proper level; t_1, initial; t_2, final.)

t', wet-bulb temperature, deg F.

t_d, dew-point temperature, deg F.

t_m, temperature of heating medium, deg F (Arts. 8.4 and 8.9).

t_n, temperature in a space *not* conditioned, deg F.

t_s, temperature of saturated steam, deg F.

U, over-all coefficient of heat transmission, Btuh per sq ft per deg F. (U_c, ceiling; U_r, roof; U_{cr}, combination ceiling and roof.)

V, velocity, fpm. Also, volume in cu ft in $PV = MRT$ and in $\dfrac{P_1 V_1}{T_1} = \dfrac{P_2 V_2}{T_2}$.

v, velocity, fps. (v_m, main; v_b, branch.)

v_g, specific volume of dry saturated steam, cu ft (Tables 1.6 and 1.7).

V_P, piston displacement, cfm (Art. 17.12).

V_{sp}, specific volume, cu ft per lb (any fluid except dry saturated steam).

w, humidity ratio for any specific combination of dry-bulb and wet-bulb temperature, lb per lb of dry air. (w_g when in grains; w_1 or w_{g1}, initial; w_2 or w_{g2}, final.)

w', humidity ratio, lb per lb dry air when saturated at the wet-bulb temperature. (wg' when in grains.)

w_s, humidity ratio of saturated air, grains per lb of dry air. w_{sg} when in grains.)

wbt, wet-bulb temperature.

x, thickness of material, in. Also, quality of wet steam (Art. 1.17).

Conversion Equations

Heat, Power, and Work
  1 Btu = 778 ft-lb
  Latent heat of ice = 143.4 Btu per lb

ABBREVIATIONS, SYMBOLS, AND CONVERSION EQUATIONS

    1 ton refrigeration = 12,000 Btuh or 200 Btu per min
    1 kw = 1.341 hp
    1 hp = 0.7455 kw or 33,000 ft-lb per min
    1 kwhr = 3413 Btu
    1 hp-hr = 2545 Btu
    1 boiler hp = 33,475 Btuh

Weight and Volume
    1 gal (U.S.) = 231 cu in. or 0.1337 cu ft
    1 gal (British Imperial) = 277.42 cu in.
    1 cu ft = 7.481 gal or 1728 cu in.
    1 cu ft water at 60 F = 62.37 lb. (Other values for various temperatures may be obtained from specific volume data in Table 1.7.)
    1 gal water at 60 F = 8.338 lb
    1 gal water at 212 F = 7.998 lb
    1 lb (avoirdupois) = 16 oz or 7000 grains
    1 short ton = 2000 lb

Pressure
    1 psi = 144 lb per sq ft or 2.0422 in. of mercury at 62 F or 2.309 ft of water at 62 F
    1 oz per sq in. = 0.0625 psi or 1.732 in. of water at 62 F
    1 standard atmosphere = 14.696 psia or 29.921 in. of mercury at 32 F or 30.01 in. of mercury at 62 F
    1 ft water at 62 F = 0.4330 psi or 62.37 lb per sq ft
    1 in. of mercury at 62 F = 0.490 psi or 13.57 in. of water at 62 F

Viscosity
    viscosity in lb/(ft $\times$ sec) = viscosity in centipoises $\times$ 0.000672

1

Definitions, Laws, and Properties of Water and Steam

1.1. Definitions. The terms used throughout this book are in general the same as those in the ASHAE Guide. Only such terms as are employed repeatedly in this volume are defined here; those pertaining to special equipment or processes or equations will be explained in the text. Information about any expressions not given here may be obtained by referring to the index.

Air conditioning involves the control of the temperature, humidity (moisture content), and motion of the air in an enclosure. Air conditioners usually incorporate filters for removing lint and dust and may include equipment for services such as removing odors or killing bacteria.

Winter air conditioning always involves heating and may include humidification (the addition of water vapor).

Summer air conditioning always involves cooling and generally includes dehumidification (the removal of water vapor).

Atmospheric pressure is measured by a barometer, Fig. 1.1, and may be expressed either in inches of mercury or in pounds per square inch. Standard atmospheric pressure is 29.921 in. of mercury.

Gage pressure is the difference between the pressure of a fluid in a container such as a pipe or tank and that of the atmosphere. Gage

1

pressures may be either positive or negative. A negative gage pressure is usually called a vacuum.

Absolute pressure is the algebraic sum of the atmospheric (or barometric) pressure and the gage pressure. Both pressures must be expressed in the same units before the addition is made.

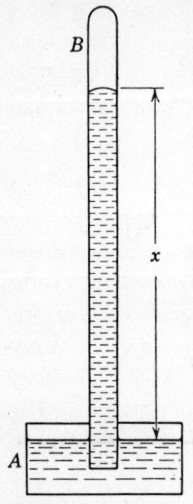

Fig. 1.1. Mercurial barometer.

Example. A gage on a steam boiler reads 5.5 psig when a barometer in the same location reads 29.6 in. of mercury. The temperature of the mercury column is 70 F. Find the absolute pressure in pounds per square inch absolute.

Solution. From Table 1.1 the density of mercury at 70 F is found to be 0.49 lb per cu in. The atmospheric pressure equals $0.49 \times 29.6 = 14.5$ psi and the absolute pressure equals $14.5 + 5.5 = 20$ psia.

Temperature is a measure of heat intensity or the ability of a body to transmit heat to a cooler body. Temperature measurements may be made by thermometers or thermocouples calibrated in terms of either degrees Fahrenheit or degrees centigrade. Temperatures, in deg F, will be used in all discussions and problems in this book. Temperature in deg F is $\frac{9}{5} \times$ temperature in deg C $+ 32$.

Absolute temperature is reckoned from the datum level where energy is lacking in a material. The volume of a perfect gas becomes zero at the absolute zero of temperature. Absolute temperature, in deg F, often referred to as temperature in degrees Rankine, is ordinary temperature in deg F $+ 459.6$.

TABLE 1.1. **Density of Mercury at Various Temperatures**

Temperature, deg F	Weight, lb per cu in.	Temperature, deg F	Weight, lb per cu in.
15	0.4920	110	0.4880
32	0.4912	140	0.4860
50	0.4910	160	0.4850
70	0.4900	180	0.4840
90	0.4890	200	0.4835

Note: Barometric pressure is often expressed in inches of mercury at 32 F regardless of the actual temperature of the column.

1.2. Force, Mass, and Weight. A *force*, measured in pounds (lb_f), represents the effect produced when it changes or tends to change the motion of a body upon which it acts. The pound is a unit of either

DEFINITIONS, LAWS, AND PROPERTIES OF WATER 3

force (lb$_f$) or mass (lb$_m$) and should be properly designated where necessary. The pound force (lb$_f$) is capable of giving an acceleration of 32.174 ft per sec^2 to a pound mass. *Mass* (lb$_m$) is a measure of the inertia or the resistance of matter to acceleration. *Weight* is a force (lb$_f$) which varies with geographical location as it is due to the gravitational attraction for a mass. Although there is a difference between pounds mass and pounds weight, the variation is not great and weight and mass will be considered as equivalent in this work.

1.3. Energy, Work, and Power. The ability to produce an effect is *energy;* it may be either stored or in a state of transition. Stored energy includes atomic, chemical, internal, kinetic, and potential forms. *Heat* is energy in transition due to thermal potential (temperature difference) between bodies, portions of bodies, or various media. Energy can neither be created nor destroyed but its form may be changed under proper conditions. The common unit of heat energy is the *British thermal unit* (Btu). That of mechanical energy is the *foot-pound* which represents the ability to lift a weight of 1 lb through a vertical distance of 1 ft.

Work results when a force acting through a distance overcomes a resistance. The unit of work is the foot-pound. Work is independent of the time required for its accomplishment. *Power* defines the rate at which work is done.

1.4. Heat. The motions of the molecules of a material increase when heat energy flows into it and decrease as it is cooled. At the absolute zero of temperature a body has no molecular motion. Heat energy may be derived from chemical action, the combustion of fuels, friction, nuclear fissions, and the resistance offered to electrical flow in a closed circuit.

Heat which produces a change of temperature in a body is known as *sensible heat;* that which is necessary to produce a change of state of a material at a constant temperature is known as *latent heat*. Latent heat has two forms: *heat of fusion* when a material passes from a solid to a liquid condition at a constant melting temperature, and *heat of evaporation* when passage from a liquid to a vaporous state occurs under the existent conditions.

The amount of heat energy possessed by different substances at a given temperature varies with the material, its quantity, and its temperature. The basis of reckoning may be above any fixed reference point. The change in the amount of sensible heat of a body is dependent upon the weight of the substance involved, its thermal capacity (specific heat), and the temperature change. Heats of fusion and also heats of evaporation vary with the material and with the temperatures and pressures involved.

In this book heat measurements are expressed in terms of the Btu, which is $\frac{1}{180}$ of the heat energy necessary to raise the temperature of 1 lb of pure water from 32 F to 212 F when held under an absolute pressure of 14.696 psi.

1.5. Specific Heat. The thermal capacity or specific heat of a substance is the amount of heat added to or taken from a unit of weight of a material to produce a change of 1 deg in its temperature. In this work specific heats will be used as Btu per lb per deg F.

TABLE 1.2. Mean Specific Heats
Solids, Liquids, and Gases

Material	Temperature Range, Deg F	Specific Heat	Material	Temperature Range, Deg F	Specific Heat
Air					
Constant pressure	32–212	0.2401	Glass	32–212	0.163
Constant volume	32–212	0.171	Gypsum	32–212	0.259
Aluminum	61–579	0.225	Mercury	32–212	0.033
Asbestos	32–212	0.200	Oak	32–212	0.570
Brass	68–212	0.104	Petroleum	32–212	0.500
Brick	32–212	0.220	Pine	32–212	0.670
Cast iron	68–212	0.119	Soil	32–212	0.440
Cinders	32–212	0.180	Steel	32–212	0.117
Concrete	32–212	0.156	Stone	68–208	0.208
Copper	59–540	0.095	Water	32–212	1.00
Cork	32–212	0.485	Zinc	68–212	0.093

All specific heats vary with a change of temperature, and distinctions must be made between the true or instantaneous and the mean specific heat. The instantaneous specific heat is the amount of heat that must be added or abstracted at any definite temperature to produce a temperature change of 1 deg per unit weight of the material. The mean specific heat, over a given temperature range, is the average amount of heat required to produce a change of 1 deg of temperature per unit weight of the substance.

Specific heats of gases and vapors are also dependent upon the conditions of maintenance, i.e., either constant pressure or constant volume. The specific heat at constant pressure c_p is greater than the specific heat at constant volume c_v. When volume changes are produced at constant pressure, by temperature changes, work is done, and the heat equivalent of the work is reflected in the specific heat c_p at constant pressure.

DEFINITIONS, LAWS, AND PROPERTIES OF WATER

The specific heats of a few materials for various temperature ranges are given by the data of Table 1.2.

1.6. Enthalpy. The total energies of fluids such as water, steam, and others are somewhat difficult to reckon from a base of zero of temperature absolute. Most calculations involve a change of heat energy; therefore, a property known as enthalpy, based on some other suitable temperature (32 F for water and steam), is satisfactory.

1.7. Dissemination of Heat. The passage of heat from regions of higher to those of lower temperature may be either by conduction, by convection, by radiation, or by combinations of the foregoing ways.

1.8. Conduction. The transfer of heat by conduction is accomplished by heat flow from one particle to another particle of a material and occurs in any direction where a lower temperature is being maintained. An example is furnished by hot and cold bodies in contact with each other. The rate of flow varies with the character of the material, the distance through which the flow takes place, and the temperature difference maintained. Some materials permit the rapid transmission of heat by conduction. Other materials, which serve as insulators, resist the flow of heat by conduction.

The total heat transmitted by conduction per hour is expressed as

$$H = \frac{k}{x}(t_h - t_c)A \tag{1.1}$$

Inspection of equation 1.1 reveals that the thicker the material, of a given conductivity, the smaller will be the amount of heat conducted at a unit area with a fixed temperature difference. Values of k are given in Table 1.3 for several materials.

TABLE 1.3.* **Approximate Unit Thermal Conductivities**[†]

Material	k	Material	k
Air	0.168	Lead	240.0
Aluminum	1416.0	Nickel	408.0
Brass (70–30)	720.0	Soil	2.4–12.0
Cast iron	336.0	Steel, mild	312.0
Copper	2640.0	Water, liquid	4.08
Glass	3.6–7.32		

* From "Heating Ventilating Air-Conditioning Guide 1957." Used by permission.

† Thermal conductivities depend to some extent on temperature. The above magnitudes are approximate only.

Note: Some tables give conductivity values in Btuh per sq ft per deg F per *ft* instead of per *in.* as in this table.

1.9. Convection. A transfer of heat takes place as a fluid (liquid or gaseous) flows over the surface of a colder or a hotter body. Heat transfer by conduction takes place between the molecules of the moving agent and the material having the surfaces where the two are in contact. The transfer of heat by the carrying medium then becomes the process of convection. The amount of heat transferred per unit of time is affected by the velocity and nature of the moving medium, the area and form of the surfaces rubbed, and the difference between the temperature of the moving fluid and that of the surfaces rubbed.

1.10. Radiation. Heat dispersed by radiation passes through space in straight lines, following the laws of the travel of light. Air and other gases, unless they contain moisture or dust particles, are not warmed directly by radiant heat. Some materials are absorbers, and others are reflectors, of radiant heat. Materials generally have the power of emitting the same amount of radiant heat that they will absorb under reversed conditions of temperature difference. A few materials have the property of allowing radiant heat to pass through them without absorbing any of the heat transmitted.

Transmission of radiant heat varies inversely as the square of the distance between the source of radiation and the absorbing surface. The heat transmission varies as the difference of the fourth powers of the absolute temperatures of the radiating and the receiving bodies.

The general expression for the transmission of radiant heat per unit of surface per hour according to the Stefan-Boltzmann law is

$$H = K(T_h^4 - T_c^4) \tag{1.2}$$

where $K =$ a coefficient which includes a radiation constant, a relative-position factor for the emitting and the receiving surfaces, and absorption and emission characteristics for the two.

1.11. Expansion and Contraction. The linear dimensions and volumes of nearly all materials are affected by changes of temperature. The coefficient of linear expansion in in. per in. of length of a material per deg of temperature change varies with the temperature; therefore mean values are usually stated for given temperature ranges.

Table 1.4 gives the coefficients of linear expansion for several metals used in heating and air-conditioning plants.

The linear expansion or contraction is

$$E_x = C_e L(t_h - t_c) \tag{1.3}$$

Volumetric expansion is of importance in fluids, such as water in hot-water heating systems, where compensation must be provided

DEFINITIONS, LAWS, AND PROPERTIES OF WATER

either in the form of expansion tanks or relief valves. For water between 32 and 212 F the coefficient of cubical expansion is 0.000,243 per deg F; for fuel oil 0.000,4 may be used.

TABLE 1.4. Coefficients of Linear Expansion

Metal	Coefficient, In. per Deg F per In.	Metal	Coefficient, In. per Deg F per In.
Aluminum	0.000,012,33	Steel, soft-rolled	0.000,006,30
Brass, rolled	0.000,010,72	Wrought iron	0.000,006,73
Cast iron	0.000,005,89	Zinc	0.000,016,53
Copper, drawn	0.000,009,26		

1.12. Gas Laws. Perfect gases are those that can be liquefied only with great difficulty and which closely follow the laws of Boyle and Charles. Air, oxygen, hydrogen, nitrogen, and helium are often considered as nearly perfect gases.

Boyle's law states that, when the temperature is held constant, the volume of a given weight of a perfect gas varies inversely as the absolute pressure. *Charles' law* specifies that, when a perfect gas receives heat at constant volume, the absolute pressure varies directly as the absolute temperature. The following single equation satisfies both of the foregoing statements.

$$PV = MRT \tag{1.4}$$

where R is the gas constant in ft-lb per lb per deg.

Equation 1.4 may also be written as $MR = PV/T$ and for a constant weight M of a given gas

$$\frac{PV}{T} = \frac{P_1 V_1}{T_1} = \frac{P_2 V_2}{T_2} \tag{1.5}$$

Equations 1.4 and 1.5 are useful in heating and air-conditioning calculations where pressure, volume, and temperature changes occur in air and other gases.

The volume of a mole weight (a weight, in pounds, equal to the molecular weight M_m of any gas) is 358.65 cu ft when its pressure is 14.7 psia and its temperature 32 F. When M is equal to M_m based on equation 1.4, the following statements are true: $R = PV/MT = PV/M_m T = B/M_m$, in which B, a constant, is PV/T or for the conditions stated $(14.7 \times 144 \times 358.65) \div (459.6 + 32) = 1544$ ft-lb per deg F. The molecular weight of oxygen, O_2, is $16 \times 2 = 32$, and the value of R for oxygen is $1544 \div 32 = 48.25$ ft-lb per deg F. Values of R for a few gases are given in Table 1.5.

TABLE 1.5. Gas Constants
GAS CONSTANTS

Gas	R	Gas	R
Air	53.35	Nitrogen	55.01
Carbon dioxide	35.10	Oxygen	48.25
Hydrogen	765.36	Water vapor	85.71
Methane	96.31		

For air at atmospheric pressure equation 1.4 may be changed to the following form to give the weight of 1 cu ft as the density d_a of dry air, at any temperature and pressure

$$d_a = \frac{P}{RT}$$

$$= \frac{144 \times 0.4912 \times \text{barometric pressure, in. of mercury at 32 F}}{53.35(459.6 + t)}$$

Hence,

$$d_a = 1.3258 \times \frac{\text{barometric pressure, in. of mercury at 32 F}}{(459.6 + t)} \qquad (1.6)$$

Example. Find the density of dry air at a temperature of 60 F when the barometric pressure is 28.5 in. of mercury at 32 F.
Solution.

$$d_a = 1.3258 \times \frac{28.5}{(459.6 + 60)} = 0.07272 \text{ lb per cu ft}$$

The specific volume, V_{sp}, or cubic feet per 1 lb of gas, is the reciprocal of the density, $1/d_a$, and is equal, in the foregoing case, to $1/0.07272 = 13.75$ cu ft per lb.

1.13. Water. Pure water (H_2O) is the result of the chemical combination of hydrogen and oxygen in the ratio of 2 parts of the former to 16 parts of the latter by weight.

The maximum density of pure water occurs at 39.2 F when under an absolute pressure of 14.696 psi. A sufficiently rapid transfer of heat to water produces increased vibration of its molecules, a rise in its temperature, and a decrease in its density. The temperature at which boiling or vaporization takes place is controlled by the absolute pressure exerted upon it (see Table 1.6).

DEFINITIONS, LAWS, AND PROPERTIES OF WATER

1.14. Saturated Steam. Steam which exists at the temperature corresponding to its absolute pressure is defined as saturated. Saturated steam may be completely free from unvaporized water particles, or it may carry water globules in suspension. Thus, saturated steam may be in either a dry or a wet condition.

In the production of saturated steam the heat added to each pound of water is utilized in two stages. The first stage raises the temperature of the water from that of its initial condition to that which corresponds to the boiling point at the existing absolute pressure. The heat necessary to accomplish this is the *enthalpy of the liquid*. This stage is reached with a comparatively small increase of volume. The second stage in the process of steam formation is the addition of an amount of heat necessary to convert the liquid at its boiling point into a vapor. The latter quantity is the *enthalpy of evaporation or latent heat of vaporization*. The vaporization takes place at a constant temperature corresponding to the boiling point.

Properties of dry and saturated steam as compiled by Professors J. H. Keenan and F. G. Keyes are given by permission in abridged form in Tables 1.6 and 1.7. The steam table data are based on a weight of 1 lb of water or steam, and the enthalpies are reckoned above a base temperature of 32 F.

1.15. Total Enthalpy of Dry Saturated Steam. The total enthalpy of 1 lb of dry saturated steam is the sum of the enthalpy of the liquid and the enthalpy of evaporation:

$$h_g = h_f + h_{fg} \tag{1.7}$$

The enthalpy of dry saturated steam is dependent upon its pressure as shown by the numerical quantities given in Tables 1.6 and 1.7 under "Enthalpy of Saturated Vapor, h_g."

1.16. Specific Volume and Density of Dry Saturated Steam. The specific volumes v_g, in cubic feet, occupied by 1 lb of dry saturated steam are given in the fourth column of Tables 1.6 and 1.7. The volume of 1 lb of steam varies inversely as the pressure.

The density d_s of dry saturated steam is the weight of steam in pounds per cubic foot. The density is the reciprocal of the specific volume.

1.17. Quality of Steam. Moisture present in steam may be due to several causes. Steam conveyed by piping may suffer a reduction in heat content through losses so that part of the latent heat of the steam is given up and water particles are formed. The generation

TABLE 1.6. Properties of Dry Saturated Steam: Pressure Table*

Pressure, Psia	Temperature F	Specific Volume Sat. Liquid	Specific Volume Sat. Vapor	Enthalpy Sat. Liquid	Enthalpy Evaporation	Enthalpy Sat. Vapor	Pressure, Psia
p	t	v_f	v_g	h_f	h_{fg}	h_g	p
0.20	53.14	0.01603	1526.0	21.21	1063.8	1085.0	0.20
0.40	72.86	0.01606	791.9	40.89	1052.7	1093.6	0.40
0.60	85.21	0.01608	540.0	53.21	1045.7	1098.9	0.60
0.80	94.38	0.01612	411.7	62.36	1040.4	1102.8	0.80
1.0	101.74	0.01614	333.6	69.70	1036.3	1106.0	1.0
2.0	126.08	0.01623	173.73	93.99	1022.2	1116.2	2.0
3.0	141.48	0.01630	118.71	109.37	1013.2	1122.6	3.0
4.0	152.97	0.01636	90.63	120.86	1006.4	1127.3	4.0
5.0	162.24	0.01640	73.52	130.13	1001.0	1131.1	5.0
6.0	170.06	0.01645	61.98	137.96	996.2	1134.2	6.0
7.0	176.85	0.01649	53.64	144.76	992.1	1136.9	7.0
8.0	182.86	0.01653	47.34	150.79	988.5	1139.3	8.0
9.0	188.28	0.01656	42.40	156.22	985.2	1141.4	9.0
10.0	191.23	0.01659	38.42	161.17	982.1	1143.3	10.0
11.0	197.75	0.01662	35.14	165.73	979.3	1145.0	11.0
12.0	201.96	0.01665	32.40	169.96	976.6	1146.6	12.0
13.0	205.88	0.01667	30.06	173.91	974.2	1148.1	13.0
14.0	209.56	0.01670	28.04	177.61	971.9	1149.5	14.0
14.696	212.00	0.01672	26.80	180.07	970.3	1150.4	14.696
16.0	216.32	0.01674	24.75	184.42	967.6	1152.0	16.0
18.0	222.41	0.01679	22.17	190.56	963.6	1154.2	18.0
20.0	227.96	0.01683	20.089	196.16	960.1	1156.3	20.0
22.0	233.07	0.01687	18.375	201.33	956.8	1158.1	22.0
24.0	237.82	0.01691	16.938	206.14	953.7	1159.8	24.0
26.0	242.25	0.01694	15.715	210.62	950.7	1161.3	26.0
28.0	246.41	0.01698	14.663	214.83	947.9	1162.7	28.0
30.0	250.33	0.01701	13.746	218.82	945.3	1164.1	30.0
35.0	259.28	0.01708	11.898	227.91	939.2	1167.1	35.0
40.0	267.25	0.01715	10.498	236.03	933.7	1169.7	40.0
45.0	274.44	0.01721	9.401	243.36	928.6	1172.0	45.0
50.0	281.01	0.01727	8.515	250.09	924.0	1174.1	50.0

* Abridged from *Thermodynamic Properties of Steam*, by J. H. Keenan and F. G. Keyes, John Wiley and Sons, Inc., New York, 1936.

DEFINITIONS, LAWS, AND PROPERTIES OF WATER

TABLE 1.6 (Continued)

Pressure, Psia	Temperature F	Specific Volume Sat. Liquid	Specific Volume Sat. Vapor	Enthalpy Sat. Liquid	Enthalpy Evaporation	Enthalpy Sat. Vapor	Pressure, Psia
p	t	v_f	v_g	h_f	h_{fg}	h_g	p
60.0	292.71	0.01738	7.175	262.09	915.5	1177.6	60.0
70.0	302.92	0.01748	6.206	272.61	907.9	1180.6	70.0
80.0	312.03	0.01757	5.472	282.02	901.9	1183.1	80.0
90.0	320.27	0.01766	4.896	290.56	894.7	1185.3	90.0
100.0	327.81	0.01774	4.432	298.40	888.8	1187.2	100.0
110.0	334.77	0.01782	4.049	305.66	883.2	1188.9	110.0
120.0	341.25	0.01789	3.728	312.44	877.9	1190.4	120.0
130.0	347.32	0.01796	3.455	318.81	872.9	1191.7	130.0
140.0	353.02	0.01802	3.220	324.82	868.2	1193.0	140.0
150.0	358.42	0.01809	3.015	330.51	863.6	1194.1	150.0
160.0	363.53	0.01815	2.834	335.93	859.2	1195.1	160.0
170.0	368.41	0.01822	2.675	341.09	854.9	1196.0	170.0
180.0	373.06	0.01827	2.532	346.03	850.8	1196.9	180.0
190.0	377.51	0.01833	2.404	350.79	846.8	1197.6	190.0
200.0	381.79	0.01839	2.288	355.36	843.0	1198.4	200.0
210.0	385.9	0.01844	2.183	359.77	839.2	1199.0	210.0
220.0	389.86	0.01850	2.087	364.02	835.6	1199.6	220.0
230.0	393.68	0.01854	1.9992	368.13	832.0	1200.1	230.0
240.0	397.37	0.01860	1.9183	372.12	828.5	1200.6	240.0
250.0	400.95	0.01865	1.8438	376.0	825.1	1200.1	250.0
260.0	404.42	0.01870	1.7748	379.76	821.8	1201.5	260.0
270.0	407.78	0.01875	1.7107	383.42	818.5	1201.9	270.0
280.0	411.05	0.01880	1.6511	386.98	815.3	1202.3	280.0
290.0	414.23	0.01885	1.5954	390.46	812.1	1202.6	290.0
300.0	417.33	0.01890	1.5433	393.84	809.0	1202.8	300.0

of steam may have been imperfect because of poor design of the boiler or too high rate of evaporation.

The dry steam content per pound of wet steam is the *quality* of the steam. The quality x of steam may be expressed either as a percentage or as the weight of dry steam existing within 1 lb of wet steam.

1.18. Enthalpy of Wet Saturated Steam. When the water is not completely vaporized, the heat carried as the heat of evaporation is less. The quality affects only the enthalpy of evaporation and does

TABLE 1.7. Properties of Dry Saturated Steam: Temperature Table*

Temperature F t	Pressure, Psia p	Specific Volume Sat. Liquid v_f	Specific Volume Sat. Vapor v_g	Enthalpy Sat. Liquid h_f	Enthalpy Evaporation h_{fg}	Enthalpy Sat. Vapor h_g	Temperature F t
32	0.08854	0.01602	3306.0	0.00	1075.8	1075.8	32
34	0.09603	0.01602	3061.0	2.02	1074.7	1076.7	34
36	0.10401	0.01602	2837.0	4.03	1073.6	1077.6	36
38	0.11256	0.01602	2632.0	6.04	1072.4	1078.4	38
40	0.12170	0.01602	2444.0	8.05	1071.3	1079.3	40
42	0.13150	0.01602	2271.0	10.05	1070.1	1080.2	42
44	0.14199	0.01602	2112.0	12.06	1068.9	1081.0	44
46	0.15323	0.01602	1964.3	14.06	1067.8	1081.9	46
48	0.16525	0.01603	1828.6	16.07	1066.7	1082.8	48
50	0.17811	0.01603	1703.2	18.07	1065.6	1083.7	50
52	0.19182	0.01603	1587.6	20.07	1064.4	1084.5	52
54	0.20642	0.01603	1481.0	22.07	1063.3	1085.4	54
56	0.2220	0.01603	1382.4	24.06	1062.2	1086.3	56
58	0.2386	0.01604	1291.1	26.06	1061.0	1087.1	58
60	0.2563	0.01604	1206.7	28.06	1059.9	1088.0	60
62	0.2751	0.01604	1128.4	30.05	1058.8	1088.9	62
64	0.2951	0.01605	1055.7	32.05	1057.6	1089.7	64
66	0.3164	0.01605	988.4	34.05	1056.5	1090.6	66
68	0.3390	0.01605	925.9	36.04	1055.5	1091.5	68
70	0.3631	0.01606	867.9	38.04	1054.3	1092.3	70
72	0.3886	0.01606	813.9	40.04	1053.2	1093.2	72
74	0.4156	0.01606	763.8	42.03	1052.8	1094.1	74
76	0.4443	0.01607	717.1	44.03	1050.9	1094.9	76
78	0.4747	0.01607	673.6	46.02	1049.8	1095.8	78
80	0.5069	0.01608	633.1	48.02	1048.6	1096.6	80
82	0.5410	0.01608	595.3	50.01	1047.5	1097.5	82
84	0.5771	0.01609	560.2	52.01	1046.4	1098.4	84
86	0.6152	0.01609	527.3	54.00	1045.2	1099.2	86
88	0.6556	0.01610	496.7	56.00	1044.1	1100.1	88
90	0.6982	0.01610	468.0	57.99	1042.9	1100.9	90
92	0.7432	0.01611	441.3	59.99	1041.8	1101.8	92
94	0.7906	0.01612	416.2	61.98	1040.7	1102.6	94
96	0.8407	0.01612	392.8	63.98	1039.5	1103.5	96
98	0.8935	0.01613	370.9	65.97	1038.4	1104.4	98

* Abridged from *Thermodynamic Properties of Steam*, by Joseph H. Keenan and Frederick G. Keyes, John Wiley and Sons, Inc., New York, 1936.

DEFINITIONS, LAWS, AND PROPERTIES OF WATER

TABLE 1.7 (Continued)

Temperature F	Pressure, Psia	Specific Volume Sat. Liquid	Specific Volume Sat. Vapor	Enthalpy Sat. Liquid	Enthalpy Evaporation	Enthalpy Sat. Vapor	Temperature F
t	p	v_f	v_g	h_f	h_{fg}	h_g	t
100	0.9492	0.01613	350.4	67.97	1037.2	1105.2	100
105	1.1016	0.01615	304.5	72.95	1034.3	1107.3	105
110	1.2748	0.01617	265.4	77.94	1031.6	1109.5	110
115	1.4709	0.01618	231.9	82.93	1028.7	1111.6	115
120	1.6924	0.01620	203.27	87.92	1025.8	1113.7	120
125	1.9420	0.01622	178.61	92.91	1022.9	1115.8	125
130	2.2225	0.01625	157.34	97.90	1020.0	1117.9	130
135	2.5370	0.01627	138.95	102.90	1017.0	1119.9	135
140	2.8886	0.01629	123.01	107.89	1014.1	1122.0	140
145	3.281	0.01632	109.15	112.89	1011.2	1124.1	145
150	3.718	0.01634	97.07	117.89	1008.2	1126.1	150
155	4.203	0.01637	86.52	122.89	1005.2	1128.1	155
160	4.741	0.01639	77.29	127.89	1002.3	1130.2	160
165	5.335	0.01642	69.19	132.89	999.3	1132.2	165
170	5.992	0.01645	62.06	137.90	996.3	1134.2	170
175	6.715	0.01648	55.78	142.91	993.3	1136.2	175
180	7.510	0.01651	50.23	147.92	990.2	1138.1	180
185	8.383	0.01654	45.31	152.93	987.2	1140.1	185
190	9.339	0.01657	40.96	157.95	984.1	1142.0	190
200	11.526	0.01663	33.64	167.99	977.9	1145.9	200
210	14.123	0.01670	27.82	178.05	971.6	1149.7	210
220	17.186	0.01677	23.15	188.13	965.2	1153.4	220
230	20.780	0.01684	19.38	198.23	958.8	1157.0	230
240	24.969	0.01692	16.32	208.34	952.2	1160.5	240

not change the enthalpy of the liquid. The enthalpy of 1 lb of wet steam is

$$h_g = h_f + xh_{fg} \tag{1.8}$$

1.19. Superheated Steam. When the steam temperature is above saturation temperature it is superheated. The enthalpy of 1 lb of superheated steam is given by

$$h = h_g + c_{ps}(t_{sh} - t) \tag{1.9}$$

The specific heat of superheated steam varies with the pressure and the temperature. Values for a wide range in both may be obtained from any complete set of steam tables. The mean specific heat of superheated steam at pressures ranging from 0.001 to 2.22 psia and at temperatures less than 200 F may be taken as 0.45 Btu per lb per deg F.

PROBLEMS

1. A closed tank confined water vapor at an absolute pressure of 20 in. of mercury measured at a temperature of 70 F when the barometric pressure was 27.8 in. of mercury at 70 F. What was the actual reading of the tank vacuum gage in inches of mercury at 32 F?

2. A low-pressure steam-heating system operates with radiator pressures of 1.5 psig when the barometric pressure is 28.5 in. of mercury measured at 80 F. Find the existent steam pressure in pounds per square inch absolute.

3. A gage indicated a steam pressure of 178.5 psi when the barometric pressure was 29.6 in. of mercury measured at 32 F. Twenty-four hours later the gage reading was unchanged at 178.5 psi but the absolute pressure had decreased to 192 psi. What reduction of the barometric pressure, in inches of mercury at 32 F, had taken place?

4. Express 90 F as degrees Fahrenheit absolute.

5. The continuous supply of power to fan motors is 30 kw. For each hour of time find the input equivalent in watts, horsepower, and Btu (see the conversion equations).

6. A steam engine, when operating at 75 per cent of its rating, develops 45 bhp of output. If the engine were loaded to 125 per cent of its rating, what would be the heat equivalent of the engine output in Btuh?

7. One hundred pounds of dry air were heated to increase its temperature from 32 F to 212 F. Find the heat in Btu that was added to the air when the operation was one of constant volume and also when one of constant pressure.

8. Nine thousand pounds of copper were cooled from a temperature of 500 F to 100 F. Find the total heat in Btu that was lost by the copper.

9. A wall 12 in. thick has a surface area of 1500 sq ft and transmits 90,000 Btuh by conduction when the inside-surface temperature is 60 F and the outside-surface temperature is 0 F. Find the thermal conductivity of the wall in Btuh per sq ft of wall area per deg F per in. of thickness.

10. A wall 16 in. thick has a surface area of 2000 sq ft and a thermal conductivity of 5.2. Calculate the heat transmitted by conduction in Btuh when the inside-surface temperature is 72 F and the outside-surface temperature is 2 F.

11. A steel pipe 200 ft long had a change of temperature from 55 F to 220 F. Find the increase of the pipe length in inches.

12. Calculate the weight of dry air in pounds per cubic foot and its volume in cubic feet per pound when its temperature is 100 F and pressure is 45 psia.

13. A tank having a volume of 100 cu ft is filled with dry air at a pressure of 2.5 psig when the barometric pressure is 29.6 in. of mercury at 50 F. Find the weight, in pounds, of 80 F air confined within the tank.

DEFINITIONS, LAWS, AND PROPERTIES OF WATER

14. A heating boiler operates to convert 8000 lb of water per hr into steam at 20 psia and 0.96 quality. Find the heat, in Btuh, added to the water if its supply temperature is 80 F.

15. How much heat must be added to 15,000 lb of water at 120 F to convert it into steam of 0.99 quality and 150 psia pressure?

16. Dry saturated steam at a pressure of 1.47 psia is brought to a temperature of 195 F without a change of pressure. Find the heat, in Btu per pound, that must be added.

17. Find the volume in cubic feet of 500 lb of dry saturated steam held at a pressure of 60 psia.

2

Psychrometric Properties of Air—Humidification and Dehumidification

2.1. Air. Pure dry air is a mechanical mixture of the following gases: oxygen, nitrogen, carbon dioxide, hydrogen, argon, neon, krypton, helium, ozone, and xenon. The oxygen and the nitrogen make up the major portion of the combination. The other gases range from a few parts per 10,000 to a mere trace. Therefore, dry air is generally considered as being made up in the following proportions: oxygen 20.91 per cent by volume and 23.15 per cent by weight, together with nitrogen 79.09 per cent by volume and 76.85 per cent by weight. Completely dry air does not exist in nature because water vapor in varying amounts is diffused through it. Air may be polluted by dust, dirt, bacteria, odor-bearing materials, and toxic gases.

Air, being one of the so-called perfect gases, approximately follows their laws, and the equations as stated in Art. 1.12 are applicable in many calculations where its weights and volumes are required at different pressures and temperatures. For ordinary computations the mean specific heat of dry air at constant pressure may be used as 0.24 Btu per lb per deg F.

2.2. Mixtures of Air and Water Vapor. In psychrometric calculations the mixture considered is that of dry air and the water vapor (steam) mixed with it. The maximum amount of water vapor which

PSYCHROMETRIC PROPERTIES OF AIR 17

can be mixed with dry air is dependent upon the absolute pressure and the temperature of the mixture. Air is saturated when it has diffused through it the maximum amount of water vapor that is possible under the conditions of temperature and absolute pressure.

When the temperature of an air-vapor mixture is held constant, the amount of water vapor required to saturate 1 lb of dry air increases as the absolute pressure of the air is diminished and decreases as the pressure is increased. This fact is of importance at high altitudes. At a given pressure the temperature of the mixture has a very marked effect on the amount of water vapor that may be carried by a given quantity of air; the maximum amount increases as the air temperature rises. Under conditions of saturation the absolute partial pressure of the water vapor is that of saturated steam at the same temperature.

When air is not saturated at given conditions of pressure and temperature the vapor held is less than that possible at the temperature of the mixture; the vapor is superheated, and its partial pressure is less than that of saturated vapor at the same temperature. With conditions other than those of saturation the relative humidity, Art. 2.8, is less than 100 per cent, and the partial pressure of the vapor is that existing at the dew-point temperature of the mixture.

2.3. Weight of Water Vapor Required to Saturate Air. Exact determinations[1] of the weight of water vapor to saturate air are somewhat involved, probably because of the effects of molecular interferences, forces of attraction and repulsion between molecules, and the solution of gas molecules in the water vapor. Neither water vapor nor air follows exactly the laws of perfect gases. However, at the temperatures and pressure involved in most psychrometric calculations they may be considered as perfect gases. Psychrometric data calculated by use of Dalton's law will be taken as having sufficient commercial accuracy through the temperature ranges involved in the discussions and problems of this book.

According to Dalton's law, air and water vapor each completely occupy a space filled by the mixture of the two as though the other were not present, and the total pressure of the mixture is the sum of the partial pressures of the water vapor and the dry air. Thus any mixture of water vapor and dry air, at any temperature, exists at an absolute pressure p which is the sum of p_w, the absolute partial pressure of the vapor, and p_a, the absolute partial pressure of the dry air. Then $p = p_w + p_a$, where each pressure is expressed in pounds per square inch absolute. Also, $p_a = p - p_w$. From Art. 1.12 the fundamental gas-law equation is $PV = MRT$. The volume V_w cu ft of water vapor at an absolute pressure of p_w psi accruing from M_w lb of moisture necessary to saturate 1 lb of dry air at an absolute temperature of T deg F is equal to the volume V_a cu ft of dry air at an absolute pressure of p_a psi at an absolute temperature of T deg F. Using $R_w = 85.71$ for water vapor and $R_a = 53.35$ for dry air (Table 1.5),

$$V_w = V_a = \frac{M_w R_w T}{144 p_w} = \frac{1 \times R_a T}{144(p - p_w)}$$

$$M_w = \frac{1 \times R_a p_w}{R_w(p - p_w)} = \frac{53.35 p_w}{85.71(p - p_w)} = 0.622 \frac{p_w}{(p - p_w)} \quad (2.1)$$

Use of the foregoing equations may be made in building up data, Table 2.1, relative to the properties of both dry and saturated air. The following examples are used to illustrate the items involved.

Example. (a) Find the volume of 1 lb of dry air existing at a barometric pressure of 14.696 psia and a temperature of 68 F. (b) Find the volume of the mixture of 1 lb of dry air and the water vapor necessary to saturate it under the same total pressure. (c) Calculate the weight of water vapor necessary to saturate 1 lb of dry air at the temperature and pressure given in (a). (d) Compute the density of the mixture.
Solution.

(a) $\quad V = \dfrac{MRT}{144p} = \dfrac{1 \times 53.35(459.6 + 68)}{144 \times 14.696} = 13.30$ cu ft

(b) $\quad p_w = 0.339$ psia (Table 1.7) for steam at 68 F

$\quad p = 14.696$ psia

$\quad V_a = \dfrac{M_a R_a T}{144(p - p_w)} = \dfrac{1 \times 53.35(459.6 + 68)}{144(14.696 - 0.339)} = 13.61$ cu ft

(c) $\quad M_w = 0.622 \dfrac{p_w}{(p - p_w)} = \dfrac{0.622 \times 0.339}{(14.696 - 0.339)} = 0.01469$ lb per lb dry air.

$\quad 0.01469 \times 7000 = 102.83$ grains per lb of dry air

(d) $\quad d_m = \dfrac{1 + M_w}{v_a} = \dfrac{1 + 0.01469}{13.61} = 0.07455$ lb per cu ft

The numerical effect of reduced air pressure is illustrated by the following example.

Example. A saturated air-vapor mixture exists at a pressure of 13.905 psia and a temperature of 68 F. Find the volume of the mixture in cubic feet per pound of dry air involved and the weight of water vapor required to saturate 1 lb of dry air.
Solution. The pressure p is 13.905 psia, and the vapor pressure at 68 F is 0.339 psia. The volume of the mixture is $1 \times 53.35 \times (459.6 + 68)/144(13.905 - 0.339) = 14.409$ cu ft. The weight of water vapor to saturate

PSYCHROMETRIC PROPERTIES OF AIR

1 lb of dry air under the conditions stated is $(0.622 \times 0.339)/(13.905 - 0.339)$ = 0.01554 lb or 108.78 grains per lb.

Comparison of the solutions of the two prior examples involving the calculation of the weight of water vapor required to saturate 1 lb of dry air indicates that for a given temperature (1) the vapor pressure for saturated water vapor remains constant irrespective of the barometric pressure under which the mixture exists; (2) the partial pressure of the dry air decreases as the barometric pressure decreases; (3) the volume of the mixture per pound of dry air and its contained water vapor increases as the barometric pressure decreases; and (4) the weight of water vapor required to saturate 1 lb of dry air increases as the barometric pressure decreases. Later in this chapter it will become apparent that the greater the amount of water vapor per pound of dry air the greater will be the total enthalpy of the mixture.

2.4. Humidity. The water vapor mixed with dry air is termed humidity.

Absolute humidity is the actual weight of water vapor in grains (1 lb equals 7000 grains avoirdupois) or pounds per cubic foot of a mixture of air and water vapor.

Humidity ratio, often designated as specific humidity, is the weight, either in pounds or in grains of water vapor diffused through 1 lb of dry air.

2.5. Enthalpy of Saturated Air. The heat held by a mixture of dry air and water vapor is generally reckoned from 0 F for the air and from 32 F for the vapor. The enthalpy of saturated water vapor may be obtained from Tables 1.6 and 1.7. A convenient empirical expression for the enthalpy, in Btu per pound, of low-pressure, low-temperature saturated water vapor is

$$h_g = 1060.8 + 0.45t \qquad (2.2)$$

Example. Find the enthalpy of 1 lb of saturated water vapor at 70 F.
Solution. From Table 1.7, $h_g = 1092.3$, or by equation 2.2, $1060.8 + (0.45 \times 70) = 1092.3$ Btu per lb.

Equation 2.2 gives the correct total enthalpy of saturated water vapor at 70 F. For temperatures above 70 F computed total enthalpies are slightly high and below it they are somewhat low. Equation 2.2 should not be used for saturated vapor temperatures above 100 F.

The enthalpy of the mixture of 1 lb of dry air and the moisture to saturate it is

$$h_s = c_{pa}t + w_s h_g \qquad (2.3)$$

Example. Find the enthalpy of 1 lb of dry air and the vapor to saturate it when the mixture exists under a barometric pressure of 14.696 psia and 68 F dbt.

Solution. By a prior calculation of part c in the example shown in Art. 2.3, the weight of water vapor for saturation is 0.01469 lb per lb. A calculated value of h_g is 1060.8 + (0.45 × 68) = 1091.4 Btu per lb of water vapor. The total enthalpy h_s = (0.24 × 68) + (0.01469 × 1091.4) = 32.35 Btu per lb.

2.6. Tabular Data Pertaining to Air-Vapor Mixtures. Several important psychrometric properties of saturated air-vapor mixtures are given in Table 2.1. The data of Table 2.1 are for dry-bulb temperatures ranging from −50 F to +130 F. Similar data based on more recent calculations[1] and covering a range from −160 F to +200 F are given in the ASHAE Guide. Where precision is essential it is recommended that the latest available data be used. However, the maximum variance of any item in Table 2.1 from the latest data for the same item is less than 0.5 per cent.

2.7. Dry-Bulb and Wet-Bulb Temperatures. The dry-bulb temperature of an air-vapor mixture is the temperature indicated by a thermometer with a clean, dry sensing element that is shielded from radiation effects. The wet-bulb temperature of air (thermodynamic wet-bulb) for all practical purposes is the temperature of adiabatic saturation as defined later in Art. 2.14. It is the lowest temperature indicated by a moistened thermometer bulb when evaporation of the moisture takes place in a current of the air-vapor mixture as it moves rapidly over the instrument.

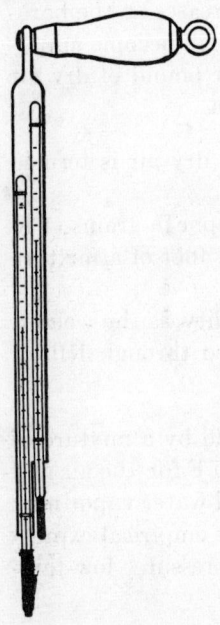

Fig. 2.1. Sling psychrometer.

Dry- and wet-bulb temperatures are determined by the use of aspirating and sling psychrometers. Aspirating psychrometers usually have shields about the thermometer bulbs to prevent errors due to radiant energy and also some means of producing rapid motion of the air-vapor mixture over the thermometer bulbs. For the best results the velocity of air motion should be from 1000 to 2000 fpm. The whirling psychrometer, illustrated by Fig. 2.1, consists of two accurate mercurial thermometers mounted upon a frame which has a handle that permits the assembly to be rapidly rotated to produce the neces-

PSYCHROMETRIC PROPERTIES OF AIR

sary air motion. One thermometer is fixed with its bulb projecting beyond the bulb of the other. This projecting bulb is covered with a very light gauze which is moistened with pure water. The other bulb is kept absolutely dry. After rapid rotation, readings are taken of both thermometers when the minimum wet-bulb temperature is indicated. The lower temperature given by the wet-bulb thermometer, in partially saturated air, is due to the evaporation of moisture from the gauze and the transformation of sensible into latent heat. When the air is saturated there is no evaporation at the moistened bulb, and the two thermometers read alike. At any given dry-bulb temperature the greater the depression of the wet-bulb temperature reading below that of the dry-bulb thermometer the smaller is the amount of water vapor held in the mixture. Hygrometers which have their wet bulbs standing in still air may give very erroneous results.

2.8. Relative Humidity and Per Cent of Saturation. The amount of water vapor diffused through dry air in an air-vapor mixture may vary from practically nothing to that necessary for saturation conditions. As an indication of the degree of saturation of the mixture, the terms relative humidity and per cent of saturation are employed.

Relative humidity, usually expressed as a percentage, is preferably defined as the ratio of the actual partial pressure of water vapor of the air-vapor mixture to the pressure of saturated water vapor at the dry-bulb temperature. *Per cent of saturation* is 100 times the ratio of the weight of water vapor actually held per pound of dry air to that necessary to saturate the pound of dry air at the same dry-bulb temperature and absolute total pressure.

2.9. Calculation of Vapor Pressures and Relative Humidities. The vapor pressure of an air-vapor mixture is always the pressure corresponding to that of saturated water vapor existing at its dew-point temperature. Since the barometric pressure and the dry- and wet-bulb temperatures are readily obtainable, vapor pressures may be secured by calculations from measured data. Equation 2.4, which was developed by Dr. Willis H. Carrier, affords a satisfactory means of making such computations. In the equation all pressures are expressed in pounds per square inch.

$$p_w = p' - \frac{(p - p')(t - t')}{(2831 - 1.43t')} \tag{2.4}$$

TABLE 2.1. Properties of Mixtures of Air and Saturated Water Vapor, −50 to 130 F*†
Barometric pressure 29.92 in. of mercury

Air Temperature, deg F	Pressure of Saturated Vapor, psia	Volume 1 Lb of Dry Air, cu ft	Volume 1 Lb of Dry Air and Vapor to Saturate, cu ft	Weight of Water Vapor for Saturation of 1 Lb of Dry Air, Grains	Weight of Water Vapor for Saturation of 1 Lb of Dry Air, Pounds	Total Enthalpy 1 Lb of Dry Air Above 0 F, Btu	Total Enthalpy Water Vapor to Saturate 1 Lb of Dry Air, Btu above 32 F	Total Enthalpy 1 Lb of Dry Air Saturated, Btu
−50	0.0009819	10.32	10.32	0.29092	0.00004156	−12.05	+0.040	−12.010
−40	0.001868	10.57	10.57	0.55349	0.00007907	− 9.629	+0.082	− 9.547
−30	0.003443	10.82	10.82	1.0206	0.0001458	− 7.216	+0.158	− 7.064
−20	0.0061989	11.07	11.07	1.8375	0.0002625	− 4.807	+0.276	− 4.531
−10	0.010856	11.33	11.34	3.2186	0.0004598	− 2.402	+0.485	− 1.917
0	0.01853	11.58	11.59	5.50	0.0007852	0.000	+0.8317	0.8317
5	0.02400	11.70	11.72	7.12	0.001017	1.200	1.080	2.280
10	0.03090	11.83	11.85	9.18	0.001311	2.400	1.395	3.795
15	0.03963	11.96	11.99	11.77	0.001682	3.599	1.783	5.392
20	0.0539	12.09	12.13	16.03	0.002290	4.802	2.447	7.249
22	0.0587	12.14	12.19	17.46	0.002494	5.282	2.667	7.949
24	0.0638	12.19	12.24	18.99	0.002713	5.762	2.903	8.665
26	0.0693	12.24	12.30	20.64	0.002949	6.243	3.158	9.401
28	0.0752	12.29	12.35	22.42	0.003203	6.723	3.433	10.156
30	0.0816	12.34	12.41	24.35	0.003479	7.203	3.731	10.934
32	0.0886	12.39	12.47	26.43	0.003776	7.683	4.053	11.736
34	0.0961	12.44	12.52	28.67	0.004096	8.163	4.400	12.563
36	0.1041	12.49	12.58	31.07	0.004439	8.644	4.773	13.417
38	0.1126	12.54	12.64	33.64	0.004806	9.124	5.174	14.298
40	0.1217	12.59	12.70	36.41	0.005201	9.604	5.603	15.207
42	0.1315	12.64	12.76	39.38	0.005626	10.084	6.062	16.146
44	0.1420	12.69	12.82	42.55	0.006079	10.564	6.553	17.117
46	0.1532	12.74	12.88	45.93	0.006561	11.045	7.081	18.126
48	0.1652	12.79	12.94	49.55	0.007079	11.525	7.647	19.172

PSYCHROMETRIC PROPERTIES OF AIR

50	0.1780	12.84				
52	0.1918	12.89				
54	0.2063	12.95				
56	0.2219	13.00				
58	0.2384	13.05				
60	0.2561	13.10	13.00	53.42	0.007632	12.005	8.256	20.261
62	0.2749	13.15	13.06	57.57	0.008224	12.485	8.909	21.394
64	0.2949	13.20	13.13	62.01	0.008859	12.965	9.606	22.571
66	0.3162	13.25	13.19	66.77	0.009539	13.446	10.350	23.796
68	0.3388	13.30	13.26	71.85	0.01026	13.926	11.143	25.069
70	0.3628	13.35	13.33	77.28	0.01104	14.406	11.991	26.397
72	0.3883	13.40	13.40	83.07	0.01187	14.886	12.899	27.785
74	0.4153	13.45	13.47	89.25	0.01275	15.366	13.869	29.235
76	0.4440	13.50	13.54	95.83	0.01369	15.847	14.905	30.752
78	0.4744	13.55	13.61	102.84	0.01469	16.327	16.009	32.336
80	0.5067	13.60	13.69	110.30	0.01576	16.807	17.187	33.994
82	0.5409	13.65	13.76	118.25	0.01689	17.287	18.441	35.728
84	0.5772	13.70	13.84	126.72	0.01810	17.767	19.778	37.545
86	0.6153	13.75	13.92	135.75	0.01939	18.248	21.203	39.451
88	0.6556	13.80	14.00	145.36	0.02077	18.728	22.724	41.452
90	0.6980	13.85	14.09	155.61	0.02223	19.208	24.346	43.554
92	0.7429	13.90	14.17	166.52	0.02379	19.69	26.07	45.76
94	0.7902	13.95	14.26	178.11	0.02544	20.17	27.91	48.08
96	0.8403	14.00	14.35	190.41	0.02720	20.65	29.86	50.51
98	0.8930	14.05	14.45	203.44	0.02906	21.13	31.94	53.07
100	0.9487	14.10	14.54	217.27	0.03104	21.61	34.14	55.75
105	1.1009	14.23	14.64	231.98	0.03314	22.09	36.48	58.57
110	1.274	14.36	14.75	247.62	0.03532	22.57	38.97	61.54
115	1.470	14.48	14.85	264.25	0.03775	23.05	41.62	64.67
120	1.691	14.61	14.96	281.92	0.04027	23.53	44.43	67.96
125	1.941	14.73	15.08	300.70	0.04296	24.01	47.43	71.44
130	2.221	14.86	15.38	352.9	0.0504	25.21	55.78	80.99
			15.72	413.6	0.0591	26.41	65.51	91.92
			16.09	484.3	0.0692	27.61	76.85	104.46
			16.51	567.1	0.0810	28.81	90.21	119.02
			16.98	663.8	0.0948	30.01	105.75	135.76
			17.51	777.2	0.1110	31.21	124.05	155.26

* Reproduced by permission of the General Electric Company.
Values for −50 F to +15 F are from data prepared by Professor W. M. Sawdon for the ASHVE Guide, 1938.

The relative humidity, expressed in equation form, is

$$\phi = \frac{p_w}{p_s} 100 \tag{2.5}$$

Likewise, an expression for per cent saturation is

$$s = \frac{w}{w_s} \times 100 \quad \text{or} \quad \frac{w_g}{w_{gs}} \times 100$$

Example. Find the relative humidity and per cent saturation for an air-vapor mixture at 14.696 psia, 80 F dbt and 60 F wbt.

Solution. Table 1.7 indicates that at 60 F wbt the saturation pressure, p', for water vapor is 0.2563 psia, and p_s at 80 dbt is 0.5069 psia. Then by equation 2.4 the actual vapor pressure is found to be: $p_w = 0.2563 - (14.696 - 0.2563)(80 - 60)/[2831 - (1.43 \times 60)] = 0.2563 - 0.1052 = 0.1511$ psia. The relative humidity is $(0.1511 \times 100)/0.5069 = 29.8$ per cent. The actual weight of water vapor required to saturate 1 lb of dry air, when the vapor pressure is 0.1511 psia according to equation 2.1, is $(0.622 \times 0.1511)/(14.696 - 0.1511) = 0.00646$ lb per lb. The weight of water vapor required to saturate 1 lb of dry air at 80 F dbt is $(0.622 \times 0.5069)/(14.696 - 0.5069) = 0.0222$ lb per lb. Thence by definition the per cent of saturation is $(0.00646 \times 100)/0.0222 = 29.1$ per cent.

The dew point is the temperature at which water vapor at its partial pressure in the atmosphere condenses. Hence, it is the saturation temperature corresponding to the partial pressure of the vapor.

2.10. Enthalpy of Air-Vapor Mixtures for Any Condition. Air may not be saturated when in combination with water vapor. Under such a condition its total enthalpy includes three components instead of the two of equation 2.3 for saturated air. These are: (1) sensible heat $c_{pa}t$, Btu per pound of dry air between 0 F and the dry-bulb temperature; (2) the total enthalpy, wh_g, of the contained water vapor, at the saturation or dew-point temperature t_d; and (3) the heat of superheat, $wc_{ps}(t - t_d)$ of the contained water vapor with $(t - t_d)$ deg F of superheat. The total enthalpy h in Btu per pound of dry air which is partially saturated by w pounds of moisture is

$$h = c_{pa}t + w[h_g + c_{ps}(t - t_d)] \tag{2.6}$$

Equation 2.2 indicates that the change of total enthalpy of saturated water vapor is 0.45 Btu per lb per deg F of change at low temperatures. Inasmuch as the change of enthalpy per degree Fahrenheit of superheat is also 0.45 Btu per lb per deg F for air temperatures less than 200 F, the enthalpy of superheated vapor equals that of saturated vapor at the same temperature and the total enthalpy h of 1 lb of dry air and the water vapor w that it contains with either a condition of saturation or partial saturation is

$$h \text{ or } h_s = c_{pa}t + wh_g \qquad (2.7)$$

where h_g is the enthalpy of saturated steam at a temperature equal to the dry-bulb temperature of the air-water vapor mixture.

Example. Find the total enthalpy of 1 lb of dry air and the moisture which it holds under the following conditions: 14.696 psia, 80 F dbt; and 60 F wbt; (a) by use of equation 2.6 and (b) by use of equation 2.7.

Solution. (a) For the conditions stated, the solution of the example of Art. 2.9 gave a vapor pressure of 0.1511 psia and a humidity ratio of 0.00646 lb of water vapor per lb of dry air. By interpolation of data of Table 1.7 the total enthalpy of saturated water vapor at a temperature of 45.6 F (the dew-point temperature) is 1081.7 Btu per lb. According to equation 2.6, $h = (0.24 \times 80) + 0.00646 [1081.7 + 0.45(80 - 45.6)] = 19.2 + 0.00646 [1081.7 + 15.48] = 19.2 + (0.00646 \times 1097.18) = 19.2 + 7.08 = 26.28$ Btu per lb. (b) The total enthalpy of saturated water vapor at 80 F (Table 1.7) is 1096.6 Btu per lb. Therefore, by equation 2.7, $h = (0.24 \times 80) + (0.00646 \times 1096.6) = 19.2 + 7.08 = 26.28$ Btu per lb.

2.11. Psychrometric Charts. These are diagrams which show graphically properties of air-vapor mixtures for conditions of saturation are of great convenience in solving problems of air conditioning. One form of psychrometric chart, as devised by the General Electric Company and reproduced by permission, is illustrated by Fig. 2.2. In this particular chart dry-bulb temperatures serve as abscissas and weights of water vapor per pound of dry air along with the corresponding vapor pressures are ordinates. Inasmuch as relative humidities represent vapor-pressure ratios, curves representing these are placed on the diagram with dry-bulb temperatures as abscissas and actual vapor pressures together with the corresponding humidity ratios as ordinates. At saturation conditions (100 per cent relative humidity) the air dry- and wet-bulb temperatures and the dew-point temperature are identical for a definite mixture. Consequently these values are placed on the saturation curve directly above the dry-bulb temperatures. Lines representing constant wet-bulb temperatures are drawn diagonally from the saturation curve to fit conditions of higher dry-bulb temperatures and lesser amounts of moisture per pound of dry air. Although a line of constant wet-bulb temperature does not exactly represent a line of constant total enthalpy, on this particular chart they are so used for 1 lb of dry air and its vapor content at any dry-bulb temperature. (See Art. 2.14.) Vapor pressures for any conditions shown on the chart are indicated at its left side, and a value can be obtained by passing horizontally from the point representing any condition to the vapor-pressure scale. Volumes of 1 lb of dry air and its contained water vapor also appear on the chart.

2.12. Use of the Psychrometric Chart. The following example serves to indicate the use of the psychrometric chart.

Example. An air-vapor mixture has 70 F dbt and 61 F wbt when the barometric pressure is 29.92 in. of mercury. From the psychrometric chart of Fig. 2.2 find the volume of 1 lb of dry air and the moisture to saturate it, the relative humidity, the humidity ratio, the dew-point temperature, the vapor pressure, and the total heat or enthalpy of the mixture.

Solution. The intersection of the vertical line through 70 F dbt and the diagonal line representing 61 F wbt fixes 13.54 cu ft as the volume of the mixture per pound of dry air and its contained water vapor and also 60 per cent as the relative humidity for the stated conditions. Passing from the point of intersection to the right, the humidity ratio is found to be 65.5 grains of moisture per lb of dry air. Moving horizontally to the left from the point of intersection, the dew-point temperature is read at the saturation curve as 55.5 F, and the vapor-pressure scale at the edge of the chart shows 0.215 psia. The total enthalpy is found by following the 61 F wbt line to the total enthalpy scale where 27.10 Btu per lb of dry air are indicated.

Whenever the barometric pressure does not differ greatly from 14.696 psia, data taken from a chart drawn for such a pressure may be used with small significant errors. However, when the barometric pressure is low, the errors accruing from the use of a standard chart may amount to more than those ordinarily permissible. When such a situation arises the psychrometric data can be computed.

2.13. Calculated Psychrometric Data Versus Data from a Standard Chart. The following example with its tabular data will show the difference between chart and calculated values when the barometric pressures are not the same.

Example. Find by calculation the psychrometric properties of an air-vapor mixture for the temperatures stated in the example of Art. 2.12 when the barometric pressure is 25 in. of mercury at 50 F. Tabulate and compare the data thus obtained with those secured by the method of Art. 2.12.

Solution. From the equations of prior articles the following may be found $p = 0.491 \times 25 = 12.275$ psia; $p_w = 0.2655 - [(12.275 - 0.2655)(70 - 61)]/[2831 - (1.43 \times 61)] = 0.226$ psia; volume of the mixture $V_a = 1 \times 53.35 (459.6 + 70)/144(12.275 - 0.226) = 16.29$ cu ft; the vapor pressure p_w from Table 1.7 is 0.3631 psia at 70 F; the relative humidity $\phi = (0.226 \times 100)/0.3631 = 62.2$ per cent; the dew-point temperature from either Table 1.7 or Table 2.1 is 56.48 F, which is the saturation temperature of water vapor corresponding to 0.226 psia; the humidity ratio $w = (0.622 \times 0.226)/(12.275 - 0.226) = 0.01167$ lb per lb or 81.69 grains; and the total enthalpy $h = (0.24 \times 70) + (0.01167 \times 1092.3) = 16.8 + 12.75 = 29.55$ Btu per lb of dry air.

Table 2.2 summarizes the data.

PSYCHROMETRIC PROPERTIES OF AIR

TABLE 2.2. Psychrometric Properties of an Air-vapor Mixture

Item	By Standard Chart	By Calculation
Barometer, in. of mercury	29.92	25.00
Dry-bulb, deg F	70	70
Wet-bulb, deg F	61	61
Vapor pressure, psia	0.215	0.226
Volume, cu ft per lb	13.54	16.29
Relative humidity, per cent	60	62.2
Dew point, deg F	55.5	56.48
Humidity ratio, grains	65.5	81.69
Enthalpy, Btu per lb	27.10	29.55

2.14. Adiabatic Saturation of Air. This fundamental term[2] was defined by Dr. Willis H. Carrier and is applicable, under certain conditions, in the humidification of air and in the drying of materials. Air may be either completely or partially adiabatically saturated with vapors from water and other liquids. In this discussion water vapor only will be considered.

Experimentally the process can be carried out in a well-insulated chamber, Fig. 2.3, where heat is neither received from nor given to outside bodies (adiabatic conditions) and where work is not done upon the air as it passes through the apparatus. The moving air stream is brought into intimate contact with a large surface area of water. The process is one involving a condition of constant enthalpy with sensible heat being transformed to latent heat. For the purposes of simplicity the heat equivalent of any work done upon the water by a pump will be ignored.

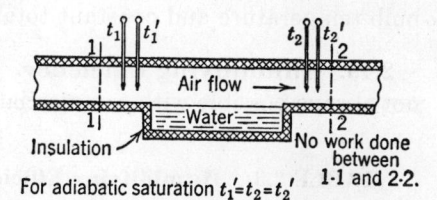

Fig. 2.3. Apparatus for adiabatic saturation of air with water vapor.

When constant conditions of operation have been established, theoretically the water temperature and the final dry- and wet-bulb temperatures of the saturated air-vapor mixture leaving the apparatus are the same as the wet-bulb temperature of the entering air-vapor mixture.

Under the conditions just specified, the process of evaporating water in air to saturate it, by having a part of the sensible heat of the entering air transformed into latent heat, is that of *adiabatic saturation*.

For air *initially dry* the heat transformation is expressed in equation form as

$$h_{fg}'w' = c_{pa}(t_1 - t') \tag{2.8}$$

where t_1 is the initial dry-bulb temperature and t' is the initial wet-bulb temperature.

The heat of transformation for air having an initial moisture content w_1, insufficient to saturate it, is indicated by

$$(w' - w_1)h_{fg}' = c_{pa}(t_1 - t') + c_{ps}w_1(t_1 - t')$$

or (2.9)

$$(w' - w_1)h_{fg}' = (c_{pa} + c_{ps}w_1)(t_1 - t')$$

where w_1 = the initial weight of water vapor per pound of dry air, as it enters the apparatus, pounds.

A large-scale psychrometric chart developed under the auspices of ASHAE indicates the deviation of lines of constant wet-bulb temperature from those of constant enthalpy. The variation of enthalpy along a line of constant wet-bulb temperature does not involve more than 0 to 0.7 Btu per lb of dry air and its vapor. For a psychrometric chart of the size of Fig. 2.2 a practical indication of the deviation is impossible. Therefore, in subsequent calculations adiabatic saturation will be shown along a straight line indicating both constant wet-bulb temperature and constant total enthalpy.

2.15. Humidifying Efficiency. Complete adiabatic saturation is not always possible with commercial air washers. Therefore the term

TABLE 2.3. Humidifying Efficiency of Commercial Washers*

Number of Spray Banks	Direction of Water Spray	Length of Washer, ft	Humidifying Efficiency, per cent
1	With air flow	4	50–60
1	With air flow	6	60–75
1	Against air flow	6	65–80
2	With air flow	8–10	80–90
2	Opposing each other	8–10	85–95
2	Against air flow	8–10	90–98

* From reference 3.

humidifying efficiency is used to express the percentage of reduction of the dry-bulb temperature actually obtained in such equipment to that theoretically possible with complete adiabatic saturation. The maximum possible reduction of the dry-bulb temperature by the process of adiabatic saturation is equal to the initial wet-bulb temperature depression below that of the dry bulb of the entering air. Sometimes the expression 1 minus the humidifying efficiency, expressed decimally,

PSYCHROMETRIC PROPERTIES OF AIR

is referred to as the *bypass factor*. Humidifying efficiency is stated in equation form as

$$E_h = \frac{t_1 - t_2}{t_1 - t_1'} 100 \qquad (2.10)$$

Table 2.3 gives typical humidifying efficiencies for commercial washers with one or two banks of spray nozzles and with three different arrangements of the nozzles in the two-bank units.

2.16. Evaporative Cooling. When conditions are favorable the partial adiabatic saturation of air can be used to reduce its dry-bulb temperature in hot weather. As moisture must be added to the air in evaporative cooling, the lowest dry-bulb temperature possible is that equal to the wet-bulb temperature of the entering air. Consequently, when the initial relative humidity is high the reduction of dry-bulb temperature possible is less than when the air is initially drier.

Example. Find the final dry-bulb temperature and relative humidity of air washed with recirculated spray water if the air is initially at 95 F dbt with 50 per cent relative humidity as it enters an air washer which has a humidifying efficiency of 85 per cent.

Solution. Figure 2.4 shows graphically the process. Air at 95 F dbt has 79 F wbt when the relative humidity is 50 per cent. The

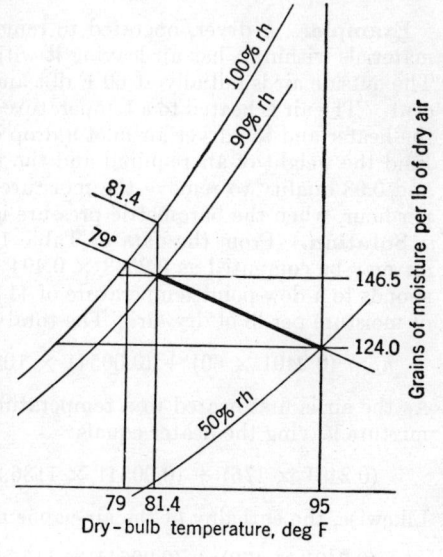

Fig. 2.4. Evaporative cooling.

maximum reduction of the dry-bulb temperature possible is $95 - 79 = 16$ F. The actual reduction of dry-bulb temperature will be $16 \times 0.85 = 13.6$ F so that the final dry-bulb temperature will be $95 - 13.6 = 81.4$ F. Following the wet-bulb temperature line, line of constant enthalpy, from 95 F dbt and 50 per cent relative humidity to 81.4 F dbt, indicates that the final relative humidity will be 90 per cent. The final conditions shown by this example are decidedly unsatisfactory from the comfort standpoint. Except in localities where air to be cooled has a low initial relative humidity, evaporative cooling in air-conditioning plants is limited in its applications. (See Art. 19.11.)

2.17. Drying of Materials. The removal of moisture from materials which are being processed is an important item in many manufacturing operations. Abstraction of moisture from materials to be dried can be effected by the process of adiabatic saturation. In a dryer, contact of the heated air with the moisture to be removed is not so readily made as is the case in apparatus where the water to be evaporated is sprayed into the heated-air stream or where the air is passed over wetted surfaces.

The weight of dry air required per hour depends upon the weight of moisture to be removed per hour and the moisture pickup per pound of dry air. The hourly weight of air M_a lb equals $M_w \div (w_2 - w_1)$. To arrive at a value for w_2, the relative humidity of the air leaving the dryer must be assumed. With the heat and the final relative humidity of the air known, the psychrometric chart gives the final humidity ratio of the air.

Example. A dryer, operated to remove 8000 lb of moisture per hr from materials within it, has air leaving it with a relative humidity of 80 per cent. The outside air is initially at 60 F dbt and has a relative humidity of 49.4 per cent. The air is heated to a temperature of 175 F by steam coils, and between the heater and the dryer air inlet a drop of 5 F occurs in the air temperature. Find the weight of air required and the necessary amount of steam at 5 psig and 0.98 quality to remove the moisture from the materials, both in pounds per hour, when the barometric pressure is 14.696 psia.

Solution. From the data of Table 1.7 the vapor pressure of the outside air may be computed as $0.2563 \times 0.494 = 0.1266$ psia. This pressure corresponds to a dew-point temperature of 41 F and a humidity ratio of 0.00541 lb of moisture per lb of dry air. The total enthalpy of the mixture at 60 F is

$$h_1 = (0.2401 \times 60) + (0.00541 \times 1088) = 20.29 \text{ Btu per lb of dry air}$$

As the air is first heated to a temperature of 175 F the total enthalpy of the mixture leaving the heater equals

$$(0.2401 \times 175) + (0.00541 \times 1136.2) = 48.16 \text{ Btu per lb of dry air}$$

Likewise, the enthalpy of the air-vapor mixture entering the dryer equals

$$(0.2401 \times 170) + (0.00541 \times 1134.2) = 46.95 \text{ Btu per lb of dry air}$$

From the psychrometric chart, air having a total enthalpy of 46.95 Btu per lb of dry air and a relative humidity of 80 per cent has a dry-bulb temperature of 88.5 F and a humidity ratio of 164 grains or 0.02343 lb of moisture per lb of dry air. The moisture pickup per pound of dry air is $0.02343 - 0.00541 = 0.01802$ lb. The weight of dry air required is $8000 \div 0.01802 = 443{,}950$ lb per hr. The heat added to the air-vapor mixture in the coils is $443{,}950(48.16 - 20.29) = 12{,}372{,}885$ Btuh. If only the latent heat of steam at 19.7 psia is available per pound of steam, the hourly weight of steam required is $12{,}372{,}885 \div (0.98 \times 960.6) = 13{,}143$ lb.

2.18. Air Humidification. Whenever water vapor is added to air, heat energy from some source is required in the process to effect a change of state of the necessary water from a liquid to a vaporous condition. With spray equipment, Fig. 20.2, the usual schemes employed are to: (1) preheat the air prior to its contact with the water, (2) heat only the spray water, and (3) moderately preheat the air and then bring it in contact with heated spray water. Method 2 is never used where the initial temperature of the air may be below 32 F as there is danger of ice formation in the humidifier. Some usage is made of an air humidification process whereby the air is preheated and water vapor produced by the heating of water by steam coils in open pans (Fig. 2.9).

In the first method of humidification mentioned the water is simply recirculated, and heat is not added to it by a separate heater. Preheating the air increases both its dry- and wet-bulb temperatures. The air passing through water sprays or brought in contact with wetted surfaces tends to become adiabatically saturated as some of the spray water evaporates. The completeness of the process is dependent upon the thoroughness with which the air is brought in contact with the water. Factors involved are the number of banks of spray nozzles used, the effectiveness of the nozzles in breaking up the water into a fine mist, the ratio of air to water handled, the velocity of the air through the spray chamber, and the distance the air travels through the spray. Under favorable conditions the air can be made to leave a washer practically saturated and with a wet-bulb temperature very near to that of the entering air. The spray water also assumes a temperature practically the same as the wet-bulb temperature of the entering air. Therefore, the amount of moisture added to air under the conditions of method 1 is dependent upon the wet-bulb temperature and the design of the washer. To secure the final necessary dry-bulb temperature and relative humidity, the air is passed through reheater coils after it leaves the spray chamber.

2.19. Numerical Examples of Humidification by Preheating Air. The various stages of preheating, humidification by adiabatic saturation, and reheating of air are illustrated by the following examples. Standard barometric pressure is assumed in all cases.

Example 1. Air at 20 F dbt and 100 per cent relative humidity is to be brought to the condition of 70 F dbt and 40 per cent relative humidity by preheating, complete adiabatic saturation with recirculated spray water held at the initial wet-bulb temperature of the air, and reheating after it leaves the water-spray chamber. Find the moisture and the heat added per pound of dry air handled. Use the psychrometric chart.

Solution. First locate on the psychrometric chart, Fig. 2.5, point D representing 70 F dbt and 40 per cent relative humidity. Pass to the left to the saturation curve, and determine C, the final dew-point temperature of 44.6 F, which is the dry- and wet-bulb temperature of the saturated air as it leaves the sprays. Follow the wet-bulb temperature line CB representing 44.6 F until it intersects at B with the horizontal line AB drawn through 20 F dbt and 100 per cent relative humidity. The processes involved are preheating from 20 to 62 F, along A to B; humidification from B to C, with decrease of the dry-bulb temperature from 62 to 44.6 F; and reheating from C to D to

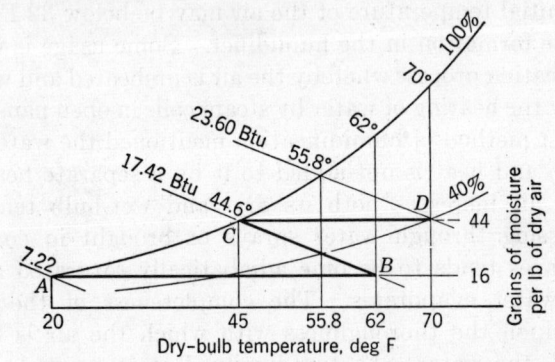

Fig. 2.5. Air humidification.

give the final temperature of 70 F dbt with 40 per cent relative humidity. The chart shows 44 grains of moisture and 23.60 Btu per lb of dry air at the final conditions and 16 grains of moisture and 7.22 Btu per lb of dry air initially. The moisture added per pound of dry air is $44 - 16 = 28$ grains or 0.004 lb, and the heat or enthalpy change is $23.60 - 7.22 = 16.38$ Btu per lb.

When the design and operation of the humidifier are such that complete adiabatic saturation is not possible the air must be preheated to a higher temperature, if the required final conditions are to be obtained.

Example 2. The initial and final conditions are to be the same as in Example 1, with the exception that the relative humidity of the air entering the reheater is 85 per cent, the highest obtainable with the equipment used. Find the temperature to which the air must be preheated and also the humidifying efficiency of a washer as used.

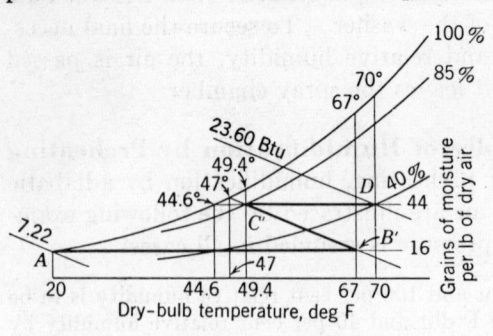

Fig. 2.6. Air humidification.

Solution. The intersection of the constant weight of vapor line for 70 F dbt and 40 per cent relative humidity and the 85 per cent relative humidity curve of Fig. 2.6 fixes the temperatures of the air leaving the humidifying chamber as 49.4 F dbt and 47 F wbt at point C'. Moving along the wet-bulb

PSYCHROMETRIC PROPERTIES OF AIR

temperature line of 47 F to the constant weight of vapor line at B', it will be noted that the preheated air must have 67 F dbt. Hence the path of the process in preheating from 20 F dbt and 100 per cent relative humidity is along AB' to 67 dbt; humidification along $B'C'$ to give a condition of 85 per cent relative humidity; and reheating along $C'D$ to give the desired final state. The theoretical reduction of the dry-bulb temperature as water vapor is added is $67 - 47 = 20$ F. The actual reduction of the dry-bulb temperature is $67 - 49.4 = 17.6$ F. Hence the humidifying efficiency is $(17.6 \times 100) \div 20 = 88$ per cent. The final dew-point temperature and the amounts of moisture and heat added are as in Example 1.

If the relative humidity of the air leaving the washer is not known, the data of Table 2.3 may be used in a simple trial-and-error method to determine the temperature to which the air must be preheated prior to humidification by partial adiabatic saturation.

Example 3. Initial and final conditions again are the same as Example 1. A washer with two spray banks opposing each other is to be used but no information is available about its performance. Find the temperature to which the air must be preheated.

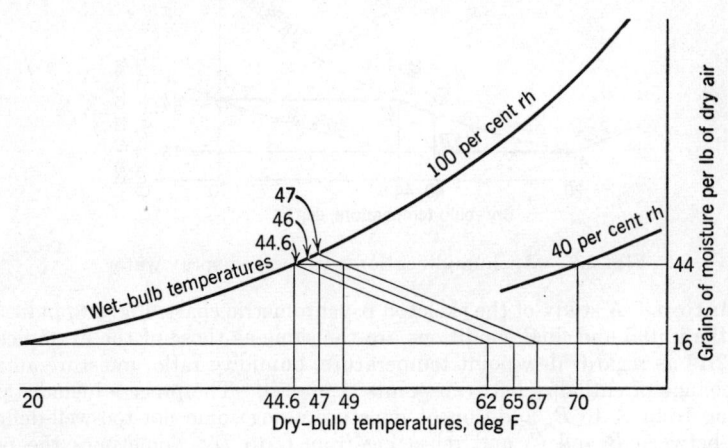

Fig. 2.7. Skeleton psychrometric chart illustrating the method of Example 3, Art. 2.19.

Solution. The median value of 90 per cent shown in Table 2.3 will be taken as the humidifying efficiency. Draw horizontal lines on the psychrometric chart, Fig. 2.7, representing the initial and final humidity ratios. Draw the adiabatic humidification line for a perfect washer as in Fig. 2.5, wet-bulb temperature 44.6 F. As was shown in Example 2, the wet-bulb temperature at which humidification occurs must be higher in the actual washer. Try a wet-bulb temperature of 46 F. From Fig. 2.7 it is seen that the humidifying efficiency with this wet-bulb temperature must be $(65 - 47)/(65 - 46) = 0.948$. Try a wet-bulb temperature of 47 F. In this case the humidifying efficiency needed is $(67 - 49)/(67 - 47) = 0.90$ or 90 per cent. Since the washer to be used may be expected to give a humidifying efficiency of 90 per cent, humidification at a constant wet-bulb temperature of 47 F will be satisfactory and the air must be preheated to a dry-bulb temperature of 67 F.

2.20. Humidification of Preheated Air by Use of Heated Water or by Steam. With conditions of thorough water and air contact, spray water heated by an independent heater is maintained at a temperature corresponding to the dew point of the air at the final desired conditions.

Example. Air at 20 F dbt and 100 per cent relative humidity after being preheated to 40 F dbt is to be humidified and heated to give the final conditions of 70 F dbt and 40 per cent relative humidity when the barometric pressure is 29.92 in. of mercury. Find the heat supplied by the air and water heaters and the moisture added per pound of dry air, if the make-up water enters the water heater at a temperature of 40 F.

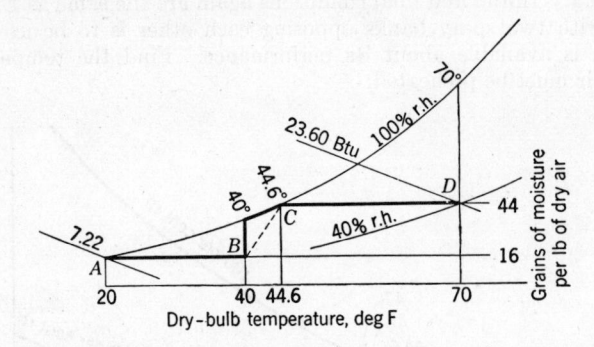

Fig. 2.8. Air humidification with heated spray water.

Solution. A study of the skeleton psychrometric chart, Fig. 2.8, indicates that the initial and final conditions are the same as those of the examples of Art. 2.19 as regards dew-point temperature, humidity ratio, moisture added, and change of enthalpy between points A and D. The process includes preheating from A to B, addition of moisture along some not-too-well-defined path between B and C, and reheating from C to D. Sometimes the path between points B and C is shown as the dashed line joining the two. By the chart, under initial conditions, each pound of dry air has a humidity ratio w_{g1} of 16 grains and an enthalpy h_1 of 7.22 Btu per lb. For the final conditions, w_{g2} is 44 grains and h_2 equals 23.60 Btu per lb of dry air. Therefore in the completed process the moisture increase per pound of dry air is 44 − 16 = 28 grains or 0.004 lb, and the change of enthalpy is 23.60 − 7.22 = 16.38 Btu per lb. Make-up water entering the system has an enthalpy of the liquid which amounts to 8.05 Btu per lb at 40 F. Since the value of 16.38 Btu per lb includes heat from all sources, the net heat supplied by the heaters is 16.38 − (0.004 × 8.05) = 16.35 Btu per lb of dry air and its contained vapor.

Thus far the discussion of air humidification has involved some form of spray equipment. A simple steam-coil-heated open-pan humidifier,

PSYCHROMETRIC PROPERTIES OF AIR

as illustrated by Fig. 2.9, permits the vaporization of water at 212 F (assuming standard atmospheric pressure). With such equipment the operations are as indicated by the skeleton chart of Fig. 2.10 for the data of the following example.

Example. Saturated air at 20 F is to be preheated, humidified by means of open-pan equipment and then reheated to give a relative humidity of 40 per cent and a dry-bulb temperature of 70 F. Determine the necessary procedure if the water vapor is supplied at 212 F. Assume standard barometric pressure.

Solution. The foregoing solution indicated that under the initial conditions the air has 16 grains (0.00228 lb) of water vapor and 7.22 Btu per lb of dry air. The final conditions of 40 per cent relative humidity and 70 F dbt involve a dew-point temperature of 44.6 F, 44 grains (0.00628 lb) of water vapor, and 23.6 Btu per lb of dry air. The enthalpy of dry saturated water vapor at 212 F is 1150.4 Btu per lb; therefore the heat added per pound of dry air as the vapor mixes with it is 1150.4 (0.00628 − 0.00228) = 4.6 Btu. The humidifying process may start at B, B', B'', or at any intermediate condition between B and B'' of Fig. 2.10. The enthalpy at point C (the minimum temperature for the humidified air) from the psychrometric chart is 17.5 Btu per lb. The minimum enthalpy of the air entering the humidifier is then 17.5 − 4.6 = 12.9 Btu per lb. In order to satisfy this condition the air must be preheated along line AB until it intersects the inclined line representing the required enthalpy (point B). The required minimum temperature of the preheated air is read as 43.5 F. After the air has received all of the required water vapor between B and C, additional heat amounting to 23.6 − 17.5 = 6.1 Btu must be added as reheat along line CD.

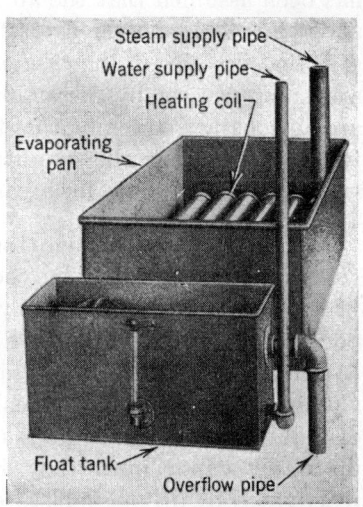

Fig. 2.9. Pan humidifier. (Courtesy Johnson Service Co.)

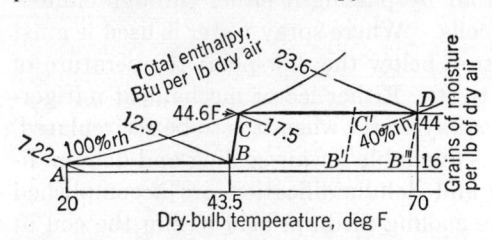

Fig. 2.10. Air humidification with pan-type unit.

The maximum enthalpy of the air entering the humidifier equals 23.6 − 4.6 = 19.0 Btu per lb. This condition on Fig. 2.10 is represented by point B'' where the temperature from the psychrometric chart is found to be 68.8 F (not shown on Fig. 2.10). Air preheated to this temperature will be at 70 F after 44 − 16 = 28 grains of saturated water vapor at 212 F are added to each pound. No reheating of the air is required in this example if the air is preheated to 68.8 F.

In all of the foregoing examples of air heating and humidification it has been assumed that the air would be delivered to the space served at 70 F. This delivered air temperature is generally representative of actual practice in the case of split systems in which the heat losses are taken care of by disseminators such as radiators. Where there are no separate disseminators to take care of heat losses, the final air temperature from the conditioner must exceed 70 F during cold weather except in interior areas from which there is no heat loss.

2.21. Air Dehumidification. Moisture can be removed from air containing water vapor by cooling the mixture to a temperature at which the required amount of vapor is condensed or by using sorbents. Sorbents are further classified as *absorbents*, such as calcium chloride and lithium chloride, and *adsorbents*, such as silica gel and activated alumina. There is no chemical reaction with adsorbents, but because of a difference of vapor pressure within and without them they pick up water vapor until they become inactive. Such materials are reactivated by driving moisture from them by the application of heat. When sorbents are used to remove moisture from air, the dry-bulb temperature of the air rises as a transformation of latent heat to sensible heat occurs. Adsorption and absorption are further discussed in Arts. 19.6 and 20.9.

2.22. Dehumidification of Air by Cooling Processes. Water vapor may be removed from air by passing it either through chilled-water sprays or over cooling coils. Where spray water is used it must be maintained at a temperature below the dew-point temperature of the air coming in contact with it. Either ice or mechanical refrigeration is necessary to cool the spray water when it is to be recirculated. Occasionally cold well water is available for air cooling and dehumidification. Where air cooling and dehumidification are accomplished by means of coil surfaces, the cooling medium held within the coil or coils must be maintained at a temperature well below the dew-point temperature of the cooled and dehumidified air.

2.23. Air Cooling and Dehumidification with Water Sprays. The theoretical performance of a spray dehumidifier is shown in Fig. 2.11.

Pieces of spray equipment, which are properly designed and carefully operated, give results which are close to those shown. In Fig. 2.11 the removal of sensible heat is along the constant weight of vapor

PSYCHROMETRIC PROPERTIES OF AIR

line AB until the dew-point temperature is reached at B, where condensation of water vapor begins. As condensation occurs along the saturation curve from B to C, both sensible and latent heat are removed. The tota heat removed is the sum of the sensible and the latent heats. As condensation takes place the ratios of the sensible to the total heat change. Enthalpies of the mixture for conditions A, B, C, and D are h_A, h_B, h_C, and h_D. The sensible heat removed is $h_D - h_C$, and the latent heat abstracted is $h_A - h_D$. The sensible-heat ratio or sensible-heat factor (SHF) is $(h_D - h_C) \div (h_A - h_C)$. The line, CA is a load-ratio line, and parts CD and DA represent proportionally the amounts of sensible and latent heat removed.

Example. Air at 100 F dbt and 60 per cent relative humidity is to be cooled and dehumidified by passing it through a cold-water spray when the barometric pressure is 14.696 psia. The spray water, at a suitable temperature, is adequate in quantity for air-washer operation, and the velocity of air flow is not excessive. Find the final relative humidity of the air, the moisture removed, the heat removed, and the ratio of sensible to total heat removed when the final temperature is 60 F.

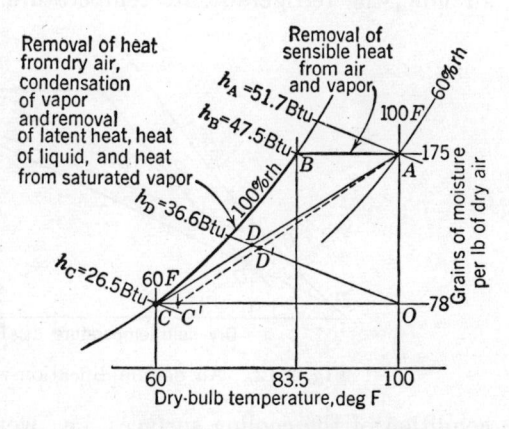

Fig. 2.11. Air dehumidification with a spray cooler.

Solution. The psychrometric chart, Fig. 2.11, indicates w_{gA} as 175 grains of moisture and h_A as 51.7 Btu per lb of dry air and its contained moisture at the initial state. Sensible-heat removal along line AB with a constant weight of vapor w_{gA} equal to 175 grains occurs until the initial dew-point temperature is produced at 83.5 F. The air remains saturated with 100 per cent relative humidity as it is cooled, with the removal of both sensible and latent heat, to a final temperature of 60 F. Under the final conditions, with air at 100 per cent relative humidity, w_{gC} is 78 grains and h_C is 26.5 Btu per lb of dry air. The moisture removed is $w_{gA} - w_{gC} = (175 - 78) = 97$ grains or 0.0139 lb per lb of dry air, and the total abstraction of heat h_T is $h_A - h_C$ or $(51.7 - 26.5) = 25.2$ Btu per lb. The sensible heat removed is $h_D - h_C$ or $(36.6 - 26.5) = 10.1$ Btu per lb. The ratio of sensible heat to total heat removed is $10.1 \div 25.2 = 0.40$.

A commercial spray dehumidifier may not give the performance required to produce state C in Fig. 2.11. When the terminal condition is at C' the humidity ratio will be the correct one but the air will not be saturated and its dry-bulb temperature will be higher than at C. The path of the process between A and C' will not be as previously assumed. Irrespective of the path followed the total heat removed will be $h_A - h_{C'}$. The load-ratio line will be $C'A$. With a higher dry-bulb temperature at C' than at C, the SHF will be different and a greater amount of air will be required to produce the desired conditions within a space.

2.24. Air Cooling and Dehumidification with Coils. When coil surfaces are used a number of factors affect their performance,[4,5] such as surface temperatures; areas; depth of sections; velocities of air flow; the refrigerant, its temperature, and direction of flow; the

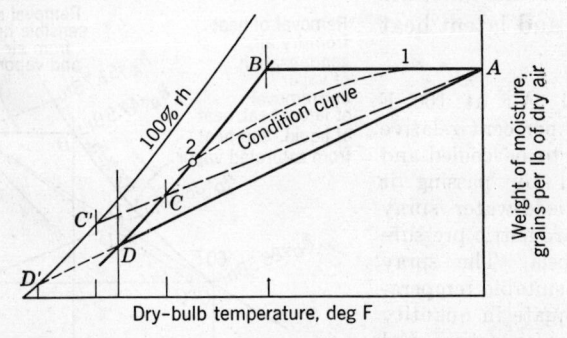

Fig. 2.12. Air dehumidification with a cooling coil.

condition of the cooling surfaces, i.e., wet or dry; and the bypassing of air over the coil surfaces.

The following discussion, dealing with finned cooling coils, involves consideration of the humidity method of determining their performances. Figure 2.12 is a skeleton representation of the use of a psychrometric chart as described in this book. When the inlet and the required exit-air conditions are as indicated by A and D, the load-ratio line AD makes an angle with the horizontal. The slope of the line is dependent upon the existing conditions.

The heat and the moisture removals between the conditions of points A and D are the same irrespective of the path of the process. Theoretically only sensible heat is removed, with a constant humidity ratio, along line AB; from B to D, with constant relative humidity, both sensible and latent heat as well as water vapor are abstracted to produce the final requirements at D. All points located on a relative humidity curve of a psychrometric chart are uniformly horizontally distant from the saturation curve as shown by BD of Fig. 2.12.

When the initial and the terminal conditions are represented by points A and D, the actual path of the dehumidifying and cooling

PSYCHROMETRIC PROPERTIES OF AIR

process falls somewhere within the area $ABCDA$. Thus if point 2 were the desired objective, the condition or pursuit curve 1-2, shown by the dashed line, might represent the removal of both sensible and latent heat along with water vapor to secure the desired condition 2. To locate points C and D similar curves may be superimposed upon the figure.

Lines AC and AD, when extended to the 100 per cent relative humidity curve, locate C' and D', the apparatus dew-point temperatures. Apparatus dew-point temperatures are discussed in Art. 19.13. The following example illustrates the amounts of heat and moisture to be removed when the exit-air-vapor-mixture conditions have been fixed for a dehumidifier.

Example. Air at 95 F dbt and 78 F wbt is to be cooled and dehumidified by passing it over a refrigerant-filled coil which has sufficient depth with the temperature of the refrigerant used to give a final condition of 60 F dbt and a relative humidity of 90 per cent. Find the heat and the amount of moisture removed per pound of dry air. What is the apparatus dew-point temperature?

Solution. The skeleton psychrometric chart, Fig. 2.13, indicates that the moisture is 118 grains per lb of dry air with a total enthalpy of 41.5 Btu per lb, a relative humidity of 47.8 per cent, and a dew-point temperature of 72 F.

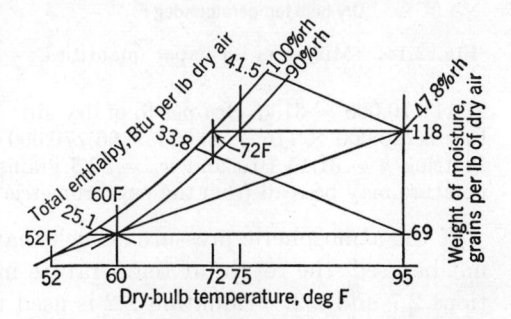

Fig. 2.13. Dehumidification of an air-vapor mixture with a cooling coil.

Sensible heat is shown to be removed at a constant humidity ratio of 118 grains until 90 per cent relative humidity is reached at 75 F dbt. Following the 90 per cent relative humidity line until 60 F dbt is reached, we find the final humidity ratio is 69 grains with a total enthalpy of 25.1 Btu per lb of dry air and contained moisture. The moisture removal is $118 - 69 = 49$ grains, and the heat removal is $41.5 - 25.1 = 16.4$ Btu per lb of dry air. The ratio of sensible heat to total heat removed is $(33.8 - 25.1) \div (41.5 - 25.1) = 0.53$. The apparatus dew-point temperature is 52.0 F.

2.25. The Mixing of Quantities of Air or Air-Vapor Mixtures.

Quite often in air-conditioning calculations it is desirable to be able to find the resultant temperature, the humidity ratio, and the enthalpy of the mixture when two or more quantities of air are combined. The resultant enthalpy can easily be found by a simple energy balance. The resultant humidity ratio may be similarly determined by equating the sum of the masses of moisture in the combining streams to that in the combined stream. If a psychrometric chart is used, the resultant temperature may be read from it as the state of the mixture is determined by its enthalpy and humidity ratio. The following example will illustrate the procedure.

Example. Twenty thousand pounds of dry air having 100 F dbt, a humidity ratio of 116 grains per lb of dry air, and an enthalpy of 42.32 Btu per lb of dry air and its contained water vapor, Fig. 2.14, are mixed with 50,000 lb of dry air having 70 F dbt, 66 grains, and 27.11 Btu. Find the resultant enthalpy, humidity ratio, and dry-bulb temperature. The barometric pressure is 29.92 in. of mercury.

Solution. The total weight of dry air involved is 20,000 + 50,000 = 70,000 lb and the enthalpy of the mixture is (20,000 × 42.32 + 50,000 × 27.11)/70,000 = 31.45 Btu per lb of dry air. The humidity ratio of the mixture is (20,000 × 116 + 50,000 × 66)/70,000 = 80.3 grains per lb of dry air. Plotting h = 31.45 Btu and w_g = 80.3 grains, Fig. 2.14, the resultant temperature may be read from the psychrometric chart as 78.6 F.

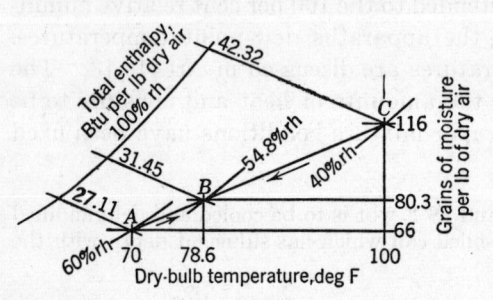

Fig. 2.14. Mixing of air-vapor quantities.

If the atmospheric pressure is such that a psychrometric chart cannot be used, the resultant temperature may be computed with equations 2.7 and 2.2. Equation 2.2 is used to express h_g of equation 2.7 in terms of the resultant temperature. In this example 31.45 = $0.24t_r + (80.3/7000)(1060.8 + 0.45t_r)$. The resultant temperature is computed as 78.6 F.

Although a slight error is made by assuming that point B, Fig. 2.14, lies on a straight line connecting points A and C, when using the type of psychrometric chart shown in Fig. 2.2, approximate values of h_r, w_r, and t_r may be found graphically by plotting the state point of each constituent, A and C in Fig. 2.14, then dividing a straight line connecting them in the proper proportions according to the respective weights of dry air. The distance BC in this example is $AC \times 50,000/70,000$ and the distance AB is $AC \times 20,000/70,000$.

2.26. Weight of Air Required as a Carrier of Heat. Generally speaking, the weight of air required is dependent upon the amount of heat to be delivered and the amount of sensible heat given up per pound of dry air which is $0.24(t_h - t_c) + 0.45w(t_h - t_c)$, where w is the humidity ratio in pounds per pound of dry air. During cold weather the humidity ratio of air which has not been humidified is

PSYCHROMETRIC PROPERTIES OF AIR

very low, and no significant error is introduced by assuming that the air processed is dry, in which case the second term of the above expression may be neglected.

Example. Heat losses amounting to 100,000 Btuh from a space to be maintained at 70 F are to be balanced by heated air introduced at 140 F. All of the air heated is to be taken from outside and the outside design temperature is 0 F. Find the exact weight of dry air to be heated and that based on calculations which assume that the air is dry. Assume that the relative humidity of the outdoor air is 100 per cent.

Solution. From Table 2.1 the humidity ratio of saturated air at 0 F is 0.0007852 lb per lb of dry air and the exact weight of dry air required is $100,000/[0.24(140 - 70) + 0.0007852 \times 0.45(140 - 70)] = 100,000/(16.8 + 0.0247) = 5940$ lb. If the effect of the small amount of moisture is neglected, the weight of dry air required is $100,000/16.8 = 5950$ lb.

REFERENCES

1. "Thermodynamic Properties of Moist Air," J. A. Goff and S. Gratch, *ASHVE Trans.*, **51**, 125 (1945).
2. "Rational Psychrometric Formulae," W. H. Carrier, *ASME Trans.*, **33** (1911).
3. "Air Washer Performance—Analyzed on the Psychrometric Chart," H. M. Hendrickson, *Heating and Ventilating*, **50**, No. 7, 92 (July 1953).
4. "Performance of Surface Coil Dehumidifiers for Comfort Conditions," G. L. Tuve and J. Seigel, *ASHVE Trans.*, **44** (1938).
5. "Air Cooling Coil Problems and Their Solutions," L. G. Seigel, *ASHVE Trans.*, **51** (1945).

PROBLEMS

1. Find by calculation the volume and the density of 1 lb of dry air when it exists at 78 F dbt and a barometric pressure of 24.5 in. of mercury at 32 F.
2. Calculate the weight of water vapor required to saturate 1 lb of dry air when at 78 F dbt and under a barometric pressure of 24.5 in. of mercury at 32 F. Find the volume of the mixture in cubic feet per pound of dry air and the density of the mixture in pounds per cubic foot.
3. Calculate the total enthalpy of saturated air at 72 F and a barometric pressure of 29.5 in. of mercury at 32 F. Express as Btu per pound of dry air.
4. Compute the enthalpy of saturated air at 72 F and a barometric pressure of 26.5 in. of mercury at 32 F. Express as Btu per pound of dry air.
5. Air has 88 F dbt and 72 F wbt when the barometric pressure is 14.14 psia. Find the actual vapor pressure and the relative humidity of the mixture.
6. Find the relative humidity, the percentage of saturation, and the dew-point temperature of air which has 90 F dbt, 70 F wbt, and is under a barometric pressure of 14.24 psia.
7. Calculate the humidity ratio and the enthalpy per pound of dry air for the conditions of Problem 5.

8. Calculate all of the psychrometric properties of saturated air for the conditions of 92 F and a barometric pressure of 14.47 psia.

9. By the use of the psychrometric chart, Fig. 2.2, find all the properties of an air-vapor mixture existing under a barometric pressure of 14.696 psia, 88 F dbt, and 76 F wbt.

10. Show a comparison of the actual psychrometric properties of an air-vapor mixture existing at a barometric pressure of 24 in. of mercury at 32 F, 80 F dbt, and 60 F wbt with those read from a standard psychrometric chart when the temperatures are the same in both cases.

11. Air under a barometric pressure of 14.696 psia has 80 F dbt and 60 F wbt. Determine by calculations the increase of the humidity ratio, and the heat transformation per pound of dry air if the air is adiabatically saturated. Check the calculations by use of a standard psychrometric chart.

12. Calculate the increase in the humidity ratio and the heat transformation per pound of dry air for the temperatures mentioned in Problem 11 if the air is adiabatically saturated when the barometric pressure is 27.3 in. of mercury at 32 F.

13. A piece of spray apparatus was used to partially adiabatically saturate air, in processing an air-vapor mixture which entered it at 80 F dbt and 65 F wbt, to give a final 68 F dbt condition. Find the humidifying efficiency and the bypass factor of the apparatus.

14. Find the humidifying efficiency and the bypass factor for spray equipment, using unheated recirculated spray water, which gives a final condition of 77 F dbt with air which enters the apparatus at 88 F dbt and 76 F wbt.

15. Air initially at 100 F dbt and 70 F wbt is washed with recirculated spray water in passing through a commercial washer having two banks of spray nozzles both spraying with the air flow. Estimate the final dry-bulb temperature and the relative humidity. Assume standard barometric pressure.

16. Ten thousand pounds of moisture are to be removed each hour from materials which are to be dried by passing heated air over them. Air from outdoors at 70 F dbt and a relative humidity of 40 per cent is heated to 160 F and passed to the dryer at 150 F dbt. The final relative humidity of the partially saturated air leaving the dryer is 80 per cent. Steam at 20 psia and 0.97 quality is used in the air heater with the condensate leaving it at steam temperature. Find the weights of wet steam and dry air required per hour if the barometric pressure is 14.696 psia.

17. Air initially at 25 F dbt with a dew-point temperature of 20 F is to be humidified by complete adiabatic saturation. The air is to be preheated to the proper temperature, humidified, and reheated to give a final condition of 68 F dbt and 52 F wbt. Find the amount of water vapor and the heat added during the process. The barometric pressure is 14.696 psia. What is the dry-bulb temperature of the preheated air?

18. Air is to be brought to a final condition of 71 F dbt and 35 per cent relative humidity from an initial condition of 25 F dbt and 90 per cent relative humidity by the processes of preheating, partial adiabatic saturation to give a relative humidity of 95 per cent, and reheating to the terminal condition when the barometric pressure is 14.696 psia. Find the moisture and the heat added during the operations. What is the dry-bulb temperature of the preheated air?

19. Air initially at 25 F dbt and 20 F wbt is to be preheated to the necessary temperature, then humidified by partial adiabatic saturation in passing through a commercial washer 6 ft long containing one bank of spray nozzles which spray with the air flow. The final condition after reheating is to be 75 F dbt and 30 per cent relative humidity. Find the temperature to which the air must be preheated in degrees Fahrenheit dry-bulb temperature. The barometric pressure is 29.92 in. of mercury at 32 F. Use the median value from Table 2.3. Find the equilibrium spray water temperature.

20. Air at an initial condition of 22 F dbt and 90 per cent relative humidity is to be brought to 72 F dbt and 35 per cent relative humidity by being preheated to 35 F dbt and then humidified by being washed with heated spray water held at a temperature corresponding to that of the dew point at the final conditions. Reheating is to be used to give the final desired temperature and relative humidity. Find the heat and moisture added per pound of dry air and the ratio of the sensible to the total heat added. The barometric pressure is 29.92 in. of mercury at 32 F.

21. Air initially at 30 F dbt and 40 per cent relative humidity is to be preheated, humidified with steam from the type of apparatus shown in Fig. 2.9, then reheated if necessary to produce a final condition of 75 F and 30 per cent relative humidity. Determine the minimum and maximum temperature of the preheated air. The barometric presure is 14.696 psia.

22. Air at 83 F dbt and 63 F wbt is to be cooled to produce saturated air at 50 F when the barometric pressure is 14.696 psia. The spray water is at a temperature below the final dew-point temperature. Calculate the amount of moisture and the heat removed per pound of dry air.

23. A spray-type air cooler reduces the dry-bulb temperature of an air-vapor mixture from 95 to 75 F when the initial relative humidity is 60 per cent and the final relative humidity is 95 per cent. The barometric pressure is 28.75 in. of mercury at 32 F. Calculate the heat and the moisture removed per pound of dry air.

24. A cooling coil is supplied with an air-vapor mixture at 90 F dbt and 40 per cent relative humidity. The apparatus discharges the air with a relative humidity of 90 per cent and 55 F dbt. Find the amount of moisture and heat removed per pound of dry air and the SHF when the barometric pressure is 29.92 in. of mercury at 32 F.

25. A cooling coil operates to give at its outlet 65 F dbt together with a relative humidity of 85 per cent when the air-vapor mixture under initial conditions has 95 F dbt and a relative humidity of 50 per cent; the barometric pressure is 14.696 psia. Calculate the moisture and the heat removed per pound of dry air and the SHF.

26. Ten thousand pounds of dry air at 10 F are mixed with 30,000 lb of dry air at 70 F. Find the temperature of the mixture.

27. Eighty thousand pounds of dry air at 80 F dbt and 60 F wbt are mixed with 25,000 lb of dry air at 50 F dbt and 40 F wbt. Calculate the humidity ratio, and the enthalpy in Btu per pound after mixing has taken place if the barometric pressure is 14.696 psia. Determine the resultant dry-bulb temperature by two methods.

28. A room has heat losses amounting to 150,000 Btuh when its air temperature is maintained at 70 F. Air is introduced into the room at 175 F dbt when the barometric pressure is 14.696 psia. Find the weight of dry air required per hour to serve as a carrier of heat and the volume of the air supplied per minute measured at 175 F.

3

Factors Affecting
Human Comfort

3.1. Foreword. If the desired results are to be obtained with either winter or summer air-conditioning systems, the designer of such plants must be cognizant of a number of factors which physiologically affect human comfort. The factors include effective temperature, the production and regulation of heat in the human body, heat and moisture losses from the human body, air motion, the effects of cold and hot surfaces within the spaces considered, and the stratification of air.

3.2. Effective Temperature. Comfort conditions for individuals, excluding radiation, air odors, and cleanliness, are dependent upon the dry- and wet-bulb temperatures of the air and its rate of motion.

The arbitrary index which combines in a single value the degree of warmth or cold felt by the human body in response to the air temperature, moisture content, and motion is termed *effective temperature*. Effective temperature cannot be measured directly but is fixed as the temperature of saturated still air (velocity 15 to 25 fpm) which induces the same sensations of warmth or coolness as those produced by the air surrounding a person.

Figure 3.1 illustrates the ASHAE Comfort Chart[1-5] (first developed by Houghton, Yaglou, and Drinker) in its present form, as later

FACTORS AFFECTING HUMAN COMFORT

modified by others. The chart is for still-air conditions. The diagonal temperature lines represent constant effective temperature. Examination of the chart reveals that several combinations of wet- and dry-bulb temperatures with different relative humidities will pro-

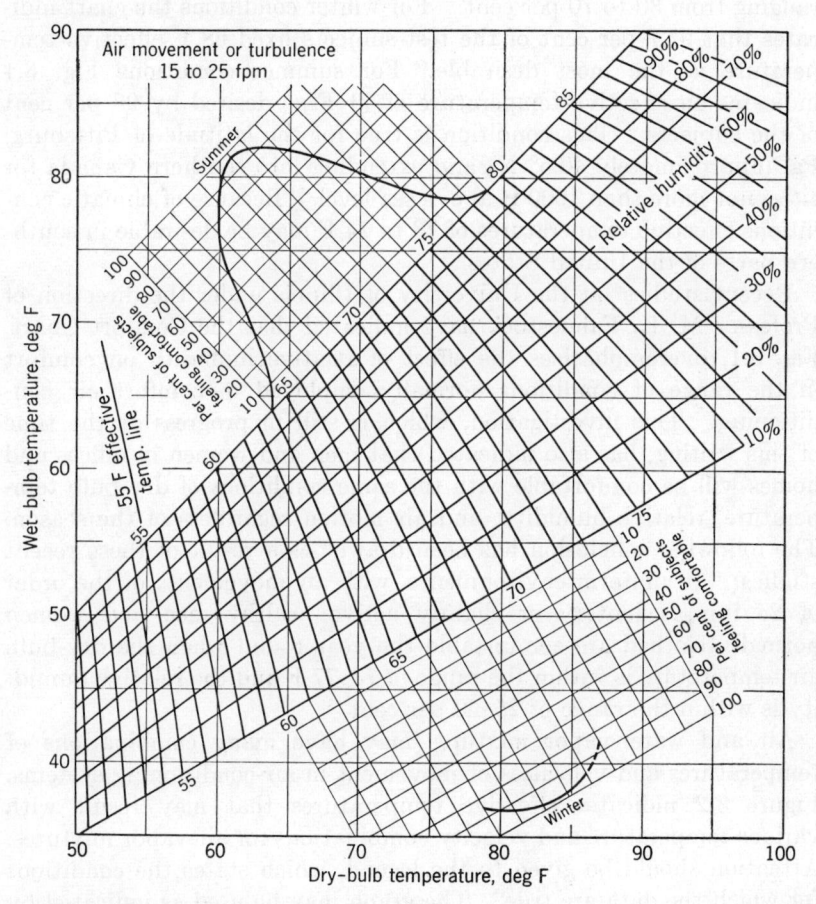

Fig. 3.1. ASHAE Comfort Chart for still air (velocities of 15 to 25 fpm). The dashed portion of the winter comfort curve represents extrapolations beyond test data. (From "Heating Ventilating Air-Conditioning Guide 1957." Used by permission.)

duce the same effective temperature. However, all points located on a given effective-temperature line do not indicate conditions of equal comfort or discomfort. Either extremely high or low relative humidities may produce conditions of discomfort regardless of the existent effective temperature.

Curves of the top and bottom of Fig. 3.1 indicate the percentages of subjects, participating in tests, who found various effective temperatures satisfactory for comfort. The tests covered both winter and summer conditions and for the latter involved relative humidities ranging from 30 to 70 per cent. For winter conditions the chart indicates that 97.7 per cent of the test subjects fixed 68 F effective temperature as the most desirable. For summer conditions Fig. 3.1 indicates an effective temperature of 71 F as desired by 98 per cent of the subjects. This condition is true for the latitude of Pittsburg, Pa. (approximately 40.5° N) and extending into southern Canada for cities not more than 1000 ft above sea level. Because of climatic conditions effective temperatures of 73 to 74 F may be desirable in southern parts of the United States.

Recent studies[6] at the University of Illinois under the direction of Professor M. K. Fahnestock have indicated that the Comfort Chart, Fig. 3.1, overemphasizes the effect of relative humidity on comfort in the range of conditions normally employed in comfort air conditioning. This investigation, which is still in progress at the time of this writing, has also indicated that men and women in offices and homes will be comfortable with the same conditions of dry-bulb temperature, relative humidity, and air motion regardless of the season. The following conclusion has been drawn[6] as a result of these recent studies: "In uniform environments, with air movements of the order of 25 fpm, sedentary or slightly active healthy men and women normally clothed are comfortable the year round when the dry-bulb air temperature is within the range of 73–77 F and the relative humidity is within the range of 25–60 per cent."

Air and water-vapor mixtures may have many combinations of temperatures and velocities of movement in air-conditioning systems. Figure 3.2 indicates effective temperatures that may result with various temperature and velocity combinations for air-vapor mixtures. Attention should be given to the legend, which states the conditions for which the data are true. The graph may be used as indicated by the dashed line which shows the effective temperatures for certain wet- and dry-bulb temperatures and for various air velocities.

The human body cannot lose heat from its surfaces, by the process of convection, when its blood temperature is less than the dry-bulb temperature of the air flowing over it. The body surface areas may receive heat from the air by convection if the temperature of the air is above 98.6 F (see top of Fig. 3.2). However, for the major portion of the cases involved in air conditioning, the effects of increased air motion are reflected by reduced effective temperatures.

FACTORS AFFECTING HUMAN COMFORT

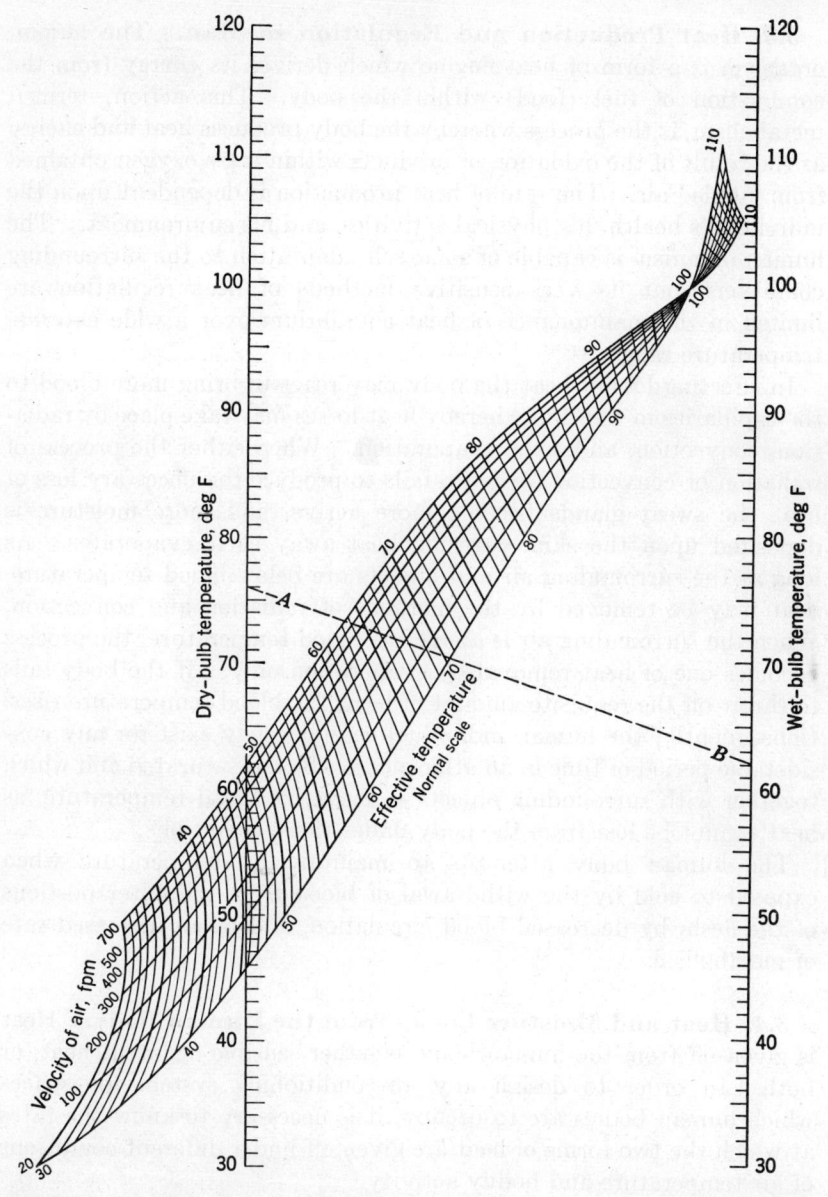

Fig. 3.2. Thermometric or effective temperature chart. Applicable to inhabitants of the United States engaged in light muscular or sedentary work and wearing the customary indoor clothing where warm-air, direct steam, hot-water, or plenum heating systems are used. (From "Heating Ventilating Air-Conditioning Guide 1957.") (Used by permission.)

3.3. Heat Production and Regulation in Man. The human organism is a form of heat engine which derives its energy from the combustion of fuel (food) within the body. This action, termed metabolism, is the process whereby the body produces heat and energy as the result of the oxidation of products within it by oxygen obtained from inhaled air. The rate of heat production is dependent upon the individual's health, his physical activities, and his environment. The human organism is capable of some self-adaptation to the surrounding conditions, but its very sensitive methods of heat regulation are limited in the maintenance of heat equilibrium over a wide external temperature range.

In effecting loss of heat the body may react to bring more blood to the capillaries in the skin whereby heat losses may take place by radiation, convection, and some evaporation. When either the process of radiation or convection, or both, fails to produce the necessary loss of heat the sweat glands become more active, and more moisture is deposited upon the skin, carrying heat away as it evaporates. As long as the surrounding air and objects are below blood temperature, heat may be removed by the methods of radiation and convection. When the surrounding air is above the blood temperature, the process becomes one of heat removal by evaporation only. If the body fails to throw off the requisite amount of heat, the blood temperature rises. Consequently, the human individual cannot safely exist for any considerable period of time in an atmosphere which is saturated and which together with surrounding objects is above his blood temperature, as heat cannot be lost from the body under those conditions.

The human body attempts to maintain its temperature when exposed to cold by the withdrawal of blood from the outer portions of the flesh, by decreased blood circulation, and by an increased rate of metabolism.

3.4. Heat and Moisture Losses from the Human Body. Heat is given off from the human body as either sensible or latent heat, or both. In order to design any air-conditioning system for spaces which human bodies are to occupy, it is necessary to know the rates at which the two forms of heat are given off under different conditions of air temperature and bodily activity.

Figures 3.3, 3.4, and 3.5, which were developed as results of research at the ASHVE laboratory[7] and which are reproduced by permission, supply information relative to total heat, sensible heat, and latent heat respectively. Figure 3.3 is based on effective temperatures; Figs. 3.4 and 3.5 employ dry-bulb temperatures. Attention is called to the

FACTORS AFFECTING HUMAN COMFORT

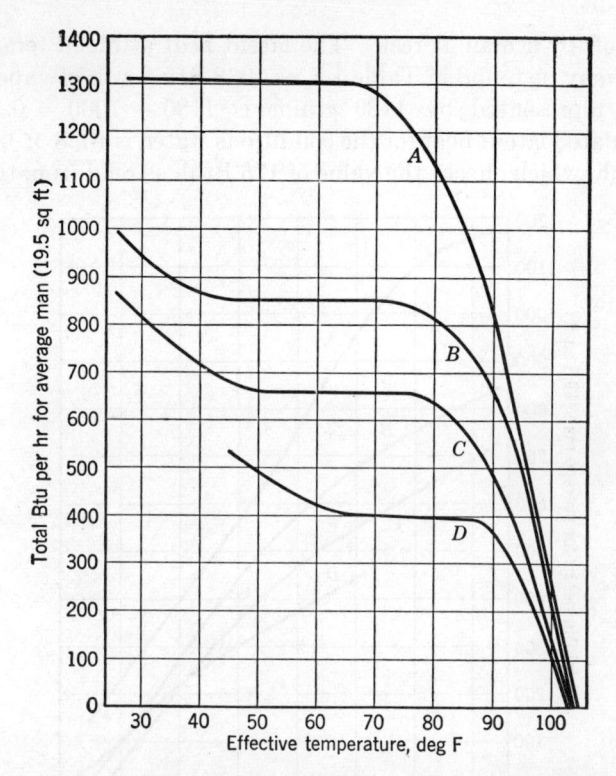

Fig. 3.3. Relation between total-heat loss from the human body and effective temperature for still air.

Curve A, men working 66,150 ft-lb per hr, metabolic rate 1310 Btuh. Curve B, men working 33,075 ft-lb per hr, metabolic rate 850 Btuh. Curve C, men working 16,538 ft-lb per hr, metabolic rate 660 Btuh. Curve D, men seated at rest, metabolic rate 400 Btuh. Curves B and D are based on test data which cover a wide temperature range. Curves A and C are based on test data at an effective temperature of 70 F and extrapolation of curves B and D which are from data at many temperatures. All curves are averages of data for high and low relative humidities; variations due to humidity are small.

[Copyright American Society of Heating and Ventilating Engineers. *ASHVE Trans.*, **37** (1931).]

practically uniform conditions of heat losses over a considerable range of effective temperature for the various rates of physical activity as shown by Fig. 3.3. Figure 3.5 shows the grains of moisture given off per hour by an individual when engaged in different rates of physical activity.

Curve D of Fig. 3.5 shows that, at a dry-bulb air temperature of 80 F, 1180 grains of moisture and 175 Btu of latent heat per hr will be given off by a man at rest. The latent heat at blood temperature of 98.6 F may be found in Table 1.7, as 1038 Btu per lb of vapor. The moisture represented by 1180 grains is 1180 ÷ 7000 = 0.1685 lb. The calculated latent heat for the conditions stated is 1038 × 0.1685 = 174.9 Btuh, which checks the value of 175 Btuh as read from the chart.

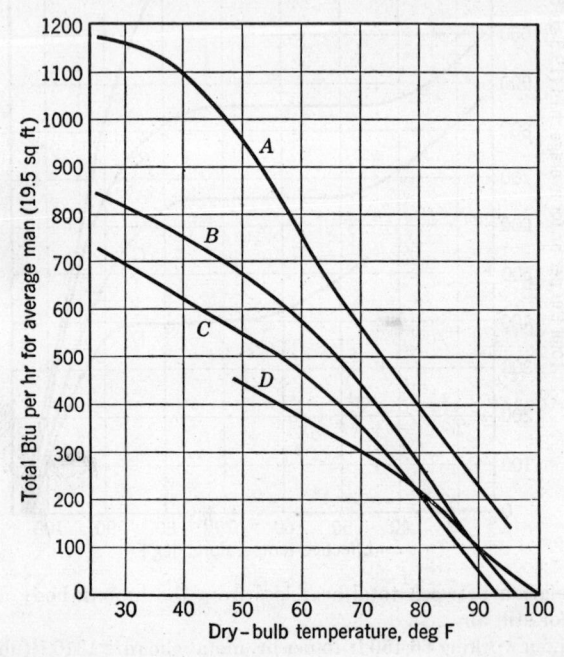

Fig. 3.4. Relation between sensible-heat loss from the human body and dry-bulb temperature for still air. Curves A, B, C, and D same as in Fig. 3.3. [Copyright American Society of Heating and Ventilating Engineers. *ASHVE Trans.*, **37** (1931).]

3.5. Moisture Content of Air. The discussion of effective temperature brought out the fact that air dry-bulb temperature, relative humidity, and air motion are interrelated. The moisture content of outside air may be low during cold weather and above the average during hot weather because the capacity of the air to carry moisture is dependent upon its dry-bulb temperature. This means that, in the winter, in-leakage of cold outside air, having a low-moisture content, will cause a low relative humidity in heated spaces unless moisture is added to the air by the process of humidification. In the summer the reverse is likely to occur unless moisture is removed from the inside air by a dehumidification process. In selecting the proper dry-bulb air temperature for either summer or winter conditions the

FACTORS AFFECTING HUMAN COMFORT

designer must be influenced by the practical consideration of relative humidities which are feasible. In general, for winter conditions in the average residence, relative humidities above 35 to 40 per cent are not practical. In summer comfort cooling, the air of the occupied space should not have a relative humidity above 60 per cent. With these limitations as to relative humidity the designer may determine

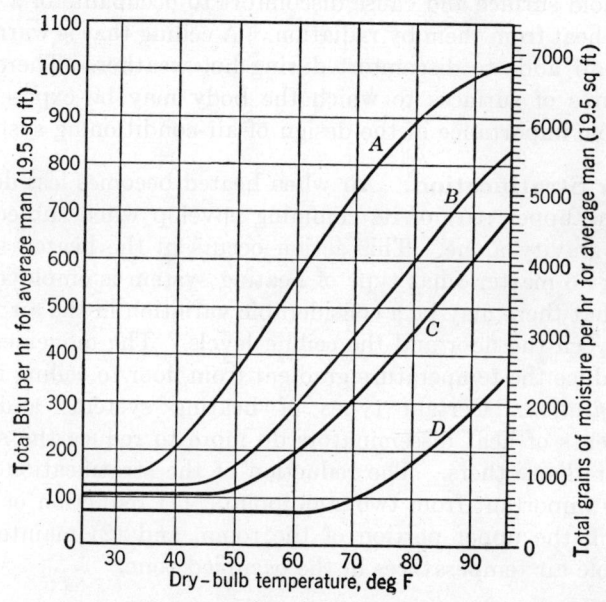

Fig. 3.5. Latent heat and moisture loss from the human body by evaporation in relation to dry-bulb temperature for still-air conditions. Curves A, B, C, and D, same as in Fig. 3.3. [Copyright American Society of Heating and Ventilating Engineers. *ASHVE Trans.*, **37** (1931).]

from the Comfort Chart the necessary dry-bulb temperatures for the air. In industrial air conditioning where relative humidities either higher or lower than those mentioned for residence work are imperative, it will be necessary to design and equip the structure so that the required relative humidites are feasible and practical.

3.6. Air Motion. No system of air conditioning is satisfactory unless the air handled is properly circulated and distributed. The air velocity in the occupied zone ordinarily should not exceed 25 to 40 fpm. The air velocities in the space above the occupied zone may be any amount which is necessary to produce good distribution of the air in the space, provided of course that the air in motion does not produce an objectionable noise. Whenever possible the flow of air should be toward the faces of the individuals in the occupied zone rather than from the rear.

3.7. Cold and Hot Surfaces. Discomfort may be occasioned by the presence of either cold or hot objects in a space. A large area of single glass exposed to the outdoor air during cold weather may constitute a cold surface and cause discomfort to occupants of a room by absorbing heat from them by radiation. A ceiling that is warmer than the room air adds to discomfort during hot weather. Therefore the temperatures of surfaces to which the body may be exposed are of considerable importance in the design of air-conditioning systems.

3.8. Air Stratification. Air when heated becomes less dense and rises to the upper part of its confining envelop when subject to the action of gravity alone. This action occurs in the heated spaces of structures no matter what type of heating system is employed. The result is that there may be a considerable variation in the air temperatures between the floor and the ceiling levels. The movement of the air to produce the temperature gradient from floor to ceiling is termed *air stratification*. Certain types of heating systems and certain arrangements of heat disseminators do more to reduce the stratification of air than others. The reduction of the stratification of air in a space is important from two standpoints: (1) reduction of the heat losses from the upper portion of the room, and (2) maintenance of comfortable air temperatures in the occupied zone.

3.9. Appraisal of Comfort Conditions. An evaluation of comfort conditions should include a study of the effects of radiation, air dry-bulb temperatures, relative humidities, and air motion. The ASHAE Comfort Chart, Fig. 3.1, is based on the last three items enumerated, the data for which may be obtained by the use of a suitable psychrometer and a means of determining air velocities. When thermal interchanges take place the following instruments[8] are of use in the appraisal of conditions as they affect the comfort of the human body: globe thermometer, eupatheoscope, and the thermointegrator.

A globe-thermometer unit[9] consists of an ordinary mercurial thermometer, Fig. 3.6, placed so that its bulb is at the center of a blackened hollow sphere which has a diameter of 8 in. The thermometer receives radiant heat from all directions when the instrument is properly suspended and provides a means of measuring radiant effects.

The eupatheoscope is a device designed to measure the effects of both air temperature and radiation upon the human body. The instrument consists of a blackened hollow copper cylinder of $7\frac{1}{2}$-in.

diameter and 22-in. height, which gives the proper surface area. The cylinder surface is maintained at 80 F (the approximate mean surface temperature of the human body) by two thermostatically controlled electric bulbs. A portion of the lamp current is passed through a coil wound around a mercury thermometer which furnishes a measure of the heat necessary to maintain the surface temperature of the cylinder at the required value. This instrument has been used in connection with studies of the performance of direct steam radiators, as a control in radiant-heating systems, and as a measure of cooling effects in occupied spaces. The device has limitations when the room air is warmer than the desired surface temperature and when it is too cool for the lamps to maintain the proper surface temperature.

The thermointegrator[10] consists of an electroplated hollow copper cylinder made of metal about 0.05 in. thick. The cylinder has a diameter of 8 in., a length of 24 in., and hemispherical ends. The device is equipped with an internal heater and can be operated so as to dissipate 17.5 Btuh per sq ft of surface. Temperatures of the ambient air and the surfaces of the cylinder are measured after equilibrium conditions have been established. The surface temperature of the thermointegrator is affected by radiation together with room-air temperature and movement. The effects of the relative humidity of the room air are not included. The thermointegrator sums up all radiation effects in a room. When the instrument is properly calibrated it can be used to measure them by subtracting from its measured total-heat losses those occasioned by the room-air temperature and velocity.

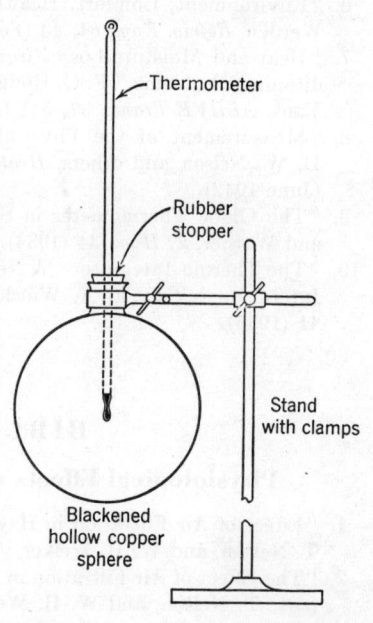

Fig. 3.6. Globe thermometer.

REFERENCES

1. "Heating Ventilating Air Conditioning Guide," ASHAE, New York, 126 (1957)

2. "Determination of the Comfort Zone," F. C. Houghten and C. P. Yaglou, *ASHVE Trans.*, **29**, 163 (1923).
3. "The Summer Comfort Zone: Climate and Clothing," C. P. Yaglou and Phillip Drinker, *ASHVE Trans.*, **35**, 269 (1929).
4. "How to Use the Effective Temperature Index and Comfort Charts," Report of Technical Advisory Committee, *ASHVE Trans.*, **38**, 411 (1932).
5. "Effective Temperature with Clothing," C. P. Yaglou and W. E. Miller, *ASHVE Trans.*, **31**, 89 (1925).
6. "Environment, Comfort, Health and People," M. K. Fahnestock and J. E. Werden, *Refrig. Eng.*, **64**, 43 (Feb. 1956).
7. "Heat and Moisture Losses from Men at Work and Application to Air-Conditioning Problems," F. C. Houghten, W. W. Teague, W. E. Miller, and W. P. Yant, *ASHVE Trans.*, **37**, 541 (1931).
8. "Measurement of the Physical Properties of the Thermal Environment," D. W. Nelson and others, *Heating, Piping, Air Conditioning*, **14**, No. 6, 382 (June 1942).
9. "The Globe Thermometer in Studies of Heating and Ventilating," Bedford and Warner, *J. Hyg.*, **34** (1934).
10. "The Thermo-Integrator—A New Instrument for the Observation of Thermal Interchanges," C. E. A. Winslow and Leonard Greenburg, *ASHVE Trans.*, **41** (1935).

BIBLIOGRAPHY

Physiological Effects of Atmospheric Environment

1. "Effect of Air Filtration in Hay Fever and Pollen Asthma," B. Z. Rappaport, T. Nelson, and W. H. Welker, *J. Am. Med. Assoc.*, **98**, 1861 (May 28, 1932).
2. "The Effect of Air Filtration in Hay Fever and Pollen Asthma," B. Z. Rappaport, T. Nelson, and W. H. Welker, *J. Am. Med. Assoc.*, **100**, 1385 (May 6, 1933).
3. "Filtered Air Relieves Hay Fever," W. H. Welker, B. Z. Rappaport, and T. Nelson, *Heating, Piping, Air Conditioning*, **5**, No. 7, 348 (July 1933).
4. "Air Conditioning in the Treatment of Pollen Asthma," T. Nelson, B. Z. Rappaport, A. G. Canar, and W. H. Welker, *Heating, Piping, Air Conditioning*, **6**, No. 8, 329 (Aug. 1934).
5. "The Effect of Low Relative Humidity at Constant Temperature on Pollen Asthma," B. Z. Rappaport, Tell Nelson, and W. H. Welker, *J. Allergy*, **6**, No. 2, 111 (Jan. 1935).
6. "Physiologic Response of Man to Environmental Temperature," F. K. Hick, R. W. Keeton, and Nathaniel Glickman, *ASHVE Trans.*, **44**, 143 (1938).
7. "Cardiac Output, Peripheral Blood Flow, and Blood Volume Changes in Normal Individuals Subjected to Varying Environmental Temperatures," F. K. Hick, R. W. Keeton, Nathaniel Glickman, and H. C. Wall, *ASHVE Trans.*, **45**, 123 (1939).
8. "The Peripheral Type of Circulatory Failure in Experimental Heat Exhaustion," R. W. Keeton, F. K. Hick, Nathaniel Glickman, and M. M. Montgomery, *ASHVE Trans.*, **46**, 157 (1940).
9. "The Influence of Physiological Research on Comfort Requirements." R. W.

Keeton, F. K. Hick, Nathaniel Glickman, and M. M. Montgomery, *ASHVE Trans.*, **47**, 159 (1941).
10. "Blood Volume Changes in Men Exposed to Hot Environmental Conditions for a Few Hours," Nathaniel Glickman, F. K. Hick, R. W. Keeton, and M. M. Montgomery, *Am. J. Physiol.*, **134**, No. 2 (Sept. 1941).
11. "The Tolerance of Man to Cold as Affected by Dietary Modifications: Carbohydrate Versus Fat, and the Effect of the Frequency of Meals," H. H. Mitchell, Nathaniel Glickman, E. H. Lambert, R. W. Keeton, and M. K. Fahnestock, *Am. J. Physiol.*, **146**, No. 1 (Apr. 1946).
12. "The Tolerance of Man to Cold as Affected by Dietary Modifications: Proteins Versus Carbohydrates, and the Effect of Variable Protective Clothing," R. W. Keeton, E. H. Lambert, Nathaniel Glickman, H. H. Mitchell, J. H. Last, and M. K. Fahnestock, *Am. J. Physiol.*, **146**, No. 1 (Apr. 1946).
13. "The Tolerance of Man to Cold as Affected by Dietary Modifications: High Versus Low Intake of Certain Water-Soluble Vitamins," Nathaniel Glickman, R. W. Keeton, H. H. Mitchell, and M. K. Fahnestock. *Am. J. Physiol.*, **146**, No. 4 (July 1946).
14. "Physiological Adjustments of Human Beings to Sudden Change in Environment," Nathaniel Glickman, Tohru Inouye, S. E. Telser, R. W. Keeton, F. K. Hick, and M. K. Fahnestock, *Heating, Piping, Air Conditioning*, **19**, No. 7, 101 (July 1947).
15. "Comparison of Physiological Adjustments of Human Beings During Summer and Winter," Nathaniel Glickman, Tohru Inouye, R. W. Keeton, and M. K. Fahnestock, *Heating, Piping, Air Conditioning*, **20**, No. 8, 113 (Aug. 1948).
16. "Physiologic Adjustments of Normal Subjects and Cardiac Patients to Sudden Change in Environment," Nathaniel Glickman, Tohru Inouye, R. W. Keeton, I. R. Callen, F. K. Hick, and M. K. Fahnestock, *Heating, Piping, Air Conditioning*, **21**, No. 2, 105 (Feb. 1949).
17. "Thermocouples and the Measurement of Body Temperatures in the Laboratory and in Flight," M. K. Fahnestock, *J. Aviation Med.*, **17**, No. 2 (Apr. 1946).
18. "Physiologic Examination of the Effective Temperature Index," Nathaniel Glickman, Tohru Inouye, R. W. Keeton, and M. K. Fahnestock, *Heating, Piping, Air Conditioning*, **22**, No. 1, 147 (Jan. 1950).
19. "The Physical Environment Unit at the University of Illinois," M. K. Fahnestock, *Heating, Piping, Air Conditioning*, **22**, No. 9, 109 (Sept. 1950).
20. "Physical Environment Control Essential in Nutrition Studies," H. H. Mitchell, *Heating, Piping, Air Conditioning*, **22**, No. 10, 87 (Oct. 1950).
21. "Stress of the Physical Environment," R. W. Keeton, *Heating, Piping, Air Conditioning*, **22**, No. 12, 96 (Dec. 1950).
22. "Physiological Adjustments of Clothed Human Beings to Sudden Change in Environment—First Hot Moist and Later Comfortable Conditions," F. K. Hick, Tohru Inouye, R. W. Keeton, Nathaniel Glickman, and M. K. Fahnestock, *Heating, Piping, Air Conditioning*, **24**, No. 2, 107 (Feb. 1952).
23. "A Comparison of Physiological Adjustments of Clothed Women and Men to Sudden Changes in Environment," Tohru Inouye, F. K. Hick, R. W. Keeton, J. Losch, and Nathaniel Glickman, *Heating, Piping, Air Conditioning*, **25**, No. 5, 125 (May 1953).
24. "Controlling the Environment for Clinical Research," M. K. Fahnestock, E. P. Heckel, Nathaniel Glickman, and Tohru Inouye, *Heating, Piping, Air Conditioning*, **25**, No. 1, 123 (Jan. 1953).
25. "Effect of Relative Humidity on Heat Loss of Men Exposed to Environments

of 80, 76 and 72 F," Tohru Inouye, F. K. Hick, S. E. Telser, and R. W. Keeton, *Heating, Piping, Air Conditioning*, **25**, No. 8, 109 (Aug. 1953)
26. "Learning the Why and How of Comfort," F. K. Hick, Tohru Inouye, and M. K. Fahnestock, *Refrig. Eng.*, **62**, No. 1, 48 (Jan. 1954).
27. "Physiological Responses to Sudden Changes in Atmospheric Environment—Studies of Normal Subjects, Obese, Hyperthyroid and Hypothyroid Patients," Tohru Inouye, F. K. Hick, R. W. Keeton, and Lionel Berstein, *Heating, Piping, Air Conditioning*, **26**, No. 7, 131 (July 1954).
28. "Things We Need to Know About Air Conditioning and Environmental Physiological Research," M. K. Fahnestock, *Heating, Piping, Air Conditioning*, **26**, No. 12, 134 (Dec. 1954).

PROBLEMS

1. What is the effective temperature of still air having dry- and wet-bulb temperatures of 75 F? Are conditions satisfactory for comfort in either summer or winter?

2. Still air is at 72 F and 50 per cent relative humidity. What effective temperature exists? What must be the dry-bulb temperature if the effective temperature is increased 2 F with the relative humidity remaining at 50 per cent?

3. Air at 70 F has a velocity of 15 fpm, due to natural turbulence, when its wet-bulb temperature is 50 F. Find the effective temperature. Find the change of effective temperature if the wet-bulb temperature is made 60 F with the other conditions unchanged.

4. Air in a room is to be maintained with a dry-bulb temperature of 82 F and an effective temperature of 71 F. If the air velocity is 20 fpm, what must be the wet-bulb temperature?

5. Air moving at a velocity of 100 fpm has wet- and dry-bulb temperatures of 70 F and 80 F. What is its effective temperature?

6. Air having dry- and wet-bulb temperatures of 75 F and 65 F moves with a velocity of 20 fpm. Find the effective temperature of the air by the use of both Figs. 3.1 and 3.2.

7. Saturated air at 80 F is moved with a velocity of 200 fpm. What effective temperature exists? If the air velocity is made 400 fpm, what decrease of effective temperature occurs?

8. Find the effective temperatures for air at 115 F dbt and 103 F wbt for velocities of 20 fpm and 300 fpm.

9. What total heat is given off, in Btuh, by a man working at the rate of 16,538 ft-lb per hr in an atmosphere having an effective temperature of 70 F?

10. What total sensible heat is given off, in Btuh, by an individual at rest in still air having a dry-bulb temperature of 80 F? What amount of latent heat and how many grains of moisture are given off per hour under the same conditions?

11. Find the sensible heat, the latent heat, and the moisture given off per hour by an individual working at the rate of 33,075 ft-lb per hr when in still air having a dry-bulb temperature of 60 F.

12. Calculate the percentages of the total heat given off as sensible and latent heats for the conditions of Problem 11.

4

Heat Transmission, Vapor Transmission, and Heat Losses from Buildings

4.1. Foreword. The discussion of heat transfer in this book will be confined to a simplified treatment of the subject applicable to the estimation of either heating or cooling loads for buildings. Several excellent publications[1-7] which deal with the subject are available. These include discussions of basic fundamentals and give procedures, equations, data, and graphs for the solution of nearly all of the problems involved in the design of heat-transfer apparatus.

4.2. Over-all Coefficient of Heat Transmission. In equation form the expression of the transmission of heat through each component portion of the enclosing envelope of a space is

$$H = UA(t - t_o) \qquad (4.1)$$

The over-all coefficient of heat transmission for a wall section may be determined either experimentally[8] or by calculations involving the use of known data for the materials included.

The commonly used method of determination of the coefficient U is that of the guarded hot box. The Nicholls heat meter[9] has been developed for use in the determination of the over-all coefficients of heat transmission of actual building walls in place and subjected to actual weather conditions. The apparatus consists of a piece of

Bakelite $\frac{1}{8}$ in. thick and 2 ft square fitted with a number of commonly connected thermocouples embedded flush with its surfaces.

4.3. Calculation of U. Heat passing through a building wall composed of a single homogeneous material, as in Fig. 4.1, is received at the wall surface exposed to the region of higher air temperature by three methods: radiation, convection, and conduction. The flow then takes place by conduction through the material to the surface exposed at the region of lower air temperature where the heat is dispersed through the processes of radiation, convection, and conduction.

The amount of heat H entering the wall per hour per unit of area is fixed by the temperature drop $t - t_1$ and the combined coefficient of radiation, convection, and conduction f_i which is effective at the warmer surface of the wall. The hourly quantity of heat H received per unit area (1 sq ft) is $f_i(t - t_1)$. The foregoing quantity of heat then flows by conduction through the material from the surface at a temperature of t_1 to the surface having the lower temperature of t_2. The heat transfer by conduction is $H = (k/x)(t_1 - t_2)$ Btuh per sq ft.

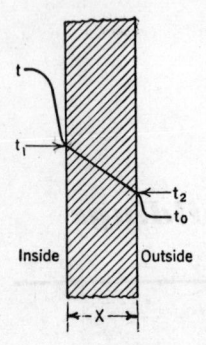

Fig. 4.1. Simple wall.

The symbol x designates the material thickness in inches. k is the thermal conductivity; it is the amount of heat in Btuh passing by conduction through a section of the material 1 in. in thickness and 1 sq ft in area, per 1 F difference between the temperatures of the surfaces. Finally, the loss of heat from the surface having the lower temperature of t_2 is $H = f_o(t_2 - t_o)$, where f_o is the combined effect of radiation, convection, and conduction for the conditions which prevail at the wall surface of lower temperature. The items f_o and f_i usually designated as *surface conductances*, are expressed as Btuh per sq ft of surface per deg F.

As the same quantity of heat is involved in each of the examples of transmission cited for Fig. 4.1 and is also equal to $U(t - t_o)$, then

$$H = f_i(t - t_1) = \frac{k}{x}(t_1 - t_2) = f_o(t_2 - t_o) = U(t - t_o)$$

from which the following may be derived:

$$U = \frac{1}{\dfrac{1}{f_i} + \dfrac{1}{f_o} + \dfrac{x}{k}} \qquad (4.2)$$

HEAT LOSSES FROM BUILDINGS, VAPOR TRANSMISSION

The fractions of the denominator of equation 4.2 are resistances to heat flow. The film or surface resistances are $1/f_i = R_i$ and $1/f_o = R_o$; the resistance to heat transfer by conduction $x/k = R_c$. The sum of R_i, R_o, and R_c is equal to R_t, the total resistance for a simple wall. U, the over-all coefficient of heat transmission, is equal to the reciprocal of R_t and is

$$U = \frac{1}{R_t} = \frac{1}{R_i + R_o + R_c} \qquad (4.3)$$

The flows of electrical energy and heat energy are analogous in that a difference of motive potential is required in each case to overcome the resistances encountered. Electrical energy must have a difference of voltage and heat involves the potential produced by a difference of temperature. Ohm's law for electric circuits indicates that the current flow is directly proportional to the difference in potential and inversely proportional to the resistance to be overcome. When heat flows through resistances placed in series, as in Fig. 4.1, the resistances are added as in equation 4.3.

4.4. Transmission Coefficients for Compound Walls. When building walls are constructed of layers of different materials, and when air spaces are placed between materials, as shown by Fig. 4.2, the additional resistances must be considered.

A conductance C is defined as the amount of heat in Btuh passing through 1 sq ft of area of any material of the thickness and arrangement stated, per 1 F difference of the material surface temperatures. *A conductance is always written into the equation for U as 1/C, and never otherwise.* The subscript a is used to designate the conductances of air spaces of various widths and various air temperatures. Air-space conductances are discussed in Art. 4.7.

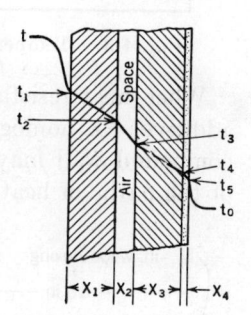

Fig. 4.2. Wall with air space.

The expression of U for the compound wall of Fig. 4.2, which includes several materials and an air space, is

$$U = \frac{1}{R_t} = \frac{1}{\dfrac{1}{f_i} + \dfrac{1}{f_o} + \dfrac{x_1}{k_1} + \dfrac{1}{c_a} + \dfrac{x_3}{k_3} + \dfrac{x_4}{k_4}} \qquad (4.4)$$

Equation 4.4 may be modified to include any number of materials of different thickness and any number of and width of air spaces.

Typical walls of wood-frame construction are shown in Fig. 4.3. Actual temperature gradients[10] through the walls are given for both wind and still-air conditions at their exteriors.

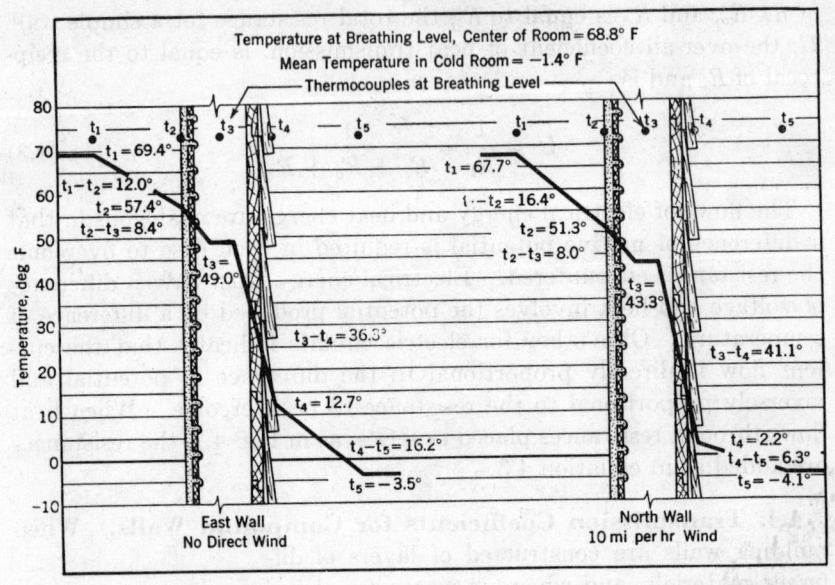

Fig. 4.3. Temperature gradients through walls of frame construction.

When heat resistances are in parallel the *conductances* involved are additive. A homogeneous material has a conductance $C = k/x$. A compound wall may have some of its materials placed so that some of the paths of heat flow are in parallel, as in the wood studding and

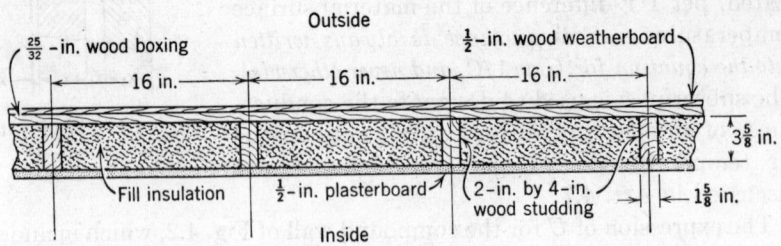

Fig. 4.4. Wood-frame wall construction with insulation.

air spaces of the example of wood-frame construction of Fig. 4.3. Another case of the same type of wall has the air spaces between studding filled with insulation, as indicated in Fig. 4.4. Where a portion of a wall is not of a single homogeneous material, the conductance for that portion may be found by multiplying each fractional part of

HEAT LOSSES FROM BUILDINGS, VAPOR TRANSMISSION

1 sq ft by its proper conductance and then totaling all such proportional conductances.

Example. Find the total resistance to heat flow offered by each square foot of the stud space in the wall of Fig. 4.4 which has standard 2-in. by 4-in. wood studding placed the customary distance apart of 16 in. center to center. The spaces between the studding are filled with mineral wool. The conductivities of the wood and the mineral wool are, respectively, 0.80 and 0.27 Btuh per sq ft per deg F per in.

Solution. The combination of wood studding and the mineral-wool filling between them produces resistances of the materials which are in parallel. The actual dimensions of the wood studding are $1\frac{5}{8}$ in. by $3\frac{5}{8}$ in. For each material the linear distance through which flow of heat may occur is $3\frac{5}{8}$ in. The studding thickness produces $1.625/16 = 0.1016$ sq ft as the proportional part of 1 sq ft of the wall surface. The amount of surface per square foot of wall represented by the insulation-filled spaces is $1.0000 - 0.1016 = 0.8984$ sq ft. The conductance of the wood per square foot of wall area is $(0.1016 \times 0.80)/3.625 = 0.0224$ and the conductance of the mineral wool per square foot of surface is $(0.8984 \times 0.27)/3.625 = 0.0668$. The total conductance per square foot of wall area is $0.0224 + 0.0668 = 0.0892$, and the resistance of each square foot of the wall section having two parallel paths of heat flow is $1/0.0892 = 11.21$ hr per deg F per Btu.

4.5. Surface Conductances. The data[11] of Fig. 4.5 indicate that air movement over the surface of a material increases the surface conductance. For fixed conditions the increase of the surface conductance is nearly directly proportional to the increase of the air velocity. Surface conductances for all materials at conditions of constant air velocity increase with an increase of mean temperature. The mean temperature is the average of the surface temperature and the temperature of adjacent air which may be measured at a distance of 1 in. from the surface.

Under service conditions the air movement may be either parallel to the wall surface or at some angle to it. Angular incidence of the air at the surface tends to reduce the surface conductance some for otherwise fixed conditions. However, the general practice is to assume air flow parallel to the surface.

Commonly accepted average values for the surface conductances of all materials are: f_i as 1.65 for still-air conditions and f_o as 6.0 when the average wind velocity over their surfaces is 15 mph. More exact values of f_i and f_o are given in Table 4.1. (These appear as C values in Table 4.1.)

4.6. Thermal Conductivities and Conductances. The thermal conductivities k for building materials vary greatly and are dependent upon density, moisture content, age and proportions of the mix in the

TABLE 4.1. Conductivities k, Conductances C, and Resistances R of Building and Insulating Materials—Design Values[a]

These constants are expressed in Btuh per sq ft per deg F temperature difference. Conductivities k are per inch thickness and conductances C are for thickness or construction stated, not per inch thickness.

Material	Description			Density, (lb per cu ft)	k	C	R Per Inch Thickness $\left(\frac{1}{k}\right)$	R For Thickness Listed $\left(\frac{1}{C}\right)$
Air spaces[b]	Position	Heat Flow	Thickness					
	Horizontal	Up	¾–4 in.	—	—	1.18	—	0.85
	Sloping (45°)	Up	¾–4 in.	—	—	1.11	—	0.90
	Vertical	Horizontal	¾–4 in.	—	—	1.03	—	0.97
	Sloping (45°)	Down	¾–4 in.	—	—	0.97	—	1.03
	Horizontal	Down	¾ in.	—	—	0.98	—	1.02
	Horizontal	Down	8 in.	—	—	0.80	—	1.25
Air surfaces[c]	Position		Heat Flow					
Still air	Horizontal		Up	—	—	1.63	—	0.61
	Up Sloping (45°)		Up	—	—	1.60	—	0.62
	Vertical		Horizontal	—	—	1.46	—	0.68
	Sloping (45°)		Down	—	—	1.32	—	0.76
	Horizontal		Down	—	—	1.08	—	0.92
15-mph wind	Any position—any direction			—	—	6.00	—	0.17
7½-mph wind	Any position—any direction			—	—	4.00	—	0.25
Building board[d]								
Boards,	Gypsum or plasterboard		⅜ in.	50	—	3.10	—	0.32
panels,	Gypsum or plasterboard		½ in.	50	—	2.25	—	0.45
sheathing, etc.	Plywood			34	0.80	—	1.25	—
Building paper	Vapor—permeable felt			—	—	16.70	—	0.06
	Vapor—seal, 2 layers of mopped 15-lb felt			—	—	8.35	—	0.12
	Vapor—seal, plastic film			—	—	—	—	Negl
Flooring	Asphalt tile		⅛ in.	120	—	24.80	—	0.04
Materials	Ceramic tile		1 in.	—	—	12.50	—	0.08
	Cork tile		⅛ in.	—	—	3.60	—	0.28
	Plywood subfloor		⅝ in.	—	—	1.28	—	0.78
	Rubber of plastic tile		⅛ in.	110	—	42.40	—	0.02
	Terrazzo		1 in.	—	—	12.50	—	0.08
	Wood subfloor		25/32 in.	—	—	1.02	—	0.98
	Wood, hardwood finish		¾ in.	—	—	1.47	—	0.68
Insulating	Cotton fiber[e]			0.8–2.0	0.26	—	3.85	—
Materials	Mineral wool, fibrous form, processed							
Blanket and	from rock, slag, or glass[e]			1.5–4.0	0.27	—	3.70	—
batt	Wood fiber[e]			3.2–3.6	0.25	—	4.00	—
Board	Glass fiber			9.5	0.25	—	4.00	—
	Acoustical tile[f]		½ in.	—	—	0.84	—	1.19
	Sheathing (impreg. or coated)			20.0	0.38	—	2.63	—
Board and	Cellular glass			9.0	0.40	—	2.50	—
slabs	Plastic (foamed)			1.62	0.29	—	3.45	—
Loose fill	Mineral wool (glass, slag, or rock)			2.0–5.0	0.30	—	3.33	—
	Vermiculite (expanded)			7.0	0.48	—	2.08	—
Roof insulation	All types[g]							
	Preformed, for use above deck							
	Approx.		1 in.	—	—	0.36	—	2.78
	Approx.		2 in.	—	—	0.19	—	5.26
	Approx.		3 in.	—	—	0.12	—	8.33
Masonry	Cement mortar			116	5.00	—	0.20	—
Materials	Lightweight aggregates, including ex-			120	5.2	—	0.19	—
Concretes	panded shale, clay, or slate; ex-			80	2.5	—	0.40	—
	panded slags; cinders; pumice;			40	1.15	—	0.86	—
	perlite; vermiculite; also cellular concretes			20	0.70	—	1.43	—
	Sand and gravel or stone aggregate (not oven dried)			140	12.00	—	0.08	—

HEAT LOSSES FROM BUILDINGS, VAPOR TRANSMISSION

TABLE 4.1 (Continued)

Material	Description		Density, (lb per cu ft)	k	C	R Per Inch Thickness $\left(\dfrac{1}{k}\right)$	R For Thickness Listed $\left(\dfrac{1}{C}\right)$
Masonry Units	Brick, common		120	5.00	—	0.20	—
	Brick, face		130	9.00	—	0.11	—
	Clay tile, hollow						
	1 cell deep	4 in.	—	—	0.90	—	1.11
	2 cells deep	8 in.	—	—	0.54	—	1.85
	3 cells deep	12 in.	—	—	0.40	—	2.50
	Concrete blocks, three oval core						
	Sand & gravel aggregate	4 in.	—	—	1.40	—	0.71
		8 in.	—	—	0.90	—	1.11
		12 in.	—	—	0.78	—	1.28
	Cinder aggregate	4 in.	—	—	0.90	—	1.11
		8 in.	—	—	0.58	—	1.72
		12 in.	—	—	0.53	—	1.89
	Gypsum partition tile:						
	3 × 12 × 30 in. 4-cell		—	—	0.74	—	1.35
	4 × 12 × 30 in. 3-cell		—	—	0.60	—	1.67
Plastering Materials	Cement plaster, sand aggregate		116	5.00	—	0.20	—
	Gypsum plaster:						
	Sand aggregate		105	5.60	—	0.18	—
	Sand aggregate on metal lath ¾ in.		—	—	7.70	—	0.13
	Lightweight aggregate		45	1.50	—	0.67	—
	Lightweight agg. on metal lath ¾ in.		—	—	2.13	—	0.47
Roofing	Asphalt roll roofing		70	—	6.50	—	0.15
	Asphalt shingles		70	—	2.27	—	0.44
	Built-up roofing	⅜ in.	70	—	3.00	—	0.33
	Sheet metal		—	400+	—	Negl	—
	Wood shingles		—	—	1.06	—	0.94
Siding Materials (On flat surface)	Shingles						
	Wood, 16-in. 7½-in. exposure		—	—	1.15	—	0.87
	Siding						
	Wood, drop, 1 × 8 in.		—	—	1.27	—	0.79
	Wood, bevel, ½ × 8 in., lapped		—	—	1.23	—	0.81
	Wood, bevel, ¾ × 10 in., lapped		—	—	0.95	—	1.05
	Wood, plywood, ⅜ in., lapped		—	—	1.59	—	0.59
Woods	Maple, oak, and similar hardwoods		45	1.10	—	0.91	—
	Fir, pine, and similar softwoods		32	0.80	—	1.25	—

[a] Representative values for dry materials at 75 F mean temperature, selected by the ASHAE Technical Advisory Committee on Insulation. They are intended as design (not specification) values for materials of building construction in normal use. For conductivity of a particular product, the user may obtain the value supplied by the manufacturer or secure the results of unbiased tests.

[b] Air-space resistance values shown here are based on a temperature difference of 20 F and a mean temperature of 50 F for spaces faced both sides with ordinary *nonreflective* materials.

[c] Surface resistance values shown here are for ordinary *nonreflective* materials.

[d] See also Insulating Materials, Board.

[e] Includes paper backing and facing if any.

[f] Insulating values of acoustical tile vary depending on density of the board and on the type, size, and depth of the perforations.

[g] The U. S. Department of Commerce "Simplified Practice Recommendation for Thermal Conductance Factors for Preformed Above-Deck Roof Insulation," No. R 257-55, recognizes the specification of roof insulation on the basis of the C values shown. Roof insulation is made in thicknesses to meet these values. Therefore, thickness supplied by different manufacturers may vary depending on the k value of the particular material.

case of concrete, and the mean temperature. In general, the materials having the smaller weights per cubic foot (density) have lower unit rates of heat transfer by conduction. The same is true with respect to moisture content. Conductivities and conductances for representative building and insulating materials are given in Table 4.1, which also includes data on conductances of air spaces and air surfaces.

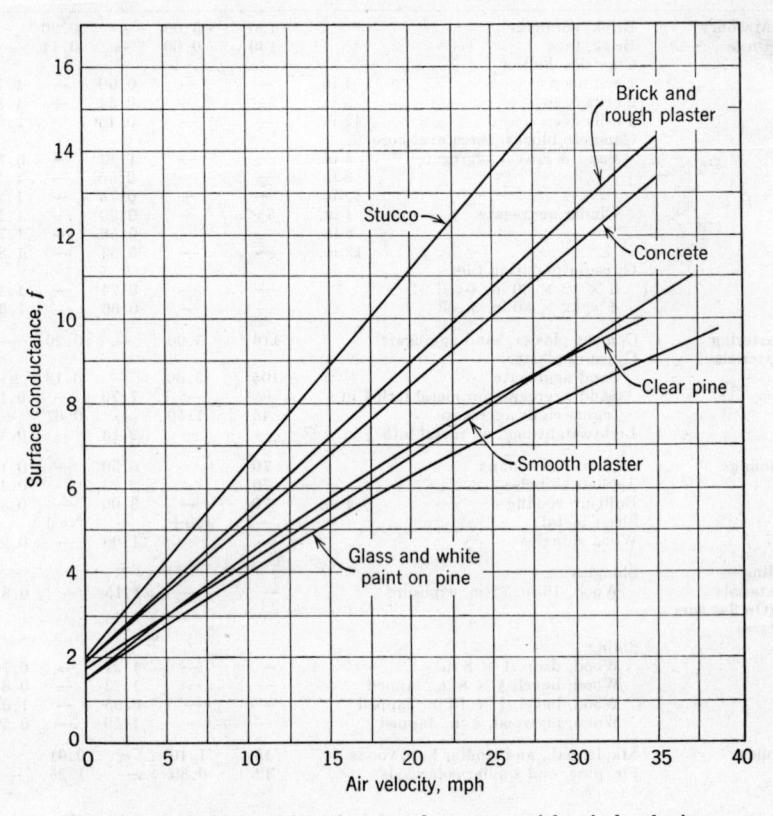

Fig. 4.5. Variation of surface conductances with wind velocity.

4.7. Conductances of Air Spaces.[12] Heat is transferred across an air space by the processes of radiation, convection, and conduction. The rate of transfer by radiation is affected by the temperatures of the two bounding surfaces and their respective emissivities but not by either the distance apart of the surfaces or by their orientation. The rate of heat transfer for the remaining portion, due to the combined effects of convection and conduction, is affected by the direction of heat flow, and the following items pertinent to the air space: depth or

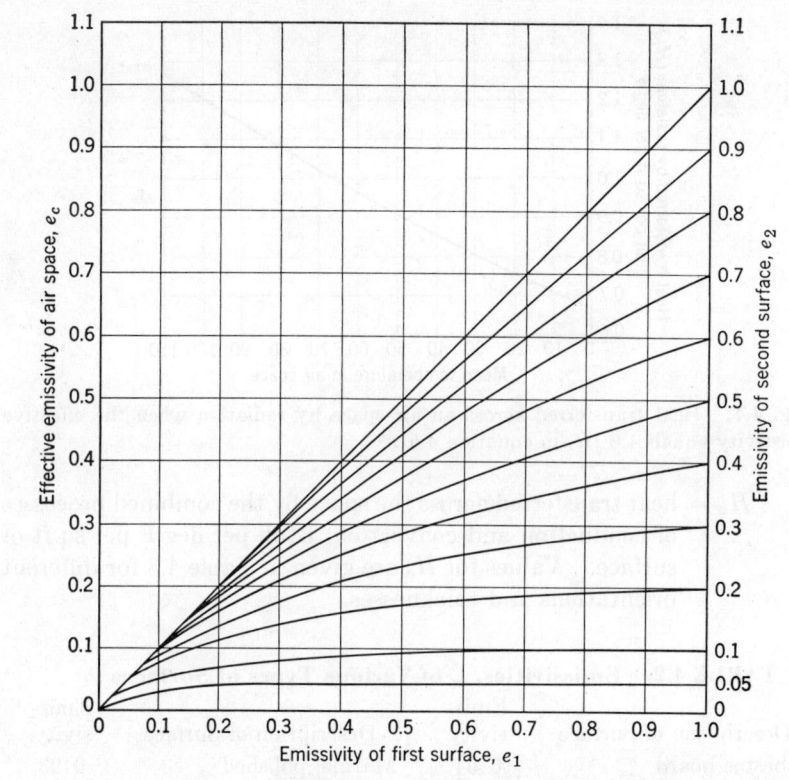

HEAT LOSSES FROM BUILDINGS, VAPOR TRANSMISSION

Fig. 4.6. Effective emissivity of an air space.[12]

distance across, mean temperature of its air, and the temperature difference maintained between the surfaces. The total conductance C_a of an air space is

$$C_a = e_c H_r + H_c \qquad (4.5)$$

where e_c = the combined emissivity of the two bounding surfaces calculated as $1/e_c = (1/e_1 + 1/e_2) - 1$. Various combinations of e_1 and e_2 give the data of Fig. 4.6 for e_c. Emissivities of various types of materials are given in Table 4.2.

H_r = heat transfer by radiation across the air space, provided both bounding surfaces have emissivities of 1.0 Btuh per deg F per unit of surface. $H_r = 0.172 \times 10^{-8}(T_1^4 - T_2^4)/(T_1 - T_2) = 0.00686(T_m/100)^3$ where all temperatures are in degrees Fahrenheit absolute. Data for H_r are shown in Fig. 4.7.

AIR CONDITIONING AND REFRIGERATION

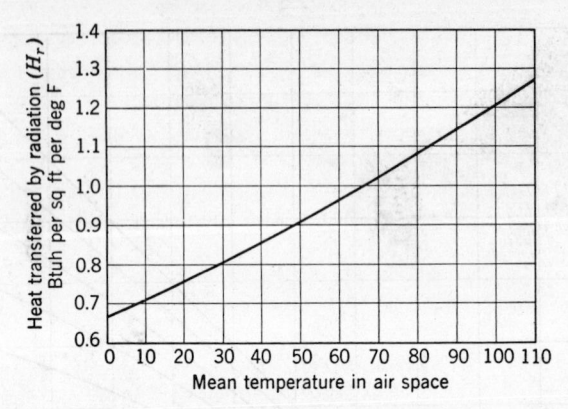

Fig. 4.7. Heat transferred across an air space by radiation when the effective emissivity equals 1.0 (H_r in equation 4.5).[12]

H_c = heat transferred across the space by the combined processes of conduction and convection, Btuh per deg F per sq ft of surface. Values for H_c are given in Table 4.3 for different orientations and thicknesses.

TABLE 4.2. Emissivities, e, of Various Types of Surfaces

Description of Surface	Emissivity	Description of Surface	Emissivity
Asbestos board	0.94	Marble, polished	0.93
Aluminum coated paper, polished	0.20	Masonry, brick, concrete, mortar, stone	0.90
Aluminum, dull	0.22	Oil paint	0.94
foil, average	0.05	Paper	0.93
polished	0.04	Plaster	0.91
Black body	1.00	Roofing paper	0.91
Brass, dull	0.22	Rubber, hard	0.91
polished	0.03	soft	0.86
Glass	0.92	Tile	0.91
Iron and steel, dull	0.82	Water	0.95
Lead, dull	0.28	Wood	0.90

The data of Table 4.3 are correct only when the mean temperature within the air space is 50 F; in general, they may be used without correction in the calculation of any air-space conductance. Whenever a correction of the data of Table 4.3 is necessary the following equation applicable to all cases except for the downward flow of heat may be used:

$$H_c' = H_c[1 - 0.001(t - 50)] \tag{4.6}$$

HEAT LOSSES FROM BUILDINGS, VAPOR TRANSMISSION

where H_c' = actual heat transfer by conduction and convection, Btuh per sq ft per deg F.

H_c = heat transfer by conduction and convection when the mean temperature within an air space is 50 F (Table 4.3).

When heat flows downward through an air space the pertinent values for H_c of Table 4.3 may be corrected by use of the following equation:

$$H_c' = H_c[1 - 0.0017(t - 50)] \qquad (4.7)$$

TABLE 4.3. Heat, H_c, Transferred Across Air Spaces by Convection and Conduction*†

H_c, Btuh per deg F temp. diff.

Air Space	Temperature Difference, deg F	\multicolumn{7}{c}{Air-Space Thickness, in.}
		0.25	0.50	0.75	1.00	1.50	2.50	3.50
Horizontal heat flow upward	10	0.670	0.434	0.405	0.390	0.370	0.340	0.330
	20	0.670	0.528	0.498	0.479	0.450	0.417	0.401
	30	0.670	0.586	0.553	0.532	0.504	0.468	0.449
	40	0.670	0.634	0.600	0.580	0.548	0.506	0.489
	60	0.670	0.716	0.677	0.651	0.617	0.575	0.550
At 45 deg, heat flow upward	10	0.675	0.366	0.314	0.323	0.321	0.297	0.290
	20	0.675	0.407	0.404	0.410	0.386	0.366	0.357
	30	0.675	0.452	0.466	0.460	0.436	0.413	0.410
	40	0.675	0.500	0.516	0.500	0.470	0.451	0.449
	60	0.675	0.583	0.583	0.560	0.534	0.510	0.507
Vertical, heat flow horizontal	10	0.680	0.340	0.240	0.220	0.231	0.248	0.245
	20	0.680	0.350	0.280	0.280	0.308	0.310	0.307
	30	0.680	0.360	0.312	0.340	0.345	0.355	0.345
	40	0.680	0.372	0.348	0.373	0.395	0.390	0.374
	60	0.680	0.406	0.410	0.440	0.450	0.440	0.420
At 45 deg, heat flow downward	10	0.690	0.345	0.240	0.190	0.165	0.176	0.182
	20	0.690	0.350	0.250	0.213	0.210	0.230	0.225
	30	0.690	0.355	0.262	0.242	0.246	0.265	0.257
	40	0.690	0.360	0.278	0.263	0.280	0.290	0.278
	60	0.690	0.370	0.307	0.302	0.323	0.327	0.309
Horizontal heat flow downward	10	0.690	0.350	0.240	0.187	0.125	0.080	0.065
	20	0.690	0.350	0.240	0.188	0.127	0.082	0.071
	30	0.690	0.350	0.240	0.189	0.129	0.085	0.077
	40	0.690	0.350	0.240	0.190	0.135	0.087	0.081
	60	0.690	0.350	0.240	0.191	0.140	0.090	0.089

* Extracted from reference 12.
† All data are for a mean temperature of 50 F.

A variation of 10 F in the mean temperature of the air space produces a correction factor of 1.0 − 0.01 or 0.99 for equation 4.6.

The following example illustrates the use of the data of Figs. 4.6 and 4.7 and also of Tables 4.2 and 4.3.

Example. Find the conductance C_a and the resistance R_a of an air space 1.5 in. thick which is bounded on one side by bright aluminum foil and on the other by masonry work. The air space is vertical with heat flow in a horizontal direction across it. The mean temperature of the air in the space is estimated to be 50 F and the temperature drop across the space is assumed to be 20 F.

Solution. The total conductance is obtained from $C_a = e_c H_r + H_c$. Table 4.2 gives an emissivity of 0.05 for aluminum foil and 0.90 for masonry surface. e_c from Fig. 4.6 is 0.05; H_r is 0.91 as indicated by Fig. 4.7 for a mean air temperature of 50 F; and in Table 4.3 H_c is found to be 0.308. Therefore, $C_a = (0.05 \times 0.91) + 0.308 = 0.354$ Btuh per sq ft per deg F and the resistance of the air space is $1/0.354 = 2.82$.

If the aluminum foil of the foregoing example were eliminated and if one boundary of the air space were a plaster surface (e_1) and the other boundary a masonry surface (e_2), then $\overset{\ast}{e_c}$, H_r, H_c, and C_a would be 0.83, 0.91, 0.308, and 1.063 respectively. *Many estimators use 1.10 as the conductance C_a for all air spaces. This practice is justified when both sides of the air space are bounded by ordinary building materials and when the resistance of the air space is a small portion of the total wall resistance.*

Wood-frame building construction may have open air spaces between the wall studding and between the joists of either floors or ceilings. The resistance to heat flow offered by the combination of the wood structural members and the accompanying air spaces may be computed as in the example of Art. 4.4 in which resistances in parallel were involved.

4.8. Numerical Calculations of U. The examples of Fig. 4.8 illustrate the procedure used in calculating over-all coefficients of heat transmission U. The examples of Fig. 4.8 have all of the resistances to heat flow, including that of the outside-air film, considered in the order of their respective positions in the path of the transmission. The concrete wall I of Fig. 4.8 has still air at one surface and air in motion at the opposite one. Wall II has still-air conditions at both surfaces. Wall III includes 4-in. hollow tile (1 cell deep) for which a conductance $C = 0.90$ is given in Table 4.1; therefore, the tile thickness is not used in the calculation of U for the wall. The resistance $\frac{1}{5}$ is for the effect of two $\frac{1}{2}$-in. thicknesses (1 in. total) of cement mortar for which a k

HEAT LOSSES FROM BUILDINGS, VAPOR TRANSMISSION 69

I
Concrete (stone aggregate)
Inside 70 F — Outside 0 F, Wind 15 mph
$f_i = 1.46$, $f_o = 6.0$, 12″

$$U = \frac{1}{\frac{1}{1.46} + \frac{12}{12} + \frac{1}{6.0}} = \frac{1}{1.852} = 0.540$$

II
Concrete (stone aggregate)
Inside 70 F — Inside 30 F (unheated room)
$f_i = 1.46$, $f_i = 1.46$, 12″

$$U = \frac{1}{\frac{1}{1.46} + \frac{12}{12} + \frac{1}{1.46}} = \frac{1}{2.370} = 0.422$$

III
Inside 70 F — Outside 0 F, Wind 15 mph
4″ hollow tile, ½″ cement mortar, 8″ common brick
$f_i = 1.46$, $f_o = 6.0$

$$U = \frac{1}{\frac{1}{1.46} + \frac{1}{5} + \frac{1}{0.9} + \frac{8}{5} + \frac{1}{6.0}} = \frac{1}{3.763} = 0.266$$

IV
Inside 70 F — Outside 0 F, Wind 15 mph
8″ concrete (stone aggregate), Furred to give 1½″ air space, $f_i = 1.46$, ¾″ total metal lath and gypsum plaster (sand aggregate), 3¾″ face brick, ¾″ air space, $f_o = 6.0$

$U = 0.2435$ (see example of Art. 4.8)

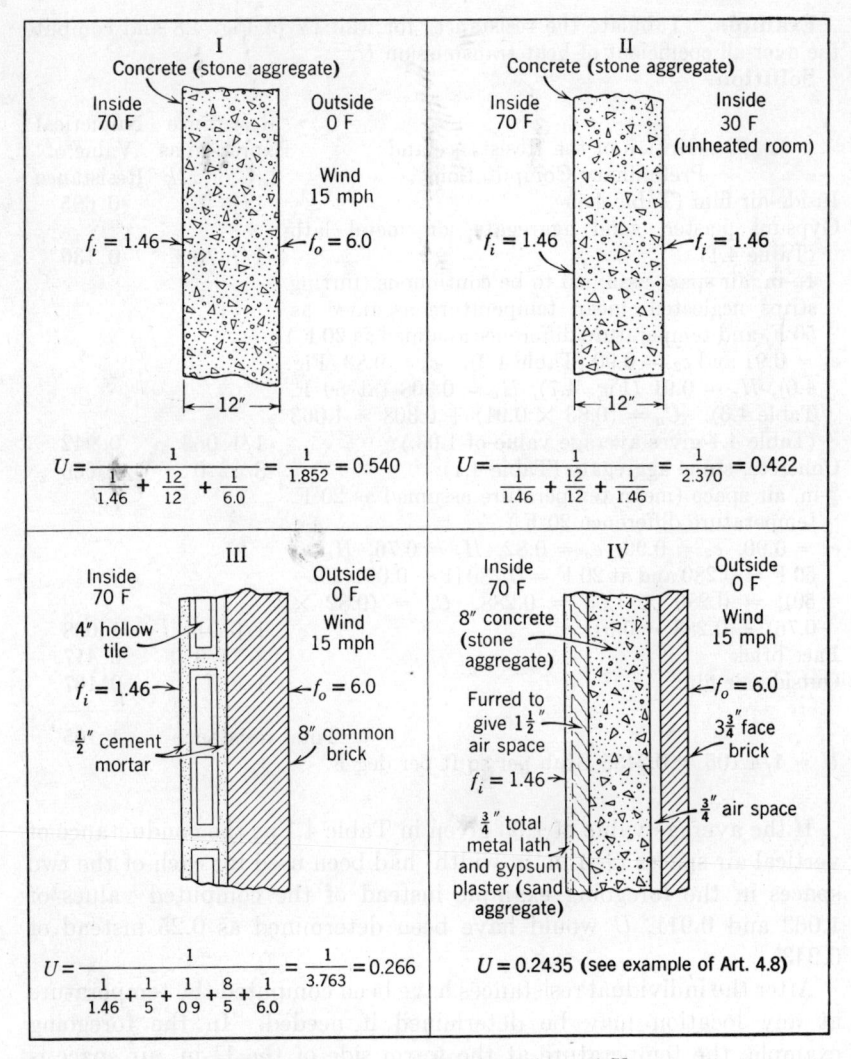

Fig. 4.8. Numerical calculations of U.

value of 5 is given by Table 4.1. The method of handling the two air spaces of wall IV is shown in the solution of the example given in the following paragraph. In calculating the over-all coefficient of heat transmission U, it may be more convenient to tabulate the separate resistances instead of setting up the solution as in the examples of Fig. 4.8. This method will be used in the following example.

Example. Tabulate the resistances for wall IV of Fig. 4.8 and compute the over-all coefficient of heat transmission U.

Solution.

Description of the Resistance and Preliminary Computations	Resistance Expressed as $1/C$ or x/k	Numerical Value of Resistance
Inside-air film (Table 4.1)	1/1.46	0.685
Gypsum plaster, sand aggregate, on metal lath (Table 4.1)	1/7.70	0.130
$1\frac{1}{2}$-in. air space assumed to be continuous (furring strips neglected, mean temperature assumed as 50 F, and temperature difference assumed as 20 F.) $e_1 = 0.91$ and $e_2 = 0.90$ (Table 4.2). $e_c = 0.83$ (Fig. 4.6), $H_r = 0.91$ (Fig. 4.7), $H_c = 0.308$ (at 50 F, Table 4.3). $C_a = (0.83 \times 0.91) + 0.308 = 1.063$ (Table 4.1 gives average value of 1.03.)	1/1.063	0.942
Concrete, stone aggregate (Table 4.1)	8/12.0	0.666
$\frac{3}{4}$-in. air space (mean temperature assumed as 20 F, temperature difference 20 F.) $e_1 = 0.90$, $e_2 = 0.90$, $e_c = 0.82$, $H_r = 0.76$, H_c at 50 F = 0.280 and at 20 F = 0.280 [1 − 0.001(20 − 50)] = 0.280 × 1.03 = 0.288. $C_a = (0.82 \times 0.76) + 0.288 = 0.911$	1/0.911	1.098
Face brick	3.75/9.0	0.417
Outside-air film	1/6.0	0.167
Total resistance		4.105

$U = 1/4.105 = 0.2435$ Btuh per sq ft per deg F.

If the average value of 1.03 given in Table 4.1 as the conductance of vertical air spaces ($\frac{3}{4}$–4 in. in width) had been used for each of the two spaces in the foregoing example instead of the computed values of 1.063 and 0.911, U would have been determined as 0.25 instead of 0.243.

After the individual resistances have been computed the temperature at any location may be determined if needed. In the foregoing example, the temperature at the warm side of the $1\frac{1}{2}$-in. air space is equal to $70 − (0.685 + 0.130)(70 − 0)/4.105 = 70 − 13.9 = 56.1$ F, while the temperature at the cooler side of the air space is $70 − (0.685 + 0.130 + 0.942)(70 − 0)/4.105 = 70 − 28.2 = 41.8$ F. The mean temperature in this air space is $(56.1 + 41.8)/2 = 48.9$ F, which is close to the assumed temperature of 50 F.

The assumed temperature drop used when obtaining H_c from Table 4.3 may also be checked. *The actual temperature drop across any resistance equals (the amount of the resistance times the over-all drop of temperature) divided by the total resistance.* The actual temperature

HEAT LOSSES FROM BUILDINGS, VAPOR TRANSMISSION 71

drop for the 1½-in. air space in the example is $(0.942 \times 70)/4.105 = 16.1$ F, instead of 20 F which was assumed. The actual temperature drop across the ¾-in. air space amounts to $(1.098 \times 70)/4.105 = 18.8$ F. The computed coefficient U is slightly in error because the H_c data from Table 4.3, in each case, are for temperature drops of 20 F. However, the computed U may be assumed to be sufficiently accurate for practical purposes.

The over-all coefficient U for any wall, floor, ceiling, or roof may be computed by the method illustrated from data given in Tables 4.1, 4.2, and 4.3 and in Figs. 4.6 and 4.7. Computed U values for many different constructions which are commonly employed may be found in the ASHAE Guide and in many other publications used throughout the heating industry.

4.9. Combination Coefficients. When an attic space is unheated a combined coefficient of heat transmission for the roof above and the ceiling below it may be estimated. This combined coefficient is used with the ceiling area and the difference between the temperature of the air below the ceiling and that of the outside air. The following expression may be used to calculate a combined coefficient for a ceiling and a roof.

$$U_{cr} = \frac{U_r \times U_c}{U_r + \dfrac{U_c}{r}} \qquad (4.8)$$

where r is the ratio of roof area to ceiling area.

When still-air conditions prevail in the attic the radiation of heat through the air of the space is compensated by increasing the surface conductance of the roof and ceiling areas within the space to 2.50 when computing U_r and U_c. The roof area used in the computation of r should include any wall areas which may be present in the attic space. If the wall areas are large and contain windows of appreciable size, the sum of the coefficients U of each individual section, i.e., roof, walls, and windows, multiplied by its fractional part of the total area should be used as U_r. When the attic air motion is rapid because of open vents the *ceiling area*, the *ceiling coefficient* of heat transmission the *air temperature beneath the ceiling*, and the *outdoor-air temperature* must be used in calculating the heat losses which occur at ceiling areas.

4.10. Air Temperature of Unheated Spaces. The air temperature in an unheated space in a heated building may be determined by equating the heat flowing in to that which flows out.

Example. Estimate the attic temperature of a 24-ft by 32-ft house constructed with a roof pitched downward in each direction from its longitudinal ridge which is located 8 ft above the ceiling. The design conditions are: outside-air temperature, -10 F; air temperature below ceiling underneath the attic, 75 F; and the U values for the ceiling, roof, and the gable walls 0.61, 0.48, and 0.20 respectively.

Solution. The ceiling area is $24 \times 32 = 720$ sq ft The roof area is $2 \times 32 \times \sqrt{12^2 + 8^2} = 922$ sq ft. The area of the two gable walls is $(2 \times 24 \times 8)/2 = 192$ sq ft. The heat flow through the ceiling to the attic is $0.61 \times 720 \times (75 - t_a)$. The heat flow from the attic to the outdoor air is $0.48 \times 922 \times [t_a - (-10)] + 0.20 \times 192 \times [t_a - (-10)]$. Equating heat flow in to heat flow out and simplifying, $439(75 - t_a) = 443(t_a + 10) + 38.4(t_a + 10)$ and $t_a = 30.5$ F.

4.11. Heat Transmission Losses from Basements and Crawl Spaces and Through Concrete Slabs Poured on the Ground.

The following suggested method of estimating the heat losses of concrete floor slabs laid upon the ground is based upon the results of experimental work[13] done at the University of Illinois. The heat losses are estimated in two parts: an edge loss, which is directly proportional to the inside-outside-air temperature difference, and an interior loss which is independent of the outdoor weather.

The edge loss is given by equation 4.9:

$$H_{fo} = FL(t_f - t_o) \tag{4.9}$$

where H_{fo} = heat losses from all of the floor area within 3 ft of any exposed edges, Btuh.
F = edge loss factor which varies with the amount of insulation provided between the floor slab and the foundation wall as given in Table 4.4.
L = length of exposed edge of floor, feet.
t_f = inside air temperature at floor level, degrees Fahrenheit.

TABLE 4.4. Edge Loss Factors

Edge Construction	F when $t_f - t_o = 70$	F when $t_f - t_o = 80$
No edge insulation	0.58	0.55
1-in. edge insulation at edge only	0.46	0.43
2-in. insulation at edge and extending under slab for a distance of 2 ft or down the inside surface of the foundation wall for the same distance	0.32	0.31

HEAT LOSSES FROM BUILDINGS, VAPOR TRANSMISSION

The interior loss is given by equation 4.10:

$$H_{fi} = U_f \times A_i(t_f - t_g) \tag{4.10}$$

where H_{fi} = heat loss through the interior floor area, Btuh.
U_f = over-all coefficient of heat transmission for heat flow from the room air through the concrete floor to deep soil, Btuh per sq ft per deg F. *Usually taken as 0.10 based on studies with heat meters.*
A_i = interior floor area, square feet, which is the total floor area less the area of a strip 3 ft wide along all exposed edges. A_i for a slab exposed on all four sides is (length $-$ 6) $\times$ (width $-$ 6).
t_g = stable deep ground temperature, degrees Fahrenheit, which may be taken as the well-water temperature that is typical of the locality.

If U_f is assumed to be 0.1, t_f is taken as 70 F, and t_g as 50 F, then $H_{fi} = 2A_i$ Btuh.

The heat loss from a basement floor in Btuh per square foot is the same as that through the interior area of a slab on top of the ground. The heat loss, in Btuh per square foot of wall area below the gradeline level, of a basement wall is usually taken as 0.10 times the average difference between the temperature of the basement air and that of the soil on the opposite side regardless of the construction used. The transmission rate for any portion of a basement wall which is below grade level is often assumed to be 4.0 Btuh per sq ft.

Estimations of the heat transmission losses through portions of basement walls which are above the earth grade line and through windows in basement walls involve no special considerations.

Exactly the same procedures are involved in estimating the heat losses from the crawl spaces below the first floors of structures without basements as are employed when a basement is in existence, except where the crawl-space ventilators are not closed during the heating season. When the latter condition exists the outdoor-air design temperature is presumed to exist in the cool region.

4.12. Air Infiltration. The leakage of air into and out of a building may be the result of the action of wind or of the differential in temperature between the inside and the outside air with a resultant difference in their densities. These factors may operate separately or in combination. Outward leakage is also the result of maintaining the air under pressure in the structure. In any event, the air leaking out of the space is replaced by an equal weight of air which comes from the outside. Infiltration with the corresponding exfiltration produces an additional load on either the heating or the cooling plant of a building.

4.13. Equivalent Wind Velocities. For low buildings the chimney or stack effect on air infiltration is generally negligible, but for tall structures consideration should be given to it. A scheme of making allowances for the stack effect of a building is to compute equivalent wind velocities for use in estimating the air infiltration at different levels.

The mid-height of a building is usually taken as the location of the neutral zone[14] where there is no air-pressure differential between the outside and the inside spaces due to stack effect. When the neutral zone is used as a datum level, equivalent wind velocities can be estimated for use in determining air leakages. Equations which may be used for finding equivalent wind velocities at various distances above and below the neutral zone are

$$v_e = \sqrt{v^2 - 1.75a} \tag{4.11}$$

$$v_e = \sqrt{v^2 + 1.75b} \tag{4.12}$$

where v_e = equivalent wind velocity, miles per hour.
v = wind velocity existent at the considered location, miles per hour.
a = distance of location above mid-height of building, feet.
b = distance of location below mid-height of building, feet.

Equations 4.11 and 4.12 are based on an analysis of pressure variations due to stack effect in a building which permits relatively free circulation of air from the bottom to the top. A more exact method is needed for estimating the infiltration rate of different levels in tall buildings having different arrangements of floors, stairways, and elevator shafts.

4.14. Air Infiltration Rates. A number of investigations[15-21] have been made in laboratories and in actual buildings at various places in an endeavor to ascertain the probable leakage of air through the cracks about window sashes when subjected to different wind velocities. Laboratory investigations have also been carried on to determine the leakage of air through various wall sections.

The leakage of air into a structure increases rapidly with an increase of wind velocity. The air leakage through brick walls which are unfinished on the inside may be a considerable amount but it is reduced

HEAT LOSSES FROM BUILDINGS, VAPOR TRANSMISSION

to a negligible figure by the addition of plastering to the inside faces. Leakage through walls which incorporate a layer of building paper is negligible.

Air leakage between window frames and masonry walls can be diminished by calking the cracks with plastic materials. The leakage about window frames is dependent upon the wind velocity, the kind and type of wood or steel sash, and the crack width together with the sash clearance. Numerical quantities are given in Table 4.5 for the infiltration of air through cracks about various kinds of windows. The volumes in cubic feet as given are per linear foot of window crackage. Attention is called to the reduction of air leakages by the use of weather stripping which is discussed in the following article.

The leakage of air at doors varies with their fit and warpage. For heating-load estimations, well-fitted doors may have about the same leakage per foot of crack considered as poorly fitted double-hung sash, and poorly fitted doors should have an allowance of twice the amount used for well-fitted doors. The foregoing values may be reduced by one-half if weather stripping is used.

Revolving doors are very helpful in reducing the infiltration of either cold or hot air into buildings which have considerable in-and-out traffic. Simpson[22] states that the infiltration through a revolving door that is maintained in good condition is only 18.5 cu ft per person passing through under average design conditions.

Some heating engineers obtain the total volume of infiltration air in cubic feet per hour by multiplying the volume of the space to be heated by an assumed number of air changes per hour. This method involves little calculation but its accuracy depends entirely upon the judgment of the estimator.

4.15. Weather Stripping and Storm Sash. Air leakage about window sash and doors may be materially reduced by the installation of weather stripping. Such building equipment may be of felt and rubberized strips installed to seal cracks, pieces of thin spring bronze attached to window sash and doors to bear against their enclosing frames, and some form of tongue-and-groove construction fabricated from metal strips. Data pertinent to the leakages of air at various wind velocities are given in Table 4.5 for three different cases of window construction operating without and with weather stripping.

Storm windows fall under two general types: those which fill the window opening and which bear against the blind stops of the window frames, and those which are clamped to the window sash and cover the glass areas. When this latter type of installation is made the storm

TABLE 4.5. Air Infiltration Through Windows*
Expressed in cubic feet per ft of crack per hour†

Type of Window	Remarks	\multicolumn{5}{c}{Wind Velocity, MPH}
		5	10	15	20	25	30
Double-hung wood sash windows (unlocked)	Around frame in masonry wall—not calked‡	3	8	14	20	27	35
	Around frame in masonry wall—calked‡	1	2	3	4	5	6
	Around frame in wood frame construction‡	2	6	11	17	23	30
	Total for average window, non-weather-stripped, $\frac{1}{16}$-in. crack and $\frac{3}{64}$-in. clearance.§ Includes wood frame leakage‖	7	21	39	59	80	104
	Ditto, weatherstripped‖	4	13	24	36	49	63
	Total for poorly fitted window, non-weather-stripped, $\frac{3}{32}$-in. crack and $\frac{3}{32}$-in. clearance.¶ Includes wood frame leakage‖	27	69	111	154	199	249
	Ditto, weatherstripped‖	6	19	34	51	71	92
Double-hung metal windows**	Non-weather stripped, locked	20	45	70	96	125	154
	Non-weather stripped, unlocked	20	47	74	104	137	170
	Weather stripped, unlocked	6	19	32	46	60	76
Rolled section steel sash windows¶¶	Industrial pivoted, $\frac{1}{16}$-in. crack††	52	108	176	244	304	372
	Architectural projected, $\frac{1}{32}$-in. crack‡‡	15	36	62	86	112	139
	Architectural projected, $\frac{1}{16}$-in. crack‡‡	20	52	88	116	152	182
	Residential casement, $\frac{1}{64}$-in. crack§§	6	18	33	47	60	74
	Residential casement, $\frac{1}{32}$-in. crack§§	14	32	52	76	100	128
	Heavy casement section, projected, $\frac{1}{64}$-in. crack‖‖	3	10	18	26	36	48
	Heavy casement section, projected $\frac{1}{32}$-in. crack‖‖	8	24	38	54	72	92
Hollow metal, vertically pivoted window**		30	88	145	186	221	242

* From "Heating Ventilating Air-Conditioning Guide 1957." Used by permission.
† The values given in this table, with the exception of those for double-hung and hollow metal. windows, are 20 per cent less than test values to allow for building up of pressure in rooms and are based on test data reported in the papers listed in chapter references.
‡ The values given for frame leakage are per foot of sash perimeter as determined for double-hung wood windows. Some of the frame leakage in masonry walls originates in the brick wall itself and cannot be prevented by calking. For the additional reason that calking is not done perfectly and deteriorates with time, it is considered advisable to choose the masonry frame leakage values for calked frames as the average determined by the calked and non-calked tests.
§ The fit of the average double-hung wood window was determined as $\frac{1}{16}$-in. crack and $\frac{3}{64}$-in. clearance by measurements on approximately 600 windows under heating season conditions.
‖ The values given are the totals for the window opening per foot of sash perimeter and include frame leakage and so-called *elsewhere* leakage. The frame leakage values included are for wood frame construction but apply as well to masonry construction assuming a 50 per cent efficiency of frame calking.
¶ A $\frac{3}{32}$-in. crack and clearance represent a poorly fitted window, much poorer than average.
** Windows tested in place in building.
†† Industrial pivoted window generally used in industrial buildings. Ventilators horizontally pivoted at center or slightly above, lower part swinging out.
‡‡ Architecturally projected made of same sections as industrial pivoted except that outside framing member is heavier, and it has refinements in weathering and hardware. Used in semi-monumental buildings such as schools. Ventilators swing in or out and are balanced on side arms. $\frac{1}{32}$-in. crack is obtainable in the best practice of manufacture and installation, $\frac{1}{16}$-in. crack considered to represent average practice.
§§ Of same design and section shapes as so-called *heavy section casement* but of lighter weight. $\frac{1}{64}$-in. crack is obtainable in the best practice of manufacture and installation, $\frac{1}{32}$-in. crack considered to represent average practice.
‖‖ Made of heavy sections. Ventilators swing in or out and stay set at any degree of opening. $\frac{1}{64}$-in. crack is obtainable in the best practice of manufacture and installation, $\frac{1}{32}$-in. crack considered to represent average practice.
¶¶ With reasonable care in installation, leakage at contacts where windows are attached to steel framework and at mullions is negligible. With $\frac{1}{64}$-in. crack, representing poor installation, leakage at contact with steel framework is about one-third and at mullions about one-sixth of that given for industrial pivoted windows in the table.

sash does not reduce the infiltration of air at the window opening. In such cases the windows should be fitted with weather stripping. Storm windows fitted with felt gaskets and drawn tightly against the blind stops of window frames are of value in reducing air infiltration. However, if a storm sash is fitted loosely, no reduction of air infiltration may take place. For tight-fitting storm sashes the reduction of air infiltration is about the same as when weather stripping is used.

Storm sashes are quite effective in reducing the amount of heat transmitted by conduction through window-glass areas. *The coefficient of heat transmission U for two panes of glass separated by an air space 1 in. wide is 0.53 Btuh per sq ft per deg F.*

Double glazing of window sashes is also used to reduce heat losses and gains through glass areas. These units employ two or more panes of glass placed a small distance apart and carefully sealed about their edges by metallic or other materials. The air of the space is partially evacuated and the assembling is done in a room where the humidity ratio is maintained at a low level. *The U value for a single thickness of glass is 1.13 Btuh per sq ft per deg F whereas that for two panes and three panes spaced $\frac{1}{4}$ in. apart is 0.61 and 0.41 respectively.* Weather stripping should be used with multiple-glazed sashes to reduce the air leakages about their perimeters.

4.16. Linear Feet of Crack Used in Computing Air Leakage. The amount of air infiltration is dependent upon the wind velocity, the width of the cracks, and the linear feet of cracks. The number of linear feet of window and door cracks to be used may be empirically fixed as: rooms with one exposure, all the linear feet of cracks in the outside wall; rooms with two exposures, the linear feet of crack in the outside wall having the greater amount of cracks; and rooms with three or four exposures, the length of the cracks in the wall having the greatest amount; but in no case less than one-half the total crackage.

4.17. Heat Losses due to Air Infiltration. Unless building walls are of poor construction air leakage through them may usually be ignored. The data of Table 4.5 and the assumptions of Art. 4.16 provide means for estimating heat loss due to infiltration through window cracks.

$$H_i = c_{pa} q_i d_a L(t - t_o) \tag{4.13}$$

where q_i is from Table 4.5. For convenience d_a is often assumed to be 0.075 lb per cu ft for all ordinary conditions. H_i is in Btuh.

The crackage of a double-hung window is three times its width plus two times its height.

Example. Find the heat loss due to infiltration from a room which has one double-hung wood sash window in each of two exposed walls. The windows are of average fit and are not weather-stripped. Each sash measures 24 in. by 24 in. The design conditions are: $t = 70$ F, $t_o = -10$ F, barometric pressure = 14.7 psia, and the wind velocity = 15 mph.

Solution. The infiltration rate, from Table 4.5, is 39 cu ft per ft of crack per hr. The density will be taken as 0.075 lb per cu ft. The effective crackage L is $3 \times 2 + 2 \times 4 = 14$ ft. Then $H_i = 0.24 \times 39 \times 0.075 \times 14 \times [70 - (-10)] = 787$ Btuh.

Heating engineers generally base their calculations on the average wind velocity during the 3 months of the most severe winter weather. A commonly used value for wind velocity is 15 mph.

4.18. Thermal Storage of Building Materials. Walls and roofs have the ability to store variable amounts of heat. This property is dependent upon the method of construction, the kinds of materials used, and the thickness of the individual component parts. The item of thermal storage is of importance in problems both of winter heating and of summer cooling.

Obviously, thick walls will store more heat than thin walls of the same material. During the winter the heat stored in heavy walls will tend to maintain the desired inside-air temperature during periods when the heating plant is not actively operating. On the other hand, when the building is intermittently heated, heavy thick walls which have become cold may maintain a low inside-surface temperature for a considerable period after the inside-air temperature has reached that value deemed necessary for comfort.

4.19. Building Insulation. Basic materials from which building insulation is derived are asbestos, bagasse (crushed sugar-cane stalks), cork, cornstalks, cotton, eelgrass, gypsum, glass, hair, jute, kapok, limestone, metals, moss, mica, paper pulp, rubber, straw, and wood fiber in several forms.

Insulation may be divided into the following types: rigid, semirigid, flexible, and fill. *Rigid* insulation is made in panels of various sizes and thicknesses, and some of it has considerable structural strength. *Semirigid* insulations embrace those of the order of felts which have some flexibility and are available in panels of different widths, lengths, and thicknesses. *Flexible* insulation embodies the quilts and blankets having paper or fabric on each side with loosely packed material between the coverings. Included in this category are the metal foils which are used either in crumpled form as filling material or as thin sheets suspended in air spaces and spaced a slight distance apart. *Fill* insulation is produced in shredded or granulated form and is used to fill spaces in walls, floors, and ceilings.

Each of the different insulations just mentioned depends upon numerous air cells for its effectiveness as a resistance against the flow of heat. A special type of insulation is now available which is designed to create additional air spaces. The air spaces are obtained by the use of sheets of bright aluminum foil which have a low emissivity factor and which are separated from each other by the use of corrugated paper. Insulation of this type comes in two- and three-sheet forms and it is competitive with other insulations both in cost and effectiveness. The principal advantages are low shipping cost, small space requirements for storage before applications, and low heat-storage capacity when in place. The latter item may be important during the evenings and nights of hot weather. This type of insulation is most effective against the downward flow of heat. When such insulation is installed in the spaces between studs in vertical walls, it is extremely important that it be continuous from the bottom to the top of each space with its joints at both the top and the bottom securely sealed against air movement.

The use of sufficient amounts of an insulating material[23] in building construction is justified from the standpoints of economy and comfort, and sometimes its use is a necessity. Part of the expenditure for insulation can be saved in the reduced first cost of either a heating or a cooling plant. A plant of smaller capacity is required in either case, if the heat losses or heat gains can be reduced by the use of building insulation, and the operating costs are less.

Each wall, roof, floor, or ceiling is an individual problem in which an attempt should be made to reduce economically the over-all coefficient of heat transmission to a low value. There is an economic thickness in each case beyond which the cost of additional material cannot be justified. The actual placement of the insulating material in the walls has much to do with the final results obtained. Where insulators of the rigid, semirigid, and flexible types are used the best results are obtained when the insulation is so placed that the number of air spaces in the walls is increased.

4.20. Outside-Air Temperature. Table 4.6 gives data for weather conditions of several cities of the United States. The cities have been selected so that representative information necessary for the design of heating plants is available for all sections of the country. Similar data for a greater number of cities are included in the

TABLE 4.6. Heating Season Climatic Data
Compiled from Records of the U. S. Weather Bureau and Other Sources

State	City	Average Temperature, Oct. 1– May 1	Lowest Temperature	Design Temperature Suggested by TAC	Average Wind Velocity, Dec., Jan., Feb., mph	Direction of Prevailing Wind, Dec., Jan., Feb.	Normal Degree Days, Total for Year
Ala.	Birmingham	53.9	−10	21	8.6	N	2618
Ariz.	Phoenix	59.5	16	31	3.9	E	1446
Ark.	Little Rock	51.6	−12	21	9.9	NW	3005
Calif.	Los Angeles	58.6	28	32	6.1	NE	1390
	San Francisco	54.3	27		7.5	N	3143
Colo.	Denver	39.3	−29	0	7.4	S	5863
D. C.	Washington	43.2	−15	14	7.3	NW	4598
Fla.	Jacksonville	61.9	10	31	8.2	NE	1161
Ga.	Atlanta	51.4	−8	22	11.8	NW	3002
Idaho	Lewiston	42.5	−13		4.7	E	4924
Ill.	Chicago	36.4	−23	−3	17	SW	6287
Ind.	Indianapolis	40.2	−25	2	11.8	S	5487
Iowa	Sioux City	32.1	−35		12.2	NW	6909
Kan.	Dodge City	40.2	−26		10.4	NW	5077
Ky.	Louisville	45.2	−20	9	9.3	SW	4428
La.	New Orleans	61.5	7	36	9.6	N	1208
Mass.	Boston	37.6	−18	8	11.7	W	5943
Mich.	Detroit	35.4	−24	4	13.1	SW	6580
Minn.	Minneapolis	29.6	−33	−15	11.5	NW	7989
Mo.	St. Louis	43.3	−22	3	11.8	NW	4610
Mont.	Billings	34.7	−49	−17	12.4	W	7119
Neb.	Lincoln	37.0	−29	−2	10.9	N	6010
N. M.	Santa Fe	38.0	−13		7.3	NE	6124
N. Y.	Buffalo	34.7	−21	3	17.7	W	6935
	New York	40.3	−14		13.3	NW	5306
N. C.	Raleigh	49.7	−2	20	7.3	SW	3281
N. D.	Bismarck	24.5	−45	−21	9.1	NW	8969
Ohio	Cleveland	36.9	−18	6	14.5	SW	6171
Okla.	Oklahoma City	48.0	−17	14	12.0	N	3698
Oreg.	Portland	45.9	−2	22	6.5	S	4379
Pa.	Philadelphia	41.9	−11		11.0	NW	4749
S. C.	Charleston	56.9	7	26	11.0	N	1870
Tenn.	Knoxville	47.0	−16		6.5	SW	3665
Tex.	El Paso	53.0	−2		10.5	NW	2538
	San Antonio	60.7	4	32	8.2	N	1424
Utah	Salt Lake City	40.0	−20	7	4.9	SE	5637
Va.	Lynchburg	45.2	−7		5.2	NW	4082
Wash.	Seattle	45.3	3	24	9.1	SE	4864
Wis.	Milwaukee	33.0	−25	−6	11.7	W	7086
Wyo.	Cheyenne	33.9	−38	−3	13.3	NW	7549

HEAT LOSSES FROM BUILDINGS, VAPOR TRANSMISSION

ASHAE Guide. Local Weather Bureau data may be used when such are available.

In the design of heating plants it is not customary to base the estimate of the heating requirements on the lowest outside-air temperature on record in any locality because this low temperature may prevail for only a few hours during the winter, if at all. Past practice has been to select the outside-air temperature on the basis of 10 to 15 deg above the lowest temperature on record in the locality where a heating plant is to be installed. However, a Technical Advisory Committee (TAC) of the ASHAE has made a very careful study of the design temperature to be used and has recommended that it be selected on a different basis. The selection is made on the basis of the maximum hourly outdoor-air temperature which has been exceeded 97.5 per cent of the time during the months of December, January, February, and March for the period of record. This information is given in Table 4.6 for all of the cities listed where the necessary data have been obtained. In every case the data for this temperature have been collected at an airport. In general the TAC design temperature which is given may be safely used not only for the city but also for surrounding suburban and rural areas.

4.21. Inside-Air Temperatures. The usual specified air temperatures in heating are either the breathing-line temperature, taken 5 ft above the floor, or the temperature at the 30-in. level and at a location not nearer than 3 ft from an outside wall in either case. Common specifications in different room usages are: homes 70–72, classrooms 70–72, playrooms 60–65, ballrooms 65–68, foundries 50–60, and paint shops 80. The specified temperatures are dry-bulb values.

4.22. Air Temperatures at the Proper Level. The increase of air temperature above that of the breathing level is not directly proportional to the distance above it. In view of the fact that authentic data are lacking for all conditions, an approximation is necessary. The assumption is made that the increase of air temperature per foot of height above the 5-ft level is at the rate of 1 per cent of the breathing-line temperature up to ceiling heights of 15 ft. Beyond the level of 15 ft above the floor the increase is assumed to be 0.10 F per ft. In equation form the rule is

$$t_h = t + 0.01(h - 5)t \qquad (4.14)$$

where t_h is the air temperature at the level h, and t is that at the breathing level. In a case where the point of interest is more than 15 ft above the floor the figure 15 is used for h in equation 4.14 and $0.1(h - 15)$

is added to the result obtained. The rule is satisfactory for rooms heated by direct radiators and for the majority of rooms heated with forced warm air but does not hold so well for gravity warm-air heating or for those systems of heating and ventilation which partially project heated air downward against its natural tendency to rise. When the air temperature at the breathing line is not less than 55 F the air temperature at the floor may be taken as 5 F less than that 5 ft above the floor, in all rooms where equation 4.14 is applicable. Application of equation 4.14 should not be made in the calculations of rooms heated either by means of panels located in floors, walls, and ceilings or by use of radiant baseboard units. In cases of rooms heated by these systems and in all cases where the ceiling height is less than 10 ft, breathing-level to ceiling and breathing-level to floor differentials may be neglected in estimating heat losses.

4.23. Procedure in Making Heat-Loss Calculations. Certain data are necessary before the estimated losses of heat from a room or a structure may be calculated. Therefore, the following items must be either computed or fixed.

1. The separate net areas of the walls, glass, ceiling or roof, and floors through which the transmission of heat will occur.

2. The over-all coefficients of heat transmission for the component parts of the building listed in item 1. These are to be taken from tables of data or calculated.

3. The inside-air temperature, at the breathing level 5 ft above the floor, which is considered to be necessary when severe winter weather exists; also the air temperatures at the proper levels for the calculation of heat losses through the various exposed building areas (Art. 4.22).

4. The outside-air temperature which is to be used for design purposes.

5. The computed losses of heat occurring at the various areas listed in item 1, based on the proper coefficients of heat transmission and the difference in temperature between the inside and the outside air.

6. An estimate of the heat required to warm the inleaking cold air based on the methods given in Art. 4.17.

7. The total of the individual transmission losses and those due to air leakage. This final summation gives the estimated heat losses from the space considered, for the air temperature conditions chosen.

The heat loss from a building consisting of several rooms is usually taken as the sum of the individual room totals. A more exact determination for the purpose of sizing a heating plant is obtained by the summation of the individual total conduction losses to which is added the infiltration loss computed for the building as a whole.

HEAT LOSSES FROM BUILDINGS, VAPOR TRANSMISSION

4.24. Intermittent Heating. Economy of heating-plant operation is sometimes sought by reducing the output of the plant during the night or periods of time when the building is unoccupied. Such a mode of operation will effect a saving in the cost of fuel, but it may also result in uncomfortable conditions in the structure when occupancy again takes place after a period of reduced air temperatures. The storage of heat in heavy building walls causes a lag in the drop of

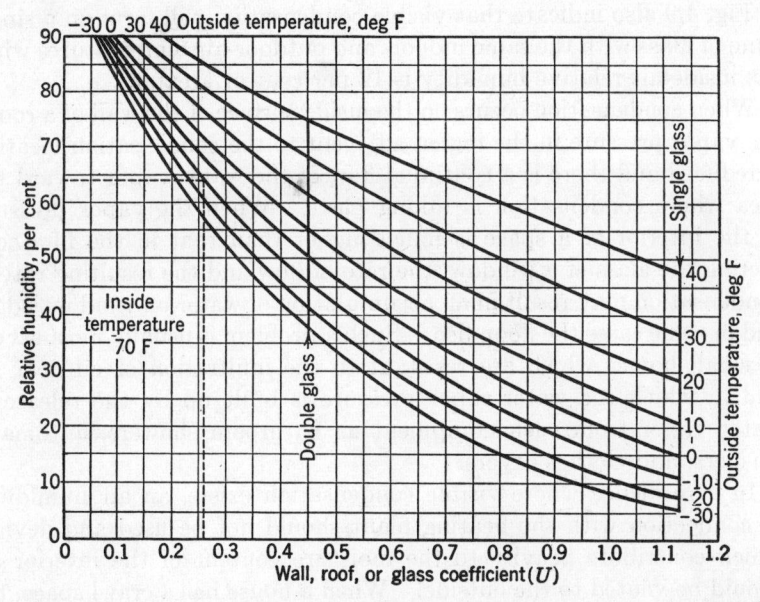

Fig. 4.9. Relative humidities at which visible condensation of water vapor occurs on inside surfaces.

air temperature as the heating-plant output is reduced. Once the walls are cooled it may take a considerable period of time to warm them again to achieve comfort conditions. Therefore the savings theoretically made by intermittent heating may be largely wiped out. The additional allowances to be made in estimating the size of a heating plant operated intermittently depend upon local conditions and the heating engineer's judgment.

4.25. Visible Condensation. When an air water-vapor mixture is cooled to a temperature below that of its dew point, vapor condensation occurs. During periods of cold weather the inside-surface temperature of single-thickness glass windows is always below the dew-point value for air at 70 F unless its relative humidity is very low. Walls that are poorly constructed may have condensation on their inside surfaces during severely cold weather. The data of Fig. 4.9

from reference 24 are based on an inside-air temperature of 70 F and indicate the relative humidities either at or above which condensation will occur on inside surfaces. The example illustrated by the broken line indicates that the inside-air relative humidity would have to be greater than 64 per cent in order for visible condensation to form on a wall having a U coefficient of 0.26 when the outdoor-air temperature is -10 F and the inside-air temperature is 70 F. However, the data of Fig. 4.9 also indicate that visible condensation will form on a single pane of glass with the same indoor- and outdoor-air temperatures when the inside-air relative humidity is 12 per cent or greater.

When condensation occurs on the inside surface of the wall of a room, the vapor pressure in the region adjacent to the condensation location is reduced and there is a continual flow of the water vapor toward the area where condensation is taking place. When the vapor pressure in the interior of a space is much higher than that in the film contacting the glass of a window, the rate of flow and the resulting rate of condensation may result in an accumulation of water on window ledges and in some cases the floor also. Such a problem is usually most severe in small homes which are resistant to the infiltration of outdoor air and in which the water-vapor pressure is built up by the release of water vapor from such equipment as bathroom showers, automatic washers, and clothes dryers.

In a residence where visible condensation exists, an air humidifier in connection with the heating plant should not be used and devices which contribute heavily to the moisture content of the interior air should be vented to the outside. When a house has a crawl space, the surface of the soil beneath it should be covered with 55-lb rolled roofing or an equivalent vapor barrier. When all practical steps have been taken to reduce the vapor content of the air without beneficial results, the only solution of the problem may be the further introduction of outside air to the space either by opening one or more windows slightly or by operating an exhaust fan to induce a greater rate of air infiltration.

4.26. Invisible Condensation. Most building materials permit the passage of water vapor through them when the pressure on one side is greater than that on the opposite side.[25–28] Special papers and thin sheets of metal which are very resistant to the passage of water vapor are called *vapor barriers*. During periods of cold weather the water-vapor pressure in outside air is always lower than that within a heated building if there are any interior sources from which water vapor may be released. Generally there is a vapor-pressure differ-

HEAT LOSSES FROM BUILDINGS, VAPOR TRANSMISSION 85

ential across all building walls, windows, ceilings, and floors which separate heated air from cold air. Consequently, a continuous flow of water vapor exists through all segments of the enclosure which do not include completely effective vapor barriers such as panes of glass and sheets of metal.

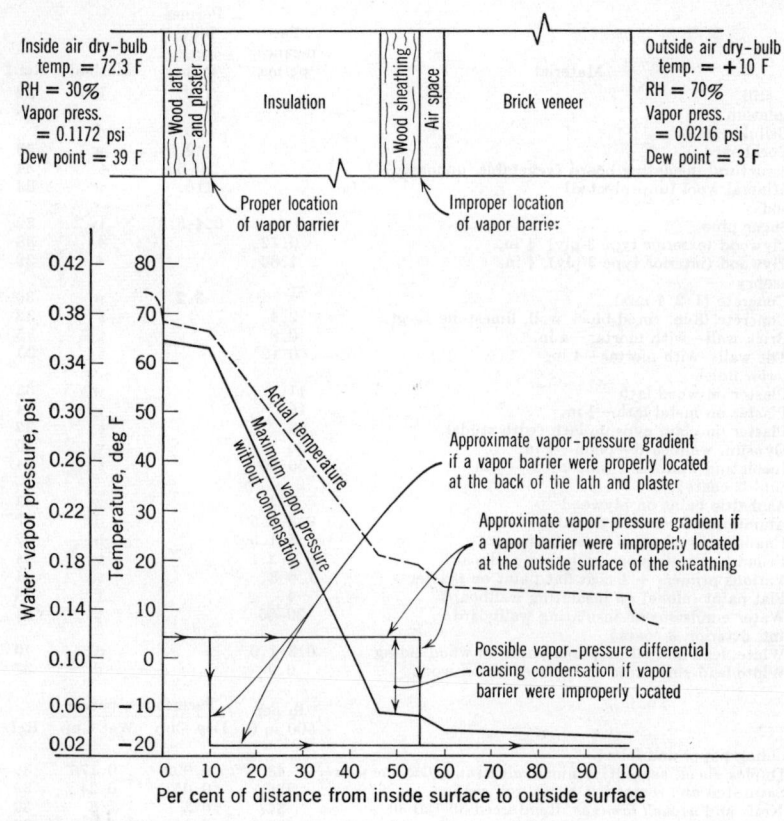

Fig. 4.10. Diagrammatic representation of possible temperature and vapor-pressure gradients in building walls.

The vapor pressure at any point within a wall, floor, or ceiling will always be intermediate between that existing at the inside and that at the outside. The decrease is according to some gradient determined by the relative resistances of the various materials incorporated in the construction. Figure 4.3 indicates that the temperature of the air in a wall, varies from the inside to the outside in a gradient which is determined by the relative resistances of the various components. The dashed line of Fig. 4.10 shows the actual temperature gradient as measured by thermocouples located in a wall of a research residence at the University of Illinois. The inside- and outside-air temperatures at the time the data were taken are indicated as well as the wall construction. Invisible condensation will occur at any point within either a wall, floor, or ceiling where the vapor pressure is such that the

TABLE 4.7. Permeance and Permeability of Materials to Water Vapor*

Material	Permeance, perms	Permeability, perm-inches	Method†	Ref.‡
Air, still		120.	b	32
Insulation				
Cellular glass		0.0	d	
Corkboard		9.5	w	38
Structural insulating board (vegetable, uncoated)		20–50	t	34
Mineral wool (unprotected)		116.	w	33
Wood				
Sugar pine		0.4–5.4	tv	32
Plywood (exterior type 3-ply), ¼ in.	0.72		4	38
Plywood (interior type 3-ply), ¼ in.	1.86		4	38
Masonry				
Concrete (1:2:4 mix)		3.2	w	36
Concrete (8-in. cored block wall, limestone agrgt.)	2.4		t	32
Brick wall—with mortar—4 in.	0.8		t	35
Tile wall—with mortar—4 in.	0.12		t	35
Interior finish				
Plaster on wood lath	11.		w	33
Plaster on metal lath—¾ in.	15.		t	34
Plaster on plain gypsum lath (with studs)	20.		t	32
Gypsum wallboard—plain—⅜ in.	50.		v	40
Insulating wallboard (uncoated)—½ in.	50–90		t	34
Paint—2 coats§				
Asphaltic paint on plywood	0.4		w	33
Aluminum in varnish on wood	0.3–0.5		d	37
Enamels, brushed on smooth plaster	0.5–1.5		b	32
Primers or sealers on insulating wallboard	0.9–2.1		t	34
Various primers + 1 coat flat paint on plaster	1.6–3.		t	34
Flat paint (alone) on insulating wallboard	4.		t	34
Water emulsions on insulating wallboard	30–85		t	34
Paint, exterior, 3 coats§				
White, lead and oil prepared paint on wood siding	0.3–1.0		d	40
White lead-zinc oxide and linseed oil on wood	0.9		d	37

	lb per 500 sq ft	Permeance, perms Dry Cup	Wet Cup	Ref.
Building paper and felts§				
Duplex sheet, asphalt, laminae, *aluminum foil* one side	43	0.002	0.176	39
Saturated and *coated* felt heavy roll roofing	326	0.05	0.24	39
Kraft and *asphalt laminae*, Reinforced 30-120-30	34	0.3	1.8	39
Asphalt-saturated and coated sheathing paper	43	0.3	0.6	39
Asphalt-saturated sheathing paper	22	3.3	20.2	39
15-lb asphalt felt	70	1.0	5.6	39
15-lb tar felt	70	4.0	18.2	39
Single sheet Kraft, double infused	16	30.8	41.9	39

* From "Heating Ventilating Air-Conditioning Guide 1956." Used by permission.
† d—dry cup; w—wet cup; t—two temperatures; b—special cell; v—air velocity both sides; 4—average of four methods.
‡ Reference 34 also includes *Bulletins* 22 and 25 of the *Engineering Experiment Station, University of Minnesota*.
§ Description is a guide only, and does not insure permeance.

dew-point temperature corresponding to it is equal to or greater than the actual dry-bulb temperature. The heavy solid-line curve of Fig. 4.10, which generally follows below the temperature-gradient line, indicates the maximum vapor pressures which may exist in the wall

HEAT LOSSES FROM BUILDINGS, VAPOR TRANSMISSION

described without the occurrence of water-vapor condensation. Two additional vapor-pressure curves are shown at the bottom of the figure. The upper one indicates the approximate gradient that would occur if the building paper attached to the outside of the sheathing were an effective vapor barrier. The lower curve indicates the approximate gradient that would occur if the vapor barrier were properly placed, i.e., as near as possible to the warm surface of the wall. Both gradients are based on the assumption that the resistances to vapor transmission of the other materials of the wall construction are negligible compared with that of the vapor barrier. From the foregoing reasoning and a study of Fig. 4.10 it may be noted that a building paper applied for the purpose of preventing air infiltration through the wall should *not* be an effective vapor barrier. In the case of the wall of Fig. 4.10, a vapor barrier at the usual location of the building paper would definitely cause condensation in part of the insulation and on both sides of the sheathing. If a building paper is used it should be of a type which is *not* very resistant to the passage of water vapor and another paper especially designed for use as a vapor barrier or a metal foil should be applied to the inside surface of the studding before attaching the lath and plaster.

Invisible condensation within a wall may be evidenced by the peeling of paint from the outside surfaces. The paint on the outside of a wall may constitute a vapor barrier and cause trouble when there is no effective barrier near to the warm side. In case this trouble develops in a house already erected the only practicable solution is the application of a paint to the inside-wall surfaces. Most oil paints and varnishes are quite effective for the purpose, as indicated in Table 4.7.

4.27. Permeability and Permeance.[29–31] The terms permeability and permeance, which are used in connection with the transmission of water vapor through materials subjected to a difference in vapor pressure, are comparable with the terms conductivity k and conductance C used in calculations involving heat transmission. The permeability of a material is the weight of water vapor, in grains per hour, which will pass through a representative sample 1 in. thick when the vapor-pressure differential between its two sides is 1 in. of mercury. Permeance is the vapor transmission per hour per square foot for the existent thickness. Permeance data are always given for units of building construction such as lath and plaster, hollow tile, papers, and paint films.

The weight of water vapor transmitted may be expressed as grains per hour, grains per hour per square foot, or grains per hour per square foot per inch of mercury pressure difference. The word *perm* is used to express a rate of 1 grain per hr per sq ft per in. of mercury. Permeability is expressed in perms per inch of thickness or simply as *perm-inches* and permeance in perms.

Most of the data which are available for the permeance and permeability of the various elements of building construction have been determined by the use of either the wet-cup method, the dry-cup method, or a combination of the two.[30]

The different methods of either permeability or permeance tests do not give the same results with identical materials. Consequently, considerable doubt exists as to the reliability of most of the available data when applied to the actual conditions existing in a building wall. Table 4.7 indicates the latest information available, at the time of this writing, for a number of materials commonly used in building construction. Metal foils not mentioned in Table 4.7 are practically impervious to the transmission of water vapor if there are no cracks and holes in them and if their joints are effectively sealed.

The over-all permeance of a typical wall, floor, or ceiling structure consisting of several parts is found in exactly the same manner as the over-all coefficient of heat transmission U.

$$P_{et} = \frac{1}{\dfrac{1}{P_{e1}} + \dfrac{1}{P_{e2}} + \cdots + \dfrac{1}{P_{en}}} \qquad (4.15)$$

where P_{et} equals the over-all or composite permeance, and P_{e1}, P_{e2}, and P_{en} equal the permeance of each segment of the wall, floor, or ceiling construction. In the case where permeability is given instead of permeance, x/P_e is used instead of $1/P_e$. The material thickness in inches is x and its permeability is P expressed in perm-inches.

Example. Find the permeance of a wood-frame wall consisting of plaster on metal lath, an air space $3\frac{5}{8}$ in. wide, structural insulating board $\frac{25}{32}$ in. thick, a layer of 15-lb tar felt paper, an air space $\frac{1}{2}$ in. wide, and 4 in. of brick and mortar construction.

Solution. Based on data from Table 4.7 the over-all permeance of the wall is

$$P_{et} = \frac{1}{\dfrac{1}{15} + \dfrac{3.625}{120} + \dfrac{0.781}{35} + \dfrac{1}{11.1} + \dfrac{0.5}{120} + \dfrac{1}{0.8}} = \frac{1}{1.4635} = 0.683 \text{ perm}$$

Permeability data are given in Table 4.7 for still air and the various materials used. The resistance to vapor transmission for each portion of the wall is either its thickness in inches divided by the permeability given in Table 4.7

or 1 divided by the permeance. For the insulating board of this problem Table 4.7 indicates the permeability as ranging from 20 to 50. An average of 35 perm-in. is used in this solution. In every case where a range was given the median value was used in the solution.

In the foregoing example most of the resistance to vapor transmission is provided by the brick and mortar portion of the wall, which is adjacent to the outside air. The condition in the wall approaches that of the incorrect location of a vapor barrier as illustrated in Fig. 4.10.

Because of the importance of using building papers properly they have been classified as follows: *Class A* for use where a high degree of water-vapor resistance is required, *maximum* permeability 0.567 perm. *Class B* for use where a lower degree of water vapor and water resistance is required, *maximum* permeability 0.864 perm. *Class C* employed where a moderate degree of water resistance is necessary, minimum water resistance of 8 hr, permeability not specified. *Class D* for use where a low resistance to water vapor is necessary, minimum water resistance 10 min and *minimum* permeability 5.04 perms. In typical frame-wall construction of a home a Class A paper should be applied between the lath and the studs and a Class D paper should be placed between the sheathing and the exterior finish.

REFERENCES

1. *Heat Transmission*, W. H. McAdams, McGraw-Hill Book Co., New York (1954).
2. *Introduction to Heat Transfer*, A. I. Brown and S. M. Marco, McGraw-Hill Book Co., New York (1951).
3. *Process Heat Transfer*, D. Q. Kern, McGraw-Hill Book Co., New York (1950).
4. *Heat Transfer*, Vol. I, Max Jakob, John Wiley & Sons, New York (1949).
5. "Heat Transfer by Free Convection from Heated Vertical Surfaces to Liquids," V. S. Touloukian, G. A. Hawkins, and Max Jakob, *ASME Trans.*, **70**, 13 (1948).
6. *Heat Transfer Notes*, L. M. K. Boelter, V. H. Cherry, H. A. Johnson, and R. C. Martinelli, University of California Press, Berkeley (1946).
7. *Applied Heat Transfer*, H. J. Stoever, McGraw-Hill Book Co., New York (1941).
8. "Standard Test Code for Heat Transmission Through Walls," *ASHVE Trans.*, **34**, 253 (1928).
9. "Measuring Heat Transmission in Building Structures and a Heat Transmission Meter," P. Nicholls, *ASHVE Trans.*, **30**, 65 (1924).
10. "Investigation of Heating Rooms with Direct Steam Radiators Equipped with Enclosures and Shields," A. C. Willard, A. P. Kratz, M. K. Fahnestock, and S. Konzo, *Bull. No. 192*, Eng. Exp. Sta., University of Illinois (1929).
11. "Surface Conductances as Affected by Air Velocity, Temperature, and Char-

acter of Surface," F. B. Rowley, A. B. Algren, and J. L. Blackshaw, *ASHVE Trans.*, **36,** 429 (1930).
12. "The Thermal Insulating Value of Air Spaces," H. E. Robinson, F. J. Powlitch, and R. S. Dill, *Housing Research Paper No. 32*, Housing and Home Finance Agency, U. S. Government Printing Office (1954).
13. "Temperature and Heat Loss Characteristics of Concrete Floors Laid on the Ground," H. D. Bareither, A. N. Fleming, and B. E. Alberty, *Small Homes Council Technical Report, University of Illinois* (1948).
14. "Neutral Zone in Ventilation," J. E. Emswiler, *ASHVE Trans.*, **32,** 59 (1926).
15. "Air Infiltration Through Various Types of Brick Wall Construction," G. L. Larson, D. W. Nelson, and C. Braatz, *ASHVE Trans.*, **35,** 183 (1929).
16. "Air Infiltration Through Double Hung Wood Windows," G. L. Larson, D. W. Nelson, and R. W. Kubasta, *ASHVE Trans.*, **37,** 571 (1931).
17. "Air Leakage Through the Openings in Buildings," F. C. Houghten and C. C. Schrader, *ASHVE Trans.*, **30,** 105 (1924).
18. "Air Leakage Studies on Metal Windows in a Modern Office Building," F. C. Houghten and M. E. O'Connell, *ASHVE Trans.*, **34,** 34 (1928).
19. "Aeration of Industrial Buildings," W. C. Randall, *ASHVE Trans.*, **34,** 159 (1928).
20. "The Weathertightness of Rolled Section Steel Windows," J. E. Emswiler and W. C. Randall, *ASHVE Trans.*, **34,** 527 (1928).
21. "Air Leakage Around Window Openings," C. C. Schrader, *ASHVE Trans.*, **30,** 313 (1924).
22. "Revolving Doors Cut Heating, Cooling Costs," A. M. Simpson, *Heating, Piping, Air Conditioning*, **26,** No. 11, 109 (Nov. 1954).
23. *Heat Insulation*, G. B. Wilkes, John Wiley & Sons, New York (1950).
24. "Permissible Relative Humidities in Humidified Buildings," P. D. Close, *Heating, Piping, Air Conditioning*, **11,** No. 12, 766 (Dec. 1939).
25. *Thermal Insulation of Buildings*, P. D. Close, Reinhold Publishing Corp., New York (1947).
26. "Condensation of Moisture and its Relation to Building Construction and Operation," F. B. Rowley, A. B. Algren, and C. E. Lund, *ASHVE Trans.*, **45,** 231 (1939).
27. "A Theory Covering the Transfer of Vapor Through Materials," F. B. Rowley, *ASHVE Trans.*, **45,** 545 (1939).
28. "Simultaneous Heat and Vapor Transfer Characteristics of an Insulating Material," F. G. Hechler, E. R. McLaughlin, and E. R. Queer, *ASHVE Trans.*, **48,** 505 (1942).
29. "Comparative Resistance to Vapor Transmission of Various Building Materials," L. V. Teesdale, *ASHVE Trans.*, **49,** 124 (1943).
30. "Permeance Measurement Improved by Special Cell," F. A. Joy and E. R. Queer, *ASHVE Trans.*, **55,** 377 (1949).
31. "Automatic Permeance Measurement by the Permeometer," F. A. Joy and A. W. Sherdon, *Heating, Piping, Air Conditioning*, **25,** No. 7, 125 (July 1953).
32. "Water Vapor Transfer Through Building Materials," F. A. Joy, E. R. Queer, and R. E. Schreiner, *Bull. No. 61*, Eng. Exp. Sta., Pennsylvania State College (Dec. 1948).
33. "Remedial Measures for Building Construction," L. V. Teesdale, *Rept. R 1710*, U. S. Dept. Agr., Forest Service, Forest Products Lab. (1947).
34. "Methods of Moisture Control and their Application to Building Construc-

tion," F. B. Rowley, A. B. Algren, and C. E. Lund, *Bull. No. 17*, Eng. Exp. Sta., University of Minnesota.
35. "The Relation of Wall Construction in Fill-Type Insulation," H. J. Barre, *Bull. No. 71*, Iowa State College of Agriculture and Mechanics Arts (1940).
36. "Moisture Migration—A Survey of Theory and Existing Knowledge," P. F. McDermott, *Refrig. Eng.*, **42**, No. n, 103 (Aug. 1941).
37. "Permeability of Paint Films to Moisture," R. I. Wray and A. R. Van Vorst, *Ind. Eng. Chem.*, **25**, 842 (1933).
38. "Water Vapor Transmission of Building Materials using Four Different Testing Methods," R. R. Britton and R. C. Reichel, *Tech. Bull. No. 12*, Housing and Home Finance Agency, U. S. Government Printing Office (Jan. 1950).
39. "Water Vapor Permeability of Building Papers and Other Sheet Materials," E. R. Bell, M. G. Seidl, and N. T. Krueger, *ASHVE Trans.*, **57**, 287 (1951).
40. Unpublished data of Eng. Exp. Sta., Pennsylvania State College.

PROBLEMS

1. A wall having a coefficient of heat transmission U which is equal to 0.40 Btuh per sq ft per deg F has 200 sq ft of surface exposed to inside air at a temperature of 65 F when the outside-air temperature is -15 F. Find the heat flow through the wall in Btuh.

2. Find by calculation the over-all coefficients of heat transmission U for a 16-in. solid brick wall (a) without interior finish and (b) with $\frac{1}{2}$ in. of cement plaster (sand aggregate), on the inside. The outside-wind velocity is 15 mph. Still-air conditions prevail within the enclosed space and $f_i = 1.46$ and $f_o = 6.0$; k for brick is 5.0 and for plaster 5.0.

3. A compound wall is made up of $3\frac{3}{4}$ in. of face brick, $\frac{1}{2}$ in. of cement mortar, 8 in. of hollow clay tile, and $\frac{1}{2}$ in. of gypsum plaster (lightweight aggregate). The outside-wind velocity is 15 mph and the exposed surface of the plaster is in contact with still air. Find the coefficient of heat transmission U.

4. The coefficient of heat transmission U for a 12-in. concrete wall with $\frac{1}{2}$ in. of gypsum plaster, lightweight aggregate, on its interior is 0.47 Btuh per sq ft per deg F. The wall surface conductances are: $f_o = 7.6$ and $f_i = 1.46$. Find (a) the over-all resistance of the wall, and (b) the conductivity k for the concrete when C for the plaster is 3.12 Btuh per sq ft per deg F.

5. A wall has a coefficient of heat transmission U of 0.23 Btuh per sq ft per deg F. The inside-air temperature is 70 F and the outside-air temperature -20 F. The surface conductances are: $f_i = 1.46$ and $f_o = 6.0$. Find the inside-wall-surface temperature and also the outside-wall-surface temperature.

6. A wall transmits 27 Btuh per sq ft when the inside-air temperature is 75 F and the inside-wall-surface temperature is 60 F. The outside-air temperature is -15 F and the outside-surface temperature of the wall is -10 F. Find the surface conductances f_i and f_o.

7. The data for the example of Art. 4.4, Fig. 4.4, are changed to 2-in. by 6-in. (actual dimensions $1\frac{5}{8}$-in. by $5\frac{5}{8}$-in.) wood studding and expanded-vermiculite insulation with k equal to 0.48. The conductance c for the gypsum

board is 2.25; the surface conductances f_i and f_o are 1.46 and 6.0 respectively. Find the total resistance to heat flow and the over-all coefficient of heat transmission U for each square foot of wall surface, when all other data remain unchanged. (Use $k = 0.80$ for all of the wood.)

8. The wall of Fig. 4.4, Art. 4.4, is changed so that the 2-in. by 4-in. wood studding is placed with the $1\frac{5}{8}$-in. dimension as the distance through which outward flow of heat occurs. The $12\frac{3}{8}$-in. wide spaces between studding are completely filled with foamed plastic slabs $1\frac{5}{8}$ in. thick which have a conductivity k of 0.29. The resistance R of $\frac{1}{2}$-in. gypsum board is 0.45; the total thickness of the outer wooden covering of the studding is 1.281 in.; and the surface coefficients f_i and f_o are 1.55 and 5.45 respectively. All other data of Fig. 4.4 are unchanged. Find the total resistance R_t for each square foot of wall surface and the over-all coefficient of heat transmission U for the wall section. (Use $k = 0.80$ for all of the wood.)

9. Find the conductance C_a and the resistance R_a for an air space $1\frac{5}{8}$ in. wide placed in a wall with the inner or warm side bounded by paper and the outer or cool side formed by wood sheathing. The air space is vertical with heat flow across it in a horizontal direction. The mean temperature of the air in the space is 50 F and the temperature drop across it is 20 F.

10. Find the conductance C_a and the resistance R_a for the conditions of Problem 9 except that the air space is 3.5 in. wide.

11. Find the conductance C_a and the resistance R_a of an air space 0.75 in. wide which is bounded on the cool side by brickwork and on the inner or warm side by concrete. The air space is vertical with horizontal flow of heat across it. The mean temperature and the temperature drop are as in Problem 9.

12. Find the air-space conductance C_a and the resistance R_a for the air spaces in a ceiling composed of 2 in. by 4-in. joists, wood flooring on top and paper-surfaced plasterboard underneath. Mean temperature and temperature drop as in Problem 9.

13. A compound wall of a building has its outside surface exposed to wind moving at 15 mph. The interior surface is in contact with still air. The wall construction from outside inward is: $3\frac{3}{4}$ in. of face brickwork, $\frac{1}{2}$ in. cement mortar, 12-in. hollow clay tile, and $\frac{1}{2}$ in. cement plaster (sand aggregate). Calculate the over-all coefficient of heat transmission U for the wall. Use surface conductance f_0 taken from Fig. 4.5.

14. A roof is made up of asphalt shingles, $\frac{25}{32}$-in. soft-wood decking, and wooden rafters which may be ignored. The outside-wind velocity is 25 mph and the air beneath the roof has movement such that f_i is 1.60. The outside-air temperature is -30 F and the air temperature beneath the roof may be taken as 20 F. Find the heat losses per hour through 1800 sq ft of the roof.

15. A roof with fir decking $1\frac{5}{8}$ in. thick is covered with built-up roofing $\frac{3}{8}$ in. thick. The air beneath the roof has a dry-bulb temperature of 92 F and a relative humidity of 80 per cent. What thickness of cellular glass must be placed beneath the wood to prevent vapor condensation at its surface when the outside air temperature is -60 F. Assume still air inside and a 15-mph wind outside.

16. Find the hourly losses of heat though a 6-in. concrete slab laid upon the ground for the following conditions: dimensions of the slab measured to the outside edges of the foundation walls are 30 ft by 40 ft; air temperature at floor slab, 70 F; outside-air temperature, 0 F; ground temperature, 50 F;

and insulation 2 in. thick at the slab edge and for a distance of 2 ft beneath it. The foundation walls around the slab are 8 in. thick.

17. A basement measuring 24 ft by 32 ft has two 18-in. by 30-in. residential casement steel windows ($\frac{1}{32}$-in. crack) in each of its two long sides. The ceiling height is 7 ft. The basement floor is concrete. The side walls are of 8-in. common brickwork. The grade-line level is 5 ft above the basement floor. The air temperatures involved are: basement 70 F at the 5-ft level, outside -5 F, and just above the floor 65 F. The ground temperature may be taken as 45 F. Assume that the wind velocity is 15 mph. Find the hourly heat losses from the basement if the ceiling losses are ignored.

18. An attic roof has a pitch of $\frac{1}{3}$, i.e., the total rise equals $\frac{1}{3}$ of the span. The ceiling beneath it is 20 ft by 40 ft with the slope of the roof from each long side to the center-ridge pole. The attic ends are triangular areas. The U for the unfloored ceiling below the attic is 0.61 Btuh per sq ft per deg F. The attic side walls are plain (face) brick walls 8 in. thick ($U = 0.41$). The roof consists of asbestos-cement shingles on softwood boards $\frac{25}{32}$ in. thick ($U = 0.44$). The outside-air temperature is -20 F and the room-air temperature just below the ceiling beneath the attic is 75 F. Find the air temperature of the unheated attic space and the heat loss through the ceiling in Btuh.

19. A crawl space with a dirt floor is surrounded by foundation walls for which the U valve is 0.41. The ground temperature is 50 F; the outside-air temperature, -10 F; and the air temperature above the wooden floor covering the space is 68 F. The upper 2 ft of all foundations are exposed to outside air and the dimensions of the enclosed space are 28 ft by 40 ft. The double wood floor has an over-all coefficient of heat transmission U which is 0.34 Btu per sq ft per deg F. Find the air temperature within the crawl space. (Neglect heat loss into the ground in a trial solution.)

20. An attic has a roof area of 2760 sq ft and a ceiling area beneath it of 1800 sq ft. Neither gable areas nor windows are involved. Heat transmission coefficients U are: for roof U_r is 0.52 and for ceiling U_c is 0.28 Btu per sq ft per deg F. Find the combination coefficient of heat transmission U_{cr}.

21. A building is 300 ft high. The wind velocity at a point 75 ft above the mid-height of the structure is 18 mph and at a location 75 ft above the ground it is 16 mph. Find the equivalent wind velocities for the two levels.

22. A living room 13 ft by 22 ft by $8\frac{1}{2}$ ft high has two air changes per hour measured at the room temperature of 72 F when the outside-air temperature is -5 F and the barometric pressure is 29.00 in. of mercury at 32 F. Find the heat losses due to infiltration in Btu.

23. A room has double-hung windows each having 26 ft of nonweather-stripped, $\frac{1}{16}$-in. crack and $\frac{3}{64}$-in. clearance. The windows are wood sash and are unlocked. The room is a corner one and has one window in the short side and two windows in its longer side. The inside-air temperature at the mean height of the windows is 73 F and the outside-air temperature is -3 F with a wind velocity of 10 mph. Calculate the infiltration losses. Assume an air density of 0.075 lb per cu ft.

24. A room space has north, east, south, and west exposures with a total of 20 double-hung, weather-stripped, unlocked, metal windows each measuring 30 in. wide by 60 in. high. The window placements are four in the east end, three in the west end, five in the north side, and eight in the south side. The wind velocity is 25 mph, the inside-air temperature at the mean height of the windows 65 F, the outside-air temperature 5 F. Assume that the air density is 0.075 lb per cu ft. Estimate for design purposes the hourly losses of heat from the space due to air infiltration and exfiltration.

25. A room heated by direct steam radiators has a ceiling height of 14 ft and a breathing-line air temperature of 75 F. The room windows are 3 ft wide and 6 ft high with the bottoms of their openings located 5 ft above the floor. Find the inside-air temperatures above the floor, beneath the ceiling, and at the mean heights of both the walls and the glass areas.

26. A single-story building which is to be heated with unit heaters (assume temperature gradients same as with forced warm air) has common-brick walls 16 in. thick plastered with $\frac{1}{2}$ in. of cement mortar on the inside. The building is 60 ft by 150 ft (outside measurements) exposed on all sides and has no ceiling except the roof. The 6-in. stone concrete floor is laid on the ground which has a temperature of 52 F. (There is no edge insulation.) The roof has a ridge along the longitudinal axis of the building which is 8 ft above the tops of the side walls which extend 16 ft above the floor level. The roof has $1\frac{5}{8}$-in. softwood decking covered with asphalt roll roofing. Each side wall has 15 double-hung metal windows each 4 ft wide by 8 ft high, placed with their bottoms 3 ft above the floor. Each end wall has six windows placed as are the windows of the side walls. All of the windows are nonweather-stripped and locked. Each end wall has a door opening 14 ft wide and 12 ft high. The doors in each opening are two in number and are made of softwood that is $2\frac{1}{8}$ in. thick. Each door opening has two horizontal cracks 14 ft long together with three vertical cracks 12 ft high. All cracks are $\frac{1}{16}$-in. wide and may be considered as having the same air leakage as industrial pivoted steel sash in estimations of air leakage. Assume an average wind velocity of 15 mph. Calculate for design purposes the hourly heat losses from the building based on an inside temperature of 70 F at the 5 ft level and on an outside temperature of -10 F.

27. An exterior wall consists of 4 in. of brickwork, $\frac{1}{2}$-in. air space, $\frac{25}{32}$ in. of pine-wood sheathing, $3\frac{5}{8}$ in. of mineral-wool insulation, asphalt-saturated and -coated sheathing paper on the warm side of the mineral wool bats and wood lath and plaster. Calculate the permeance of the wall. Find the weight of water vapor in grains that would pass through 1000 sq ft of this wall in 1 hr when the inside air has 75 F dbt and 40 per cent relative humidity while the outdoor temperature and relative humidity are 20 F dbt and 60 per cent. Assume standard barometric pressure.

5

Fuel Selection
and Utilization

5.1. Fuel Types, Combustible Elements, and Heating Values. Several good books[1-3] on fuels and combustion are available. Treatment of the subject here will be limited to facets of primary interest to the air-conditioning engineer. In the broadest manner fuels can be classified as solid, liquid (fuel oils), and gaseous types. The general nature of the burner to be used is determined by the type of fuel selected.

Although the physical characteristics of solid, liquid, and gaseous fuels are completely different, the same combustible elements are found in all three. These are carbon which combines with oxygen to form carbon dioxide, CO_2; hydrogen which forms water vapor, H_2O; and sulfur which produces sulfur dioxide, SO_2. Carbon is the principal combustible element in all commercial fuels, regardless of type. Hydrogen has the highest heating value per unit weight and generally contributes an important part of the total heat released. Sulfur contributes a negligible portion of the heating value in every case and is objectionable because of the damage to vegetation, metals, and painted surfaces which may result indirectly from the SO_2 formed. Gaseous fuels contain practically no sulfur, and it is present in but slight amounts in most of the grades of fuel oil. However, the sulfur in the heaviest grade of fuel oil and in bituminous coal from certain areas may be enough to warrant special consideration.

Heating values as used by air-conditioning engineers in this country are generally expressed in Btu per pound for solid fuels, Btu per gallon for fuel oils, and Btu per cubic foot for gases. In the case of fuel gases the value is usually given in Btu per cubic foot of *standard gas* measured at a temperature of 60 F (520 F abs) and under an absolute pressure of 30 in. of mercury. If the gas is metered under a condition other than standard, the heating value must be obtained as illustrated in the following example.

Example. Natural gas in a certain locality has a heating value of 1000 Btu per cu ft of standard gas. The barometric pressure is 26.5 in. of mercury, the gas line pressure is 7 in. of water, and the temperature of the gas passing through the meter is 80 F. Find the heating value of the gas as metered.

Solution. The first step is to find the volume of 1 cu ft of standard gas when under metered conditions by applying equation 1.5. The absolute pressure of the gas at the meter in inches of mercury is $26.5 + 7/13.58 = 27.014$ and its absolute temperature is $80 + 460 = 540$ F abs. (13.58 is the specific gravity 60/60 of mercury, Art. 5.9.) Representing the desired volume by V_m, $(27.014 \times V_m)/540 = (30 \times 1)/520$. $V_m = 1.148$ cu ft. The actual heating value of the gas as metered is then $1000/1.148 = 872$ Btu per cu ft.

Fuel technologists often refer to the actual heating value of a fuel as the *higher heating value* to differentiate between it and the lower heating value. The *lower heating value* is the net heat liberated per pound of fuel after deducting the latent heat of vaporization of the steam formed by the combustion of hydrogen and that formed by the evaporation of any moisture present in the fuel, as fired. Calorimeters used for determining the heating value of fuels condense practically all of this steam and give the higher heating value. The lower heating value is a better measure of the true worth of a fuel but it is necessary to have the chemical analysis (called the *ultimate analysis*) before it can be determined. Practically all calculations involving fuels are based on the higher heating value. If the ultimate analysis of a fuel is available, the higher heating value in Btu per pound may be computed by Dulong's formula.

$$F = 14{,}540\mathrm{C} + 60{,}958(\mathrm{H} - \mathrm{O}/8) + 4050\mathrm{S} \tag{5.1}$$

where C, H, O, and S are respectively the weights in pounds per pound of fuel of the carbon, hydrogen, oxygen, and sulfur as indicated by the analysis. Heating values of gaseous fuels can usually be obtained from the local utility company and that of either solid or liquid fuels from local suppliers.

FUEL SELECTION AND UTILIZATION

5.2. Anthracite Coal. Anthracite coal, often called *hard* coal and mined extensively only in Pennsylvania, may be easily handled without excessive breakage and the individual pieces do not soften in the burning process. It may be stored in large bins for long periods without danger of spontaneous combustion and when suitably prepared as to size of pieces it is well adapted to use in furnaces or boilers equipped with a hopper (*magazine*) for feeding the fire by gravity action. Several different types of automatic stokers incorporating

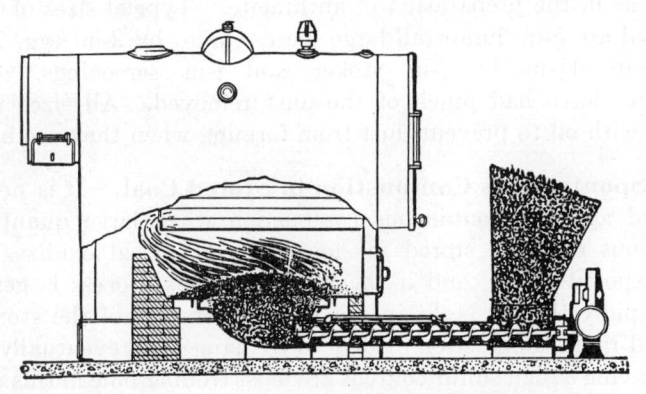

Fig. 5.1. Underfeed stoker with heating boiler.

automatic ash removal are available for burning this fuel. Commercial sizes of anthracite (all sizes in inches) are: broken ($3\frac{1}{4}$ to $4\frac{3}{8}$), egg ($2\frac{5}{16}$ to $3\frac{1}{4}$), stove ($1\frac{5}{8}$ to $2\frac{7}{8}$), nut ($1\frac{3}{16}$ to $1\frac{5}{8}$), pea ($\frac{9}{16}$ to $1\frac{3}{16}$), buckwheat ($\frac{5}{16}$ to $\frac{9}{16}$), rice ($\frac{3}{16}$ to $\frac{5}{16}$), barley ($\frac{3}{32}$ to $\frac{3}{16}$), No. 4 ($\frac{3}{64}$ to $\frac{3}{32}$), and No. 5 (smaller than $\frac{3}{64}$). The smallest sizes are least costly but more expensive equipment is required to handle them.

5.3. Bituminous Coal. Bituminous coal, often referred to as soft coal, is sold in a wide variety of types, preparations, and burning characteristics. It is mined in many different sections of the country and hence is less expensive than anthracite in most regions because of lower transportation costs. Some bituminous coals are nearly as hard as anthracite when at outdoor or room temperature, but they all soften and the lumps fuse together to a varying extent when heated in the preliminary stage of the burning process. This burning characteristic must always be kept in mind when selecting a burner to handle this fuel.

Objectional smoke results from the hand firing of bituminous coal unless special techniques[4] are used. Underfeed stokers[5] of the general type shown in Fig. 5.1 are usually employed for burning this fuel in homes and in small commercial buildings. Chain grate stokers and spreader stokers are often used where a large fuel-burning capacity is required. In central power-generating stations this fuel is generally burned in pulverized form. When either spreader stokers or burners using powdered coal are used, an excessive amount of fly ash is likely to be discharged into the atmosphere unless special removal equipment is employed in the plant.

Sizing practice in the bituminous coal industry has not been standardized as in the preparation of anthracite. Typical sizes of bituminous coal are 5-in. lump (all large pieces), 5-in. by 2-in. egg, 2-in. by 1¼-in. nut, 1¼-in. by ¾-in. stoker, and ¾-in. screenings. Washed screenings have had much of the dust removed. All sizes may be treated with oil to prevent dust from forming when they are handled.

5.4. Spontaneous Combustion in Stored Coal. It is necessary to guard against spontaneous combustion when large quantities of bituminous coal are stored for long periods. Coal oxidizes slowly when exposed to air, and if the heat from this process is generated more rapidly than it is dissipated the temperature of the stored fuel rises and rapid combustion evidenced by flames will eventually result. Coals having a high sulfur content are most troublesome in this respect, particularly when damp. Spontaneous combustion in coal piles can be avoided either by storing only carefully screened, large sizes so that the pile will be well ventilated or by storing a preparation that is so fine that air cannot enter the mound of fuel. The latter course is usually the more practicable solution, particularly if the coal is to be pulverized before it is burned. One commonly used technique is to drive a heavy tractor back and forth over the stored coal to thoroughly crush and compact the outer layer that is exposed to the air. If suitable facilities are available, coal of any preparation or analysis can be stored under water for an indefinite period without any danger of spontaneous combustion.

5.5. Proximate Analysis. A special type of analysis, termed proximate, is often applied to coals. The proximate analysis gives the percentage by weights of moisture, volatile matter, fixed carbon, and ash. The volatile matter consists of hydrogen and carbon in combination. The fixed carbon is responsible for the characteristic glow in a fuel bed after the moisture and volatile matter have been driven out. Ash is the incombustible final residue.

5.6. Coke. Ordinary coke is made from bituminous coal by heating it in a coke oven so as to drive out the moisture and volatile

FUEL SELECTION AND UTILIZATION

matter while preventing the combustion of the fixed carbon. Coke consists essentially of fixed carbon and ash. Petroleum coke is produced as a residue from the destructive distillation of crude oil. It generally has a low ash content.

5.7. Fuel Oils. The chief constituents of fuel oils are carbon and hydrogen, which are united in a series of very complex hydrocarbon

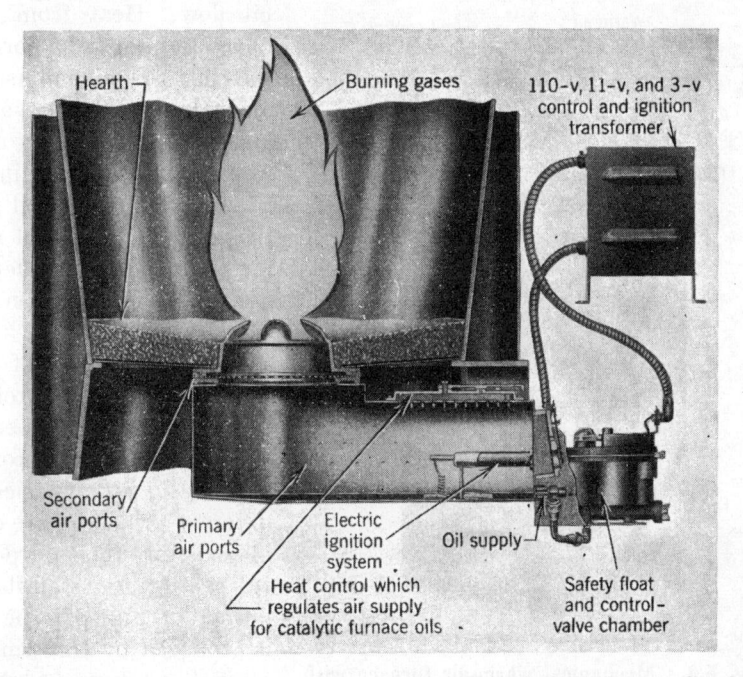

Fig. 5.2. Vaporizing or pot oil burner. (Courtesy H. C. Little Burner Co., Inc.)

compounds. Other items in the analyses are oxygen, nitrogen, sulfur, and moisture. The last is very undesirable, especially when it forms an emulsion.

Fuel oils are standardized in different grades[6,7] which are designated by numbers 1, 2, 4, 5, and 6. At the present time a No. 3 grade is not listed. The detailed requirements of the different grades of fuel oils are stated in reference 6. Oils of grades No. 1 and 2 are light ones suitable for burners found in small heating plants. Grade No. 4 has a viscosity such that preheating it is not necessary. The heaviness of oils is progressively indicated by the magnitude of the grade number. Grade No. 6, the heaviest of the fuel oils, requires preheating and has a usage in industrial furnaces and those of steam power plants. Grade

No. 5 may or may not require some preheating, and its uses are similar to those of No. 6 oil.

5.8. Oil Burners. Number 1 oil can be burned in the simple arrangement illustrated by Fig. 5.2 which is known as a vaporizing or pot-type burner. This type of oil burner consists of a pot and a control which regulates the oil flow. Heat from the burning process vaporizes the oil. The combustion air is drawn into the vapor above the oil pool by natural draft or is forced in by a small fan. Careful adjustment of both fuel and air is necessary in order to avoid the production of smoke.

Heavier grades of oil must be atomized (broken into small droplets) before they can be satisfactorily burned. Two basically different principles are employed for this purpose. In the rotary wall-flame burner of Fig. 5.3 the oil is atomized by throwing it against a flame ring by means of small tubes rotating at high speed. Fan blades rotating with the tubes draw the proper amount of air into the combustion chamber. Burners of this type function efficiently in burning No. 2 oil. Burners which atomize the oil by throwing it from the rim of a cup rotating at speeds up to 10,000 rpm can handle any grade of oil up to No. 5 without preheating and can handle No. 6 oil if it is preheated.

The gun-type burner of Fig. 5.4 atomizes the oil by forcing it under considerable pressure through a small orifice. A fan on the same shaft as the oil pump delivers the combustion air so as to thoroughly mix it with the atomized oil. Gun-type burners are made as high-pressure and low-pressure types. The low-pressure units employ larger orifices

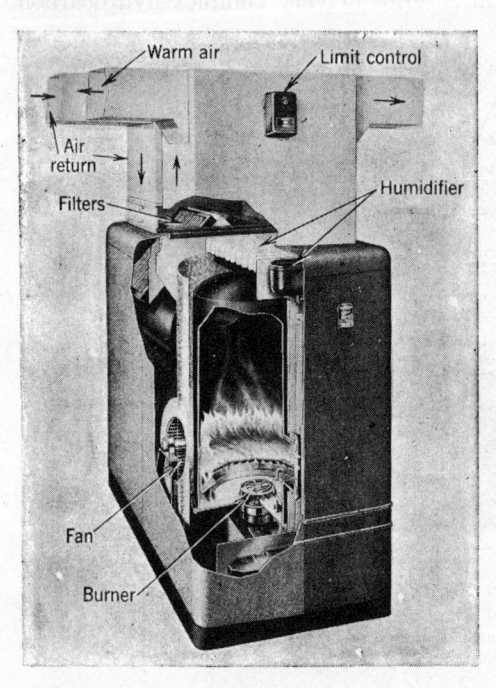

Fig. 5.3. Mechanical warm-air furnace with rotary wall-flame oil burner. (Courtesy Timken Silent Automatic Division, Timken-Detroit Axle Co.)

FUEL SELECTION AND UTILIZATION

and secure adequate atomization by passing the oil droplets together with a part of the combustion air through a second restriction. Low-pressure burners are usually more efficient but the initial cost is higher than comparable units of the high-pressure type. Large boilers burning oil are normally equipped with atomizing burners which employ either steam or high-pressure compressed air for breaking the

Fig. 5.4. Domestic gun-type conversion oil burner.

oil down into very small droplets. Installations of this type usually burn either No. 5 or No. 6 oil. If No. 6 oil is used, provision must be made to preheat it before it reaches the atomization process. Provision must also be made for keeping it warm at all times when in storage tanks.

5.9. Fuel-Oil Specifications. Fuel oils should be free from grit, acid, and fibrous materials, and they should have limitations on their sulfur, silt, and moisture contents. Specifications for fuel oils should include the items of viscosity, specific gravity, flash point, burning point, pour point, calorific value, and water, silt, and sulfur contents.

Viscosity is a measure of the internal friction of the oil or its resistance to free flowing. Viscosity is specified at definite temperatures as determined by some specific viscosimeter and is the time in seconds required for a definite quantity of the oil, at the prescribed temperature, to flow through the orifice of the viscosimeter.

Specific gravity is the ratio of the weight of a given volume of the oil to the weight of an equal volume of water at the same temperature. A temperature of 60 F is commonly used so that sp gr 60°/60° F means that oil at 60 F has been referred to water at 60 F. Specific gravities may be determined by either a hydrometer or a Westphal balance. Specific gravities are useful in computing the weights of unit volumes, as oils are usually sold either by the volume in gallons or barrels of 42 gal.

The *API* (American Petroleum Institute) *gravity*, measured by means of a hydrometer, is generally used by the oil industry. The instrument scale is calibrated to read 10 deg when that graduation on its float stem is at the surface of 60 F pure water in which it is partially immersed. The API scale fixes a reading of 10 deg as equal to a specific gravity of 1.00. Readings greater than 10 deg indicate a specific gravity less than 1.0 or an oil which is lighter than water. Actual specific gravities may be calculated from degrees API by the following equation:

$$\text{Sp gr} = \frac{141.5}{131.5 + \text{deg API}} \qquad (5.2)$$

Flash point and *burning point* are, respectively, the temperatures at which, when an oil is heated, its vapors flash or burn momentarily and burn continuously when a small flame is brought near its surface. *Pour point* is the lowest temperature at which cold oil will flow.

Actual determinations of heating value, made with a calorimeter, are the most useful in the case of fuel oils, because the calorific value may be only approximately calculated by Dulong's equation. Water and sediment contents should not exceed 0.05 per cent for No. 1 oil and 2 per cent by volume for No. 6 oil. Although sulfur is undesirable, an amount not in excess of 2 per cent is not seriously objectionable in the heavier oils.

5.10. Gaseous Fuels. The use of gas as a fuel in heating plants has become quite extensive, especially since the advent of long-distance high-pressure transmission lines for the conveyance of natural gas.

Natural gas occurs in many localities and usually consists of a high percentage of methane CH_4, some ethane C_2H_6, and small percentages of nitrogen, carbon monoxide, carbon dioxide, and other materials.

FUEL SELECTION AND UTILIZATION

Coal and *coke-oven gases* are obtainable by carbonizing high volatile bituminous coals, without air in contact with them, in suitable retorts or ovens which are heated by the combustion of gas. The heating values of the two forms of gases are not greatly different and in their production valuable by-products of the coal are generally obtained. Their heating values range from 400 to 700 Btu per cu ft.

Oil gas is obtained by distilling petroleum oil in iron or fireclay retorts similar to those used for coal gas.

Water gas, a somewhat misleading designation, is obtained by alternately passing steam and air through a fuel bed containing incandescent carbon to produce a mixture of carbon monoxide, carbon dioxide, hydrogen, and nitrogen.

Carburetted water gas results from the enrichment of water gas, which alone is not suitable for use as a domestic fuel, with gas from a petroleum oil. In the process ethylene and methane are usually added.

Producer gases fall within a general group of those obtained by firing coal or coke with an insufficient amount of air for complete combustion and are formed either with or without an admixture of steam. Due to their low heating value, 125 to 150 Btu per cu ft, they are not used in small domestic heating plants.

Other *manufactured gases* are *reformed natural gas* and *refinery oil gas*. *Mixed gases* usually contain natural gas together with one or more of the manufactured types.

Liquid-petroleum gases, by-products of oil refineries, consist of butane, C_4H_{10}, and propane, C_3H_8. The materials can be liquified at reasonable temperatures and pressures which allow them to be transported in tank cars and stored in suitable containers on the purchaser's premises. The gases are used in various ways for both industrial and domestic purposes. Each may be used alone when vaporized and mixed with air, they may be blended together and mixed with air to form a fuel gas, or they may be used with other fuel gases, in various proportions, to augment a fuel supply. The higher heating value of butane is 21,320 and of propane 21,620 Btu per lb.

5.11. Gas Burners and Gas-Fired Heating Units. A gas burner functions to convey a mixture of air and fuel gas to the combustion chamber of a furnace where its ignition and burning occur. Gas burners operate with either atmospheric or power injection of the combustion air. A combined burner-boiler unit is shown in Fig. 5.5. The *draft hood* or divertor shown at the top of the boiler should be part of every gas burner installation. It includes an inverted cone about which the flue gases pass on their way to the chimney. An annular opening in the draft hood allows air from the furnace room to be drawn into the flue connection. The objectives of the draft hood are to prevent a chimney from producing excessive draft conditions in the

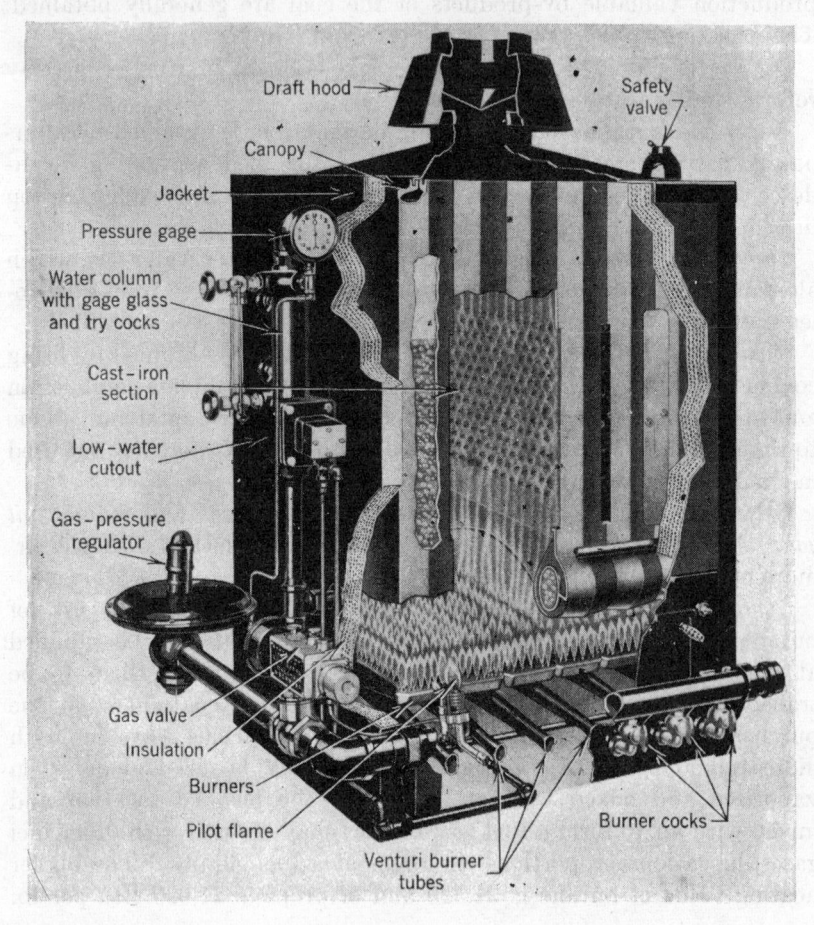

Fig. 5.5. Gas-fired steam-heating boiler. (Courtesy American Radiator and Standard Sanitary Corp.)

furnace and also to prevent downward flow of air in the chimney from extinguishing either a pilot flame or that of the main burner.

Details of a conversion burner suitable for heating boilers and warm-air furnaces are given in Fig. 5.6. The unit shown has only one large port in place of the many small ports in the burner of Fig. 5.5. A flat circular disk directs the flames toward the walls of the combustion

FUEL SELECTION AND UTILIZATION

chamber where heat is transferred from the products of combustion to the heating medium (water or air). Other arrangements used for this purpose include stainless steel cones and refractory baffles.

Gas burners are rated in terms of the hourly input of heat units. Given in this manner a rating applies to any kind of gas. The size of the orifice used or the position of the regulating screw must be adjusted to suit the type of gas to be burned.

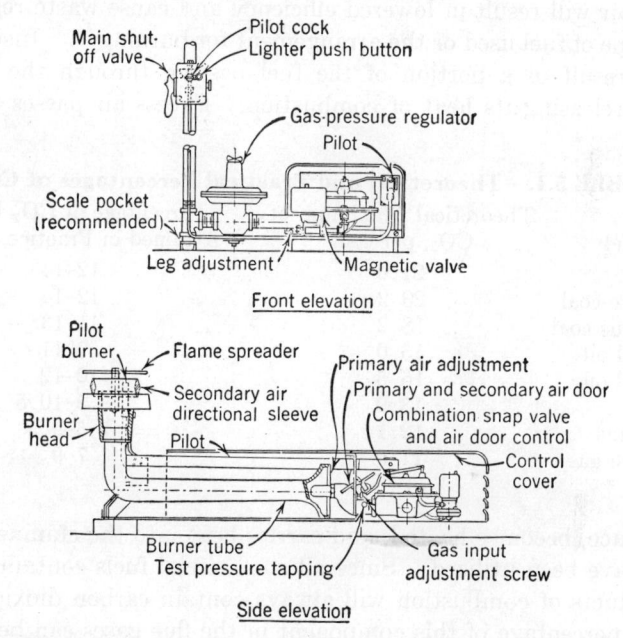

Fig. 5.6. Spreader-flame conversion gas burner. (Courtesy Bryant Heater Co.)

Small-capacity gas burners usually are designed to aspirate the required amount of primary air into the burner tube with the high-velocity jet of gas as in Fig. 5.6. The general type is called an *atmospheric burner*. Gas pressures ranging from 2 to 12 in. of water are used in supplying atmospheric burners. Large-capacity boilers or furnaces may be equipped with high-pressure burners employing gas pressures to 40 psig. High-pressure burners may use the same principle of air procurement employed in atmospheric burners. In one design a fan is operated by a turbine powered by the gas to be burned. In other types the combustion air is supplied either from a fan or from an air compressor.

5.12. Combination Oil and Gas Burners. In localities where the gas supply may be inadequate to meet the demand in severe winter weather, burner units may be installed which are able to utilize fuel oils during periods when gaseous fuel is not available. Such units are called combination oil and gas burners. Each unit of the combination burner is fitted with the necessary control devices required for the fuel preparation and the promotion of its combustion.

5.13. Burner Adjustment. An improper proportion of combustion air will result in lowered efficiency and cause waste regardless of the type of fuel used or the arrangement for burning it. Insufficient air will result in a portion of the fuel passing through the furnace without releasing its heat of combustion. Excess air passes through

TABLE 5.1. Theoretical and Practical Percentages of CO_2

Fuel	Theoretical Maximum of CO_2, per cent	Percentage of CO_2 Usually Attained in Practice, per cent
Coke	21.0	12–14
Anthracite coal	20.2	12–14
Bituminous coal	18.2	11–13
No. 2 fuel oil	15.0	9–11
No. 6 fuel oil	16.5	10–12
Propane	13.9	8.5–10.5
Natural gas	12.1	7.5–9.5
Coke-oven gas	11.0	7–9

the furnace, becomes heated, and carries heat into the chimney which might have been utilized. Since all commercial fuels contain carbon, the products of combustion will always contain carbon dioxide, CO_2, and the percentage of this component in the flue gases can be used as a guide in burner adjustment.

The percentage of CO_2 that can be obtained with perfect combustion depends on the percentage of carbon in the combustible portion of the fuel burned. The middle column of Table 5.1 gives the per cent of CO_2 that would result from perfect combustion of each of several commercial fuels if exactly the required amount of air were supplied. However, perfect mixing of the air with the combustible elements of the fuel cannot be depended upon in the actual combustion chamber so it is advisable to introduce a greater amount of air than that which is theoretically required in order to increase the chance that every particle of fuel will contact enough oxygen to burn it. An excess of air results in a percentage of CO_2 lower than that which can theoretically be achieved. The right-hand column of the table gives the range of CO_2, for each fuel listed, which is acceptable in actual practice. If an analysis of a sample of flue gas shows that the percentage

FUEL SELECTION AND UTILIZATION

of CO_2 is appreciably below the range given in the right-hand column of the table for the fuel that is burning, an adjustment of the burner or air damper should be made and another sample of flue gas analyzed. This process should be repeated until an acceptable percentage of CO_2 is attained.

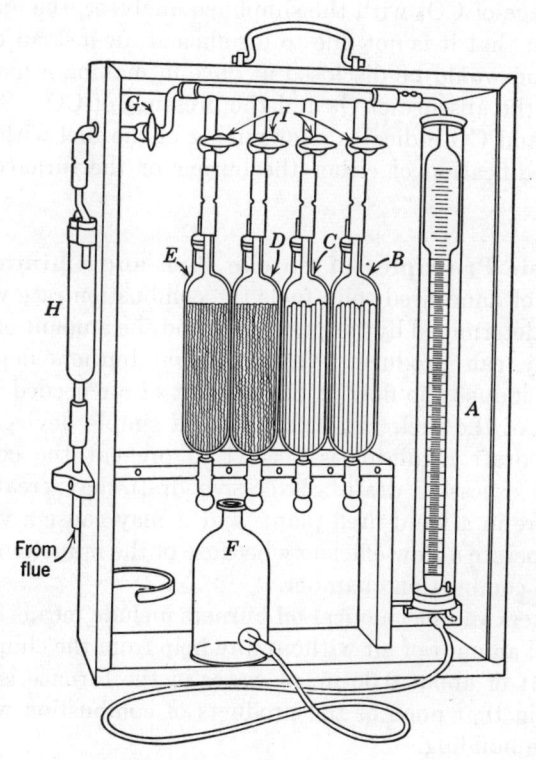

Fig. 5.7. Orsat apparatus.

A portable analyzer known as the Orsat apparatus is shown in Fig. 5.7. The essential parts are: the measuring burette A, the leveling bottle F, the gas cleaner H, the combination inlet and discharge valve G, and the absorption pipettes B, C, D, and E which are interconnected by means of a manifold. The valves I are manipulated so as to direct the sample of flue gas into one pipette at a time. Carbon dioxide is absorbed in pipette B which is partially filled with caustic potash, KOH; oxygen, O_2, is taken out in C which contains an alkaline solution of pyrogallic acid; and carbon monoxide, CO, is absorbed by an acid solution of cuprous chloride in D and E. The absorptions must occur in the order indicated and the nitrogen is obtained by difference. The fourth pipette E is for a second check on the CO and some analyzers of this type do not include it.

A simple analyzer having only the measuring burette, which in this case contains the caustic potash, can be used for burner adjustment as it measures the percentage of CO_2. However, when obtaining a low percentage of CO_2 with the simplified analyzer, the operator must make certain that it is not due to insufficient air instead of an excess. This situation would be disclosed at once in making a more complete analysis by the absence of O_2 and the presence of CO. The presence of both O_2 and CO indicates poor mixing of the fuel with the air and calls for modification of either the burner or the furnace or of both of them.

5.14. Basic Principles of Smoke Pipe and Chimney Design. In the case of hand-fired solid fuels the combustion rate which can be achieved is determined by the grate area and the amount of draft which the chimney can produce. Vaporizing oil burners depend on the chimney to induce the flow of the amount of air needed for complete combustion of the fuel. In either case, a simple device known as a barometric draft regulator is needed to prevent the chimney from creating an excessive draft. Excessive draft may create a dangerously hot fire in a hand-fired plant, and it may cause a vaporizing oil burner to operate at low efficiency because of the induction of too much air into the combustion chamber.

Coal stokers and mechanical oil burners include means for providing the required amount of air without any help from the chimney. However, a draft of about 0.03 in. of water in the furnace is desirable to make certain that none of the products of combustion will be forced out into the building.

Gas-burning boilers and furnaces are generally designed so that the required draft is created within the unit. A chimney is needed to carry the products of combustion to the outside but a draft hood, Fig. 5.5, must be provided so that varying chimney draft will have no effect on the combustion process. The draft hood also prevents a blocked flue, or a down draft in the flue, from creating a condition which might result in the formation of carbon monoxide.

Boilers and furnaces designed for coal and later converted to gas often produce too much draft within themselves, thus requiring the installation of a turn damper in the smoke pipe upstream from the draft hood. However, such a damper when used should be designed so that it cannot completely block the smoke pipe.

FUEL SELECTION AND UTILIZATION

A down draft may be caused by a point on the roof of the same building that is higher than the top of the chimney or by a nearby building, tree, or hill. Usually, a down draft condition due to wind effects can be eliminated by extending the top of the chimney until it is 2 ft higher than any nearby object. Down draft will always be found in a chimney if at any time its lining is at a lower temperature than that of the outdoor air. This condition is more likely to occur when the chimney is located on the outside of the building. Special devices employing small fans are now available for installation in smoke pipes where a down draft condition has been troublesome.

Boiler and furnace manufacturers usually indicate in their data for each unit size the proper size of the flue and a suitable chimney height (see Table 9.1). Considerable information about chimneys is given in the ASHAE Guide.

5.15. Heating Loads and Fuel Costs per Month and per Season. When either steam, gas, or electricity is purchased for heating purposes, the rate per unit used is often dependent upon the amount consumed. The estimated costs, either per month or per season, must be based on the expected heat requirements of the building per unit of time considered. The purchased commodity—steam, coal, gas, fuel oil, or electricity—must provide for the heat losses of the building as a unit and other losses such as from piping.

When heat energy is derived from electrical energy or from the combustion of any fuel, the efficiency of the heating-plant operation must be considered in arriving at the amounts necessary and their costs. The efficiency of resistance electric heaters placed within the spaces to be warmed is 100 per cent. For fuel-burning devices the percentages may range from 45 to 85. A part of the heat from the fuel which appears to be lost aids in warming the structure. Areas from which unregulated heat energy may be emitted are the external surfaces of boilers and warm-air furnaces, smoke-pipe connections, and chimney faces. Other sources of heat which help to warm a structure are lights, occupants, cooking operations, electric appliances, and independently operated water heaters. Therefore, in the estimation of fuel costs and fuel requirements for a definite period of time, consideration should be given to the seasonal efficiency of the heating plant as it operates in the structure which it serves. The seasonal efficiencies are greater than those based on heat outputs at boiler outlets and warm-air furnace bonnets. Seasonal efficiencies are affected by many items and are dependent for any given structure on the conditions pertinent to it alone. Approximate seasonal efficiencies for different types of fuel-burning equipment and for electric heaters are given in Table 5.2.

TABLE 5.2. Approximate Seasonal Heating Efficiencies

Type of Fuel-Burning Unit	Seasonal Efficiency, per cent	Type of Fuel-Burning Unit	Seasonal Efficiency, per cent
Gas, designed unit	75–80	Anthracite, hand-fired with controls	60–80
Gas, conversion unit	60–80		
Oil, designed unit	65–80	Anthracite, hand-fired without controls	50–65
Oil, conversion unit	60–80		
Bituminous coal, hand-fired without controls	40–60	Anthracite, stoker-fired	60–80
		Coke, hand-fired with controls	60–80
Bituminous coal, hand-fired with controls	50–65	Coke, hand-fired without controls	50–65
Bituminous coal, stoker-fired	50–70	Direct electric heating	100

Two methods are available for the estimation of heat or fuel consumption. The first method is based on the heat requirements of the building per unit of time involved as estimated for the average inside-outside air-temperature difference for the period. The second method requires data relative to a term known as the degree day.

By the method first mentioned the heat requirements per period considered are

$$H_m = \frac{H_b(t_r - t_a)N}{t - t_o} \qquad (5.3)$$

where H_m represents the heat losses in Btu per month or per season as required, and H_b is in Btu per hour under design conditions. The number of hours of heating service rendered per month or per season is represented by N, and the temperatures, all in deg F, are: t_r, average inside; t_a, average outside; t, design inside; and t_o design outside.

Purchased steam may be expected to yield approximately 1000 Btu per lb when used in a heating system. The weight M_s of steam per unit of time considered is $H_m \div 1000$. When fuel is burned to produce heat the required amount is

$$M_f = \frac{H_m}{F_{cu}E_s} \qquad (5.4)$$

FUEL SELECTION AND UTILIZATION

where M_f = units of fuel required per time interval, i.e., tons of coal, cubic feet of gas, or gallons of oil.
H_m = total heat losses for the period, Btu.
F_{cu} = heating value of the unit of fuel used, i.e., Btu per ton per cu ft, per gal or per therm. (A *therm* is equal to 100,000 Btu. Gas is often sold by the therm instead of by the cubic foot.)
E_s = seasonal efficiency of fuel utilization expressed decimally.

The term *degree day* was originated by the American Gas Association. It is based on the premise that heat is not required in a building maintained at 70 F when the average outside-air temperature, represented by the mean of the maximum and minimum outside-air temperatures for the day, does not fall below 65 F. Each degree that the average outside-air temperature falls below 65 F represents a degree day. Thus, if the average of the maximum and the minimum temperatures of a day is 60 F, the number of degree days is 5. The degree days thus determined are totaled and reported per month and per heating season. These totals vary with different localities; see Table 4.6. Data for the degree days of various localities.

When the number of fuel units actually used during a prior heating season is known, an estimate of the probable fuel consumption to date, during a current heating season, may be made by the following method: (Fuel used during a prior season)(degree days to date)/(total degree days for the heating season during which a definite quantity of fuel used is known). Degree-day records are also of help to a fuel user who is interested in comparing annual fuel costs when a change has been made in the type of fuel used.

If one wishes to compare the cost of heating a structure by means of one fuel with that which would result from using a different type, it is necessary to convert the commercial price quotations to a common basis. It is also necessary to take into consideration any difference in seasonal efficiency which may be involved. A convenient common unit of heat measurement to use for this purpose is the therm, (100,000 Btu). The cost of heating in cents per therm of utilized heat may be computed for any fuel after the heating value and the price are known. It is given by:

$$\frac{\text{Cost per therm of utilized heat}}{} = \frac{\text{Price per sale unit in cents} \times 100{,}000}{\text{Btu per sale unit} \times \text{efficiency}} \quad (5.5)$$

The method will be illustrated by two examples.

Example 1. Compare the cost of heating with natural gas at 8 cents per therm with that when using bituminous coal costing $15.00 per ton and having a heating value of 13,000 Btu per lb. The coal is hand-fired without automatic controls. The gas is to be used in a conversion unit.

Solution. The median seasonal efficiencies from Table 5.2 are 70 per cent for the gas and 50 per cent for the coal. The cost per therm of utilized heat in the case of gas is $(8 \times 100{,}000)/(100{,}000 \times 0.70) = 11.42$ cents. In the case of coal it is $(1500 \times 100{,}000)/(2000 \times 13{,}000 \times 0.50) = 11.53$ cents.

Example 2. It is desired to know the relative cost of heating with electric resistance panels compared with No. 2 oil burned in a heater specifically designed for this fuel. The oil costs 15 cents per gal and its heating value is 140,000 Btu per gal. The rate for electric service is 1.5 cents per kwhr.

Solution. Using the median efficiencies given for oil and electricity in Table 5.2, the heating value of the oil, and the Btu per kwhr from the table of conversion equations, the costs per therm of utilized heat are: For oil $(15 \times 100{,}000)/140{,}000 \times 0.725 = 14.8$ cents; for electricity $(1.5 \times 100{,}000)/(3413 \times 1.00) = 44.0$ cents.

In a similar manner the cost of a therm of utilized heat from any available energy source may be determined. The cost of electricity is usually relatively high, as indicated in Example 2. By generous use of insulation and other heat-conserving devices the heat loss from a home or any other building may be reduced to the point where heating with electricity is not too costly. It should also be pointed out that it is easier to reduce the amount of energy used with separately controlled electric panels than is the case when heating a building with a central plant which burns some type of fuel. The heat pump (see Chap. 17) multiplies the effectiveness of electricity when used for heating, thereby reducing the relative cost of this type of energy.

REFERENCES

1. *Fuels and Combustion*, Smith and Stinson, McGraw-Hill Book Co., New York, 1st ed. (1952).
2. *Fuels, Combustion and Furnaces*, John Griswold, McGraw-Hill Book Co., New York, 1st ed. (1946).
3. *Combustion Engineering*, Otto de Lorenzi, Combustion Engineering Co., New York (1947).
4. "Hand Firing of Bituminous Coal in the Home," A. P. Kratz, J. R. Fellows, and J. C. Miles, *Circ. No. 46*, Eng. Exp. Sta., University of Illinois (1942).
5. "Industry Standards, Recommended Practices, Technical Information," Stoker Manufacturers Association, Chicago (1944).
6. "Fuel Oil," *Commercial Standard CS12-48*, U. S. Department of Commerce (1948).
7. "Tentative Specifications for Fuel Oils," *ASTM Spec. D396-48T*, American Society for Testing Materials (1948).

FUEL SELECTION AND UTILIZATION

PROBLEMS

1. A barrel of fuel oil contains 42 gal. The oil has an API gravity of 16.8 measured at 60 F and a coefficient of cubical expansion of 0.0004. (a) Find the volume of 10,000 gal of the oil measured at 60 F, expressed in barrels, when its temperature is 80 F. (b) If the heating value of the oil is 18,840 Btu per lb, how many tons of coal having a heating value of 11,695 Btu per lb would have the heat content of a barrel of the oil when at a temperature of 60 F.

2. A storage tank contains 5000 cu ft and is filled with gas at a pressure of 25 psia and a temperature of 80 F. Under standard conditions the gas has a heating value of 450 Btu per cu ft. (a) Find the heat energy stored in the gas in the tank under the foregoing conditions. (b) If the pressure is reduced to 15 psia and the gas temperature is lowered to 40 F, find the heat energy stored in the tank.

3. Calculate the higher heating value of coal for which the following ultimate analysis gives its component parts in percentages: carbon, 66.5; hydrogen, 4.26; sulfur, 2.99; oxygen, 6.30; nitrogen, 1.04; free moisture, 7.88; and ash, 11.03.

4. Find the weight, in pounds, of 1 gal of fuel oil for which the API gravity taken at 60 F is 36 deg. If the higher heating value of the oil is 140,000 Btu per gal, how many gallons would be equivalent to a ton of the coal of Problem 3 assuming the same over-all efficiency?

5. An analysis of the flue gas from a heating plant burning No. 6 fuel oil gave a CO_2 percentage of 8.25. What is the maximum percentage attainable with the fuel? Should the burner be adjusted for a different proportion of air?

6. A building has hourly heat losses amounting to 125,000 Btu based on an inside temperature of 70 F and an outside temperature for design purposes of −10 F. The average outside-air temperature during a month of 30 days was 35 F. Find the number of tons of coal required for the month if fuel having a heating value of 10,670 Btu per lb is used with a seasonal efficiency of 65 per cent. Assume 24 hr per day for the heating service.

7. During the heating season a building having a volume of 80,000 cu ft required 1200 Btu per 1000 cu ft of volume per degree day for heating. The total degree days for a season amounted to 5400. Find the cost of heating the building with oil, which costs 12 cents per gal, that weighs 7 lb and has a heating value of 18,500 Btu per lb, if the oil is utilized with a seasonal efficiency of 70 per cent.

8. A home owner used 10 tons of bituminous coal in a hand-fired furnace without controls during a mild season when the degree days amounted to 5000. The higher heating value of the coal as fired was 11,000 Btu per lb. Estimate the gallons of oil with a heating value of 140,000 Btu that would be required during a normal season of 5500 degree days if the oil were used in a conversion burner. (Use median value from Table 5.2.)

9. During one heating season 1500 gal of oil having a higher heating value of 140,000 Btu per gal and costing 16 cents per gal were used in a boiler designed especially for this fuel. Estimate the cost of maintaining the same temperature in the same building during a season having the same number of degree days if some form of electric resistance heating is used and the rate is 1.5 cents per kwhr.

6

Fluid Flow and Pressure Losses in Ducts and Pipes

6.1. Necessity for Pressure-Loss Calculations. The successful operation of any heating, ventilating or air-conditioning system is dependent upon the adequate circulation within it of one or more fluids. These media may be either in a liquid, vaporous, or gaseous state and include water, refrigerants, steam, and air. A careful estimate of the pressure losses involved is necessary when a duct or a conveying pipe or tube is to be selected to carry the required quantity of any medium. This chapter is a presentation of the essential fundamentals involved in the estimation of friction and pressure losses accompanying flow in ducts and pipes.

6.2. Total, Static, and Velocity Pressures. The flow of fluids (liquids and gases) within a duct or piping system is produced by a pressure difference existing between different locations. The pressures involved are *total*, *static*, and *velocity*. Total and static pressures may be either positive or negative in terms of gage values. Velocity pressure is always numerically positive.

Total pressure is the algebraic sum of a static pressure and its accompanying velocity pressure. With all of the pressures mentioned the units of measurement must be consistent.

Static pressure tends to burst a duct or pipe when greater than

FLUID FLOW AND PRESSURE LOSSES

atmospheric and tends to collapse the confining envelope when its force is less than that of the atmosphere. Static pressures are used to effect increases of velocity and to overcome friction and shock losses as fluid flow occurs.

Velocity pressure is equal to the drop in static pressure necessary to produce a given velocity of flow, or conversely it is equal to the increase of static pressure possible when the velocity is reduced to zero. Figure 6.1 indicates the use of a thin-walled glass tube, which has been bent at an angle of 90 deg, as a device for the measurement of the velocity pressure of a liquid flowing in an open channel. The

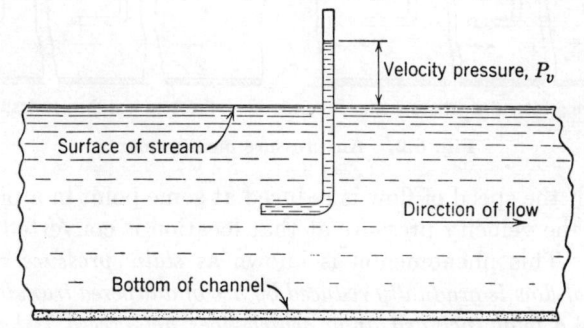

Fig. 6.1. Measurement of velocity pressure in an open stream.

moving fluid impinges against the open end of the short leg of the tube and its impact causes the level of the liquid within the tube to stand above the fluid surface of the moving stream. The elevation of the liquid in the vertical portion of the tube, above the fluid surface in the channel, is the velocity pressure P_v as static pressure is not involved. The device, as shown by Fig. 6.1, is an elementary Pitot tube.

Often because of practical considerations the fluid in pressure-measuring devices known as manometers must be something other than that flowing within either a pipe or a duct. Air cannot be used as a gage fluid; water may be satisfactory in some cases but not in others where a special light oil may be required. For pressures of greater magnitude mercury is useful; other requirements may necessitate the application of mechanical pressure gages. The manometer arrangements of Fig. 6.2 are suitable for the measurement of total, static, and velocity pressures. The choice of the manometer fluid is dependent upon the magnitude of the pressures involved.

The single static pressure connections of Fig. 6.2 are not so likely to give true indications of pressure as are obtainable when multiple openings are made about the periphery of a duct or pipe as shown in Fig. 6.3. The arrangement of Fig. 6.3 is called a *piezometer ring*.

Velocity, static, and total pressures are interrelated. When an increase of velocity of flow is produced in any part of a system, a part of the static pressure at that point is required to produce the change.

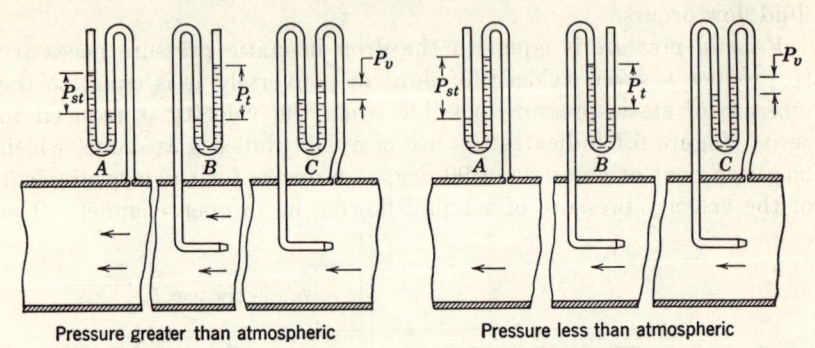

Fig. 6.2. Air-pressure measurements.

Likewise if the speed of flow is reduced at some point in a pipe or duct, a part of the velocity pressure at that location is converted into static pressure. This phenomenon is known as *static pressure regain*[1]. *If the speed of flow is gradually reduced by use of a tapered transition section, which has a total included angle which does not exceed 7 deg, the static pressure increase will be an amount that is only slightly less than the change of velocity pressure at the duct section under consideration.* The principle involved is employed in the recovery tube of a Venturi meter and it may be applied at the discharge end of any system handling a fluid at a high velocity.

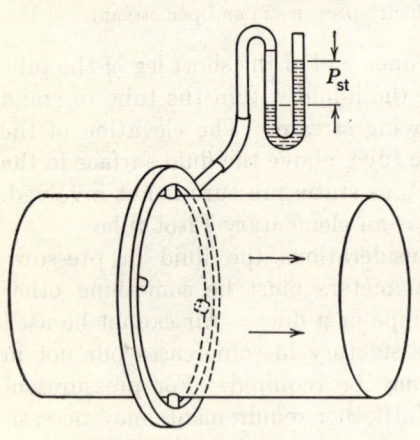

Fig. 6.3. Piezometer ring.

6.3. Pitot Tubes. A simple Pitot tube, Fig. 6.1, for the measurement of total pressure only, may be constructed of thin-walled small internal-diameter tubing. A short leg is cut squarely across at the open or upstream end and is placed parallel to the duct or pipe axis.

A Pitot tube for the measurement of both total and static pressures, more properly called a *Pitot-static tube*, consists of two tubes, one within the other as in Fig. 6.4. The center tube is open at the end of the short

FLUID FLOW AND PRESSURE LOSSES

leg and is used to measure the total pressure. The outer tube is closed at the end of its short leg which has properly located holes that are used for the measurement of static pressure. The locations of the small holes with respect to the end of the outer tube and the stem are important as well as their spacing around the tube. The ASHAE has

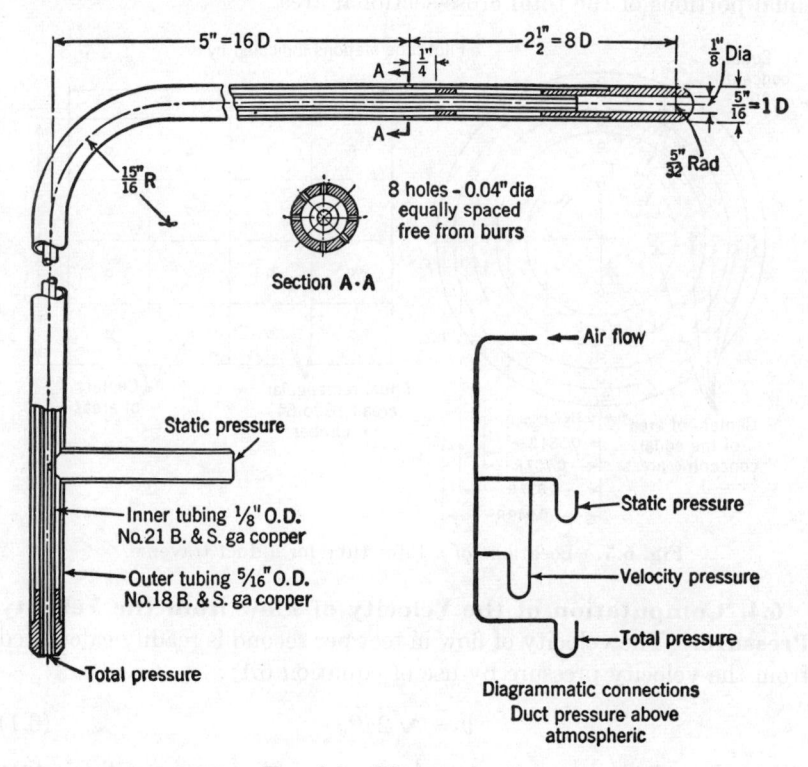

Fig. 6.4. Pitot-static tube.

adopted the definite specifications for a Pitot-static tube as shown in Fig. 6.4.

Pitot tubes are not adapted to the measurement of air velocities less than 6 fps unless used with very sensitive pressure-measuring gages. Pressure measurements are best made in a straight section of pipe at least 20 diameters in length with at least 10 diameters of straight pipe on each side of the tube location.

The velocity of flow in a duct is never uniform so that a number of observations are necessary in order to accurately determine the average. Schemes of locating a tube in circular, rectangular, or square pipes and ducts are shown in Fig. 6.5. The procedure of taking a number of readings at specific locations in a plane across the fluid stream, in order to obtain data from which the average velocity of flow may be determined, is called a *traverse*. In every case the traverse points are laid out in such a way that all of the readings represent equal portions of the total cross-sectional area.

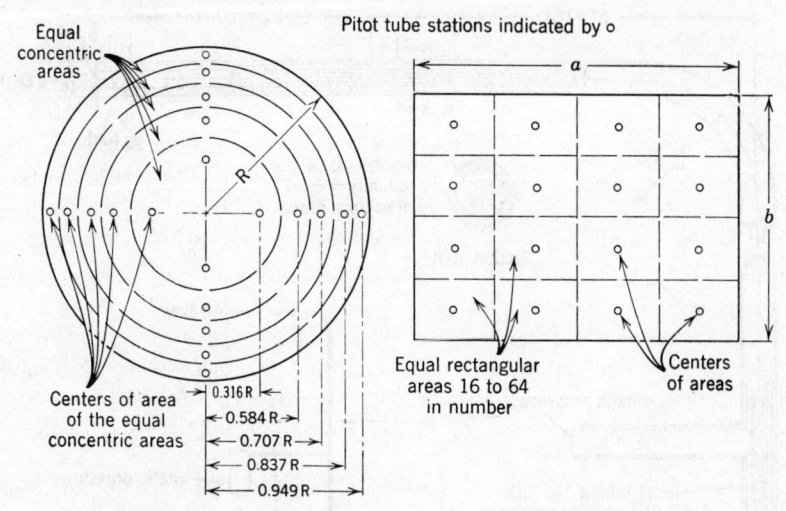

Fig. 6.5. Locations of a Pitot tube for a duct traverse.

6.4. Computation of the Velocity of Flow from the Velocity Pressure. The velocity of flow in feet per second is readily calculated from the velocity pressure by use of equation 6.1:

$$v = \sqrt{2gP_v} \tag{6.1}$$

When the velocity pressure must be measured in terms of a fluid other than that flowing, the observed gage reading must be converted into the equivalent feet head of the fluid in motion for use as P_v in equation 6.1. The following numerical case illustrates the method.

Example. The average velocity pressure of dry air at 100 F is measured as 1.2. in. of water as it flows in a duct under a static pressure of 2.0 in. of water when the barometric pressure is 29.5 in. of mercury at 70 F and the room air temperature is 70 F. Find the velocity pressure in terms of feet head of air and the velocity of air flow in feet per second.

Solution. From Table 1.1 the density of mercury is 0.490 lb per cu in. at 70 F and from Table 1.7 the density of water at 70 F may be computed as $1/0.01606 = 62.27$ lb per cu ft. The absolute pressure of the air flowing in the duct is $29.5 + [(2 \times 62.27)/(0.490 \times 1728)] = 29.5 + 0.147 = 29.647$ in. of mercury. (The factor 1728 represents the equivalent of 1 cu ft in cubic

FLUID FLOW AND PRESSURE LOSSES

inches.) By use of equation 1.6 the density of the air being handled at a temperature of 100 F is $(1.3258 \times 29.647)/(459.6 + 100) = 0.0702$ lb per cu ft. The velocity pressure in terms of feet of air is $(1.2 \times 62.27)/(12 \times 0.0702) = 88.7$ ft, and the average velocity of flow $v = \sqrt{2 \times 32.174 \times 88.7} = 75.6$ fps.

The average velocity of flow in a duct respresents the mean of all of the velocities at the section where the measurement is made. The square root of the velocity pressure is involved in the calculation of the velocity at each location; therefore, in arriving at the average velocity of flow the average of the square roots of the velocity pressures must be used. The procedure in obtaining the average velocity pressure from several readings taken in a traverse is as follows: (1) tabulate the square roots of all of the velocity pressures read, (2) find the average of the tabulated square roots, and (3) square the average square root. The result of the three operations is the velocity pressure corresponding to the average velocity and is sometimes referred to as the *root mean square*. This pressure expressed in feet of the fluid flowing when inserted in equation 6.1 permits the calculation of the average velocity of flow in feet per second. Equation 6.1 rearranged may be used to compute the velocity pressure when the velocity is known.

6.5. Causes of Pressure Losses in Pipes and Ducts. Pressure loss or *loss of head* may be caused by friction between the moving particles of a fluid and the interior surfaces of a pipe or between the adjacent particles in a stream of the moving medium. When a pressure loss occurs in a straight pipe or duct it is usually classed simply as *friction loss*. A greater pressure loss, per unit of pipe length, may be caused by the excessive friction between particles in the eddy currents following an abrupt enlargement in the pipe or following a special fitting (elbow, valve, tee, or other unit) in which case the term *shock loss*[2] is usually used.

6.6. Friction Losses in Circular Pipes or Ducts. The frictional resistance to flow is independent of the absolute pressure exerted by the fluid on the confining pipe walls; it varies almost as the square of the velocity of flow, directly as the length of the pipe or duct, and inversely as the hydraulic radius which is the cross-sectional area of the conduit divided by the wetted perimeter. The following equation, which is based on work by Wiesbach, D'Arcy and Hazen, may be used to compute friction-pressure losses.[3]

$$P_f = f \frac{L}{R_h} \frac{v^2}{2g} \qquad (6.2)$$

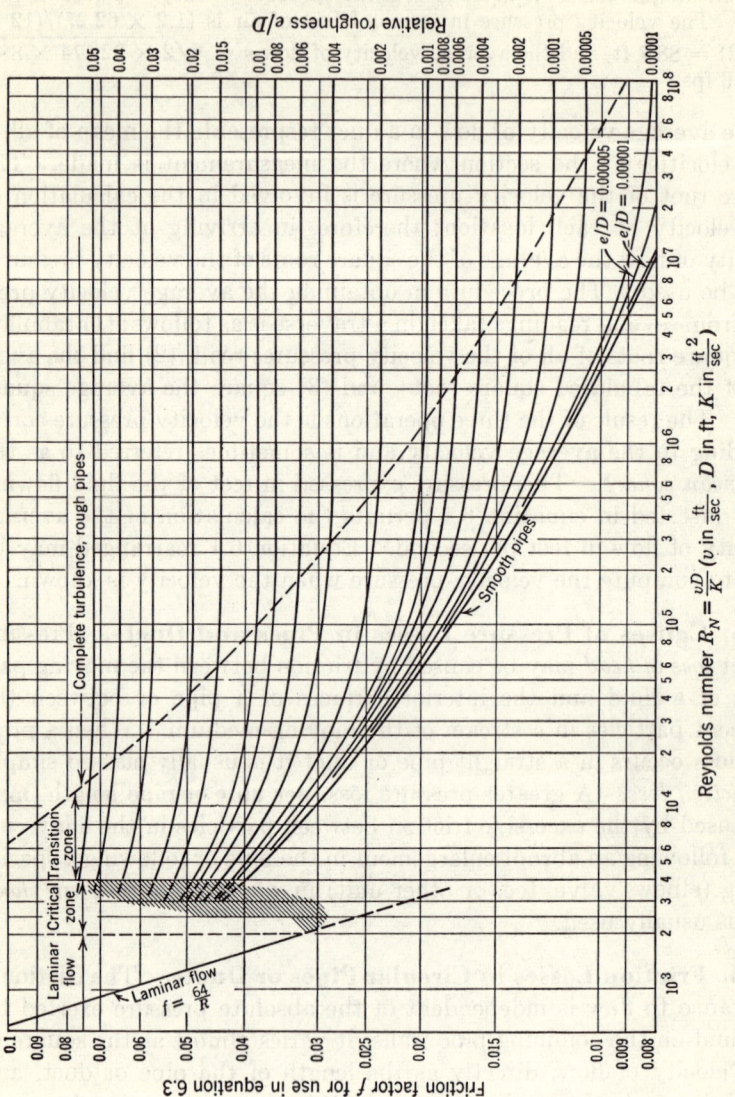

Fig. 6.6. Moody chart showing friction factors for fluid flow in circular pipes. (Reprinted by permission from *ASME Trans.* (Nov. 1944.))

FLUID FLOW AND PRESSURE LOSSES

Equation 6.2 is applicable to the flow of a fluid in any enclosed conduit whether it be circular, square, rectangular, or annular. For commonly used circular ducts and pipes, equation 6.2 is generally expressed in the form

$$P_f = f \frac{L}{D} \frac{v^2}{2g} \qquad (6.3)$$

Inasmuch as the hydraulic radius of a circular pipe is $D/4$, the numerical value of the friction factor intended for use in equation 6.2 must be multiplied by 4 before it can be used in equation 6.3. *Most charts, such as Fig. 6.6, show values of the friction factor for direct usage in equation 6.3. However, care must be taken to ascertain the intended application of such data if there is any doubt about their suitability.*

The foregoing equations are applicable to all incompressible fluids (such as water, oil, and brine) and may be applied with negligible error (less than 1 per cent) to compressible fluids if the friction-pressure loss does not exceed 10 per cent of the initial absolute static pressure.

Numerous studies, covering many fluids, have established that *the friction factor f of either equation 6.2 or 6.3 is a function of a dimensionless ratio known as Reynolds number,*[4] R_N.

$$R_N = \frac{Dv}{\mu/\rho} \qquad (6.4)$$

$$R_N = \frac{Dv}{K} \qquad (6.5)$$

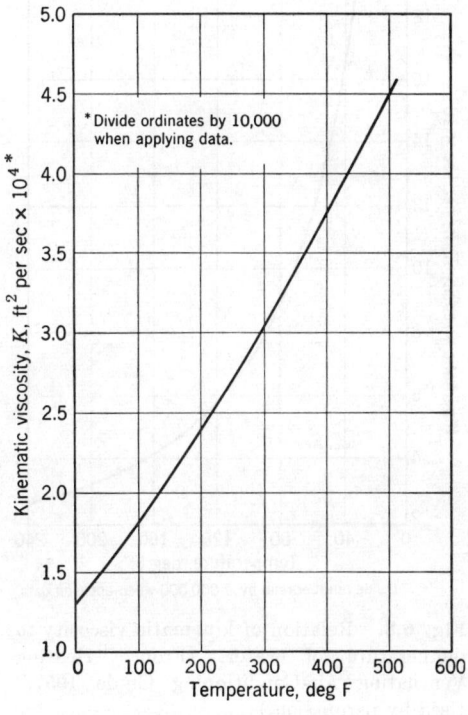

Fig. 6.7. Relation of kinematic viscosity to temperature of air. (From "Heating Ventilating Air-Conditioning Guide 1957." Used by permission.)

*Divide ordinates by 10,000 when applying data.

(Kinematic viscosity, K, ft^2 per sec $\times 10^4$ vs. Temperature, deg F)

where μ is the absolute viscosity of the fluid in lb ÷ (ft × sec) and ρ is the density. [Viscosity in centipoises × 0.000672 = viscosity in lb ÷ (ft × sec).] K is the kinematic viscosity which is μ/ρ, expressed in ft² per sec.

Kinematic viscosities of air at temperatures ranging from 0 F to 500 F are given in Fig. 6.7; the same properties for water at temperatures ranging from 20 F to 240 F are shown in Fig. 6.8. Figure 6.9 indicates the absolute viscosities of steam through wide ranges of temperature and pressure.

Experiments have shown that when the conditions of fluid flow are such that R_N is less than 2000, the fluid particles move in straight lines which are parallel to the pipe or the duct axis. This type of fluid movement is variously referred to as either *viscous*, *streamline*, or *laminar* flow. Under these conditions the maximum velocity of flow occurs at the center of the pipe and is twice the average velocity. The average velocity occurs at a distance from the pipe center that is 0.707 times its radius. Under conditions of laminar flow the friction factor is independent of the degree of roughness of the pipe wall as may be noted from Fig. 6.6, which gives friction factors versus Reynolds numbers *for use in equation 6.3*.

Fig. 6.8. Relation of kinematic viscosity to temperature of water. (From "Heating Ventilating Air-Conditioning Guide 1957." Used by permission.)

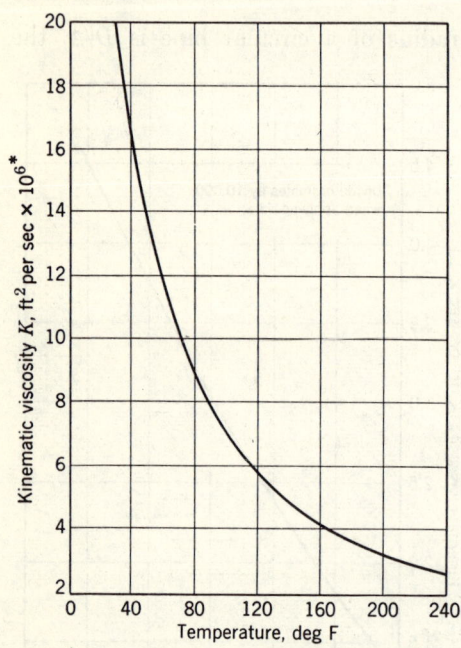

*Divide number read by 1,000,000 when applying data.

When the conditions of fluid flow are such that R_N is greater than 4000, the particles move in irregular and unpredictable paths which are parallel to the duct longitudinal axis only a very small percentage of the time. Such conditions of flow are classed as turbulent and the friction factor is greatly affected by roughness of the pipe wall. Investigations of the effects of roughness of the inner surface of a pipe when turbulent flow conditions exist indicate that the friction factor is

FLUID FLOW AND PRESSURE LOSSES

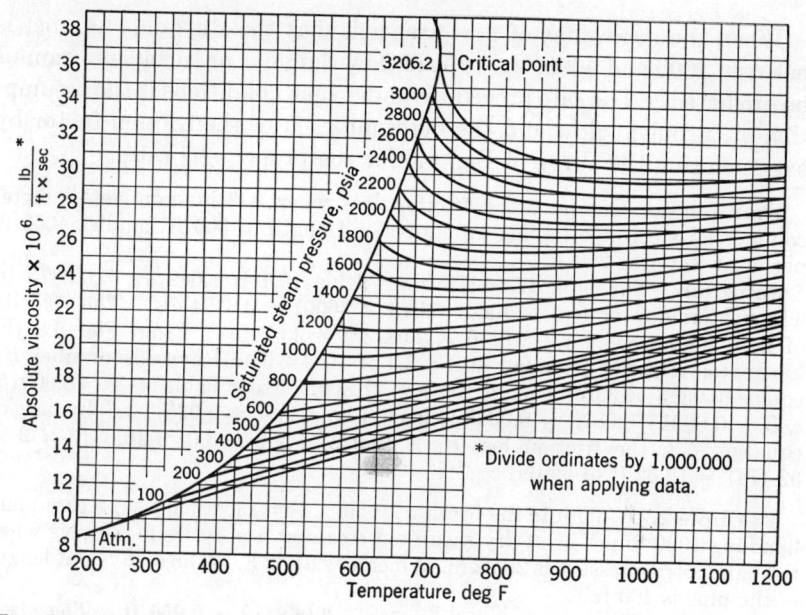

Fig. 6.9. Lieb chart for absolute viscosity of steam. [Reprinted by permission from *Combustion* (Dec. 1940).]

directly proportional to the absolute roughness of the surface (usually represented by the symbol e which is measured in feet) and inversely proportional to the pipe diameter, D, expressed in feet. Figure 6.6 includes a curve for each of several values of e/D. Values of e for several types of commercial pipe or duct work are given in Table 6.1.

TABLE 6.1. Absolute Roughness Factors e for Different Types of Pipes*

Type of Pipe	Absolute Roughness Factor e, ft
Smooth-drawn tubing (glass, brass, and lead)	0.000005
Commercial steel- or wrought-iron pipe	0.00015
Galvanized iron or steel (air ducts)	0.0005
Cast iron	0.00085
Riveted steel (light weight, small rivets)	0.003
Riveted steel (heavy weight, large rivets)	0.03
Smooth concrete (carefully troweled)	0.001
Average concrete	0.004
Very rough concrete	0.01
Brick-line conduits	0.03
Wood-stave conduits	0.003

* Extracted from reference 5.

When the conditions of flow are such that the Reynolds number is between 2000 and 4000, the type of flow (laminar or turbulent) cannot be predicted. The safe procedure under such conditions is the assumption of turbulent flow and the determination of the friction factor by extrapolating the curve of Fig. 6.6 for the proper e/D ratio.

Example 1. Compute the friction loss when a 1-in. nominal-diameter commercial steel pipe having an equivalent length of 300 ft handles 4000 lb per hr of water at 200 F.

Solution. The pipe diameter from Table 10.1 is $1.049/12 = 0.0873$ ft. The volume flowing equals $4000/(60.13 \times 3600) = 0.0184$ cfs. The velocity of flow is $(0.0184 \times 4)/(3.1416 \times 0.0873^2) = 3.07$ fps. From Fig. 6.8 the kinematic viscosity is found to be 3.4×10^{-6}. The Reynolds number R_N equals $(0.0873 \times 3.07)/(3.4 \times 10^{-6}) = 77,800$. From Table 6.1, $e = 0.00015$; $e/D = 0.00015/0.0873 = 0.00172$; from Fig. 6.6, $f = 0.025$; and by use of equation 6.3, the friction loss P_f is $(0.025 \times 300 \times 3.07^2)/(0.0873 \times 2 \times 32.174) = 12.52$ ft of water.

Example 2. Compute the friction loss in a 3-in. (nominal size) steam main handling 1000 lb per hr of dry saturated steam at a pressure of 5.3 psig when the barometric pressure is 29.92 in. of mercury at 32 F. The equivalent length of the pipe is 400 ft.

Solution. The actual pipe diameter is $3.068/12 = 0.256$ ft. The absolute steam pressure equals $(29.92 \times 0.4912) + 5.3 = 20.0$ psi. From Table 1.6 the specific volume of the saturated steam is found to be 20.089 cu ft per lb when the change of volume of the steam due to the pressure drop in the pipe is temporarily neglected. The velocity is $(1000 \times 20.089 \times 144)/(3600 \times 7.393) = 108.7$ fps. The kinematic viscosity of the steam is not available but its density equals $1/20.089 = 0.0497$ lb per cu ft, and from Fig. 6.9 the absolute viscosity is 9.2×10^{-6} lb/(ft $\times$ sec). Computation of the Reynolds number R_N, by use of equation 6.4, gives $R_N = (0.256 \times 108.8 \times 0.0497 \times 1{,}000{,}000)/9.2 = 150{,}020$. From Table 6.1 $e = 0.00015$ and $e/D = 0.00015/0.256 = 0.000586$. The friction factor $f = 0.0197$, from Fig. 6.6, and the friction loss by equation 6.3 is $P_f = (0.0197 \times 400 \times 108.7^2)/(0.256 \times 2 \times 32.174) = 5660$ ft of steam.

The solution correction necessary, because of changes of the specific volume of the steam with its pressure decrease, is made as follows: $(5660 \times 0.0497)/144 = 1.950$ psig (the equivalent of 5660 ft of steam). The average steam pressure in the pipe line is $20 - (1.950/2) = 19.025$ psia and the corresponding specific volume is 21.055 cu ft per lb which represents a density of $1/21.055$ 0.0475 lb per cu ft. The corrected average velocity is $(1000 \times 21.055 \times 144)/(3600 \times 7.393) = 114.0$ fps. The corrected Reynolds number is $(0.256 \times 114.0 \times 0.0475 \times 1{,}000{,}000)/9.2 = 150{,}500$. The friction factor $f = 0.0197$ is unchanged. The actual friction loss is $(0.0197 \times 400 \times 114.0^2)/(0.256 \times 2 \times 32.174) = 6200$ ft of steam.

6.7. Pressure Losses in Noncircular Ducts. Equation 6.2 is applicable to any conduit regardless of its cross-sectional shape. However, to use the friction factor data of Fig. 6.6 it is necessary to revise equation 6.2 by replacing the term R_h by the equivalent diameter of

FLUID FLOW AND PRESSURE LOSSES

the duct D_e. The equivalent diameter D_e is four times the hydraulic radius of the noncircular conduit under consideration. The equivalent diameter, in feet, is expressed as

$$D_e = \frac{(4 \times \text{duct cross-sectional area, sq ft})}{(\text{Wetted perimeter of duct, ft})} \qquad (6.6)$$

With a change of nomenclature equation 6.2 becomes

$$P_f = f \frac{L}{D_e} \frac{v^2}{2g} \qquad (6.7)$$

in which the value of f is the same as that of the f which is used in equation 6.3.

The equivalent diameter D_e may be used in the expressions for the Reynolds number as indicated by equations 6.8 and 6.9.

$$R_N = \frac{D_e v \rho}{\mu} \qquad (6.8)$$

$$R_N = \frac{D_e v}{K} \qquad (6.9)$$

Here all symbols are identical with those of equations 6.4 and 6.5 except that D_e is substituted for D. The equivalent diameter D_e is also used to determine the relative roughness e/D needed when a friction factor is to be found by use of Fig. 6.6.

Example. Compute the friction losses, in inches of water at 60 F, that occur when 2000 cfm of dry air at 80 F are passed through an 8-in. by 10-in. galvanized steel duct 200 ft in length when the barometric pressure is 29.92 in. of mercury measured at 32 F.

Solution. The duct dimensions are 0.666 ft and 0.833 ft. The equivalent diameter of the duct is $D_e = (4 \times 0.666 \times 0.833)/[(2 \times 0.666) + (2 \times 0.833)] = 0.741$ ft. From Table 2.1 the air density is found to be $1/13.6 = 0.0735$ lb per cu ft. The air velocity is equal to $(2000 \times 144)/(60 \times 8 \times 10) = 60$ fps. The kinematic viscosity at 80 F (Fig. 6.7) is $1.7/10,000$, and by equation 6.9, $R_N = (0.741 \times 60 \times 10,000)/1.7 = 262,000$. The absolute roughness e (Table 6.1) is 0.0005 ft and e/D_e equals $0.0005/0.741 = 0.000675$. Figure 6.6 indicates a friction factor f equal to 0.0193. By use of equation 6.7 the friction loss is $(0.0193 \times 200 \times 60^2)/(0.741 \times 2 \times 32.174) = 291$ ft of air or, in terms of inches of water at 60 F, $P_f = (291 \times 12 \times 0.0735)/62.37 = 4.1$.

6.8. Friction Charts. Calculation of the friction losses in a definite type of pipe carrying a specified fluid may be made for a range of assumed conditions covering many pipe sizes and fluid volumes and weights. These data may be plotted as in Fig. 12.5, which is a chart giving friction losses in milinches of water versus weight of water flowing in pounds per hour. A similar chart covering flow of water in copper tubing is shown in Fig. 12.6. Friction charts applying to other fluids are included in chapters which involve their use.

Friction charts are always limited in their application and are usually based on an assumed constant fluid density and viscosity.

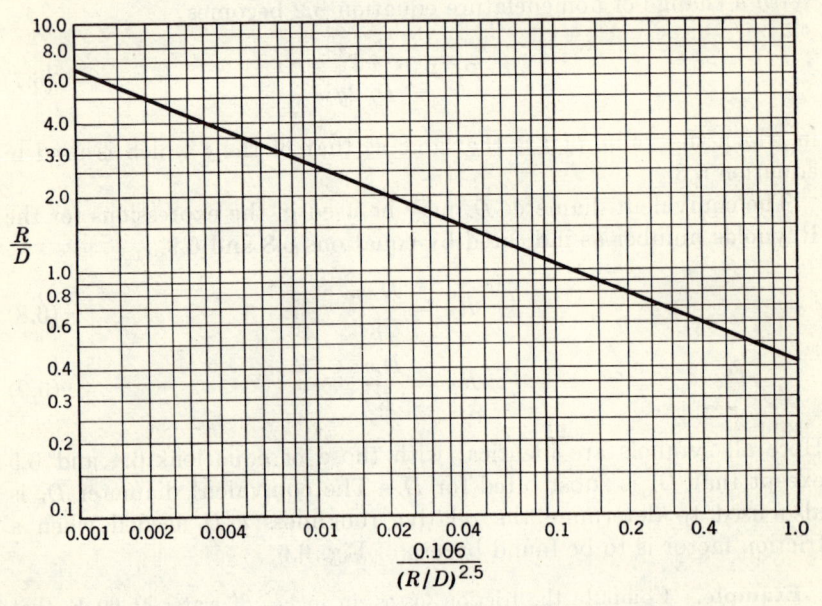

Fig. 6.10. Computed values of the first term in equation 6.11.

Whenever there is doubt, the friction losses as obtained from a chart should be checked against those computed by the method of Art. 6.6 when circular pipes are involved, and by the procedure of Art. 6.7 if the conduit is of another cross-sectional shape. Friction-pressure loss charts are usually applicable to circular pipes or ducts but their use can be extended to conduits of either square or rectangular shape by first determining the circular equivalent as indicated by Art. 14.19.

6.9. Friction Losses in Pipe and Duct Fittings. The data relative to predicting the pressure losses in fittings are much more uncertain[5, 6] than those pertinent to straight pipe because the nature of the passages through which the fluid flows is not exactly the same for all fittings. This is particularly true in the case of valves.

Two different methods of computing the pressure loss caused by a

FLUID FLOW AND PRESSURE LOSSES

fitting are in use. One method involves the velocity pressure, $v^2/2g$, existing in the fitting.

$$P_{sh} = \frac{K_f v^2}{2g} \qquad (6.10)$$

where v is the velocity in a pipe of the same nominal diameter as that of the fitting.

Constants K_f have been determined for many specific fittings used in definite applications. When such data are available for the fittings

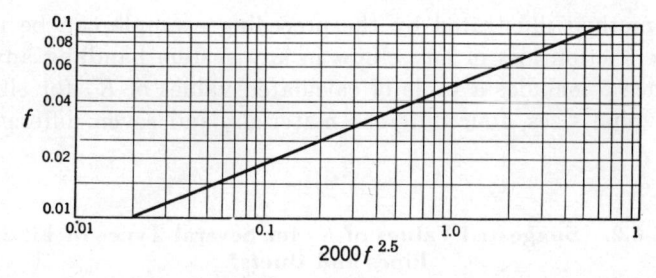

Fig. 6.11. Computed values of the second term in equation 6.11.

involved they should be used. Elbows and bends are the most commonly used fittings and Pigott[5] suggests that the total loss produced by them is likely to be made up of three elements. These are: (1) the ordinary pipe friction; (2) the minimum true bend loss for smooth conduit which depends only on R/D (ratio of the center-line radius to the pipe diameter); and (3) a turbulence loss due to the relative roughness of the pipe. This loss is affected by the friction factor which in turn is dependent upon the Reynolds number and the roughness of the surface over which fluid flow takes place. Pigott suggests that item 1 can be included by the addition of the length of the bend in the calculation of the length of straight pipe in the system. When this is done K_f, in equation 6.10, includes only items 2 and 3. Equation 6.11 has been suggested[5] after an extensive study of the results of several investigations.

$$K_f = \frac{0.106}{(R/D)^{2.5}} + 2000 f^{2.5} \qquad (6.11)$$

where K_f = a constant for use in equation 6.10. Computed data for the two terms of equation 6.11 are given in Figs. 6.10 and 6.11.

Example. Calculate the bend loss in an elbow which is part of an 8-in. round galvanized sheet-steel duct, which handles 1500 cfm of dry air at 140 F when the barometric pressure is 29.92 in. of mercury measured at 32 F. The center-line radius is 12 in.

Solution. The duct diameter in feet is 0.666, the ratio R/D equals 1.0/0.666 and is 1.50, the cross-sectional area amounts to 0.349 sq ft, and the velocity of air flow within the duct is 1500/(60 × 0.349) or 71.7 fps. Equation 6.10 will be used. $R_N = DV/K = (0.666 \times 71.7 \times 10{,}000)/2.04 = 234{,}000$. Table 6.1 indicates that e is 0.0005; $e/D = 0.0005/0.666 = 0.00075$; and, from Fig. 6.6, $f = 0.0265$. Using equation 6.11, $K_f = (0.106/1.50^{2.5}) + (2000 \times 0.0265^{2.5}) = 0.038 + 0.230 = 0.268$. The numerical data 0.038 and 0.230 were obtained from Figs. 6.10 and 6.11 respectively. By use of equation 6.10, $P_{sh} = (0.268 \times 71.7^2)/(2 \times 32.174) = 21.35$ ft of air.

The method illustrated by the preceding example can be used to find the friction loss in any elbow in any system handling any fluid. Reference 5 includes a table of calculated values of K_f for elbows of several pipe sizes, four different materials, and seven different R/D ratios.

TABLE 6.2. Suggested Values of K_f for Several Types of Fittings for Pipes and Ducts*

Type of Fitting	K_f	Authority
Globe valve	10	Crane Co.[8]
Angle valve	5	Crane Co.[8]
Swing check valve, fully open	2.5	Crane Co.[8]
Close return bend	2.2	Crane Co.[8]
Standard elbow	1.0†	Giesecke et al.[9]
Medium-sweep elbow	0.75†	Crane Co.[8]
Long-sweep elbow	0.60†	Crane Co.[8]
45-deg elbow	0.42†	Crane Co.[8]
Gate valve, fully open	0.19	Crane Co.[8]

* For use in equation 6.10. See Art. 14.20 for discussion of expanding or contracting sections.

† In the cases of all elbows, for more exact results use equation 6.11.

The quantity P_{sh} from equation 6.10, for long-radius bends of smooth pipe under conditions of flow where the Reynolds number is greater than 100,000, is very small. Such a condition indicates that the friction loss is little more than in a straight pipe of the same length.

Fittings which do not change the direction of flow but which either decrease or increase the cross-sectional area of the pipe or duct also cause a shock loss and thereby increase the flow resistance. Globe valves, Art. 10.5, involve two changes of direction of fluid flow. K_f data for valves are given in a bulletin[7] published by the University of Illinois. Average numerical K_f factors, for several types of fittings, which have been collected from various sources, are given in Table 6.2.

FLUID FLOW AND PRESSURE LOSSES

The resistance of a tee is dependent upon the condition of flow within it. Giesecke et al.[9] suggested first finding the resistance of the tee in terms of the number of standard-pipe short-radius elbows that would have the same resistance. The elbow equivalent, for flow from a main

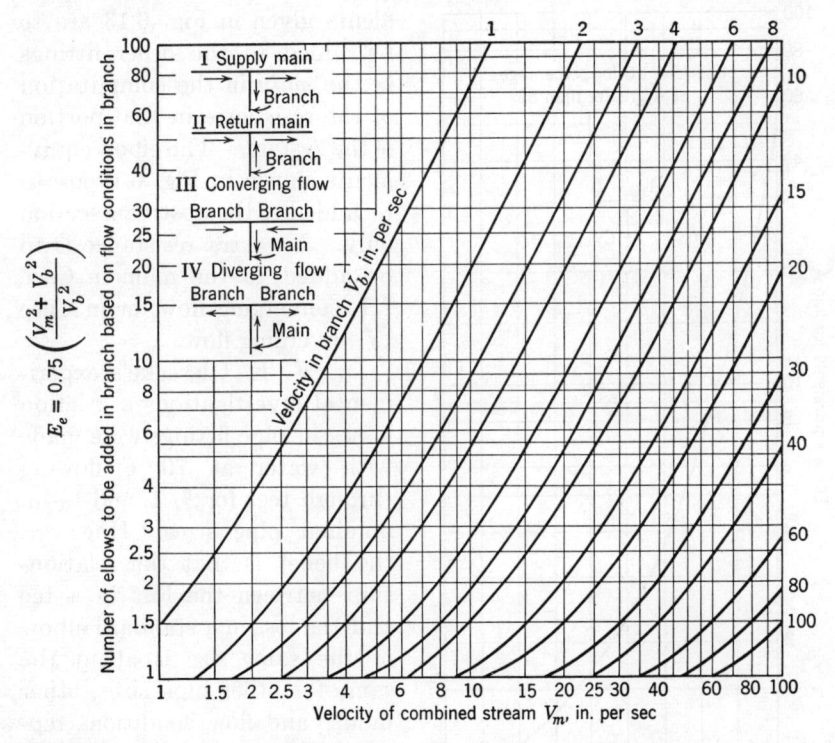

Fig. 6.12. Elbow equivalents of tees. (Applicable to the flow in a branch under any of the conditions illustrated.)

into a branch or from a branch into a main, may be found by use of equation 6.12:

$$E_e = 0.75 \left(\frac{v_m^2 + v_b^2}{v_b^2} \right) \tag{6.12}$$

where v_m and v_b are the velocities in the main and in the branch, inches per second.

Figure 6.12 gives data obtained by the use of equation 6.12 for a wide range of fluid velocities in both a main and a branch from it. *The diagrams in the upper left-hand corner of Fig. 6.12 indicate the conditions of flow to which the data apply. The resistance of a tee in a supply main, Case I, Fig. 6.12, or in any case where the flow is out from the side outlet, is negligible as far as flow in the main is concerned.* Figure 6.13, plotted from Giesecke's data,[9] gives the elbow equivalents for a tee in a return main (Case II, Fig. 6.12) when the fluid flows *into it* from a branch. The elbow equivalents given in Fig. 6.13 are to be added to the other fittings in the *main* in the computation of the resistance in that portion of the system. The elbow equivalents shown in Fig. 6.13 are to be added to the upstream section only. No extra resistance is to be added to the main in Case III, converging flow, or in Case IV, diverging flow.

Dr. F. E. Giesecke's experimental investigations of friction losses in pipe fittings were made with water at 70 F flowing through tees for $\frac{3}{4}$, 1, and $1\frac{1}{2}$-in. nominal pipe sizes. However, the belief is that the relationship between the loss in a tee and the loss in a standard elbow of the same size is about the same for other pipe sizes, other fluids, and flow conditions represented by other Reynolds numbers. Equation 6.12 and Figs. 6.12 and 6.13 can therefore be used for all cases where fluids flow through cast-iron tees in iron or steel pipes, when specific data applicable to the condition are not available. After the elbow equivalent of a tee has been found, the K_f value for a standard short-radius elbow of the same nominal pipe size, in the section of interest, may be found by use of equation 6.10. *The R/D ratio for standard cast-iron short-radius elbows of all sizes is 0.58.*

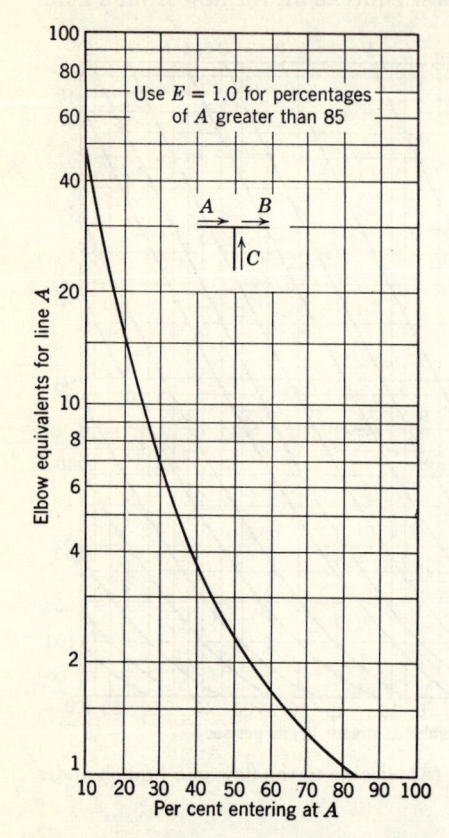

Fig. 6.13. Elbow equivalents for a return tee. (Applicable to the upstream section A of the return main.)

Example. Find the loss of head, in feet of water, which affects flow in a branch connection when 1000 lb per hr of water at 200 F flows from a $1\frac{1}{4}$-in. steel supply main, carrying 4000 lb per hr, into a $\frac{3}{4}$-in. branch pipe as in Case I of Fig. 6.12.

FLUID FLOW AND PRESSURE LOSSES

Solution. The velocity of the combined water stream, carrying 4000 lb. per hr in a $1\frac{1}{4}$-inch pipe, from Fig. 12.5 is 21 in. per sec. The water velocity in the $\frac{3}{4}$-in. branch, carrying 1000 lb per hr of water, is 14.5 in. per sec. The equivalent of the tee is 2.5 standard elbows as indicated in Fig. 6.12. For the $\frac{3}{4}$-in. elbow involved K_f is determined by use of equation 6.11. R/D for a standard elbow is 0.58. After the Reynolds number and the roughness ratio have been ascertained, the friction factor f may be obtained from Fig. 6.6. Equation 6.5 indicates that $R_N = DV/K$. From Table 10.1 D, for $\frac{3}{4}$-in. pipe, is $0.824/12 = 0.0686$ ft. $v = 14.5/12 = 1.21$ fps. K, from Fig. 6.8, is $3.4/1{,}000{,}000$. $R_N = 0.0686 \times 1.21 \times 1{,}000{,}000/3.4 = 24{,}400$. Table 6.1 indicates that $e = 0.00015$. $e/D = 0.00015/0.0686 = 0.00218$, and f from Fig. 6.6 is 0.029. $K_f = 0.42 + 0.28 = 0.70$ when Figs. 6.10 and 6.11 are used respectively for the solutions of the two terms of equation 6.11. The head loss in one standard short-radius elbow under the conditions of flow may then be computed by use of equation 6.10 as $(0.70 \times 1.21^2)/(2 \times 32.174) = 0.0160$ ft of water; therefore the head loss in the tee is $2.5 \times 0.0160 = 0.040$ ft of water.

An alternate method of making an allowance for the friction losses occurring in each pipe fitting is to estimate its effect in terms of the equivalent straight pipe. The equivalent length of a fitting is that length of a straight pipe of the same diameter that offers an equal resistance to fluid flow. The actual lengths of straight pipe and the equivalent lengths of the fittings are totaled to give the total equivalent length of the system and the friction losses are then computed on the basis of straight pipe alone. A simple test setup may be used to determine the equivalent length of any specific fitting when operating with definite conditions of fluid flow.

When the applicable equivalent length of a fitting is known, the friction losses occurring within it may be determined from the equation $P_f = (fL_e v^2)/(D2g)$ in which L_e is substituted for L in equation 6.3. When P_f is thus computed it is equivalent to P_{sh}, as determined by use of equation 6.10. When the foregoing equation is combined with equation 6.10 the following relationship is obtained: $(fL_e v^2)/(D2g) = K_f v^2/2g$ or

$$L_e = K_f D/f \qquad (6.13)$$

Reference 5 gives computed equivalent lengths of short-radius and long-radius cast elbows in pipe lines of three different materials and of 90-deg turns in brass tubing and steel pipe. Equivalent lengths of fittings commonly used in steam-heating systems are given in Table 11.1. Similar data pertaining to turns in sheet-metal ducts are given in Table 14.2 and Fig. 14.18. The method of computing an equivalent length of a turn will be illustrated.

Example. Find the equivalent length in feet of a 48-in. diameter 90-deg bend in a sheet-metal duct through which air at 150 F is flowing with a velocity of 3000 fpm. The center-line radius is 1.5 times the duct diameter.

Solution. Equation 6.13 will be used but some preliminary calculations are necessary. K for air at 150 F (Fig. 6.7) is $2.1/10{,}000$, $D = 4.0$ ft, and $V = 50$ fps. The Reynolds number $R_N = DV/K = (4.0 \times 50 \times 10{,}000)/2.1 = 952{,}000$. The absolute roughness e (Table 6.1) is 0.0005 ft, $e/D = 0.0005/4 = 0.000125$, and f from Fig. 6.6 = 0.0138. Equation 6.11 indicates that $K_f = 0.106/1.5^{2.5} + 2000 f^{2.5} = 0.038 + 0.042 = 0.080$ when Figs. 6.10 and 6.11 are employed. $L_e = K_f D/f = 0.080 \times 4.0/0.0138 = 23.2$ ft.

The sheet-metal elbow of this example is assumed to be formed in a continuous curve. With the usual form of commercial construction the equivalent length of an elbow is greater, (Fig. 14.18). Fittings other than tees are sometimes expressed first in terms of elbow equivalents before their equivalent lengths are determined. This practice is extensively used in connection with the design of the piping of hot-water heating systems. Table 12.1 includes elbow equivalents for various pipe fittings.

REFERENCES

1. "Power Saving Through Static Pressure Regain," J. R. Fellows, *Heating, Piping, Air Conditioning,* **11,** No. 4, 219 (Apr. 1939).
2. "Pressure Losses Resulting from Changes in Cross-Sectional Area in Air Ducts," A. P. Kratz and J. R. Fellows, *Bull. No. 300,* Eng. Exp. Sta., University of Illinois (1938).
3. "Practical Considerations in the Interpretation of Commonly Used Formulae Governing the Flow of Fluids in Piping, Bends, Etc.," P. L. Martin, *J. Inst. Heating Ventilating Engrs.* (*London*), **17,** No. 168, 199 (July 1949).
4. "An Experimental Investigation of the Circumstances which Determine Whether the Motion of Water Shall Be Direct or Sinuous and the Law of Resistance in Parallel Channels," Osborne Reynolds, *Phil. Trans. Royal Society London,* **174,** 935 (1883).
5. "Pressure Losses in Tubing, Pipe and Fittings," R. J. S. Pigott, *ASME Trans.,* **72,** 679 (1950).
6. "A Study of the Data on the Flow of Fluids in Pipes," Emery Kemler, *ASME Trans.,* **55,** Hy 55-2 (1933).
7. "Loss of Head in Flow of Fluids Through Various Types of 1-in. and 1½-in. Valves," W. M. Lansford, *Bull. No. 340,* Eng. Exp. Sta. University of Illinois (1943).
8. "Flow of Fluids Through Valves, Fittings, and Pipes," *Tech. Paper No. 409,* Crane Company, Chicago (1942).
9. "The Loss of Head in Cast-Iron Tees," F. E. Giesecke, W. H. Badgett, and J. R. D. Eddy, *Bull. No. 41,* Eng. Exp. Sta., A and M College of Texas (1932).

FLUID FLOW AND PRESSURE LOSSES

PROBLEMS

1. The total pressure of air flowing in a duct is 2.35 in. WG (water gage) when the velocity pressure is 0.55 in. Find the existent static pressure in inches of water.

2. Water at 80 F and a static pressure of 30 psig flows through a pipe section with a velocity of 25 fps. Find the total head in feet of water at the location of the stated pressure.

3. Oil weighing 53 lb per cu ft, at a temperature of 60 F, moves through a commercial steel pipe that has a cross-sectional area of 28.91 sq in. The static head of the oil is 200 ft of the fluid and its velocity head is 25 ft. Find: (a) the velocity of oil flow, feet per second; (b) the total pressure of the oil, pounds per square inch gage; and (c) the quantity of oil handled, cubic feet per minute.

4. Twelve thousand five hundred cubic feet per minute of dry air at 180 F flow through a circular duct which is 24 in. in diameter when the barometric pressure is 28.5 in. of mercury measured at 70 F. The static pressure of the dry air is 1.4 in. WG at 70 F. Find the total pressure of the air measured both in inches of water at 70 F and in pounds per square inch.

5. An 18-in. by 24-in. rectangular galvanized sheet-steel duct 325 equivalent feet in length has a friction-loss factor f equal to 0.0055, (equation 6.2), when the velocity of flow of the 170 F dry air within it is 900 fpm. (a) Find the loss of head in feet of the fluid flowing. (b) Find the total friction losses expressed in inches of water at 60 F. The barometric pressure is 29.0 in. of mercury measured at 70 F. The static pressure may be neglected.

6. A circular sheet-steel duct of 4-ft internal diameter, 200 equivalent feet in length, conveys 628 cfs of air measured at 40 F when the barometric pressure is 28.75 in. of mercury measured at a temperature of 50 F. (a) Find the total pressure of the air if its static pressure is 0.92 in. WG when the gage-fluid temperature is 50 F. (b) Find the head loss due to friction and turbulence effects in the duct if the factor f, (equation 6.3), is 0.022. Express the losses in feet of air. (Assume dry air.)

7. Calculate the Reynolds number for superheated steam at 600 psia and 600 F total temperature which flows with a velocity of 125 fps in an 8-in. nominal-size steel pipe which has an internal diameter of 7.948 in. The specific volume of the steam is 0.9463 cu ft per lb.

8. Water at a temperature of 190 F flows with a velocity of 5 fps in a 2-in. nominal-size steel pipe which has an internal diameter of 2.067 in. Calculate the existent Reynolds number.

9. Find the friction factor f applicable to the data of Problem 8.

10. What friction factor f is involved with the superheated steam of Problem 7?

11. Compute the friction loss when a $2\frac{1}{2}$-in. commercial steel pipe with an internal diameter of 2.469 in. and a length of 150 ft handles 12,000 lb per hr of water at 210 F. Express answer in pounds per square inch.

12. A 6-in. nominal-size steel pipe with an internal diameter of 6.065 in. has 4000 lb per hr of dry saturated steam flowing through it when the average line pressure is 22 psia. The equivalent length of the pipe is 175 ft. Compute the actual friction loss in pounds per square inch.

13. Calculate the friction losses, in inches of water at 70 F, that are pro-

duced in a 16-in. by 24-in. galvanized steel duct which has no bends and is 125 ft long. The duct handles 5000 cfm of dry air at 100 F and the barometric pressure is 29.92 in. of mercury measured at 32 F. The static pressure is negligible.

14. Determine by calculation the pressure losses in a 90-deg elbow in a 16-in. diameter galvanized iron duct. The center-line radius of the elbow is 18 in. Dry air at 150 F flows through the duct at the rate of 3500 cfm. The pressure is the same as in Problem 13. Express the loss in inches of water at 60 F.

15. Dry air at 120 F flows through a 90-deg riveted steel elbow (light weight) at the rate of 6000 cfm when the barometric pressure is 28.3 in. of mercury measured at 70 F (neglect static pressure). The duct diameter is 16 in. and the elbow center-line radius is 20 in. Compute the head losses in the elbow when measured in inches of water at 60 F.

16. Find the loss of head, in feet of water, which affects flow in a 1-in. branch connection when 2000 lb per hr of water at 160 F flows from it into a 2-in. return main which, downstream from the tee, carries 6000 lb per hr of water at the same temperature (Case II of Fig. 6.12).

17. Determine the loss of head, in feet of water, which affects flow in the branch when 1000 lb per hr of water at 200 F flows from a $1\frac{1}{4}$-in. supply main carrying 8000 lb per hr. The branch pipe is $\frac{3}{4}$ in. nominal size.

18. Find the equivalent length in feet of a 36-in. diameter, 90-deg bend in a galvanized sheet-steel duct through which dry air at 180 F is flowing with a velocity of 2400 fpm. The center-line radius is equal to the duct diameter.

19. Determine the equivalent length of the elbow of Problem 18 if the center-line radius is two times the duct diameter.

7

Heating with Warm-Air Furnaces

7.1. Applications and Classifications. Furnace systems are suitable for buildings of relatively small size. Included in this category are residences, schools, stores, and churches having dimensions which will permit the circulation of air to and from a fuel-burning heater. The circulation can be produced either by gravity flow or mechanically. In any event the temperature of the warmed air delivered to the spaces to be heated should not exceed 175 F with gravity flow and 150 F when the air is circulated by mechanical means.

7.2. Advantages of Warm-Air Furnace Heating. Warm-air furnaces are generally lower in first cost than steam and hot-water plants. They possess flexibility of operation because they begin to function quickly after a fire is started. The operating costs compare favorably with those of other systems. The temperature of the air entering rooms is adjustable to weather conditions, and recirculation of the air through the furnace is an aid in securing uniformity of air temperature. Humidification of air is more easily accomplished than in systems using either direct radiators, convectors, or radiant baseboards and this type of heating system can be arranged to continually introduce a controlled supply of ventilation air. Warm-air furnace systems take little occupied space above the basement, and there is no danger of damage to them due to freezing if left without a fire in winter.

7.3. Comparison of Gravity-Flow and Forced Circulation. Gravity systems are noiseless as far as the air circulation is concerned, and both the initial cost and the cost of maintenance are less with this type of warm-air system. However, gravity-flow systems have certain limitations which are not present when the air circulation is produced by means of a fan. Structures more than 40 ft square and above two stories in height are not suitable for gravity-flow heating. The furnace occupies considerable space in the basement and requires a central location because the leader pipes should be limited to lengths of 10 to 12 ft. Adequate spaces for stacks in the interior partitions may be difficult to locate or not obtainable. Leaders may not run through unheated spaces unless exceptionally well insulated. The most effective air filters cannot be used for cleaning because their resistance to circulation is too much to be overcome by the small gravity head. Gravity systems require larger ducts for conveying the air to and from the furnace; this interferes with the use of the basement for recreation or other purposes. Because of the lower velocity of the air circulation in gravity systems, heat transfer from the furnace to the moving air is not as good, and a greater portion of the heat absorbed from the furnace is lost between it and the warm-air registers.

Since relatively few gravity systems are currently being installed, no space will be devoted to them in this book. Information about this type of system may be obtained from Manual 5, one of a series of manuals published by the National Warm-Air Heating and Air-Conditioning Association (NWAH & ACA) and listed in the Bibliography at the end of this chapter.

7.4. Forced Warm-Air Systems. A fan furnishes the motive head required to produce the circulation of the necessary amount of air in forced warm-air systems. This type of design has greater flexibility than gravity-flow installations with respect to the furnace location, size and placement of air ducts, use of air filters, and adaptability to all-year air conditioning.

Many different piping arrangements are used in forced warm-air systems. Instructions for the design and installation of the different systems are included in NWAH & ACA manuals.

7.5. Conventional Forced-Air System. Figure 7.1 shows part of a typical piping arrangement for the conduction of heated air from the plenum chamber, at the top of the furnace, and its delivery to the

HEATING WITH WARM-AIR FURNACES

spaces to be heated. The illustration also shows some of the ducts required to collect the return air and conduct it to the blower cabinet which contains an air filter. In some mechanical systems the fan and the air filter are placed in the lower portion of the furnace casing as shown in Fig. 7.2; with another arrangement, Fig. 5.3, these parts are housed in a separate cabinet which is adjacent to the furnace casing.

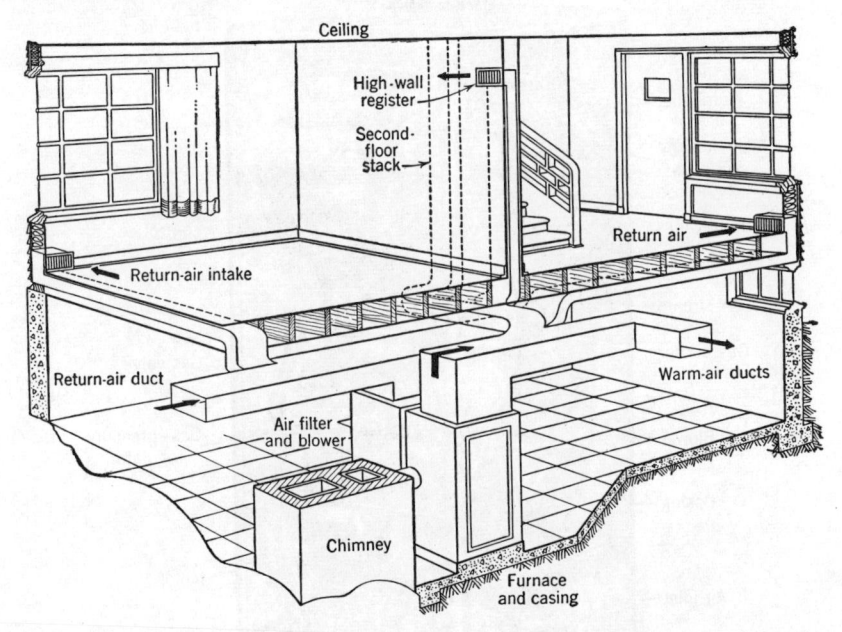

Fig. 7.1. Diagrammatic section of a typical forced-circulation warm-air heating system.

7.6. Perimeter-Loop System. The perimeter-loop system, a plan of which is shown in Fig. 7.3, was developed especially for heating residences and other small structures' which rest on floor slabs which lie upon the ground. A special counterflow furnace with the general form of Fig. 7.2 but with the fan at its top and the warm-air plenum at its bottom is used to warm the air. In the perimeter-loop system the return air is drawn from a level near the ceiling of the space to be heated and is forced downward over the heating surfaces into a plenum chamber embedded in the concrete floor slab. Figure 7.4 indicates the path of the air. (However, the plenum shown in Fig. 7.4 is suspended in a crawl space instead of being embedded in a concrete slab.) Warmed air flows from the plenum chamber through radial feeders to the perimeter duct from which it is discharged, into the space to be heated, either through floor or baseboard diffusers located beneath windows. The heated air passing through the radial feeders warms a considerable portion of the interior floor area and that passing

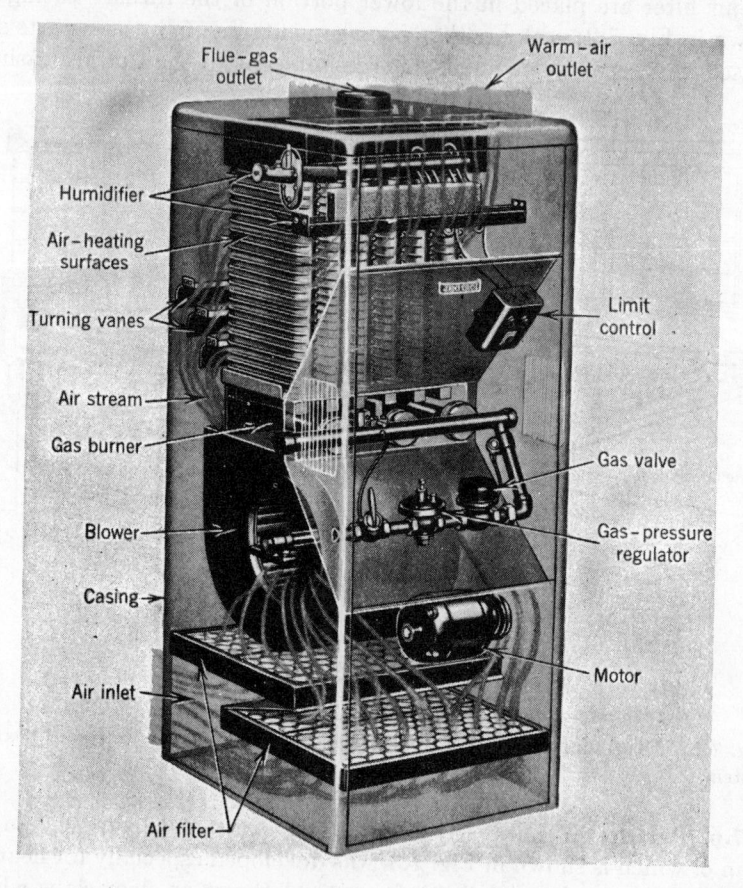

Fig. 7.2. Typical gas-fired packaged-unit forced-circulation warm-air furnace. (Courtesy Surface Combustion Corp.)

through the perimeter duct warms all parts of the outer edge. Floor-surface temperatures are not uniform when this system is used. However, studies made in a research residence indicated that the temperatures are not too low for comfort in any location when the system is properly designed and installed. The investigations revealed narrow areas directly over the feeder ducts where the floor-surface temperature exceeds that which has been considered the maximum allowable (85 to 90 F). These warm spots have not deterred home owners from

HEATING WITH WARM-AIR FURNACES

using perimeter systems. The temperature of the slab surface directly above radial feeders can be prevented from becoming excessive either by placing these ducts so that they will be farther below the surface

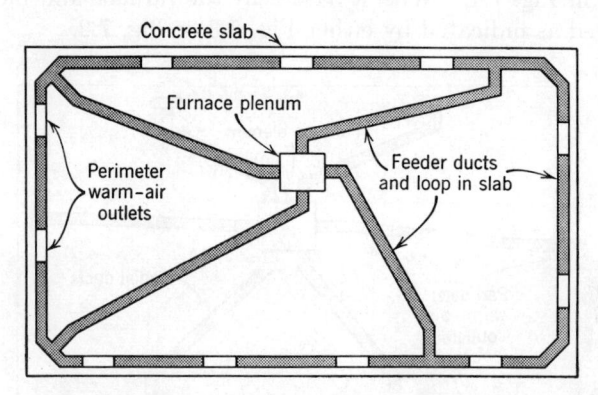

Fig. 7.3. Perimeter-loop system with feeder and loop ducts in concrete slab.

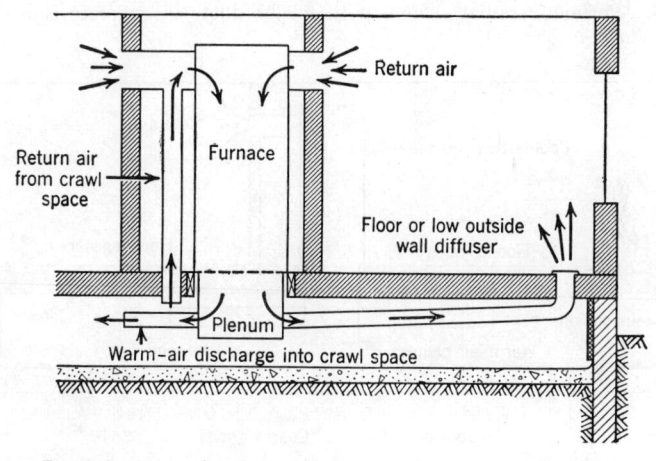

Fig. 7.4. Bottom warm-air plenum system with recirculation from the crawl space.

in the critical areas or by wrapping the critical sections with a proper amount of insulation before the concrete slab is poured.

7.7. Perimeter-Radial System. Figure 7.5 illustrates the principle of a perimeter-radial system in which the warm-air outlets are located at the ends of the radial feeders. There is no perimeter duct as shown in Fig. 7.3. A system of this type can be used for heating a residence constructed upon a concrete slab, over a crawl space, or over a conventional basement. When the distributing ducts are installed in a slab or in a crawl space, the furnace is usually located in a utility room and the circulation of air through it is from top to bottom as shown in Fig. 7.4. With a basement the furnace and blower may be arranged as indicated by either Fig. 5.3 or Fig. 7.2.

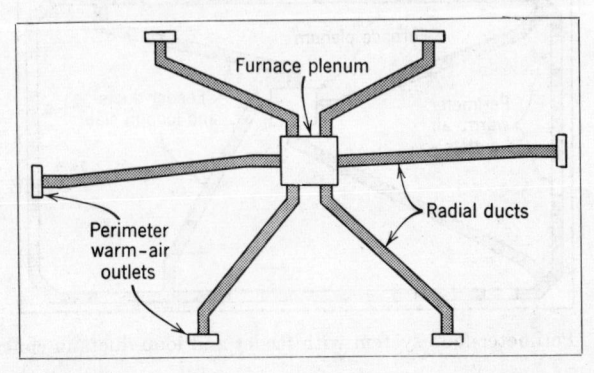

Fig. 7.5. Perimeter-radial system with feeder ducts in concrete slab or crawl space.

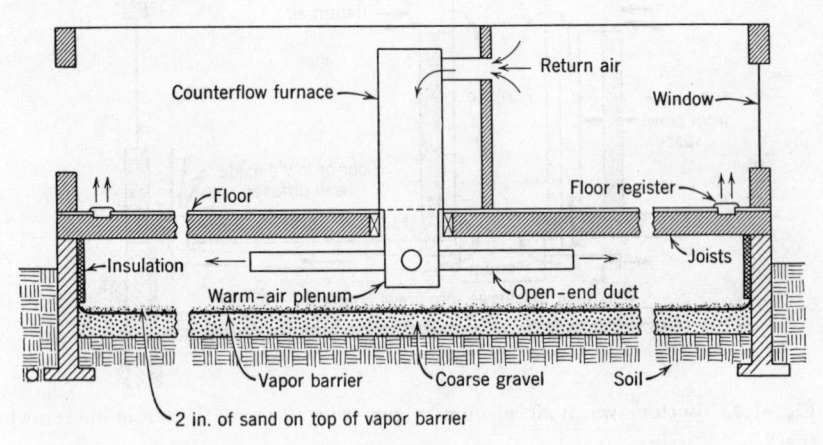

Fig. 7.6. Crawl-space plenum system.

When a perimeter-radial system is used in a house built upon a slab, certain areas of the floor, particularly those in the room corners, are likely to be too cool for human comfort during cold weather. However, when installed within a structure which has either a crawl space or a basement, the heat emitted from the plenum and from the radial ducts warms the ambient air and thus provides a uniformly comfortable floor. If the vagrant heat from the plenum and the radial ducts

HEATING WITH WARM-AIR FURNACES

is not enough, one or two stub ducts may be used to introduce air from the plenum as indicated in Fig. 7.4.

7.8. Crawl-Space Plenum System. When a crawl space has been suitably prepared it may serve as a secondary warm-air plenum from which heated air passes directly, through properly placed registers, into rooms above it. Figure 7.6 shows the arrangement required

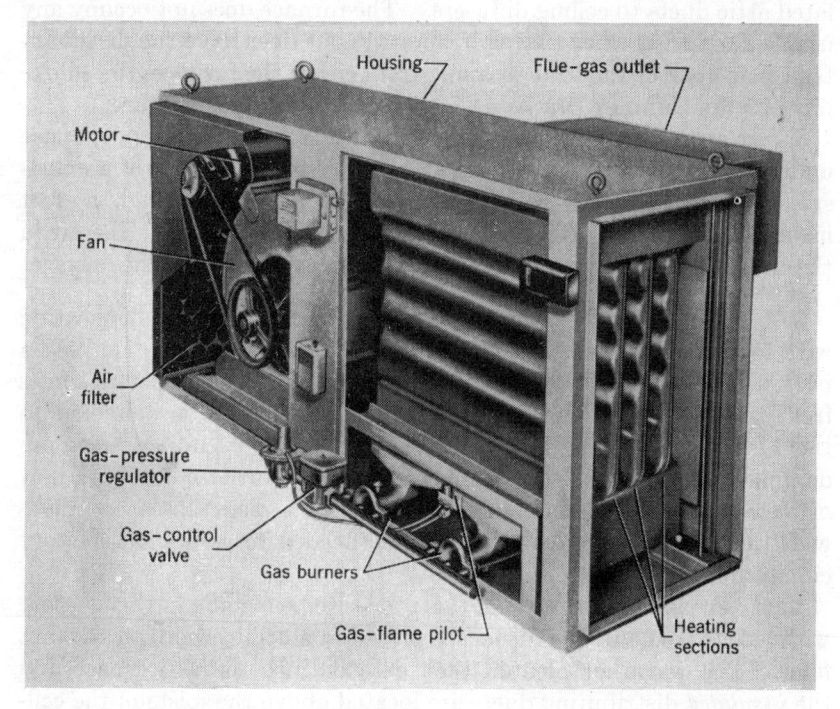

Fig. 7.7. Gas-fired horizontal furnace. (Courtesy Armstrong Furnace Co.)

and indicates the necessary treatment of the bottom and the sides of the crawl space. The system may be classed as a perimeter type since the warm air is delivered through floor registers located near the outer walls of the building served. The furnace, the return-air system, and the warm-air registers may be the same as those used in other perimeter systems. The only supply ducts needed are four straight runs from the four sides of the plenum. The purpose of the ducts is to insure a uniform temperature of the air in all parts of the crawl space.

An advantage of this type of system, over any warm-air installation that can be used in a house built upon a concrete slab, is a floor that is uniformly warm in all rooms of the structure. A disadvantage is a slightly higher fuel consumption, due to increased heat losses from the crawl space.

The National Board of Fire Underwriters in Pamphlet 90b requires 2 in. of sand on top of the vapor barrier as a precaution against fire.

7.9. Attic Stowaway Type of Heating System. The furnace of Fig. 7.7 may be located in the attic and deliver heat through insulated attic ducts to ceiling diffusers. The furnace does not occupy any usable floor area, when placed in an attic, but does have the disadvantage that none of the heat escaping from either the furnace, the smoke pipe, or the chimney can be utilized in warming the structure.

When an attic furnace is used to heat a home with a crawl space underneath it, the return air is usually collected by means of a crawl-space system of pipes which leads to the furnace through a duct installed in either a utility room, a closet, or a partition. The air in the crawl space may be warmed by one or more ducts from the furnace.

7.10. Blend-Air System. Another principle used in home heating with forced warm air is illustrated in Fig. 7.8. In this system only a part of the heated air delivered through the warm-air registers comes from the furnace. The heated air from the furnace is delivered to the blending units as shown in Fig. 7.8. The blending units are designed in such a way that the hot air from the furnace induces a flow of recirculated air in the heated spaces. This recirculated air mixes with the hot air to provide the volume needed to assure satisfactory circulation.

In the event that a system of this type is to be installed in a structure without a basement, the furnace is placed in a utility room on the first floor. The warm-air plenum then extends into the attic space and the *insulated* distributing ducts are located above the joists of the ceiling. In this arrangement the air for each room enters the top of a blending unit. However, in this application the blender is inverted with its recirculation near the ceiling and the blended air is delivered at the floor level.

7.11. Air Humidification with Warm-Air Furnaces. The necessity for air humidification arises from the small amount of moisture held by outside air when the weather is cold. Air humidification in connection with space heating is always desirable unless moisture released from other sources, such as drying clothes, bathing, and cooking, is sufficient for satisfactory environmental conditions. The addition of a humidifier to a warm-air furnace system is usually a simple matter, as shown in Fig. 7.2. The unit should be placed in the warm-

HEATING WITH WARM-AIR FURNACES

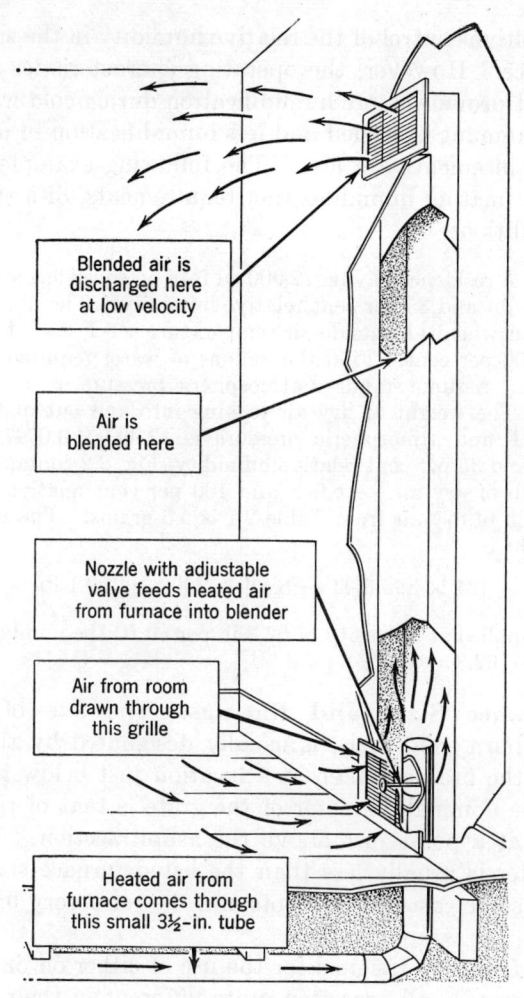

Fig. 7.8. Blend-air heating system. Sectional view of a high-register-outlet blender. The air flow may be reversed when used in a basementless house. (Courtesy The Coleman Co.)

Labels in figure:
- Blended air is discharged here at low velocity
- Air is blended here
- Nozzle with adjustable valve feeds heated air from furnace into blender
- Air from room drawn through this grille
- Heated air from furnace comes through this small 3½-in. tube

air furnace plenum where the air temperature is highest so that a sufficient amount of evaporation of water from the pan may be obtained. The water to be evaporated is drawn to the upper portions of the porous plates by capillary action and the rate of evaporation can be increased or decreased by changing the number of the plates in the rack which holds them in place.

The arrangement for humidification, shown in Fig. 7.2, does not provide a positive control of the relative humidity in the spaces heated by the furnace. However, the operating characteristics are satisfactory in that it provides more humidification during cold weather when the greatest amount is needed and less humidification in mild weather when the requirements are less. The following example indicates a method of estimating humidification requirements, of a structure, for specified conditions.

Example. A residence having 12,000 cu ft of occupied space is to be maintained at 71.5 dbt and 35 per cent relative humidity. The house has one air change per hour when the outside-air temperature is 0 F, and it has a relative humidity of 100 per cent. Find the gallons of water required per 24 hr for humidification. Assume standard atmospheric pressure.

Solution. The weight of dry air leaking into and out of the house per hour at 71.5 F and atmospheric pressure is $12,000 \times 0.07471 = 896.5$ lb. Air at 71.5 F and 35 per cent relative humidity, Fig. 2.2 contains 41 grains of moisture per lb of dry air. At 0 F and 100 per cent relative humidity the moisture of 1 lb of dry air from Table 2.1 is 5.5 grains. The moisture to be added per 24 hr is

$$[24 \times 896.5(41 - 5.5)] \div 7000 = 109.1 \text{ lb}$$

With water supplied at a density of 62.3 lb per cu ft, the number of gallons is $109.1 \times 7.48 \div 62.3 = 13.1$.

7.12. Furnace Sizes and Ratings. The size of a furnace, designed to burn solid fuels, is usually designated by the maximum diameter of the fire pot taken at a location just below the feed-neck section. The nominal diameter of the grate is that of the bottom of the fire pot at a point just above the ashpit section. The nominal grate diameter is usually less than the listed furnace size because of the taper of some cast-iron fire pots and the refractory linings of steel furnaces.

Warm-air furnaces designed for the use of either oil or gas alone do not include grates and are often quite different in their construction from those built for the use of coal. The differences in design lead to more efficient transfer of heat from the products of combustion to the circulating air in furnaces built for the exclusive use of either oil or gas, and also less consideration has to be given to the accessibility of their heating surfaces for cleaning. Special designs of furnaces for the burning of either oil or gas as a fuel are more efficient than those built for the use of coal and then later fitted with conversion burners. Furnaces designed for the exclusive use of either oil or gas are generally rated in terms of heat of the fuel input to the furnace, heat output at the furnace bonnet, or the heat delivered by the warm-air registers, **all in Btuh.**

HEATING WITH WARM-AIR FURNACES

The combined combustion and heat-transfer efficiency is called the bonnet efficiency. It is the ratio of the heat outlet at the bonnet to the heating value of the fuel consumed. Catalogues do not generally give this information, but it can easily be computed in the case of units burning either gas or oil when the input capacity is given.

Mechanical warm-air furnaces are rated in terms of Btuh, based on either register delivery or bonnet capacity. The register-delivery capacity with mechanical circulation is usually assumed as 85 per cent of the output at the bonnet. If practicable, ratings should be determined by actual test.

Many manufacturers offer oil-burning and gas-burning packaged units which are designed specifically for forced circulation. Figure 7.2 shows the interior view of such a unit designed for the use of gas as a fuel. Included within a single casing are filters, a fan, burners, and the required amount of heat-transfer surface. Rating data are given for several different sized furnaces of this type in Table 7.1.

7.13. Furnace Selection. A warm-air furnace for a home with a basement which is *not* to be heated (except for vagrant heat) may be selected either on the basis of a register capacity equal to the heat loss from the rooms above the basement or on the basis of a bonnet capacity equal to the heat loss of the entire structure. In the case of a home with a basement which *is* to be heated and in the cases of homes built on concrete slabs or over crawl spaces, the selection should be on the basis of a bonnet capacity equal to the *total* heat loss.

7.14. Registers in Mechanical Warm-Air Systems. When air circulation is produced by means of a fan, satisfactory heating of a room may be achieved with the warm-air registers located in the floor, low on a side wall, high on a side wall, or in the ceiling. The principal requirement from a comfort standpoint is that the register is so located that the air stream does not strike an occupant before it has lost its high velocity. The high side-wall location is extensively used because when the register is in this position there is little chance that the air stream will strike an occupant anywhere in the room; an additional advantage is that it does not interfere with any furniture arrangement which may be desired. The ceiling register also offers these advantages but is less commonly used because when the furnace is located in the basement more piping is required to reach a warm-air outlet in this location, and when there is no basement its use is likely to result in cold floors. A register suitable for use in either a high or low side-wall location in a forced-circulation system is shown in Fig. 7.9b.

The registers used in perimeter systems are usually a special type

TABLE 7.1. Rating Data for Packaged-Type, Gas-Fired, Forced-Circulation, Warm-Air Furnaces*

Model Number	Rating, Btuh† Input	Output at Bonnet	Output at Register‡	Cfm Delivery for Stated Deg F Temperature Rise§ 80	90	100	Gas Capacity, Cfh with Btu Heating Value of 550	800	1000	Cfm Delivery at 0.2 In. of Water Static Pressure	Approximate Shipping Weight, lb
FEC 60-45	60,000	48,000	43,200	525	465	418	109	75	60	540	312
FEC 75-45	75,000	60,000	54,000	655	580	525	136	94	75	690	325
FEC 90-45	90,000	72,000	64,800	790	700	630	164	113	90	880	356
FEC 105-45	105,000	84,000	75,600	920	815	733	191	131	105	985	380
FEC 120-45	120,000	96,000	86,400	1050	935	841	218	150	120	1030	407
FEC 150-45	150,000	120,000	108,000	1310	1170	1053	273	188	150	1510	481
FEC 180-45	180,000	144,000	129,600	1575	1400	1260	328	225	180	1690	572

* Surface Combustion Corporation.
† If specified on order for high altitude, approved units will be supplied to develop full catalogue rating per AGA high-altitude requirements.
‡ The output at the register allows for a 10 per cent heat loss from the warm-air ducts between the bonnet and the register. (15 per cent is more commonly allowed.)
§ Blowers will deliver proper cubic feet per minute for Save-Way System and for NWAH & ACA 4-in. duct system.
∥ Janitrol conditioners have sufficient blower capacity to deliver the required cubic feet per minute at a 100 F temperature rise against the static pressures of small-diameter duct systems designed in accordance with the Janitrol Save-Way Manual and the NWAH & ACE Manual 10.

HEATING WITH WARM-AIR FURNACES

of floor or low wall unit. Figure 7.10 is a cutaway view of a typical floor diffuser. The vanes for spreading the rising stream of air are clearly indicated as well as the mechanism for adjusting the position of the inlet damper. Figure 7.11 represents a segment of a perimeter-loop system and indicates alternate locations for a warm-air register below a window. Air passing through the register in either location

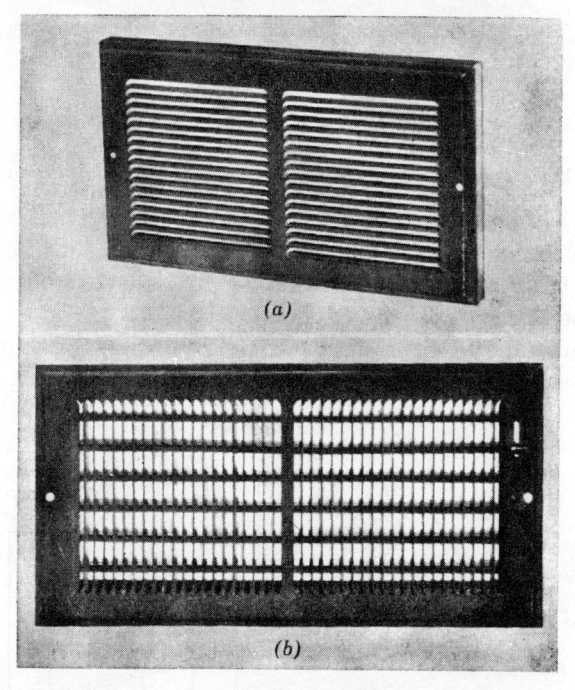

Fig. 7.9. Forced-circulation warm-air registers. (a) Baseboard return-air unit. (b) High or low side-wall register arranged to deflect one-half of air to right and one-half to left.

rises in a V-shaped current which intercepts the falling streams of cold air from windows and adjacent walls, as indicated by the arrows shown in Fig. 7.11. Either arrangement can also be used in any perimeter-radial system.

The baseboard diffuser of Fig. 7.12 may be used in place of a register for the dissemination of warm air in any perimeter type of warm-air system. Some home owners prefer the appearance of the baseboard diffuser over that of the small registers of Fig. 7.10. Units up to 8 ft in length are available and it is somewhat easier to obtain a continuous "blanketing" effect for a cold wall than when registers are used. When the disseminators of a warm-air system are also used for diffus-

148 AIR CONDITIONING AND REFRIGERATION

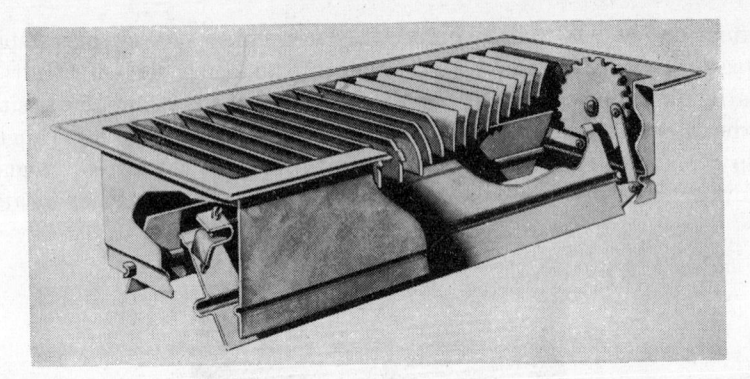

Fig. 7.10. Floor register for a perimeter system. (Courtesy Hart and Cooley Mfg. Co.)

ing cold air during the summer, it may be more difficult to obtain satisfactory distribution with the baseboard diffusers than when using the proper type of register.

In designing forced-circulation warm-air systems of the conventional type, the usual practice is to include a return-air grille in each room

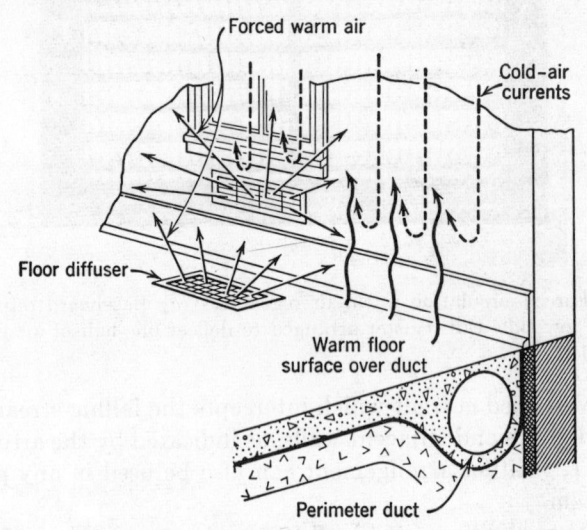

Fig. 7.11. Placement of floor or low-wall register in a warm-air perimeter-loop system.

to be heated with the exception of bathrooms, closets, kitchens, and lavatories. A vent to the outside, equipped with a small fan, should be provided in the kitchen to remove cooking odors from the house. Large living rooms are usually served by two return-air intakes.

HEATING WITH WARM-AIR FURNACES

Intake grilles for conventional mechanical systems are generally placed in an outer wall of the room with their lower edges at the floor level; they have the appearance of the side-wall outlet registers and do not include a means of regulating the flow of air. A typical return-air grille, suitable for many forced-air circulation systems, is shown in Fig. 7.9a. Floor intakes may also be used and offer less resistance to the flow of air than those placed in a vertical position. However, the side-wall type of intake is usually preferred because it does not interfere with the placement of any type of floor covering and there is less chance that small articles will accidentally drop into it.

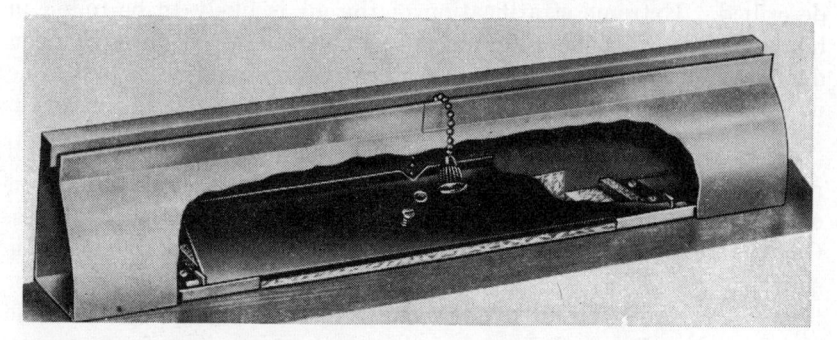

Fig. 7.12. Baseboard diffuser for warm-air perimeter systems.

Perimeter systems may include only one return-air register the placement of which is not critical. Cold air from windows and exposed walls is prevented from reaching the floor and causing discomfort to the room occupants by warm air moving upward over all of these areas. The return-air register in a system of this type is usually placed in such a way that little duct work is needed to convey air from it to the special downflow furnace (see Fig. 7.4). When the installation of a straight duct between the register and the return-air plenum is impractical a ceiling register location may be used and elbows employed as required. Return-air ducts in an attic or in any other unheated space should be insulated to conserve fuel.

7.15. Design of Warm-Air Systems. Any warm-air system can be designed by the methods discussed in Chapter 14. Special simplified procedures for designing small systems applicable to a residence, for all of the commonly used types, have been worked out through investigations sponsored by the NWAH & ACA and are published in its manuals for a list of which see the Bibliography.

7.16. Floor Furnaces. A special type of pipeless warm-air furnace, known as a floor furnace, is used extensively for heating one-story houses, built without basements, in certain sections of the country. Oil and gas are the common fuels used with floor furnaces which are set in floor openings, as near as possible to the center of the floor area of the space to be heated. One large register at the top of the unit protects its top. The air circulation is by gravity action. The products of combustion are conducted through a smoke pipe, located in a crawl space below the floor, to a chimney.

Floor furnaces are lower in first cost than any other type of warm-air furnace installation, but the results are generally inferior to those produced by any of the piped systems which have been previously described. Extreme stratification of the air is likely to be found in homes heated with floor furnaces. Circulation of air through open doorways must be depended upon for heating rooms adjacent to the one in which the register is located when this type of system is employed.

7.17. Wall Furnaces. Wall furnaces are installed in many small residences where the building budget will not permit the installation of any of the central systems that have been discussed in prior articles. One manufacturer's version of this relatively new type of heater is shown in Fig. 7.13. Gravity action alone or aided by a small fan or blower circulates the air over the heating surfaces. In general, the air circulation in a wall furnace is better than in a floor furnace even when a fan is not used.

In the design shown in Fig. 7.13 the combustion air is taken from outside and the products of combustion are discharged to the outside through a special vent. Some wall furnaces are designed for installation in a partition separating two rooms so that a unit can be located near the center of the space which it is to serve. In such a location the special vent shown in Fig. 7.13 could not be used and it would be necessary to employ a vertical flue to carry the products of combustion to the outside. A wall furnace located in an interior partition may deliver heat to two rooms.

As with floor furnaces, rooms which do not contain a part of the unit can be heated only by air circulation through open doorways when a single wall furnace is used to heat a home.

7.18. Control of Warm-Air Heating Systems. In the operation of a forced warm-air heating system the temperature in the space served is usually controlled by a thermostat which causes the fuel burner to be started or stopped as directed. The fan which circulates air through the system is started by a temperature-sensitive device known as a *fan switch* which is located in a warm-air duct near the bonnet. This switch has two set points and is designed to start the

HEATING WITH WARM-AIR FURNACES

fan when the air temperature reaches its higher setting and to stop it when it falls to a level corresponding to the lower setting. Both the set points and the temperature increment between them are usually adjustable. The cutout point should preferably be only 5 to 10 deg

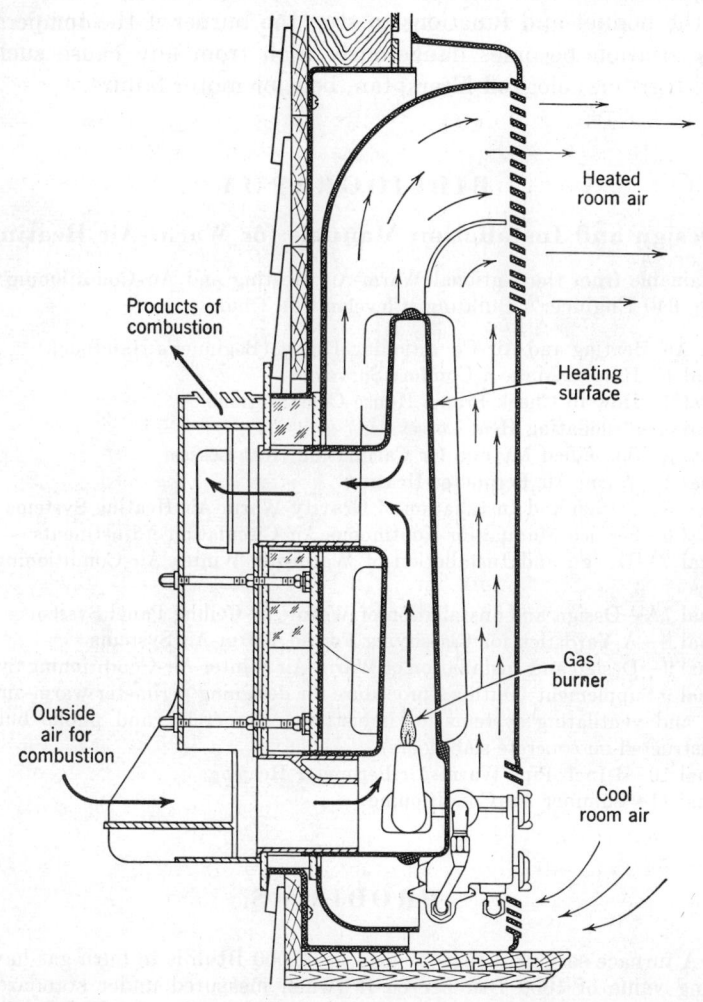

Fig. 7.13. Wall furnace. (Courtesy Stewart-Warner Corp.)

above the desired room temperature, and the differential should approximate 15 deg. With proper adjustment of the fan switch together with the proper fan capacity, the air will be circulated continuously except in mild weather. Continuous air circulation during cold weather is essential for good results with a forced warm-air system. Additional information on the adjustment procedure is given in Manual 6 (see Bibliography).

All warm-air systems which are automatically operated should be equipped with a limit switch, Art. 21.17, which is also located in or near the bonnet and functions to stop the burner if the temperature at its location becomes dangerously high from any cause such as: closed registers; clogged filters; fan, belt, or motor failure.

BIBLIOGRAPHY

Design and Installation Manuals for Warm-Air Heating

Obtainable from the National Warm-Air Heating and Air-Conditioning Association, 640 Engineer's Building, Cleveland 14, Ohio.

Warm-Air Heating and Air-Conditioning Digest (Beginner's Handbook)
Manual 1—How to Make a Comfort Survey
Manual 2—How to Check Frame House Construction
Manual 3—Calculating Heat Losses
Manual 3—Simplified Method for Calculating Heat Losses
Manual 4—Warm-Air Perimeter Heating
Manual 5—Design and Installation of Gravity Warm-Air Heating Systems
Manual 6—Service Manual for Continuous Air Circulation Adjustments
Manual 7—Design and Installation of Warm-Air Winter Air-Conditioning Systems
Manual 7A—Design and Installation of Warm-Air Ceiling Panel Systems
Manual 8—A Yardstick for Classifying Forced Warm-Air Systems
Manual 9—Design and Installation of Warm-Air Winter Air-Conditioning Systems
Manual 9 Supplement (Outlines procedure for designing perimeter warm-air heating and ventilating systems for industrial, commercial, and public buildings constructed on concrete slab floors)
Manual 10—4-Inch Pipe Warm-Air Perimeter Heating
Manual 11—Summer Air Conditioning

PROBLEMS

1. A furnace serving a register load of 60,000 Btuh is to burn gas having a heating value of 1000 Btu per cu ft (when measured under standard conditions). Assuming a bonnet efficiency of 75 per cent and a distribution efficiency of 85 per cent, find the input capacity in Btuh and in cubic feet of standard gas per hour.

2. A structure having an occupied space of 16,000 cu ft is to be maintained at a dry-bulb temperature of 72 F with a corresponding wet-bulb temperature of 57 F when the outside-air temperature is 35 F and its relative humidity is 40 per cent. The barometric pressure is 14.696 psia. What weight of water

HEATING WITH WARM-AIR FURNACES

vapor in pounds must be supplied per hour to maintain the required conditions when the volume of outdoor air entering each hour equals that of the occupied space?

3. Specify by model number from Table 7.1 the unit that should be installed in a home from which the heat losses (excluding the basement) have been estimated as 86,000 Btuh. How many cubic feet per hour of propane, heating value 2570 Btu per cu ft as metered, should be burned in the unit?

4. (a) Specify by catalogue number from Table 7.1 the gas-fired unit which would be best suited to the heating of a home from which the heat losses (excluding the basement) are estimated to be 65,000 Btuh under design conditions. (b) What would be the rated bonnet capacity of the unit you select? (c) Calculate the bonnet efficiency of the unit from the tabular data furnished. (d) How many cubic feet of 550 Btu gas are required per hour by this unit?

8

Radiators, Convectors, and Baseboard Heating Units

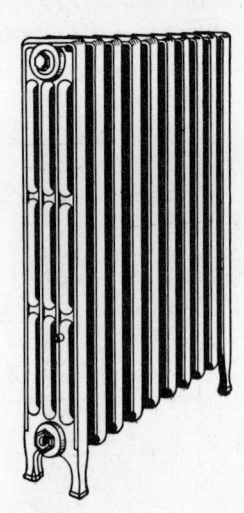

8.1. Tubular Radiators. The manufacture of column and the large tubular radiators has been discontinued; however, many of them remain in service in old buildings. Because of its pleasing appearance, low space requirements, and efficiency the small tubular radiator, as shown in Fig. 8.1, is now the common unit. Commercial sections of this form of radiation are standardized as to heights, number of tubes, rated surface, linear length, and limits for the actual width of the sections. Table 8.1 includes data for tubular radiators, as built by one manufacturer, which conform to the accepted requirements.

The required amount of surface in any radiator may be secured by fastening together a sufficient number of sections of the necessary height and width. The usual scheme of connecting the sections is by either malleable-iron screw or push nipples, Fig. 8.2. Details of the use of push nipples are shown in Fig. 8.3. Screw nipples require tapered threaded tappings in the radia-

Fig. 8.1. Tubular radiator.

RADIATORS, CONVECTORS, BASEBOARD HEATING UNITS

TABLE 8.1. Small-Tube Direct Cast-Iron Radiators—Widths, Heights, and EDR Surfaces*

Heating Surface, sq ft

	Three-Tube		Four-Tube		Five-Tube		Six-Tube	
Length 1¾ In.† per Sec	25-In. Height 1.6 Sq Ft per Sec	19-In. Height 1.6 Sq Ft per Sec	22-In. Height 1.8 Sq Ft per Sec	25-In. Height 2.0 Sq Ft per Sec	22-In. Height 2.1 Sq Ft per Sec	25-In. Height 2.4 Sq Ft per Sec	19-In. Height 2.3 Sq Ft per Sec	25-In. Height 3.0 Sq Ft per Sec	32-In. Height 3.7 Sq Ft per Sec
Distance from floor to center of top tapping, in.	23⅞	17⅞	20⅝	23⅝	20⅝	23⅝	17 11/16	23 11/16	30 11/16
Distance from floor to center of bottom tapping, in.	2½	2½	2½	2½	2½	2½	2½	2½	2½

* National U. S. Radiator Corp.
† Add ½ in. to length of all radiators for each bushing.

Width of sections, three-tube, 3⅜ in.; four-tube, 4 7/16 in.; five-tube, 5¾ in.; six-tube, 7¼ in. Width at feet is the same as width of section.

Tappings. Three-, four-, and five-tube radiators are tapped 1¼ in. at bottom both ends and 1 in. at top both ends. Six-tube tapped 1½ in. bottom both ends. 1¼ in. top both ends.

Can be furnished legless or with legs 4½ in. from floor to center of bottom tapping when ordered

To determine over-all height of three-, four-, or five-tube legless radiator, deduct 1⅜ in. from the standard heights. For six-tube deduct 1¼ in.

Radiators are not furnished in lengths exceeding 78 sections.

tor sections and do not necessitate the use of rods with nuts to hold them together.

The sections are further classified as loop and leg. The latter, of course, rest upon the floor, and the height of the radiator is based upon the nominal over-all height of the leg sections. The loop sections are assembled between the leg sections and are listed as having the height of the leg sections with which they are used. Formerly radiators were

Screw Nipple **Push Nipple**

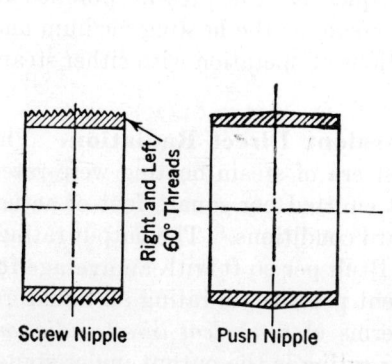

Fig. 8.2. Radiator nipples.

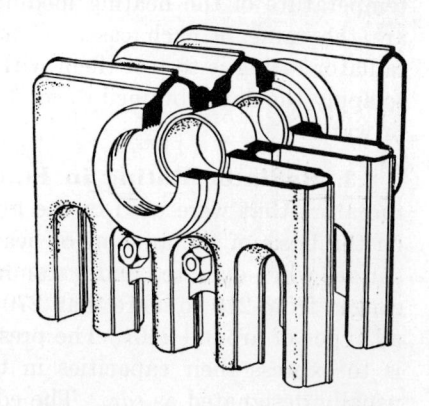

Fig. 8.3. Use of push nipples in a radiator.

also designated as steam and hot-water types. Steam-radiator sections are fastened together by nipples at their bottoms only. Such radiators cannot be used in hot-water heating because of the extreme difficulty of venting air from all the sections. Hot-water radiators have nipple connections between the sections at both the top and the bottom. As hot-water radiators may be used with any steam-heating system the manufacture of the steam-type unit has been discontinued.

Tubular radiators may be assembled without leg sections and hung on concealed brackets. Information relative to the actual lengths of loop sections is to be found in the footnotes of Table 8.1.

8.2. Standard Temperatures for Determining Radiator Performance. Definite temperature specifications[1] are necessary in the standard conditions of performance of direct radiators. Steam units are supplied, when rated for heat transmission, with steam at 215 F and are tested in a room where the air temperature, at the breathing level, is 70 F, thus giving a steam-air-temperature difference of 145 F. In all tests the radiator is to stand in practically still air with the air movement over its surfaces due to natural convection currents.

Formerly the standard conditions for testing hot-water radiators were: water entering the radiator at 180 F; water leaving the radiator at 160 F; average temperature of water in radiator, 170 F; room-air temperature, 70 F; and temperature difference, 100 F. Many modern hot-water heating systems that are equipped with closed expansion tanks operate with an average water temperature equal to that which has been adopted as standard for testing steam radiators. Furthermore, the output of a given radiator containing hot water is essentially the same as that when it contains steam, providing that the temperature of the heating medium and that of the surrounding air are the same in each case. Consequently, the present practice in radiator rating is to test them with steam as the heating medium and to apply the data obtained to conditions of operation with either steam or water.

8.3. Radiator Rating in Equivalent Direct Radiation. The radiators that were used in the first era of steam heating were rated on the basis of the amount of heat emitted per square foot of actual surface when operated under standard conditions. The output ratings ranged from 210 to more than 270 Btuh per sq ft with an average for all types of around 240. The present practice, in rating all radiators, is to express their capacities in terms of *equivalent direct radiation*, usually designated as *edr*. The edr rating is the output under standard conditions, in Btuh, divided by 240. The rating is generally

RADIATORS, CONVECTORS, BASEBOARD HEATING UNITS

given in square feet of edr per section. The output under standard conditions in Btuh is given by

$$H = M(h_f + xh_{fg} - h_{fc})\left(\frac{215 - 70}{t_s - t}\right)^{1.3} \qquad (8.1)$$

where M is the weight of steam condensed in pounds per hour. The room temperature t is measured at the 5-ft level. The last term of the equation corrects the output to standard conditions. The exponent 1.3 has been determined empirically through the study of test data.

Example. Determine the edr rating per section of a four-tube 22-in. radiator from the following test data: number of radiator sections in the test unit, 10; steam pressure, 18 psia; steam quality, 1.0; condensate temperature, 220 F; steam condensed, 4.96 lb per hr; and room-air temperature, 65 F at the 5-ft level.
Solution. The following thermal data were obtained from Tables 1.6 and 1.7: $h_f = 190.56$, $h_{fg} = 963.6$, $h_{fc} = 188.13$ Btu per lb, respectively, and $t_s = 222.41$ F. The output under standard conditions is $H = 4.96[190.56 + (1.0 \times 963.6) - 188.13][(215 - 70)/(222.41 - 65)]^{1.3} = 4310$ Btuh. The edr rating of the test unit is $4310/240 = 18$ sq ft and the edr rating per section is $18.0/10 = 1.8$.

In the foregoing example the portion of equation 8.1 which represents the enthalpy per pound of dry steam could have been obtained directly from Table 1.6 as 1154.2 Btu per lb. A radiator test arrangement may include an electric heater in the supply steam pipe so that dry steam with a fraction of a degree of superheat may be delivered to the unit. When dry steam is supplied to a radiator and the temperature difference between the steam and its condensate is small, equation 8.1 can be simplified as

$$H = Mh_{fg}\left(\frac{215 - 70}{t_s - t}\right)^{1.3} \qquad (8.2)$$

8.4. Number of Radiator Sections Required. Radiator rating data are always given in the form of Table 8.1. When the operating conditions are standard the heat output per section of any type listed is simply its edr rating times 240. The actual heat output in Btuh per sq ft of edr for conditions other than standard is called the radiation factor R_a

$$R_a = 240\left(\frac{t_m - t}{215 - 70}\right)^{1.3} \qquad (8.3)$$

When steam is used as the heating medium its temperature t_m is taken as the saturation temperature corresponding to the absolute pressure.

For hot water t_m equals the mean temperature of the water or the average of its inlet and outlet temperatures. Values of R_a computed for t_m at different temperatures with t constant at 70 F, are given in Table 8.3. The method of determining the number of sections

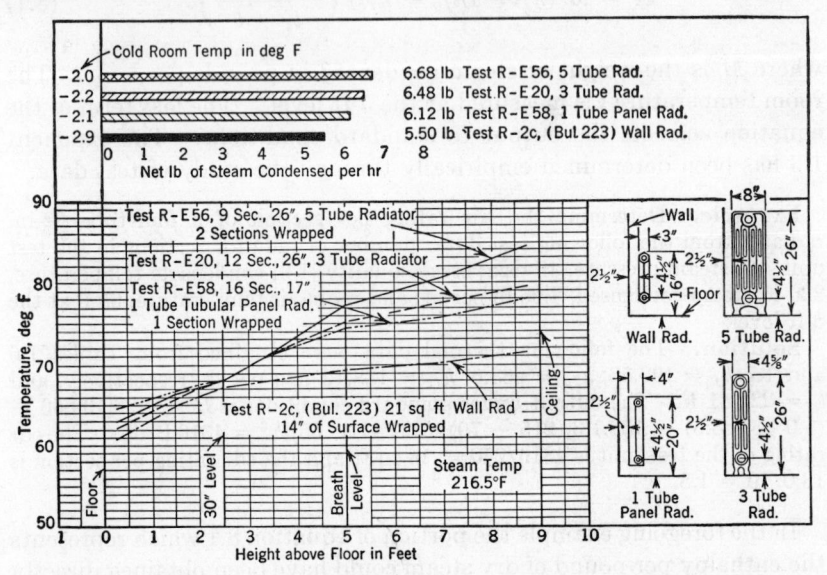

Fig. 8.4. Room-air temperature gradients and steam-condensation rates for four cast-iron direct radiators.

required for any given condition of radiator service is illustrated in the following example.

Example. Find the number of six-tube 19-in. radiator sections that would be required for maintaining the temperature in a room at 80 F when the heat losses are 7580 Btuh when the unit is served with hot water entering it at 190 F and leaving it at 170 F.

Solution. By use of equation 8.3 the actual output per hour per square foot of edr is $240[(180 - 80)/(215 - 70)]^{1.3} = 148.3$. Table 8.1 gives 2.3 sq ft, edr, as the rating of a section. Therefore, the number of sections required is $7580/(2.3 \times 148.3) = 22.2$ in which case 22 sections would be installed.

8.5. Heating Effect. The heating effect of a radiator or other disseminator is not directly proportional to the energy released by the heating medium passing through it. High radiators placed at the sides of window openings cause large differences between the air temperature measured near the ceiling and that near the floor. When the air temperature near the ceiling is considerably higher than that needed for comfort in the living zone, heating costs are increased. The heating industry has generally accepted the practice of arbitrarily adding a percentage to the actual output of certain disseminators which are capable of maintaining a small floor-to-ceiling temperature differential. This arbitrary addition to test results is called *heating effect allowance*.

Figure 8.4[2] indicates that in general as a disseminator is made lower and wider the ceiling-to-floor temperature differential is decreased. This observation led to the development of radiant baseboards and

Fig. 8.5. Cabinet convector with finned-tube heating element. (Courtesy Young Radiator Co.)

baseboard convectors which will be discussed later. Manufacturers of these devices often use heating effect allowances of as much as 15 per cent in arriving at the published rated output. Some authorities do not agree that this practice is justified and some engineers reduce published ratings by the amount of heating effect allowance indicated before using them in the selection of disseminators.

8.6. Convectors. The term convector covers a great variety of concealed heating units constructed in many forms of both ferrous and nonferrous materials. One form of such equipment functions with a heating element placed at a low level within an enclosure as shown in Fig. 8.5.

Because of the chimney action produced by the effects of stack height, Fig. 8.6, and the differences in the temperatures of air within and without the enclosure, cool air from near a room floor is circulated through the heating element of a convector. The air thus warmed is discharged through a grille either in the top of or near to the top of the enclosure.

Three general types of convector enclosures are available: cabinet, partly concealed, and totally concealed units. Thus convectors may be installed under windows either as free-standing floor cabinet units, Fig. 8.5; assemblages enclosed in wall-mounted cabinets; recess types partially set into the wall; recess types fully set into the wall; and finally those fully concealed arrangements with their heating units placed within wall spaces behind plaster coverings.

Nonferrous materials used for the construction of heating-element cores include copper, brass, and aluminum. With nonferrous materials one common form of heating element design involves inlet- and outlet-connection headers joined by tubes of either brass or copper which have attached finned extensions of either sheet copper, brass, or aluminum as shown in Fig. 8.5. Where extended air-heating surfaces are formed from sheet materials of either copper or brass, bonding between tubes and fins may be secured by soldering. Good contact between tubes and fins is often secured by heavy pressure exerted upon fin and tube parts during the assembly of the core of the heater unit.

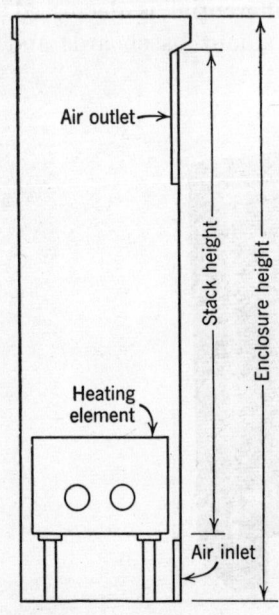

Fig. 8.6. Stack and enclosure heights of a convector.

The rate of air flow through a convector has much to do with the output obtained. The design of a convector-enclosure air inlet and also its air outlet affect its performance. The capacity of a convector is dependent upon the height of the stack for any given set of operating and design conditions. The greater the stack height the greater is the heat output per hour for a given type of heating unit. However, the high-outlet convectors tend to give poorer heating effects owing to the increased tendency for air stratification in a room. Generally it is best to avoid the use of convectors which cannot be placed beneath the usual window opening.

Long low units similar in principle to that of Fig. 8.5 are now available for placement under picture windows in modern homes. Rating

RADIATORS, CONVECTORS, BASEBOARD HEATING UNITS

data pertaining to one manufacturer's line of such units, when served with hot water, are given in Table 8.2.

TABLE 8.2. Hot-Water Capacities for Picture-Window Convectors, Thousands of Btuh*
Based on 20 F water-temperature drop and air temperature 65 F at convector inlet

Average Water Temp., deg F	Depth in.	Length, in. 40	44	48	56	64	72	80	88	96
160	6	2.4	2.6	2.9	3.5	4.0	4.6	5.1	5.7	6.2
	8	2.7	3.0	3.4	4.0	4.6	5.3	5.9	6.5	7.1
180	6	3.2	3.5	3.9	4.7	5.4	6.1	6.9	7.6	8.4
	8	3.6	4.1	4.5	5.3	6.2	7.1	7.9	8.8	9.5
190	6	3.6	4.0	4.4	5.3	6.1	6.8	7.8	8.6	9.4
	8	4.1	4.6	5.1	6.0	7.0	8.0	8.9	10.0	10.9
200	6	4.0	4.5	4.9	5.9	6.8	7.7	8.6	9.6	10.6
	8	4.6	5.1	5.7	6.7	7.8	8.9	10.0	11.1	19.1
210	6	4.4	5.0	5.5	6.5	7.5	8.6	9.6	10.6	11.7
	8	5.0	5.7	6.3	7.4	8.7	9.9	11.0	12.3	13.4
220	6	4.8	5.4	6.0	7.1	8.2	9.3	10.5	11.6	12.7
	8	5.4	6.2	6.9	8.1	9.5	10.8	12.0	13.4	14.6

* Modine Manufacturing Co.

8.7. Control of Convectors. The heat output of a convector may be regulated by either of two methods. The flow of the heating medium may be controlled by a valve in the heating-medium supply connection or the air flow may be controlled by a damper at the air outlet of the unit. Closure of the air damper stops air circulation and the unit becomes inactive. Some convectors are provided with both forms of control.

8.8. Ratings of Convectors. Convectors also may be rated on the basis of steam condensation tests. Ratings for steam operation are standardized[3] for the conditions of 215 F steam temperature within the heating element and air entering the convector enclosure at 65 F. The heat output of a steam convector operating under standard conditions, in Btuh, is

$$H = h_{fg}M \left(\frac{215 - 65}{t_s - t_i}\right)^{1.5} \qquad (8.4)$$

The assumption is made that the steam is dry and saturated and that the temperature of the condensate is that of the steam (Art. 8.3). The rating of a convector in terms of edr is $H/240$. In equation 8.4 t_i

is not equal to the room temperature t in equations 8.1 and 8.2, and also the exponent is 1.5 instead of 1.3. Sometimes an extra allowance is included in the rating for heating effect (Art. 8.5). Ratings as given should be studied critically to determine whether the allowance added for heating effect is excessive.

8.9. Heating Performance of Convectors for Conditions Other than Standard. Many convectors cannot be operated with steam at 215 F and the inlet-air temperature at 65 F. As with radiators, rating data obtained with steam may be applied to a unit when served with hot water. When the conditions of operation are other than standard, it is necessary to estimate from the rating data the performance to be expected under actual conditions of operation. The corrected radiation factor R_a is in the case of a convector rated in terms of edr

$$R_a = 240 \left(\frac{t_m - t_i}{215 - 65}\right)^{1.5} \tag{8.5}$$

When the ratings under standard conditions are given in Btuh, the corrected value for conditions other than standard becomes

$$H = H_{std} \left(\frac{t_m - t_i}{215 - 65}\right)^{1.5} \tag{8.6}$$

Table 8.3 gives computed heat emission rates in Btuh per square foot

TABLE 8.3. Heat Emission Rates of Radiators and Convectors

Average Temperature of Heating Medium, deg F	Radiators with Air Temperature at 70 F at the Breathing Level, Btuh per sq ft	Convectors with Air Temperature at 65 F at the Floor Level, Btuh per sq ft
150	111	102
160	129	121
170	148	141
180	168	161
190	188	182
200	208	206
210	229	228
215	240	240
220	250	251
230	272	277

edr for radiators and convectors. Table 8.3 can be used for air temperatures other than 70 F for radiators and 65 F for convectors by properly adjusting the medium temperature that is applied.

RADIATORS, CONVECTORS, BASEBOARD HEATING UNITS 163

8.10. Forced Convectors. As a means of adapting hot-water heating systems to space cooling in summer, the type of convector shown in Fig. 8.7 has been developed. The unit consists of a sheet-metal housing designed to fit between the joists of the usual standard wood-frame building construction, a finned-tube heat exchanger, and a small fan which circulates room air through the unit. The forced

Fig. 8.7. Forced convector. (Courtesy Penn Boiler and Burner Mfg. Corp.)

convector of Fig. 8.7 is arranged to draw air into the cabinet through the upper grille and to discharge air from the lower one. Other units have air circulation in the opposite direction. Room temperatures can be individually controlled by means of separate room thermostats which regulate the fan in each unit; the fans can also be operated continuously with the temperature of the system circulating water determined by a single thermostat placed in a strategic location, as has been common practice with more conventional types of heat disseminators. Some rebalancing of the system is usually necessary when a shift from summer-cooling operation is made to winter heating or vice versa when control of the system is maintained by a single thermostat. This situation is particularly true when the installation serves a two-story house.

8.11. Baseboard Heating Units. Developments have been made with heating units which replace the usual baseboard. Baseboard units may be divided into two general groups, those which are direct-radiant elements and those which are primarily convectors in their operation. The direct-radiant sections are made of either cast iron or steel. The convector-type units are usually made of finned tubing placed within sheet-metal enclosures. The desirable placement of

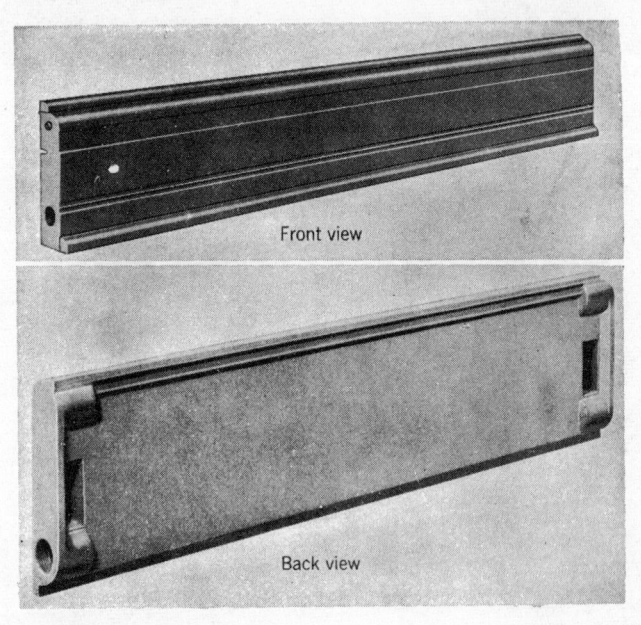

Front view

Back view

Fig. 8.8. Type-R radiant baseboard. (Courtesy Burnham Corp., Boiler Division.)

either of the two types is along the outside walls of the space to be heated. When because of inadequate wall space the required linear lengths of baseboard units cannot be placed along exterior walls, the additional amounts are located along interior walls.

Figure 8.8 shows a hollow cast-iron unit which is available in section lengths of 1 and 2 ft. The required amount of length, by 1-ft increments, which replaces the room baseboard can be obtained by joining sections together with push nipples and bolt fastenings. This particular unit is suitable for use with any type of hot-water heating system. In order to take care of inequalities of floor level such equipment is hung on wall brackets. Air leakages behind the units must be prevented by proper sealing. An allowance of about $\frac{1}{8}$ in. in 10 ft for each 180 F temperature rise should be made for expansion of the

RADIATORS, CONVECTORS, BASEBOARD HEATING UNITS

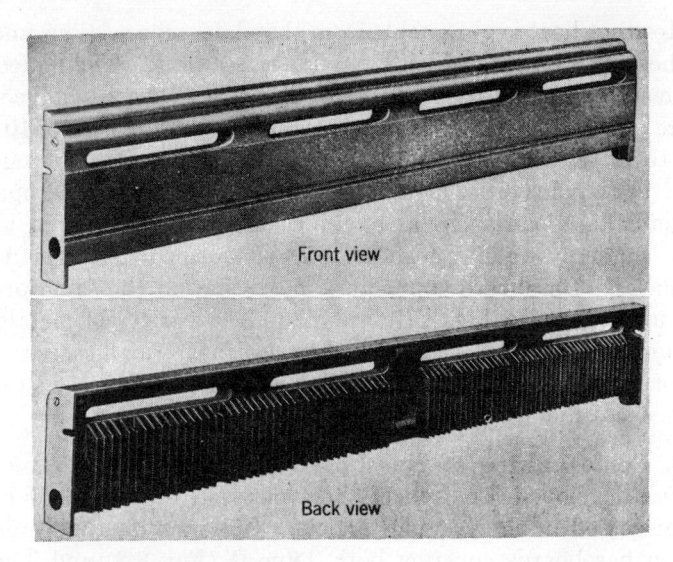

Fig. 8.9. Type-RC radiant baseboard. (Courtesy Burnham Corp., Boiler Division.)

metal. Fittings for corner connections of the sections and coverages of ends are available.

Finned radiant baseboard heaters are constructed in the form of the cast-iron unit illustrated in Fig. 8.9. These sections are fastened together in the manner stated for the units illustrated by Fig. 8.8. A baseboard unit of the convector type which is built up of sheet metal and tubing is shown in Figs. 8.10 and 8.11. Examination of Fig. 8.11

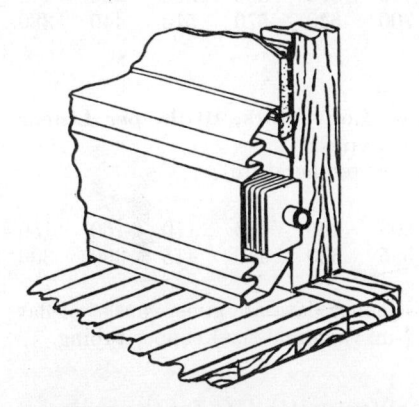

Fig. 8.10. Convector-type baseboard heating unit. (Courtesy C. A. Dunham Co.)

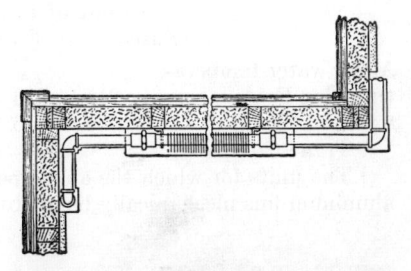

Fig. 8.11. Plan of pipe and finned sections of baseboard heating unit. (Courtesy C. A. Dunham Co.)

indicates that heat is obtainable from the plain-tube sections and from those heating elements which have finned sections. The finned units come in standard lengths of 2, 3, 4, 5, and 6 ft. A sheet-metal covering for bare pipe sections may be obtained in standard lengths of 10 ft and is cut to fit at the time of installation. The manufacturer supplies inside and outside corner fittings and those necessary at door openings.

Radiant baseboards similar to the cast-iron unit shown in Fig. 8.9 are also made by welding together suitably formed shapes made from steel plates. Baseboard convectors, operating on the same principle as the unit shown in Fig. 8.10, are made in a variety of metals, tube sizes, fin shapes, fin spacings, and cover designs. In all cases the front panel of the sheet-metal enclosure is removable so that the tubes and fins are accessible for cleaning.

8.12. Performance of Baseboard Heating Units. A code[4,5] has been developed for each of the two types of baseboard heating units discussed in the previous article. One manufacturer's data on cast-iron baseboards covering both Type-R (Fig. 8.8) and Type-RC (Fig. 8.9) construction are given in Table 8.4. Data which are applicable to a convection type similar to Fig. 8.10 but made by a different manufacturer are given in Table 8.5.

TABLE 8.4. Ratings for Cast-Iron Radiant Baseboards, Btuh per Linear Foot*

(Based on 70 F room-air temperature)

Mean water Temperature, deg F	220	210	200	190	180	170	160	150
9-in. Type R	490	450	410	370	330	290	250	210
9-in. Type RC	880	760	700	640	570	510	440	380

* Crane Company.

TABLE 8.5. Capacities of Baseboard Convectors, Btuh, per Linear Foot of Finned Length*†

(Based on 70 F room-air temperature)

Mean water temperature, deg F	215	210	200	190	180	170	160	150
$8\frac{1}{4}$-in. high unit	680	645	585	530	470	415	360	306

* Trane Company.

† The units for which the above performance data are given consist of flat aluminum fins mechanically bonded to $\frac{3}{4}$-in. (nominal-size) copper tubing.

Tables 8.6 and 8.7 indicate the manufacturer's data for the performances of bare pipe and finned heating elements made of tubing and sheet metal as shown in Fig. 8.11.

TABLE 8.6. Heat Output of 1-In. Bare Pipe in Dunham Baseboard, Btuh per Linear Foot*

Room air at 70 F

Average water temperature, deg F	160	180	200	220
Heat output	76	96	117	140

* C. A. Dunham Co.

TABLE 8.7. Heat Output of Finned Sections of Dunham Baseboard, Btuh*

Room air at 70 F

Average water temperatures, deg F

Cabinet Length, in.	Pipe Length, in.	160	180	200	220
24	20	557	750	948	1171
36	32	937	1244	1594	1967
48	44	1317	1748	2239	2762
60	56	1697	2251	2884	3558
72	68	2077	2755	3528	4354

* C. A. Dunham Co.

REFERENCES

1. "Code for Testing Radiators," *ASHVE Trans.*, **33**, 18 (1927).
2. "Steam Condensation an Inverse Index of Heating Effect," A. P. Kratz and M. K. Fahnestock, *ASHVE Trans.*, **37**, 475 (1931).
3. "Commercial Standard for Testing and Rating Convectors," *Commercial Standard CS140-47*, U. S. Department of Commerce (1947).
4. $I=B=R$ *Testing and Rating Code for Baseboard Type of Radiation*, Institute of Boiler and Radiator Manufacturers, New York, 2nd ed. (1952).
5. $I=B=R$ *Testing and Rating Code for Finned-Type Radiation*, Institute of Boiler and Radiator Manufacturers, New York, 1st ed. (1951, with addenda, 1954).

BIBLIOGRAPHY

1. $I=B=R$ *Installation Guide No. 5, Baseboard Heating Systems*, Institute of Boiler and Radiator Manufacturers, New York (1953).
2. "A Study of Radiant Baseboard Heating in the $I=B=R$ Research Home," A. P. Kratz and W. S. Harris, *Bull. No. 358*, Eng. Exp. Sta., University of Illinois (1945).
3. "The Heating Effect of Radiators," Charles Brabbée, *ASHVE Trans.*, **33**, 33 (1927).
4. "Heat Emission from Radiators," K. F. Rubert, *Bull. No. 24*, Eng. Exp. Sta., Cornell University (1937).

PROBLEMS

1. A tubular radiator operated with saturated steam of 0.99 quality and 215 F temperature in a room where the air temperature was 70 F. The steam used per hour was 35 lb when the condensate left the radiator with the same temperature as that of the steam. Find the edr rating of the unit.

2. A building having heat losses amounting to 90,000 Btuh is to be maintained at 70 F by four-tube 22-in. free-standing direct cast-iron radiators operating with saturated steam kept at a pressure corresponding to 215 F. How many radiator sections are necessary?

3. A five-tube 25-in. steam radiator gave off 12,650 Btuh when operated under standard conditions of rating. Find the number of sections in the radiator.

4. A test radiator consisting of 16 sections condensed 10.1 lb of steam per hr under the following conditions: steam pressure, 20.78 psia; steam quality, 98 per cent; condensate temperature, 220 F; and room air temperature at the 5-ft level, 80 F. Find the rating of the test radiator in square feet of edr per section.

5. How many sections of six-tube 19-in. radiation, Table 8.1, are required to emit 39,700 Btuh when operated as free-standing direct radiators in a room where the breathing-line air temperature is 60 F and the steam used for operation has a pressure of 15.65 psia?

6. Find the number of sections of three-tube 25-in. radiation that will be needed in a room from which the heat loss is 3168 Btuh which is to be maintained at 75 F at the 5-ft level if hot water is to enter the unit at 210 F and leave at 190 F.

7. Calculate the heat given off by 50 sections of three-tube 25-in. direct radiation when operated with steam at a temperature of 220 F and room air at 65 F.

8. A convector, when operating with a steam temperature of 220 F and air entering the unit at 60 F, gave off 30,000 Btuh. Find the output of the unit under standard conditions of rating convectors. What is its rating in square feet, edr?

9. A room from which the heat loss is expected to be 10,040 Btuh is to be maintained at 70 F at the 5-ft level. Assuming that a convector will be used and that it will be supplied with saturated steam at 220 F, determine the required rating of the unit in square feet, edr.

10. A livingroom in which the temperature is to be maintained at 70 F at the breathing level and from which the heat loss has been estimated as 8000 Btuh is to be heated with a convector which will be supplied with hot water having a mean temperature of 190 F as it passes through. Rating data for the type of unit to be used are given in Table 8.2. Specify the length and the depth in inches.

11. All conditions are as in Problem 10 except that the room is to be heated with a 9-in. Type-R radiant baseboard. (a) How many feet would be required? (b) How many feet would be required if the Type-RC radiant baseboard were used?

12. A baseboard heating unit is made up of 10 ft of 1-in. pipe and 56 in. of finned pipe (Table 8.7). What heat output may be expected per hour with a mean water temperature of 200 F? The room air temperature is 70 F.

9

Heating Boilers
and Appurtenances

9.1. Distinctions Between Steam Boilers and Hot-Water Heaters. The term hot-water boiler is applied to fuel-fired units for heating the water for hot-water heating systems. Such a water heater, completely filled with water, is usually identical in many respects with a steam-heating boiler of the same type of construction. Any steam-heating boiler may serve as a water heater if properly arranged and installed. Boilers made for water heating may not have enough space in the top for use as a steam boiler. The usual trimmings of a steam-heating boiler include a pressure gage, a safety valve, and a water column with a gage glass and try cocks.

The same unit when used as a water heater does not have a water column, gage glass, or try cocks. A relief valve to limit the water pressure is used in place of a safety valve. The gage attached to the heater may read either in pounds per square inch pressure or the head of water in feet. A special type of thermometer is inserted in a tapping near the hot-water outlets to indicate the temperature of the heated water.

9.2. Materials of Construction and Working Pressures. Cast iron and steel are commonly employed in heating boiler construction. Cast-iron boilers are built in sections which are assembled to produce

169

a unit of the desired size. Steel boilers are constructed with a shell of steel plates either riveted or welded together and are usually provided with either fire tubes or water tubes. All cast-iron sectional boilers and many steel heating boilers have internal fireboxes.

The maximum safe working pressures usually specified for cast-iron boilers are 15 psig for steam and 30 psig for water-heating service. Low-pressure steel heating boilers usually have the same pressure limits. Steel boilers may be built for higher pressures than those stated previously, as closed systems of hot-water heating may involve pressures greater than 30 psig.

9.3. Boiler Heating Surface. Those areas of a boiler which are in contact with the boiler water and which are exposed to either incandescent fuel, hot refractory materials, or hot flue gases are termed water-heating surfaces. Some modern cast-iron heating boilers have sections with ribbed areas which serve to increase their heating surfaces (see Fig. 5.5).

Heating surfaces may be classified as direct and indirect. Direct surfaces are those upon which the light of the fire shines; they are very effective in the transfer of heat to the boiler water. Indirect-heating surfaces are those in contact with flue gases only and are progressively less effective from the standpoint of heat transfer as the flue gases become cooled.

Additional heating surfaces are provided in boiler shells by the use of tubes. *Fire-tube boilers* are those in which the hot gases pass through tubes which are surrounded by water. The reverse is true in *water-tube boilers* where the water circulates through tubes which have their outer surfaces exposed to the travel of hot gases. Most boiler manufacturers use the outside diameters of the tubes employed in estimating the heating surfaces provided by the tubes.

9.4. Boiler Classifications. Heating boilers are built in a wide variety of designs. In the limited space here available it is impossible to describe all the different units available. One or more of the following classifications may be applied to any heating boiler:

Materials	cast-iron and steel
Settings	brick set and portable
Tubes	fire tube and water tube
Flue-gas travel	single pass and multiple pass
Draft	direct updraft and downdraft furnaces
Special	boilers with a magazine feed for coal, etc.
	oil-burning units with an integral burner
	units especially designed for the burning of gas

HEATING BOILERS AND APPURTENANCES

Cast-iron boilers are never provided with brick settings, with the possible exception of a brick base, and generally the base is of cast iron. Steel boilers of the portable type have an internal firebox surrounded by water legs, and rest either on a cast-iron or a brick base beneath the firebox, the opposite end of the boiler being supported by a pedestal. These boilers have no enclosing brickwork.

In single-pass boilers the flue gases pass directly through the flues to the smoke outlet without reversal of direction of flow. When the flue gases from the furnace are made to travel back and forth through the gas passageways of the boiler, the multiple-pass effect is secured.

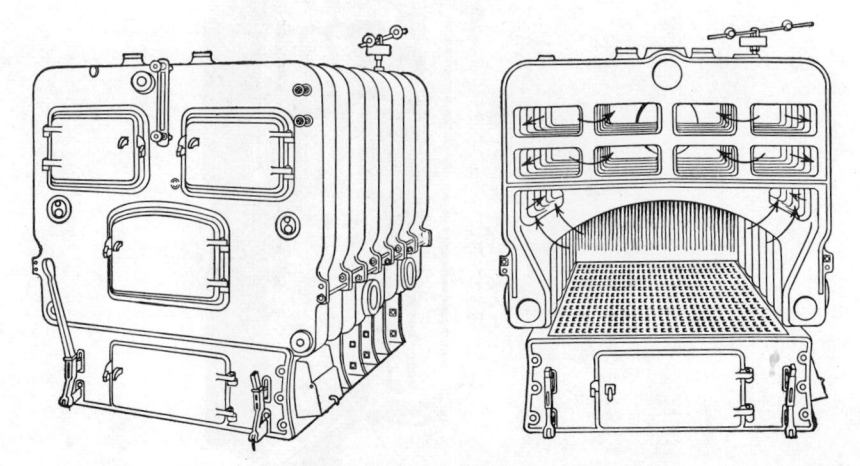

Fig. 9.1. Square sectional cast-iron steam boiler.

9.5. Cast-Iron Heating Boilers. These units are built in sizes ranging from very small up to those of moderate size. Cast-iron boilers are always constructed of sections joined together by some form of connections. Square sectional boilers in general are of the order of the unit illustrated in Fig. 9.1. The assemblage consists of front and rear sections together with a number of intermediate sections, depending upon the size of the boiler. The sections of large boilers are always connected together at their tops and at their bottoms on each side, either by nipples of the push type or by an outside header at the top and at both sides of the bottom. When push nipples are used the sections are firmly held together by rods and nuts.

Another example of a square-type cast-iron sectional boiler is the wet-bottom unit shown in Fig. 9.2. The boiler illustrated has a row of nipple connections between its sections both at their tops and their bottoms. The cast-iron boiler of Fig. 9.1 has a separate base which does not confine water and requires a floor of fireproof construction beneath it. Wet-bottom boilers are built in relatively small units for both steam and water service and may be installed upon wooden floors.

9.6. Capacity Ratings of Cast-Iron Boilers. The manufacturers of such equipment express the output of the units in terms of either Btuh or square feet of edr. The latter term may be square feet

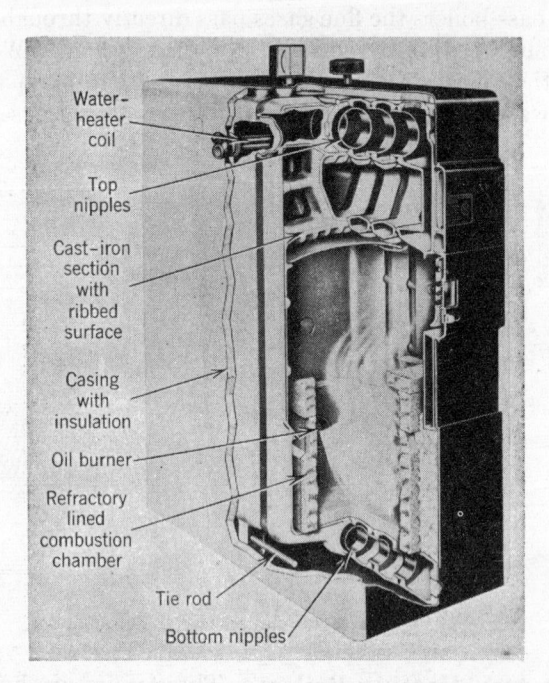

Fig. 9.2. Oil-fired wet-bottom cast-iron hot-water boiler. (Courtesy National-U. S. Radiator Corp.)

of steam radiation, 240 Btuh per sq ft, or square feet of hot-water radiation, 150 Btuh per sq ft. The hot-water rating is based on the assumption that the mean temperature of the water in the heat disseminators will always be 170 F and the air surrounding them will be maintained at 70 F. These conditions of temperature are seldom maintained with modern hot-water heating systems and such a rating is of little use.

The latest code[1] of the Institute of Boiler and Radiator manufacturers (I = B = R) calls for ratings as follows: gross output and net rating both in Btuh. The net rating is a reduced figure which allows for both piping loss and reserve capacity (pickup) for warming up the building under design conditions. For steam the net rating is also

HEATING BOILERS AND APPURTENANCES

given in square feet, edr. The code designates methods for testing and rating cast-iron heating boilers (except those of the magazine type) for conditions of both hand and automatic firing. Rating data for boilers that are automatically fired include fuel input rate in pounds per hour for stokers, in gallons per hour for oil burners, and in Btuh for gas burners. When the rating is based on the hourly heat input in the case of gas-fired units, a designation of a particular gas and its calorific value is unnecessary.

Hand-fired boilers under test must develop an over-all efficiency of 58 per cent or more with anthracite coal having a heating value of 13,000 Btu per lb on a dry basis. Limits are set for chimney area and height, draft in the stack, and the minimum time that an available fuel charge will last when it is burned with a combustion rate sufficient to produce the listed $I=B=R$ gross output.

Oil-burning boilers must operate with a minimum over-all efficiency of 68 per cent and a maximum temperature of 600 F for the flue gases leaving the boiler; the burner adjusted to give 10 per cent CO_2 (with variations of ± 0.2 per cent); heat release in the furnace not to exceed 80,000 Btuh per cu ft; and draft losses not to be greater than a designated amount. Stoker-fired units are rated with the same output as oil-fired boilers.

Boilers equipped with gas conversion burners are tested with the fuel-burning device adjusted to give a percentage of CO_2, which does not exceed 80 per cent of the ultimate possible, (Art. 5.13), for the gas used, and the CO must not be greater than 0.04 per cent on an air-free basis. The flue-gas temperature shall not exceed 600 F and the over-all efficiency shall not be less than 70 per cent. Boilers designed to burn gas only are always rated by tests conducted in the laboratories of the American Gas Association (AGA) under conditions specified in a code[2] published by that organization. In order to be eligible to bear the AGA stamp of approval the test data must indicate an efficiency (based on the higher heating value of the gas used) of at least 75 per cent.

For hand-fired units the piping losses and the pickup factor together range from 2.36 for boilers having a gross $I=B=R$ rating of 57,000 Btuh to 1.40 for boilers having a gross $I=B=R$ rating of 6,720,000 Btuh. For automatically fired boilers the combined factors range from 1.333 to 1.288 when the net capacities are from 100 to 20,000 sq ft of steam edr. Thus a hand-fired boiler having a gross $I=B=R$ output of 6,720,000 Btuh has a net $I=B=R$ rating of $6,720,000 \div 1.40 = 4,800,000$ Btuh.

Rules for the selection of heating boilers, stokers, or burners together with chimney specifications, ratings of several different makes of boilers, and other useful information are given in reference 3.

TABLE 9.1. Capacity Ratings of Weil-McLain No. 57 Series Cast-Iron Square Sectional Boilers for All Fuels

Oil-Fired

Boiler No. Steam or Water	Net I=B=R Ratings* Steam, sq ft	Steam, Btuh	Water, Btuh†	Gross I=B=R Output, Btuh	I=B=R Burner Capacity, oil gal per hr	Chimney Size, in.	Height, ft
0-157-‡	325	78,000	90,000	120,000	1.20	8 by 8	20
0-257-‡	430	103,000	118,000	157,000	1.55	8 by 8	20
0-357-‡	535	128,000	146,000	194,000	1.90	8 by 8	23
0-457-‡	640	154,000	173,000	230,000	2.25	8 by 12	25

Stoker-Fired

Boiler No. Steam or Water	Net I=B=R Ratings* Steam, sq ft	Steam, Btuh	Water, Btuh†	Gross I=B=R Output, Btuh	I=B=R Stoker Capacity coal lb per hr	Chimney Size, in.	Height, ft
K-257-‡	430	103,000	118,000	157,000	17	8 by 8	20
K-357-‡	535	128,000	146,000	194,000	21	8 by 8	23
K-457-‡	640	154,000	173,000	230,000	25	8 by 12	25

Hand-Fired

Boiler No. Steam or Water	Net I=B=R Ratings* Steam, sq ft	Steam, Btuh	Water, Btuh†	Gross I=B=R Output, Btuh	Grate Area, sq ft	Chimney Size, in.	Height, ft
157-‡	210	50,000	50,000	110,000	1.22	8 by 12	28
257-‡	320	77,000	77,000	177,000	1.81	8 by 12	30
357-‡	430	103,000	103,000	233,000	2.40	8 by 12	32
457-‡	535	128,000	128,000	283,000	3.00	8 by 12	35

* Net I=B=R ratings are based on net installed radiation of sufficient quantity for the requirements of the building and nothing need be added for normal piping and pickup.

† Exception: An additional allowance should be made for gravity hot-water systems or for unusual piping and pickup loads.

‡ Substitute S (for steam); W (for water).

9.7. Ratings of Commercial Cast-Iron Boilers. The ratings of Table 9.1 are typical of the data in catalogues supplied by the builders of cast-iron boilers.

9.8. Advantages and Disadvantages of Cast-Iron Boilers. The construction of sectional boilers allows the various parts to be

HEATING BOILERS AND APPURTENANCES

taken through ordinary doorways for assembly in the boiler room. Cast iron resists the action of corrosive agents much better than steel. Many cast-iron steam boilers have a low water line, which is a great advantage, especially where gravity-return steam-heating systems are used.

The disadvantages of cast-iron boilers include the danger of being damaged when improperly handled, small steam- and water-storage spaces, and the likelihood of discharging wet steam and slugs of water when driven hard.

9.9. Steel Heating Boilers. Heating boilers constructed of steel involve several arrangements. The categories which may be included are: brick set, portable, fire tube, water tube, downdraft, updraft, single pass and multiple passes, and vertical and horizontal.

A longitudinal view of a two-pass return-tubular boiler and setting is shown in Fig. 9.3. This type of boiler, which may be used for steam heating, is a fire-tube unit. The furnace consists of the grates and the combustion space.

Fig. 9.3. Square Type-R steel heating boiler. (Courtesy Kewanee Boiler Corp.)

The boiler of Fig. 9.3 is arranged for the hand-firing of solid fuels. However, horizontal return-tubular (HRT) boilers may be equipped with automatic burners using either coal, fuel oil, or gas.

In recent years some manufacturers have improved the performances and increased the capacities of HRT boilers by using small-diameter tubes which are practical when the fuel burned does not form excessive amounts of soot. When the assumption may be made that there will be *no* soot or flyash deposits as a result of the combustion process, a spiral-shaped strip of steel known as a spinner blade may be inserted in each fire tube to promote the transfer of heat. Each individual tube, suitably equipped with a spinner, may be fired by a small gas burner. When such operation occurs the combustion space shown in Fig. 9.3 may be completely eliminated. Special-tube designs which incorporate integral fins may be used to increase the heat-transfer

Fig. 9.4. Boiler-burner unit arranged for firing heavy fuel oil. (Courtesy Kewanee-Ross Corp. and Iron Fireman Manufacturing Co.)

HEATING BOILERS AND APPURTENANCES 177

capacity. The boiler of Fig. 9.3 is completely self-contained. Portable boilers are made in a large variety of designs and arrangements and include both fire-tube and water-tube types.

9.10. Boiler-Burner Units. Self-contained portable boilers of special design which also include all of the equipment needed for fuel burning are available from several manufacturers. Figure 9.4 shows a boiler-burner unit of the fire-tube type which is available in several

Fig. 9.5. Automatic packaged water-tube boiler capable of burning either fuel oil or gas. (Courtesy Babcock and Wilcox Co.)

sizes with either oil or gas as a fuel or as a combination oil-gas unit. The boiler-burner assembly illustrated by Fig. 9.4 is equipped to burn heavy oil (No. 5 or No. 6). The oil is automatically heated to the proper temperature by an oil heater and is automatically supplied as needed in the proper volume by a positive-displacement pump. The combustion air is supplied by a forced-draft fan and the efficiency of combustion is not dependent on the draft provided by the chimney.

Combustion takes place in the long corrugated central chamber and in the refractory-lined space at the rear end. The greater part of the heat transfer takes place in the fire tubes which are fitted with spinner blades throughout two thirds of their lengths. The unit can deliver either saturated steam or hot water at pressures up to 150 psig. When under the rated load the unit is guaranteed, by the builder, to have an over-all efficiency of 80 per cent.

Figure 9.5 shows a portable packaged unit which employs water tubes. Units of this type can burn either oil or gas and the fuel-burning equipment can be arranged as in Fig. 9.5 so that a change-over from one fuel to another can be readily made by merely manipulating certain controls and valves. Boiler-burner units of the type shown in Fig. 9.5 are available in capacities ranging from 2900 to 28,000 lb per hr of steam at pressures up to 235 psig.

Field-erected water-tube boilers[4,5] equipped to burn powdered coal, oil, or gas and capable of producing steam at pressures up to 2650 psig are available in capacities exceeding 1,500,000 lb of steam per hr.

9.11. Ratings of Steel Boilers. The Steel Boiler Institute, Inc. (SBI), a manufacturers' organization, has a code[6] which embodies specific directions for the testing of and the empirical rating of both hand-fired and mechanically fired steel heating boilers. The code classifies steel boilers into two general groups which are Table 1 and Table 2 types. The term Table 1 applies to large-size boilers and the designation Table 2 is given to those units which have less than 294 sq ft of heating surfaces and which have net ratings of not more than 4120 sq ft edr when hand-fired and 5000 sq ft edr when mechanically fired. Ratings given in square feet for either steam or water can be converted into Btuh by multiplying them by 240 or 150 as the case may be. Some manufacturers' catalogues pertinent to steel boilers include for each unit listed ratings in square feet of steam radiation, square feet of hot-water radiation, and Btuh. For large units the rating may also be expressed in terms of boiler horsepower.

The procedure used in determining the ratings of steel boilers is different for the two size classifications. The SBI rating of a Table 1 hand-fired boiler, in square feet of steam radiation, is 14 times the heating surface in square feet. The same rating for a Table 1 unit, which is mechanically fired, is 17 times the heating surface involved. The SBI net rating of Table 1 boilers is obtained by dividing the SBI rating by 1.2. In effect this method of rating allows 20 per cent of the connected load as piping losses. This allowance is made so that a boiler may be selected without making a calculation of the heat losses from the connecting piping. In cases where there is an unusual amount of piping involved, the connected load and the piping losses should be computed and totaled and the SBI rating rather than the SBI net rating used in making a boiler selection.

In the case of Table 2 boilers the SBI net rating, in square feet of

HEATING BOILERS AND APPURTENANCES

TABLE 9.2. Ratings and Specifications of Kewanee "Square Heat" Type-R Boilers Mechanically Fired

Boiler Number	3R1	3R2	3R3	3R4	3R5	3R6	3R7	3R8	3R9	3R10	3R11	3R12
SBI net rating, sq ft, steam	900	1100	1300	1500	1800	2200	2600	3000	3500	4000	4500	5000
SBI net rating, sq ft, water	1440	1760	2080	2400	2880	3520	4160	4800	5600	6400	7200	8000
Btuh ÷ 1000	216	264	312	360	432	528	624	720	840	960	1080	1200
Heating surface, sq ft	53	65	77	88	106	129	153	177	206	236	265	294
Furnace volume, cu ft	8.2	10	11.8	13.6	16.4	20	23.6	27.3	31.8	36.3	40.9	45.4
Fuel burning rates, oil gph	3.0	3.7	4.3	5.0	6.0	7.3	8.7	10.0	11.7	13.3	15.0	16.7
Gas, Btuh ÷ 1000	405	495	585	675	810	990	1170	1350	1575	1800	2025	2250
Stoker, solid fuel, lb per hr	36	44	52	60	72	88	104	120	140	160	180	200
Over-all boiler shell width, in.	30	30	30	30	34	34	34	34	41	41	41	41
Boiler shell length, in.	$30\frac{1}{2}$	$36\frac{1}{2}$	$42\frac{1}{2}$	$48\frac{1}{2}$	$42\frac{1}{2}$	$50\frac{1}{2}$	58	67	56	64	72	79
Water-line height, in.	$55\frac{1}{2}$	$55\frac{1}{2}$	$55\frac{1}{2}$	$55\frac{1}{2}$	$62\frac{1}{2}$	$62\frac{1}{2}$	$62\frac{1}{2}$	$62\frac{1}{2}$	$70\frac{1}{2}$	$70\frac{1}{2}$	$70\frac{1}{2}$	$70\frac{1}{2}$
Steam or water supply height, in.	$66\frac{1}{2}$	$66\frac{1}{2}$	$66\frac{1}{2}$	$66\frac{1}{2}$	74	74	74	74	86	86	86	86
Steam supply size, in.	4	4	4	4	4	4	4	4	6	6	6	6
Return size, in.	3	3	3	3	4	4	4	4	4	4	4	4
Smoke outlet height, in.	$44\frac{1}{2}$	$44\frac{1}{2}$	$44\frac{1}{2}$	$44\frac{1}{2}$	49	49	49	49	$54\frac{2}{3}$	$54\frac{2}{3}$	$54\frac{2}{3}$	$54\frac{2}{3}$
Breeching diameter, in.	14	14	14	14	18	18	18	18	20	20	20	20

steam radiation, is 17 times the square feet of boiler heating surface in the cases of stoker or oil firing, and 14 times the heating surface when hand firing is employed. Boiler manufacturers may not publish SBI net ratings without first submitting test data to the SBI.[6] Limiting factors in regard to CO_2 in the flue gases, flue-gas temperature, and over-all efficiency are applied when operating at 150 per cent of the SBI net ratings. In effect this procedure assures 50 per cent reserve capacity to allow for piping losses and the necessary pickup in the case of Table 2 boilers.

Table 9.2 gives SBI net ratings in three forms as well as other data relative to boilers of the type shown in Fig. 9.3. The net rating, the furnace volume, and the heating surface given for each unit in Table 9.2 correspond with these specifications for Table 2 steel boilers, oil-fired, which are given in the SBI code.[6] The table in the code also includes the specifications for boilers having other SBI net ratings.

9.12. Heating Boiler Tests. The performance of heating boilers can best be determined as the result of actual tests. The details of the necessary equipment, the methods of procedure, the observations to be made, and all the calculations involved are beyond the scope of this book. The reader is referred to the existing codes.[1,6]

The capacity of a boiler may be expressed in different units. For a steam boiler the heat output in Btuh is

$$H = (h_f + x h_{fg} - h_{ff}) M_w \qquad (9.1)$$

where M_w is the weight of feed water in pounds per hour. The capacity in square feet of edr is $H/240$.

For boilers tested as hot-water heaters the total heat output H in Btuh is

$$H = (h_{f2} - h_{f1}) M_w \qquad (9.2)$$

The over-all efficiency E_b of a boiler unit is the ratio of the heat output to the heat of the fuel burned. Thus for a steam boiler the over-all efficiency E_b in per cent is

$$E_b = \frac{(h_f + x h_{fg} - h_{ff}) M_w}{M_f \times F} \times 100 \qquad (9.3)$$

and for a hot-water heater or boiler

$$E_b = \frac{(h_{f2} - h_{f1}) M_w}{M_f \times F} \times 100 \qquad (9.4)$$

HEATING BOILERS AND APPURTENANCES

where M_f is the weight of fuel fired in pounds per hour and F is its heating valve in Btu per pound.

When either gas or oil is used as a fuel the total quantity in the denominators of equations 9.3 and 9.4 is the units of fuel used per hour times the heating value of the fuel per unit.

The over-all efficiency generally increases as the capacity developed increases until a maximum efficiency is obtained and then decreases as the load is further increased. Heating boiler over-all efficiencies generally range from 60 to 80 per cent.

Small heating boilers which are automatically fired are controlled by starting or stopping the burner as required. With this type of operation the burner capacity should be such that the boiler will operate at or near its maximum efficiency whenever fuel is being consumed.

9.13. Methods Used in Boiler Selection. The practice of boiler manufacturers in giving their units a net rating is intended to simplify the purchaser's problem of selecting the proper size. In the selection of an $I = B = R$ rated boiler the procedure is to select one which has a net rating equal to the estimated heat loss from the spaces which it is to serve. In the case of boilers intended for residences the allowances for piping losses and pickup are sufficient to take care of domestic water heating in most cases.

The SBI net ratings of Table 1 steel boilers allow 20 per cent of the connected load to care for piping losses, which is an adequate allowance for most installations; an allowance for pickup is not included in the 20 per cent extra. No allowance for pickup is needed if the spaces served are heated continuously. For special applications, such as churches, extra allowances up to 50 per cent should be made in the selection of a boiler.

SBI net ratings for Table 2 steel boilers assure satisfactory operation at actual boiler outputs up to 150 per cent of the published allowable loads. Therefore no extra allowance need be made for either piping loss or pickup except in unusual situations.

In the case of hospitals or other establishments where human life would be endangered in the case of a boiler failure, 100 per cent reserve capacity in the form of stand-by units may be necessary. In the case of large installations, of all types, consideration should be given to the desirability of installing two or more smaller boilers instead of one large one in order to produce flexibility of operation and available reserve capacity.

9.14. Safety Rules. Any boiler can fail by explosion as the result of either faulty construction or improper operation. As a service to the boiler manufacturing industry and to the public, a code[7] covering the construction, maintenance, and repair of boilers of all types was

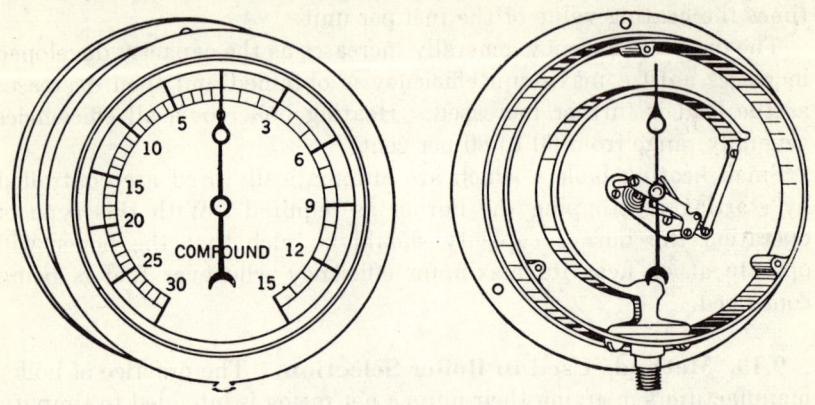

Fig. 9.6. Bourdon pressure gage.

formulated by a committee appointed by the American Society of Mechanical Engineers (ASME) in the year 1911. This body of persons, known as the Boiler Code Committee, with changing membership, has been continuously active since that time. The Committee meets at regular intervals to consider code revisions to permit progress in boiler construction and the use of new materials.

The majority of the states and many large municipalities have enacted laws which require that all boilers installed within their jurisdictions must be made according to the ASME construction code. Safety rules adopted by either cities or states usually require the regular inspection of boilers of certain specified classifications.

9.15. Boiler Accessories and Auxiliaries. Steam gages are usually of the Bourdon type as illustrated in Fig. 9.6. Where the boiler may be operated at pressures less than atmospheric, the gage should be compound in that both pressures above and below atmospheric may be measured.

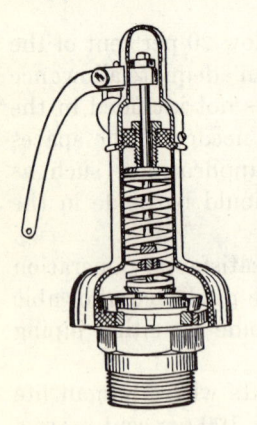

Fig. 9.7. Safety valve.

All steam boilers must have a safety valve. Safety valves are usually set to discharge steam at 5 to 10 psig above the usual pressure carried. With cast-iron boilers the safety valve must function at a

HEATING BOILERS AND APPURTENANCES

pressure not to exceed 15 psig. Figure 9.7 illustrates a spring-loaded valve—the only type that should be used. A safety valve must have a relieving capacity such that the steam pressure cannot rise more than 5 psi above the setting when it is discharging steam at the maximum rate at which it can be generated in the boiler.

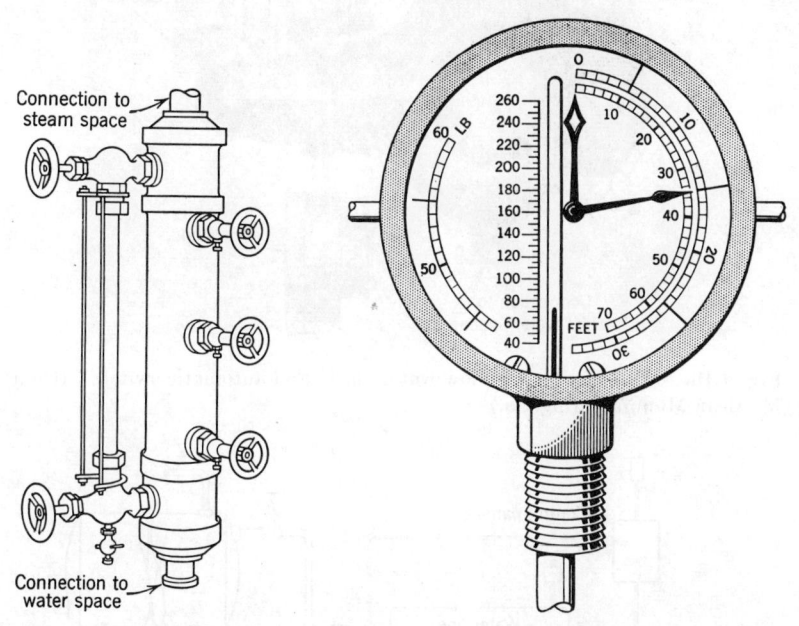

Fig. 9.8. Water column with gage glass and try cocks.

Fig. 9.9. Altitude gage and thermometer.

Steam-heating boilers may or may not be equipped with a water column. One form of water column is shown in Fig. 9.8. The functions of the gage glass and the try cocks are to indicate the position of the water level in the boiler. The try cocks check the accuracy of the indications given by the gage glass. Figure 9.9 is a combination thermometer and pressure gage designed especially for use on hot-water boilers. This type of gage usually indicates the pressure in feet of water instead of in pounds per square inch gage.

Protection should be provided for boilers in the form of equipment to stop the operation of automatic fuel-burning equipment whenever the level falls below a point where serious damage may happen. Figure 9.10 shows the details of one type of combination water feeder and low-level cutout.

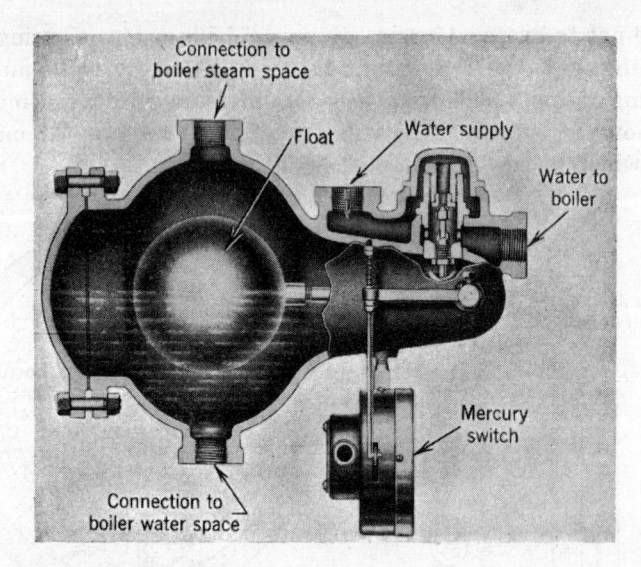

Fig. 9.10. Water feeder with low-water alarm and automatic switch. (Courtesy McAlear Manufacturing Co.)

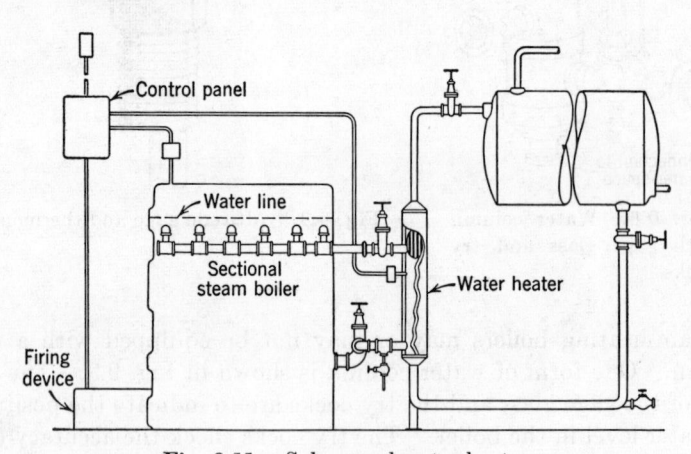

Fig. 9.11. Submerged water heater.

9.16. Water Heaters. Clean heated water is required in many buildings for domestic and other purposes. A number of operating reasons preclude the direct usage of the water of boilers. All water heaters are some form of a heat exchanger whereby the water heated is not contaminated as it passes through them. Often it is advisable to install an independently operated water heater which has as its source of heat either electricity or the combustion of a fuel.

HEATING BOILERS AND APPURTENANCES

The following water-heating devices are employed in plants where boilers are installed: (1) furnace coils, (2) coils placed within the water space of a boiler, (3) units external to the boiler which secure their supply of heat from either steam or water of the boiler, and (4) heaters which utilize exhaust steam from engines, turbines, and other sources.

Coils placed in furnaces give variable water temperatures. A typical heater coil inserted within the water space of a boiler is that

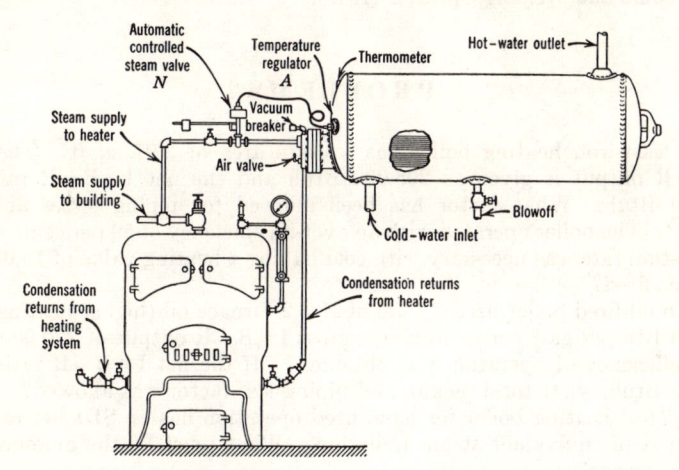

Fig. 9.12. Tank water heater.

illustrated in Fig. 9.2. Figure 9.11 shows a type of water heater which may be connected into hot-water boilers and into steam boilers below the water line. Water from the boiler flows through the outer shell and heats the domestic water supply which flows through the bank of tubes of the heater. The tank heater, Fig. 9.12, is for use with steam. The steam supply is regulated by a valve N which is under the control of a thermostatic element A placed in the heated water.

REFERENCES

1. "I=B=R Testing and Rating Code for Low Pressure Cast-Iron Heating Boilers," Institute of Boiler and Radiator Manufacturers, New York (1952).
2. "Approval Requirements for Central Heating Gas Appliances, Vol. 1, Steam and Hot-Water Boilers," American Gas Association, New York (1951 with addenda 1954).
3. "Net Load Recommendations for Heating Boilers," published semi-annually by Heating, Piping, and Air Conditioning Contractors National Association, New York.
4. *Steam, Its Generation and Use*, Babcock and Wilcox Co., New York (1955).

5. *Combustion Engineering*, Combustion Engineering Co., New York (1947).
6. "Rating Code for Steel Boilers, Steel Boiler Institute, Philadelphia, 6th ed. (July 1954)."
7. "ASME Boiler Construction Code." Section I, Power Boilers (1952 with 1954 addenda). II, Material Specifications (1949). III, Boilers of Locomotives (1949). IV, Low Pressure Heating Boilers (1952 with 1954 addenda). V, Miniature Boilers (1952). VII, Rules for Care of Power Boilers (1954). VIII, Unfired Pressure Vessels (1950). IX, Standard Qualifications for Welding Procedure and Welding Operator (1949).

PROBLEMS

1. A cast-iron heating boiler has a grate area of 3.97 sq ft. The gross $I=B=R$ output is given as 295,000 Btuh and the net $I=B=R$ rating as 134,000 Btuh. What factor has been allowed for piping losses and load pickup? The boiler operated with an over-all efficiency of 59 per cent. What combustion rate was necessary with coal having a heating value of 13,000 Btu per lb as fired?

2. An oil-fired boiler used 2.1 gal of No. 2 furnace oil (higher heating value 140,000 Btu per gal) per hr to give a gross $I=B=R$ output of 220,000 Btuh. What efficiency of operation was obtained? If the net $I=B=R$ rating was 145,000 Btuh, what total pickup and piping-loss factor was allowed?

3. A steel heating boiler for hand-fired operation has an SBI net rating of 2600 sq ft of equivalent steam radiation. What must be the amount of its heating surface?

4. A steel heating boiler intended for steam production and arranged for stoker firing has an SBI rating of 6500 sq ft, edr. What amount of heating surface is required for the rating as stated? What is the SBI net rating of the unit in Btuh?

5. A cast-iron heating boiler operated at 1.5 psig when the barometric pressure was 29.6 in. of mercury at 32 F. During a period of 8 hr 8000 lb of steam of 0.99 quality were formed from water supplied at a temperature of 200 F. Find the boiler capacity that was developed, in square feet, edr, and also in boiler horsepower. Calculate the over-all efficiency of operation if 6.5 lb of water were evaporated per each pound of 10,880 Btu per lb coal used during the test.

6. A stoker-fired cast-iron hot-water boiler heated 9000 lb of water from 170 to 195 F in 1 hr. The weight of coal used per hour was 31 lb, and its heating value as fired was 11,200 Btu per lb. Find the over-all efficiency of operation of the unit and the gross output in Btuh. Compute the net rating of the unit using a piping loss and pickup factor of 1.3.

7. A residence with heat losses of 110,000 Btuh is to be heated by a forced hot-water system using a boiler burning No. 2 furnace oil (heating value, 140,000 Btu per gal). Select an $I=B=R$ rated boiler for this installation from Table 9.1. What is the over-all efficiency of the unit?

8. A commercial building which is to be heated continuously has calculated heat losses of 950,000 Btuh. The piping will be typical for buildings of this type. Select by number the boiler best suited to this installation using the data of Table 9.2. What reserve capacity will the unit have in per cent of

HEATING BOILERS AND APPURTENANCES

the calculated load? What over-all efficiency is indicated by the tabular data when the unit is fired with gas? What efficiency is indicated when oil is burned (assume No. 4 oil, heating value of which may be taken as 150,000 Btu per gal)?

9. An oil-fired boiler is to heat a building having heat losses of 350,000 Btuh in the coldest weather. Domestic requirements of heated water are 200 gal per hr raised in temperature from 55 to 160 F. Allow 20 per cent piping losses in the heating system and 5 per cent losses between the boiler and the water heater. How many gallons of No. 2 fuel oil (140,000 Btu per gal) are required per hour if the boiler can be operated with 70 per cent efficiency?

10

Pipe, Tubing, Fittings, Coverings, and Piping Details

10.1. Pipe Materials and Their Applications. In heating and air-conditioning equipment both pipes and tubes are used for the conveyance of fluids. Materials used for pipes include wrought iron, mild steel, cast iron, aluminum, copper, and brass. For the most part nonferrous materials resist corrosion better than those made from iron. Copper tubing has advantages in that it may be bent readily and lengths of it can be easily fastened together with either solder or compression fittings. Thin-wall steel tubing has properties which allow it to be bent and welded together and it is not subject to the corrosive action of ammonia as are copper and brass.

10.2. Commercial Pipe. Butt-welded, resistance-welded, and lap-welded steel and wrought-iron pipes are manufactured in sizes which are pertinent to their diameters. For pipes up to 12 in. inclusive the sizes are based on the *nominal* inside diameters. Pipes designated as 14 in. and larger are listed on the basis of their outside diameters. All steel and wrought-iron pipes of the various sizes have external diameters which are standard for them irrespective of the thickness of the metal of their walls. This is necessary in order that the same thread-cutting equipment for each size, regardless of its wall thickness, may be used.

PIPE, TUBING, FITTINGS, AND COVERINGS

The welding of pipe sections together and the use of welding-type fittings permits the use of thinner metal sections than when threading operations are necessary. At one time pipes were designated only by the terms standard weight, extra heavy, and double extra heavy according to their abilities to withstand internal pressures. Now the American Standard for Wrought-Iron and Wrought-Steel Pipe[1] lists the various commercial sizes of pipes under different schedules depending on their wall thickness. Schedules 30 and 40 of the American Standards Association conform most closely in dimensions to those formerly classified as standard-weight pipes. Data pertinent to these schedules are given in Table 10.1 as these are suitable for most uses with the low pressures of heating installations. The dimensions for wrought-iron pipes are approximately the same as in Table 10.1, except that the wall thickness is slightly greater. Additional information

TABLE 10.1. Dimensions of Schedules 30 and 40, Welded Steel Pipe for Steam, Air, Gas, and Water*

Nominal Size, In.	Inside Diameter, In.	External Diameter, In.	Nominal Thickness, In.	Transverse Areas Internal Sq In.	Transverse Areas External, Sq In.	Transverse Areas Metal, Sq In.	Length of Pipe Per Square Foot of External Surface, Ft	Length of Pipe Containing One Cubic Foot, Ft	Nominal Weight per Foot Plain Ends, Lb	Number of Threads per Inch of Screw
1/8	0.269	0.405	**0.068**	0.057	0.129	0.072	9.431	2533.775	0.244	27
1/4	0.364	0.540	**0.088**	0.104	0.229	0.125	7.073	1383.789	0.424	18
3/8	0.493	0.675	**0.091**	0.191	0.358	0.167	5.658	754.360	0.567	18
1/2	0.622	0.840	**0.109**	0.304	0.554	0.250	4.547	473.906	0.850	14
3/4	0.824	1.050	**0.113**	0.533	0.866	0.333	3.637	270.034	1.130	14
1	1.049	1.315	**0.133**	0.864	1.358	0.494	2.904	166.618	1.678	11½
1¼	1.380	1.660	**0.140**	1.495	2.164	0.669	2.301	96.275	2.272	11½
1½	1.610	1.900	**0.145**	2.036	2.835	0.799	2.010	70.733	2.717	11½
2	2.067	2.375	**0.154**	3.355	4.430	1.075	1.608	42.913	3.652	11½
2½	2.469	2.875	**0.203**	4.788	6.492	1.704	1.328	30.077	5.793	8
3	3.068	3.500	**0.216**	7.393	9.621	2.228	1.091	19.479	7.575	8
3½	3.548	4.000	**0.226**	9.886	12.566	2.680	0.954	14.565	9.109	8
4	4.026	4.500	**0.237**	12.730	15.904	3.174	0.848	11.312	10.790	8
5	5.047	5.563	**0.258**	20.006	24.306	4.300	0.686	7.198	14.617	8
6	6.065	6.625	**0.280**	28.891	34.472	5.581	0.576	4.984	18.974	8
8†	8.071	8.625	0.277	51.161	58.426	7.265	0.443	2.815	24.696	8
8	7.948	8.625	**0.322**	50.027	58.426	8.399	0.443	2.878	28.554	8
10†	10.136	10.750	0.307	80.691	90.763	10.072	0.355	1.785	34.240	8
10	10.020	10.750	**0.365**	78.855	90.763	11.908	0.355	1.826	40.483	8
12†	12.090	12.750	0.330	114.800	127.676	12.876	0.299	1.254	43.773	8
12	12.000	12.750	0.375	113.097	127.676	14.579	0.299	1.273	49.562	8

* Nominal thickness given in bold-faced type is identical with that for schedule 40.
† Thickness same as for schedule 30.

relative to steel and wrought-iron pipes is given in two publications[2, 3] of the American Society for Testing Materials.

Brass and copper pipes from ⅛ to 12 in. are also listed with outside diameters equal to those given in Table 10.1. Such pipes can be used with fittings conforming in size to those suitable for steel and wrought-iron standard pipe. The internal diameters of brass pipe for the most part differ slightly from those of standard steel pipe. Aluminum pipe with ips, iron pipe size, welding fittings is finding use in chemical, food, soap, petroleum, and cosmetic plants. The cost is less than that of either brass or copper pipe.

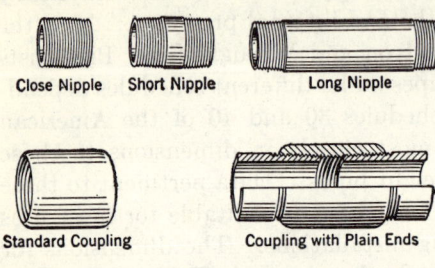

Fig. 10.1. Pipe nipples and couplings.

10.3. Pipe Threads. Pipe threads and the tappings of the various threaded fittings for commercial pipe are standardized and have a taper of 1 in 16 with the longitudinal axis of the pipe. The threads are

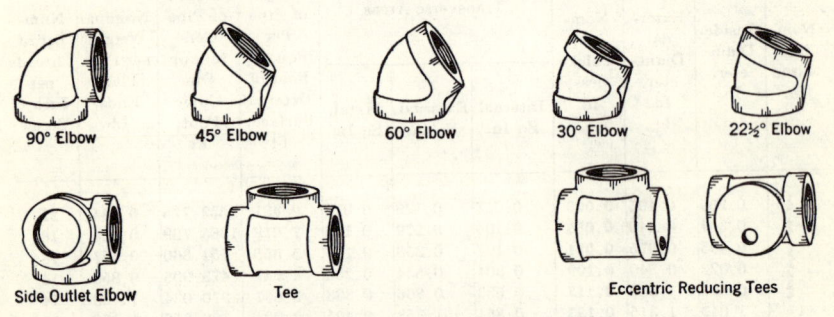

Fig. 10.2. Screwed pipe fittings.

V-shaped with a 60-deg angle between them and are flattened at their tops at the larger end of the taper. The function of the taper is to produce a very tight joint when the threads draw together the parts to be joined.

10.4. Pipe Fittings for Standard Steel and Wrought-Iron Pipe. These parts, for joining lengths of pipe together and providing outlet connections, reductions of pipe size, etc., are made in two forms. The screw type of fitting is usual for pipes less than 4 in. in diameter and the flanged type of fitting for pipes 4 in. in diameter and larger.

Materials for fittings include cast iron, malleable iron, steel, and

PIPE, TUBING, FITTINGS, AND COVERINGS 191

brass, the choice depending on the service and pressure. The weight of fittings should conform to the weight of pipe; thus standard fittings are used with standard pipe. Close and shoulder nipples (short and long) and ordinary couplings, Fig. 10.1, are made either of steel or wrought iron. Reducing nipples and double-reduction nipples, which are respectively one or two standard pipe sizes smaller at one end than

Bronze to iron seat
Nut union

Gasket type
Nut union

Flange union

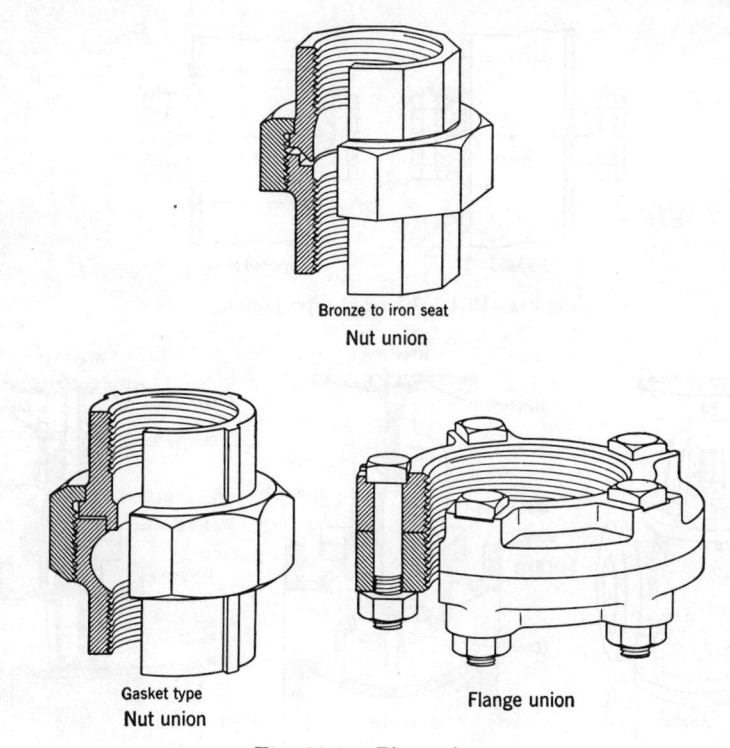

Fig. 10.3. Pipe unions.

at the other, are now available. The names of common fittings and their appearances are indicated by the various details of Fig. 10.2.

The size of fittings is always designated in accordance with the pipe with which they are to be used. The tappings of fittings, such as elbows, tees, and couplings, may be uniform in size; or one or more of the tappings may be smaller, when the term *reducing fitting* is applied.

Provision must always be made when two immobile objects are fastened together, by pipe connections, for the final joining of the last two sections of pipe. With small pipes this can be accomplished by means of couplings having a right-hand thread in one end and a left-hand thread in the other. These couplings usually have a distinguishing mark such as longitudinal ridges on their exterior. Unions or companion flanges are more satisfactory than fittings with right- and left-hand threads.

Unions, Fig. 10.3, are of two different forms, nut unions and flange unions. Nut unions consist of three parts which are the two parts

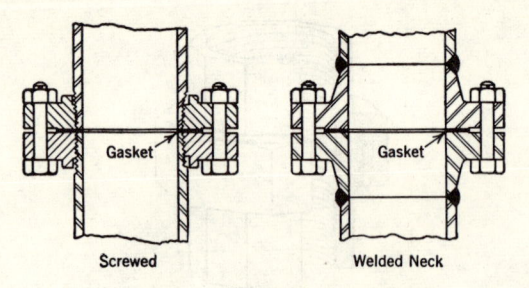

Fig. 10.4. Flanged pipe joints.

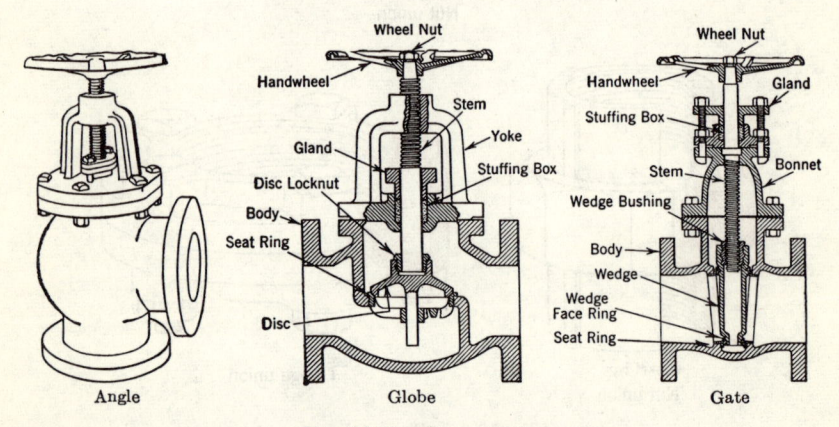

Fig. 10.5. Stop valves.

attached to the pipe ends by threads and the threaded nut which is used to draw the seat parts together. Depending upon their construction, nut unions may or may not require a gasket where their seats fit together. A flange union consists of two parts which are attached by threads to the pipe sections to be joined together. A gasket is placed between the flanges which are drawn together by bolts, or the flange union may have ground seats which do not require a gasket. Nut unions are made in all pipe sizes up to 4 in. inclusive. For pipe sizes above 2 in. flange unions are generally preferred.

Examples of pipes joined together with flanges are shown in Fig. 10.4. Flanged joints usually require a gasket of some material

PIPE, TUBING, FITTINGS, AND COVERINGS 193

between the flange faces inside of the bolt circle. Important dimensions of all of the common screwed and flanged fittings may be obtained from the ASHAE Guide.

10.5. Valves. The stop valves used in heating pipe lines may be angle, globe, or gate types, Fig. 10.5, and made of brass or cast iron. Gate valves are usually made only in the straightaway pattern; globe-valve construction may be either the straightaway or the angle type.

Swing

Horizontal　　Vertical

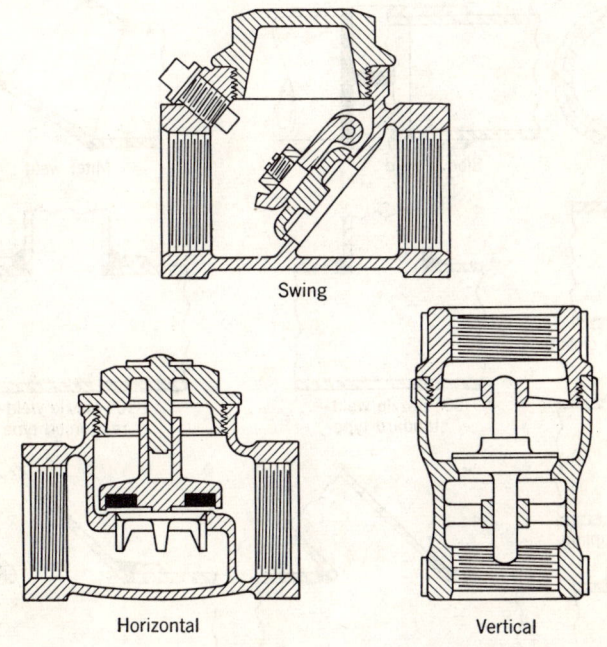

Fig. 10.6. Cross sections of check valves. (Courtesy Crane Co.)

Gate valves are desirable as they offer the least amount of obstruction to the flow of fluids. Globe valves secure a throttling action by partially closing the valve, and are consequently best where regulation of the flow is necessary. The use of a globe valve, with its stem vertical, in a horizontal steam line is not recommended as it creates an obstruction in the pipe which will cause an accumulation of water to choke the flow of steam. Furthermore, complete drainage of the line is impossible.

Gate valves may have either a rising or a nonrising stem; globe valves always have a rising stem. The rising stem requires more space but has the advantage of giving positive indication of the valve opening. All the valves of Fig. 10.5 require packing of some kind within the stuffing boxes of the bonnets and around the valve stems to prevent leakage of the fluid passing through them.

Check valves serve to prevent backward flow in a pipe line. The check valves of Fig. 10.6 are for horizontal and vertical pipes. The

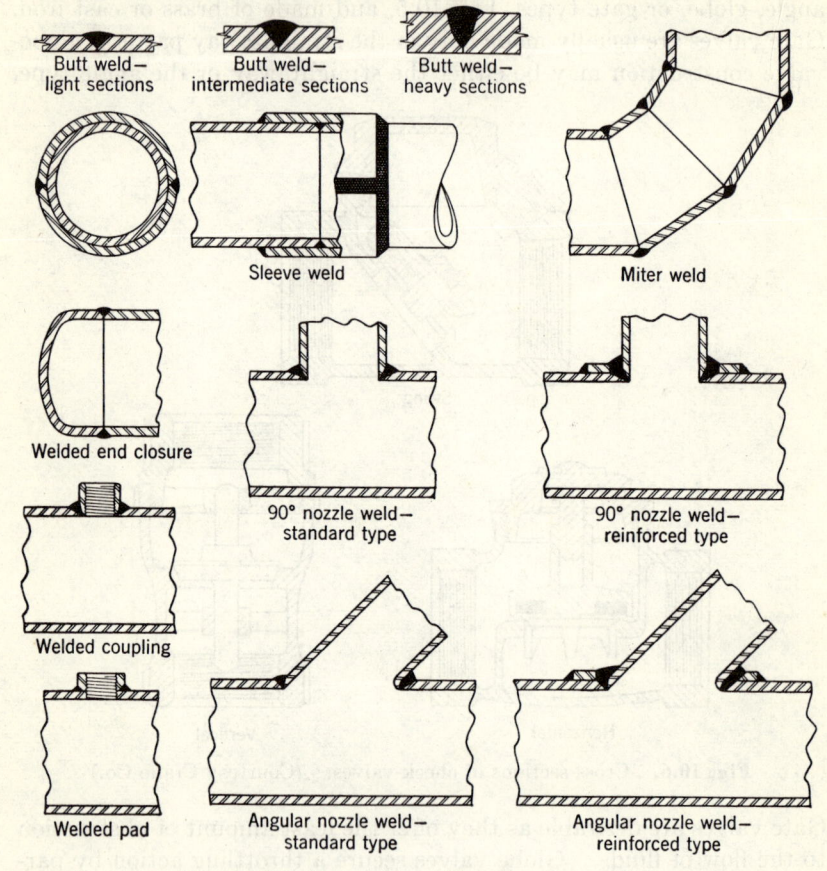

Fig. 10.7. Pipe welds. (Courtesy Crane Co.)

bodies of check valves are commonly made of brass in the smaller sizes and of cast iron in the larger ones.

Valves designed for special purposes such as blowoff and needle types may be obtained from the principal manufacturers. Gate, globe, and check valves, in which critical parts are made of special materials, are available for handling fluids that are severely corrosive and those which must be handled at high temperatures. Reference to available literature[4-6] on piping systems should be made when the selection of valves

PIPE, TUBING, FITTINGS, AND COVERINGS

to handle fluids under unusual conditions is necessary. In some cases the employment of a competent specialist may be advisable.

10.6. Welded Pipe Joints. Considerable savings in materials and in labor costs can often be made by welding together pipe sections and fittings. Figure 10.7 shows several applications of welds in fluid-handling systems. Butt joints are usually used in joining two lengths

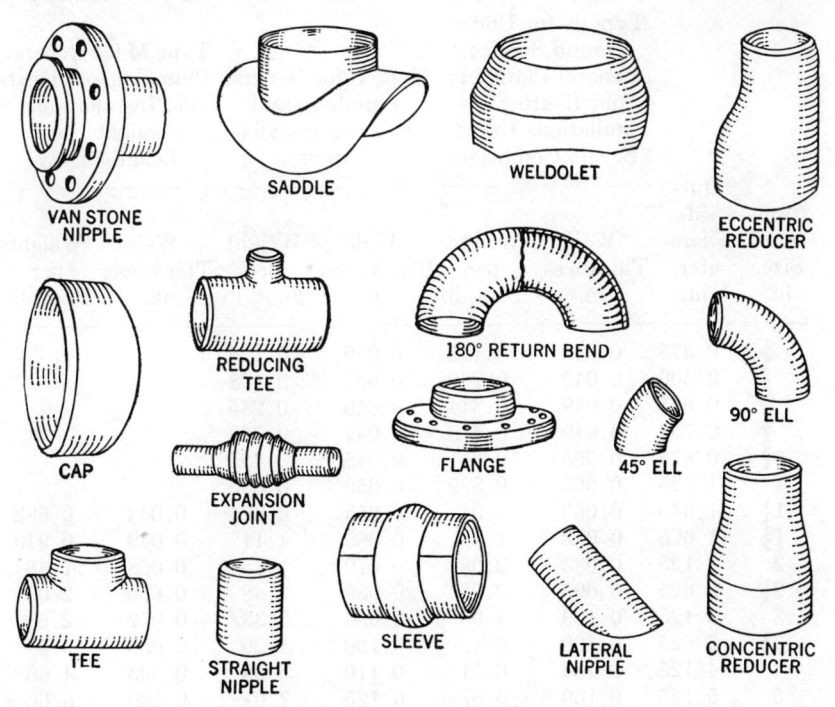

Fig. 10.8. Welding fittings.

of pipe. Commercial welding fittings may have ends provided for butt welding or the socket-weld joint may be used. Elbows may be made of short lengths of pipe, or they may be purchased fabricated as a seamless tube. Valves and flanges may also be obtained for welding into steam or other lines. Several commercial welding fittings are depicted in Fig. 10.8. Dimensions for commonly used welding fittings have been standardized and are available from references 7 and 8.

10.7. Steel Tubing. The material used in ferrous tubing may be either of carbon or alloy steel. Generally the listed sizes of steel tubes are based on their external diameters. The wall thickness of steel tubes varies with their external diameters and the pressures which hey must withstand. Tubes of small size may be readily bent, and lengths of them, having thin walls, may be joined together by welding. Applications of such pipe construction are in condensers of various kinds, flues of fire-tube boilers, and water passes of water-tube boilers.

TABLE 10.2. Copper Tubing for General Plumbing and Heating

Nominal Size, in.	Outside Diameter, in.	Type K for Underground Service, General Plumbing and Heating Installations Under Severe Conditions		Type L for General Plumbing and Heating Installations		Type M for General Plumbing and Heating Installations, Straight Hard Lengths Only	
		Wall Thickness, in.	Weight per Foot, lb	Wall Thickness, in.	Weight per Foot, lb	Wall Thickness, in.	Weight per Foot, lb
1/4	0.375	0.035	0.145	0.030	0.126		
3/8	0.500	0.049	0.269	0.035	0.198		
1/2	0.625	0.049	0.344	0.040	0.285		
5/8	0.750	0.049	0.418	0.042	0.362		
3/4	0.875	0.065	0.641	0.045	0.455		
1	1.125	0.065	0.839	0.050	0.655		
1 1/4	1.375	0.065	1.04	0.055	0.884	0.042	0.682
1 1/2	1.625	0.072	1.36	0.060	1.14	0.049	0.940
2	2.125	0.083	2.06	0.070	1.75	0.058	1.46
2 1/2	2.625	0.095	2.93	0.080	2.48	0.065	2.03
3	3.125	0.109	4.00	0.090	3.33	0.072	2.68
3 1/2	3.625	0.120	5.12	0.100	4.29	0.083	3.58
4	4.125	0.134	6.51	0.110	5.38	0.095	4.66
5	5.125	0.160	9.67	0.125	7.61	0.109	6.66
6	6.125	0.192	13.90	0.140	10.20	0.122	8.92
8	8.125	0.271	25.90	0.200	19.30	0.170	16.50
10	10.125	0.338	40.30	0.250	30.10	0.212	25.60
12	12.125	0.405	57.80	0.280	40.40	0.254	36.70

10.8. Copper Tubing. Seamless tubing for heating, plumbing, and air-conditioning installations is made of either soft and ductile or hard-drawn copper. Copper tubing is standardized[9] as Type K, L, or M, depending upon the thickness of the wall material. For a given tube size, Type K has the thickest and Type M the thinnest wall.

Data relative to the dimensions and weights of copper tubing are given in Table 10.2.

All Type-M copper tubing is hard drawn or has a hard temper.

PIPE, TUBING, FITTINGS, AND COVERINGS

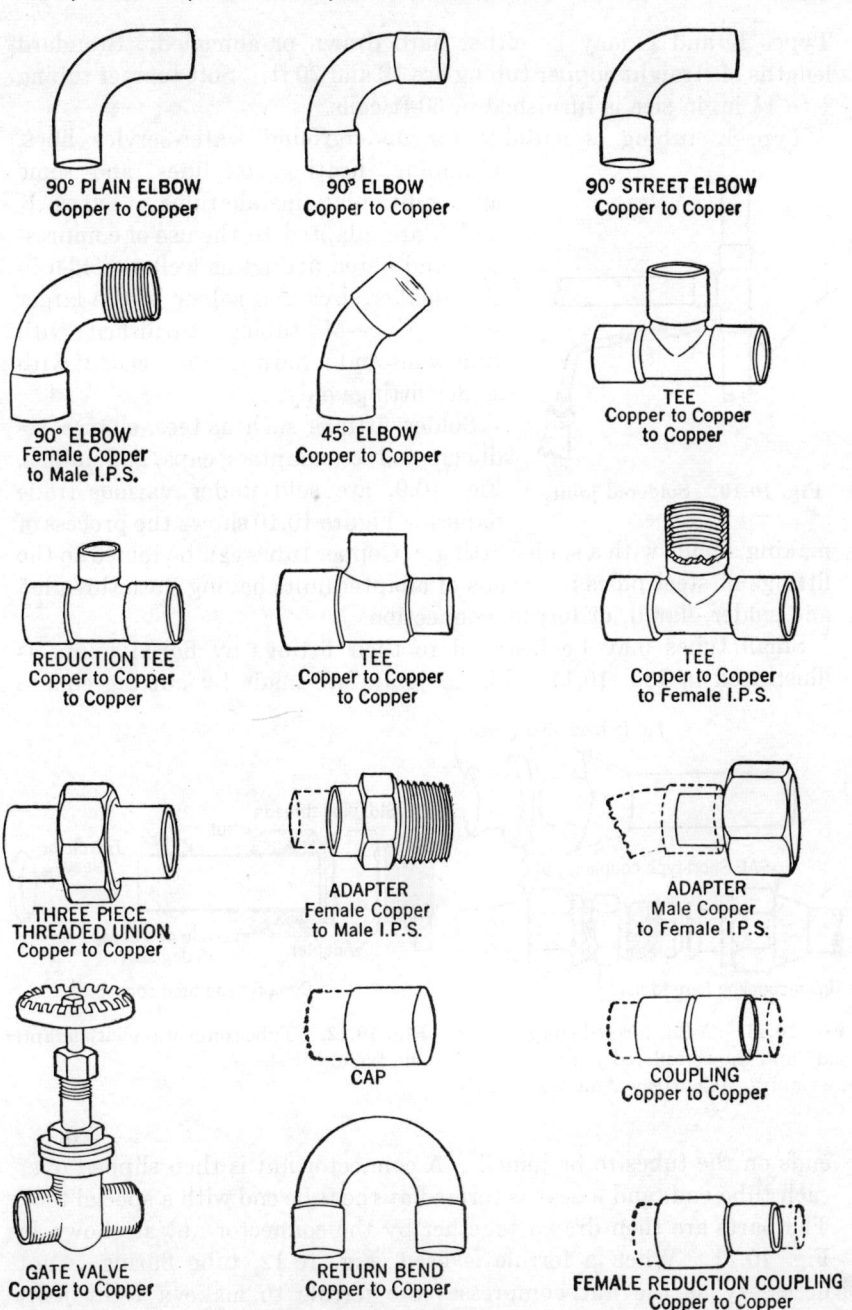

90° PLAIN ELBOW
Copper to Copper

90° ELBOW
Copper to Copper

90° STREET ELBOW
Copper to Copper

90° ELBOW
Female Copper
to Male I.P.S.

45° ELBOW
Copper to Copper

TEE
Copper to Copper
to Copper

REDUCTION TEE
Copper to Copper
to Copper

TEE
Copper to Copper
to Copper

TEE
Copper to Copper
to Female I.P.S.

**THREE PIECE
THREADED UNION**
Copper to Copper

ADAPTER
Female Copper
to Male I.P.S.

ADAPTER
Male Copper
to Female I.P.S.

CAP

COUPLING
Copper to Copper

GATE VALVE
Copper to Copper

RETURN BEND
Copper to Copper

FEMALE REDUCTION COUPLING
Copper to Copper

Fig. 10.9. Copper-tube fittings.

198 AIR CONDITIONING AND REFRIGERATION

Types K and L may be either hard drawn or annealed. Standard lengths of straight copper tubing are 12 and 20 ft. Soft temper tubing $\frac{1}{4}$ to $1\frac{1}{2}$ in. in size is furnished in 60-ft coils.

Type-K tubing is suitable for underground water-service lines, plumbing, heating, oil lines, and some air-conditioning installations. Types K and L are adapted to the use of compression and flared fittings as well as solder in the smaller sizes and solder in the larger sizes. Type-M tubing, furnished with thin walls and a hard temper, is used with solder fittings only.

Solder fittings such as tees, elbows, reducers, valves, adapters, caps, and unions, Fig. 10.9, are sold under various trade names. Figure 10.10 shows the process of making a joint with a solder fitting. Copper tubes can be joined to the fittings of steel pipes by means of adapter units having both threaded and solder, flared, or ferrule connections.

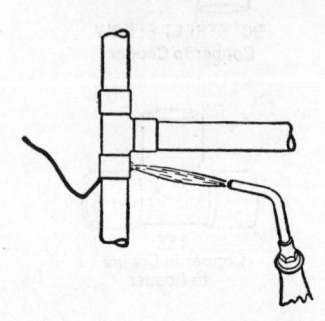

Fig. 10.10. Soldered joint.

Small tubes may be fastened to their fittings by flared joints as illustrated in Fig. 10.11. Flared joints are made by cutting square

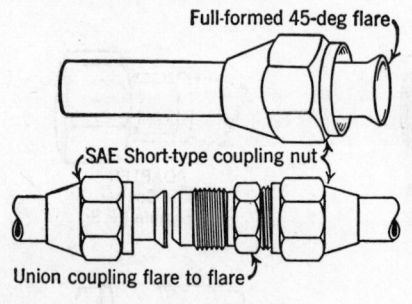

Fig. 10.11. A full-formed copper-tube flare and the completed joint ready for assembly. (Courtesy American Brass Co.)

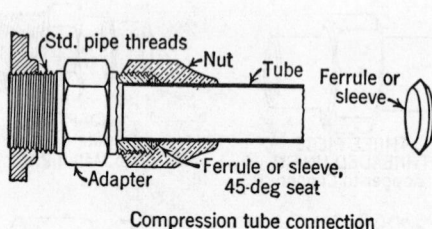

Fig. 10.12. Tube connection with adapter and ferrule or sleeve.

ends on the tubes to be joined. A connector nut is then slipped over each tube end, and a seat is formed at the tube end with a special tool. The parts are then drawn together by the connector nut as shown in Fig. 10.11. When a ferrule is used, Fig. 10.12, tube flaring is not necessary as the nut compresses the ferrule to make a tight joint between it and the tube and also where the ferrule bears on 45-deg seats of the parts joined.

PIPE, TUBING, FITTINGS, AND COVERINGS

10.9. Steam Separators and Traps. A separator is a piece of equipment for removing water particles from steam as it flows in a pipe line or for removing oil from the steam exhausted by engines. The separator illustrated by Fig. 10.13 causes a change of direction of flow to aid in the removal of the water particles. The water collects in the bottom chamber.

Traps are devices which permit the removal of collected water without the loss of steam from heaters, separators, and steam-pipe lines. Traps may be classified as float, bucket, bowl, and thermostatic. The float trap of Fig. 10.14a operates as water entering at I accumulates and causes the float C to lift to open valve F. The steam pressure then forces water through the outlet H until the float is lowered to close valve F. The trap of Fig. 10.14b has a thermostatic element at E to allow the discharge of air. The bucket trap shown in section by Fig. 10.14c has a bucket which floats and holds the valve

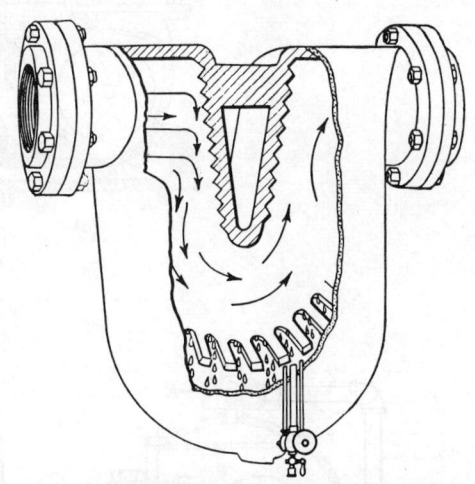

Fig. 10.13. Steam separator.

against the seat of the outlet until the water overflows into the bucket. When this occurs the bucket sinks, and steam pressure causes a discharge of water through the outlet until the bucket rises again to close the discharge opening. This trap also has a thermostatic element E which permits the discharge of air.

Other types of traps which compete with either float or bucket devices in classes of service where large volumes of condensate must be handled intermittently are: inverted traps, flash traps, impulse traps, and tilting traps. Thermostatic radiator traps are described in Chap. 11.

10.10. Pressure-Reducing Valves. When steam is drawn from high-pressure lines to heat buildings, its pressure must be reduced to the working pressure of the system. This is accomplished by means of a pressure-reducing valve, one form of which is shown in Fig. 10.15. A pressure equal to that desired is maintained beneath the diaphragm by means of the balance-pipe connection between the low-pressure main and the valve. This pressure can be controlled by adjusting the position of the spring with respect to the diaphragm. The valve should be installed with a bypass connection so that service by means of hand operation may be maintained when the pressure-reducing valve is being repaired.

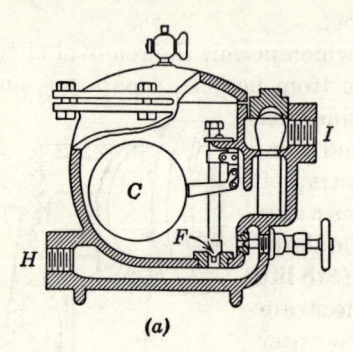

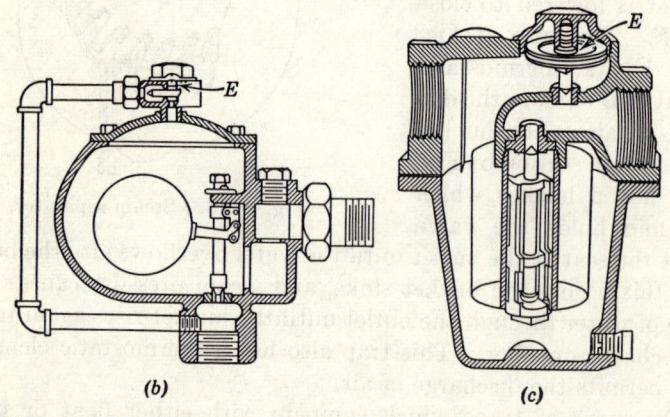

Fig. 10.14. (*a*) Float trap. (*b*) Float and thermostatic drip trap. (*c*) Bucket and thermostatic drip trap.

10.11. Pipe Insulation. The type of insulation which is proper for any particular pipe depends principally on the temperature of the fluid which flows through it. Column two in Table 10.3 indicates the maximum temperature of the fluid to be handled in the case of each of the insulation types mentioned.

Coverings which are commonly used on steam and hot-water lines include multiple corrugated asbestos paper, Fig. 10.16, and magnesium carbonate mixed with asbestos fibers which is widely known as "85 per cent magnesia."

PIPE, TUBING, FITTINGS, AND COVERINGS

Insulation placed on pipes handling cold fluids must be provided with a vapor barrier at its outer surface to prevent water vapor in the surrounding air from condensing within it.

Pipe fittings of various sorts are enclosed with special molded coverings made to fit them or with an insulating cement containing asbestos fibers. This cement is applied with a trowel, and the finished covering is wrapped with canvas to give additional strength and protection to the covering. Boilers may be insulated with magnesia blocks and an external coat of insulating cement and canvas, or they may be covered with 2 in. of plastic cement reinforced with wire netting, the final outside surface being finished with a thickness of canvas.

The efficiency of pipe covering is expressed as one minus the ratio of the heat actually lost from the pipe through the covering to that which would have been lost under the same conditions of inside-steam and outside-air temperature from uncovered pipe surface.

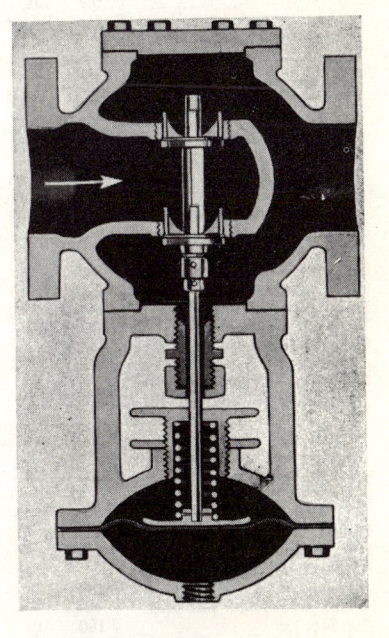

Fig. 10.15. Spring-controlled pressure-reducing valve. (Courtesy Illinois Engineering Co.)

10.12. Heat Losses from Steel Pipes. The heat losses from either bare or insulated pipes may be computed by means of fundamental heat-transfer equations[10] and the information given in a number of publications[11-14] which deal with the subject. Table 10.4 gives data for the heat losses from steel pipes carrying either hot water or steam when bare and in a horizontal position. Data covering a wider range of temperature differences and pipe sizes may be found in the ASHAE Guide.

Pipe insulations vary in effectiveness per unit of thickness and with the actual thickness that is applied. Therefore, it is not possible to give heat-loss data for insulated pipes in a single table. In the case of high-temperature steam the common practice is to use one type of insulation next to the pipe and a less expensive one adjacent to the surrounding air. However, in heating work only a single kind of insu-

TABLE 10.3. Thermal Conductivity k of Pipe Insulations (for Mean Temperatures Indicated)*

Expressed in Btuh per sq ft per deg F temperature difference per inch of thickness

Material (Composition)	Accepted Max. Temp. for Use, deg F†	Density, lb per cu ft	\multicolumn{8}{c}{Typical Conductivity k at Mean Temperature, deg F}								
			40	70	100	200	300	500	700	900
Asbestos										
Molded amosite & binder	1200	16			0.33	0.38	0.43	0.53		
Laminated asbestos paper	700	30			0.40	0.45	0.50	0.60		
Corrugated & laminated asbestos paper										
4 ply per in.	300	11–13		0.54	0.57	0.62	0.80			
6 ply per in.	300	15–17		0.49	0.51	0.59	0.69			
8 ply per in.	300	18–20		0.47	0.49	0.57	0.65			
Calcium silicate										
Calc. sil. & asbestos	1200	11			0.36	0.40	0.44	0.55		
Cellular glass	800	9	0.37	0.39	0.41	0.48	0.55			
Cork (without added binder)	200	7–10	0.27	0.28	0.29	0.30				
Diatomaceous silica	1500	22						0.64	0.66	0.71
	1900	25						0.70	0.75	0.80
85% Magnesia										
Mag. carb. & asbestos	600	11–14			0.39	0.42	0.45	0.51		
Mineral wool (rock, slag, or glass)										
Low temp. (asphalt or resin bonded)	200	15	0.28	0.30	0.33	0.39				
Low temp. (fine fiber resin bonded)	450	3	0.22	0.23	0.24	0.27	0.31			
High temp. blanket-type (metal reinforced)	1200	6–15			0.29	0.36	0.42	0.56		
Plastics (foamed)	175	1.6	0.26	0.28	0.31					
Rubber (foamed)	150	5	0.23	0.24	0.25					
Vegetable & animal fiber										
Wool felt	180	20	0.29	0.31	0.33					
Hair felt or hair felt plus jute	180	10	0.27	0.28	0.30					
85% magnesia cement (for fittings)	600	18			0.46	0.52	0.58			

* Extracted with permission from Table 6, page 657 of "Heating Ventilating Air-Conditioning Guide 1956."

† These temperatures are generally accepted as maximum. When operating temperature approaches these limits the manufacturer's recommendations should be followed.

lation is used. The hourly heat losses through a homogeneous layer of insulation placed on a round pipe are

$$H = \frac{k(t_h - t_c)}{r_o \log_e (r_o/r_i)} \quad (10.1)$$

where r_o = outer radius of the insulation, inches, and r_i = inner radius (outer radius of the pipe). Conductivity k values for pipe insulations are given in Table 10.3.

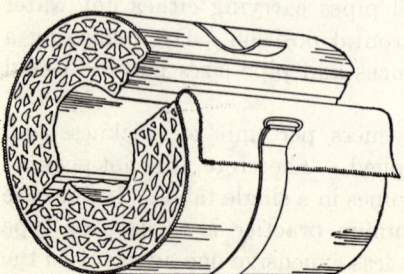

Fig. 10.16. Cellular pipe covering.

TABLE 10.4. Heat Losses from Horizontal Bare Steel Pipes*
Btuh per linear ft per deg F between the pipe
temperature and still air at 70 F surrounding it

	Hot Water				Steam		
Nominal pipe Size, in.	120 F	150 F	180 F	210 F	227.1 F, 5 psig	297.7 F, 50 psig	337.9 F, 100 psig
	Temperature Difference						
	50 F	80 F	110 F	140 F	157.1 F	227.7 F	267.9 F
$\frac{1}{2}$	0.455	0.495	0.546	0.584	0.612	0.706	0.760
$\frac{3}{4}$	0.555	0.605	0.666	0.715	0.748	0.866	0.933
1	0.684	0.743	0.819	0.877	0.919	1.065	1.147
$1\frac{1}{4}$	0.847	0.919	1.014	1.086	1.138	1.324	1.425
$1\frac{1}{2}$	0.958	1.041	1.148	1.230	1.288	1.492	1.633
2	1.180	1.281	1.412	1.512	1.578	1.840	1.987
$2\frac{1}{2}$	1.400	1.532	1.683	1.796	1.883	2.190	2.363
3	1.680	1.825	2.010	2.153	2.260	2.630	2.840
$3\frac{1}{2}$	1.900	2.064	2.221	2.433	2.552	2.974	3.215
4	2.118	2.302	2.534	2.717	2.850	3.320	3.590
5	2.580	2.804	3.084	3.303	3.470	4.050	4.385
6	3.036	3.294	3.626	3.886	4.074	4.765	5.160
8	3.880	4.215	4.638	4.960	5.210	6.100	6.610
10	4.760	5.180	5.680	6.090	6.410	7.490	8.115
12	5.590	6.070	6.670	7.145	7.500	8.800	9.530

* From "Heating Ventilating Air-Conditioning Guide 1956." Used by permission.

Example. Estimate the hourly heat losses per linear foot of 2-in. steel pipe covered with canvas-wrapped 85 per cent magnesia (magnesium carbonate and asbestos) insulation $1\frac{1}{2}$ in. thick when it conveys 200 F hot water in a basement where the temperature of the surrounding air at the pipe level is 70 F.

Solution. Equation 10.1 is applicable. By interpolation of the data of Table 10.3 k is found to be 0.40 when the temperature of the insulation is assumed as the arithmetic mean between that of the water and the estimated surface temperature (90 F will be used as a preliminary estimate). Other data for use in equation 10.1 are: $t_h = 200$ F, t_c by estimation is 90 F, from Table 10.1 r_i is 2.375/2 = 1.187 in., and r_o equals 1.187 + 1.5 = 2.687 in. Hence, $H = [0.40(200 - 90)]/[2.687 \log_e (2.687/1.187)] = (0.40 \times 110)/(2.687 \times 0.816) = 20.5$ Btuh per sq ft of outer surface. Figure 10.17 shows that the temperature difference between the canvas covering and the room air is about 13 F when the heat loss is 20.5 Btuh per sq ft. The temperature of the surface is therefore of the order of 83 F instead of 90 F. However, the heat flow through the insulation will be greater with a lower surface temperature. The surface temperature will therefore be estimated as 84 F for a second-trial solution. $H = (0.40 \times 116)/(2.687 \times 0.816) = 21.15$ Btuh per sq ft, which closely checks the loss shown by Fig. 10.17 for a surface-to-air temperature difference of 14 F. The heat losses per linear foot of pipe are $(21.15 \times 2 \times 3.1416 \times 2.687)/12 = 29.85$ Btuh.

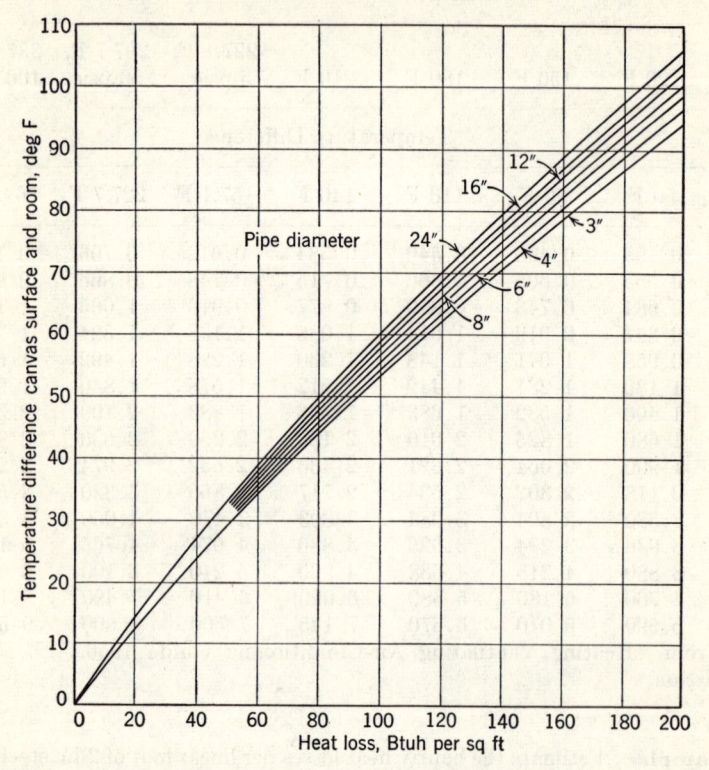

Fig. 10.17. Heat loss from canvas-covered cylindrical surfaces or various diameters. (Data and general form from "Heating Ventilating Air-Conditioning Guide 1957." Used by permission.)

10.13. Heat Losses from Pipe Fittings. In order to estimate heat losses from pipe fittings their external areas must be known. These depend upon the design of the fittings, but areas of Table 10.5 may be taken as representative. Calculations of the heat lost from an insulated fitting are possible by the method of the preceding article. An estimation of the average diameter of the fitting is required along with a determination of the average diameter of the insulation placed about it. Such preliminary steps require a careful study of the details of each fitting. For practical purposes an assumption that the heat loss per square foot of metal surface is the same as that from the insulated pipe in which the fitting is placed is practical.

PIPE, TUBING, FITTINGS, AND COVERINGS

10.14. Heat Losses from Copper Tubing. The heat losses from bare copper tubing are considerably less than those from steel pipes under the same operating conditions. Data for the heat losses from tarnished copper pipe or tubing are given in Table 10.6. The losses from copper pipe having its original bright finish are somewhat

TABLE 10.5.* Areas of Flanged Fittings, Square Feet†

Nominal Pipe Size, In.	Flanged Coupling Standard	Flanged Coupling Extra Heavy	90-Deg Ell Standard	90-Deg Ell Extra Heavy	Long Radius Ell Standard	Long Radius Ell Extra Heavy	Tee Standard	Tee Extra Heavy	Cross Standard	Cross Extra Heavy
1	0.320	0.438	0.795	1.015	0.892	1.083	1.235	1.575	1.622	2.07
1¼	0.383	0.510	0.957	1.098	1.084	1.340	1.481	1.925	1.943	2.53
1½	0.477	0.727	1.174	1.332	1.337	1.874	1.815	2.68	2.38	3.54
2	0.672	0.848	1.65	2.01	1.84	2.16	2.54	3.09	3.32	4.06
2½	0.841	1.107	2.09	2.57	2.32	2.76	3.21	4.05	4.19	5.17
3	0.945	1.484	2.38	3.49	2.68	3.74	3.66	5.33	4.77	6.95
3½	1.122	1.644	2.98	3.96	3.28	4.28	4.48	6.04	5.83	7.89
4	1.344	1.914	3.53	4.64	3.96	4.99	5.41	7.07	7.03	9.24
5	1.622	2.18	4.44	5.47	5.00	6.02	6.81	8.52	8.82	10.97
6	1.82	2.78	5.13	6.99	5.99	7.76	7.84	10.64	10.08	13.75
8	2.41	3.77	6.98	9.76	8.56	11.09	10.55	14.74	13.44	18.97
10	3.43	5.20	10.18	13.58	12.35	15.60	15.41	20.41	19.58	26.26
12	4.41	6.71	13.08	17.73	16.35	18.76	19.67	26.65	24.87	34.11

* "Heating Ventilating Air-Conditioning Guide 1957." Used by permission.

† Including areas of accompanying flanges bolted to the fitting. The area of a valve may be taken as that of a tee.

less than those indicated in Table 10.6. The original bright finish can be retained only by applying a coat of lacquer to the exposed surface. The heat losses from either copper pipes or tubing, when covered with insulation, are determined by the procedure given for steel pipe in Art. 10.12. The dimensions of copper tubing are given in Table 10.2.

10.15. The Installation of Piping. The success of steam or hot-water heating systems depends not only upon the selection of the proper sizes of pipes but also upon the care and skill used in their installation. All internal burrs caused by pipe-cutting tools should be removed by reaming. All joints must be made tight. The joint compound should not be applied to internal threads of fittings as it will be forced inside of the line to cause trouble in return lines and traps. Traps should not be put into operation until the greater portion of dirt and foreign materials has been carried out during the initial operation of the system.

TABLE 10.6. Heat Losses from a Horizontal Tarnished Copper Pipe*

Btuh per linear ft per deg F temperature difference between that of the pipe and the surrounding air at 70 F

Nominal Pipe Size, in.	Hot Water (Type-K Copper Tube)				Steam (Standard Size Pipe)		
	120 F	150 F	180 F	210 F	227.1 F, 5 psig	297.7 F, 50 psig	337.9 F, 100 psig
	\multicolumn{7}{c}{Temperature Difference}
	50 F	80 F	110 F	140 F	157.1 F	227.7 F	267.9 F
$\frac{1}{2}$	0.250	0.287	0.300	0.321	0.433	0.500	0.530
$\frac{3}{4}$	0.340	0.381	0.409	0.429	0.533	0.543	0.654
1	0.440	0.475	0.509	0.536	0.636	0.746	0.803
$1\frac{1}{4}$	0.500	0.559	0.618	0.622	0.764	0.878	0.934
$1\frac{1}{2}$	0.580	0.656	0.710	0.750	0.904	1.053	1.120
2	0.730	0.825	0.890	0.957	1.101	1.273	1.364
$2\frac{1}{2}$	0.880	1.000	1.091	1.143	1.305	1.490	1.605
3	1.040	1.175	1.272	1.343	1.560	1.800	1.940
$3\frac{1}{2}$	1.180	1.350	1.454	1.535	1.750	2.020	2.170
4	1.460	1.500	1.635	1.715	1.941	2.240	2.430
$4\frac{1}{2}$	.....	.....	.....	.....	2.131	2.465	2.650
5	1.600	1.812	1.980	2.071	2.387	2.770	2.990
6	1.840	2.125	2.270	2.430	2.740	3.210	3.440
8	2.400	2.685	2.910	3.110	3.310	4.050	4.370

* From "Heating Ventilating Air-Conditioning Guide 1957." Used by permission.

Pipe lines require adequate support so that they will not sag. Sagging causes pockets in the lines so that they cannot be completely drained. Accumulations of water interfere with the flow of steam. Suitable types of pipe hangers and supports are shown in Fig. 10.18.

All pipes must be properly pitched in order that air and water may be removed. When water accumulations interfere with steam flow the water may be carried with the steam to produce a disagreeable

PIPE, TUBING, FITTINGS, AND COVERINGS

noise known as "water hammer." Water hammer under extreme conditions may cause a rupture of either the pipe or its fittings.

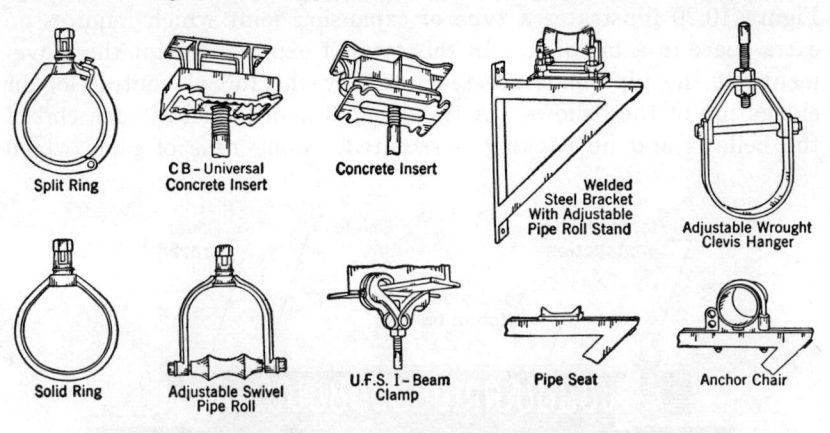

Fig. 10-18. Pipe hangers and supports. (Courtesy Grinnell Co., Inc.)

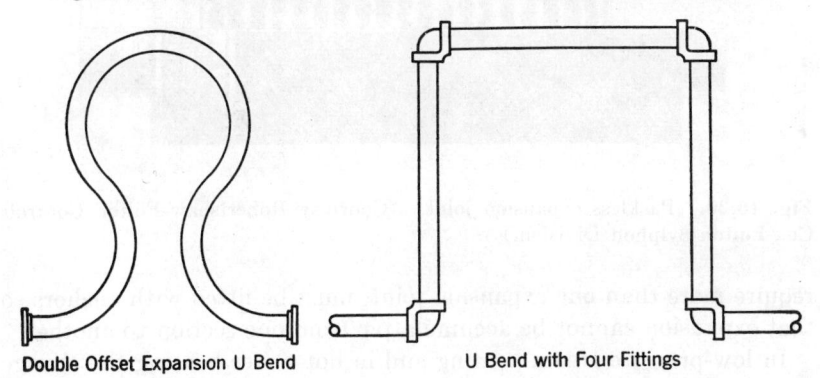

Fig. 10.19. Expansion pipe bends.

10.16. Pipe-Line Expansion. Steam and hot-water pipes are erected with the materials at the temperature of the surrounding air. When such lines convey a heated medium they change in length. In high-pressure lines, the pressure of the material conveyed also affects the length, owing to the elastic properties of the materials. In lines operating with pressures less than 50 psig, the effect of pressure other than its effect on temperature may be ignored.

Provision may be made for pipe-line expansion (1) by the turning of certain fittings on their threads, (2) by the placement of bends in certain pipes, (3) by the use of expansion loops which may be constructed of short pieces of pipe and elbows, and (4) by the use of any of the types of expansion joints.

Dependence is often placed on method 1 in low-pressure heating installations. The scheme illustrated by Fig. 10.19 is self-explanatory. Figure 10.20 illustrates a type of expansion joint which requires no extra space in a building. In this type of expansion joint the movements of the pipe ends are taken up by the forced contraction or elongation of the bellows. A tight seal is maintained at each end of the bellows and no packing is required. Long runs of pipe, which

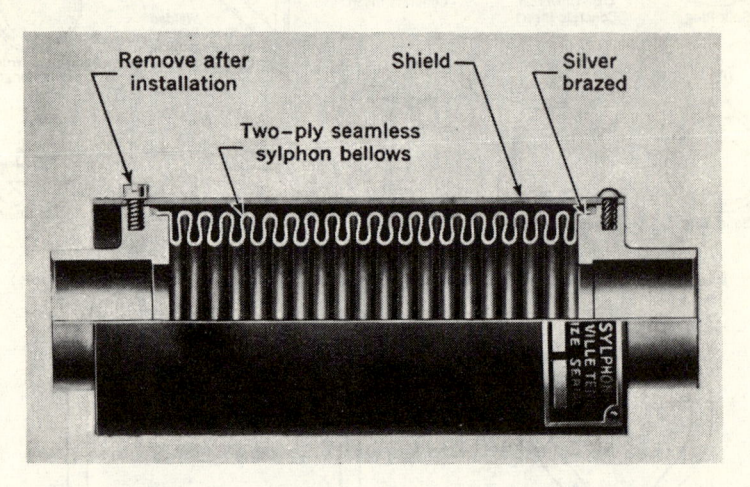

Fig. 10.20. Packless expansion joint. (Courtesy Robertshaw-Fulton Controls Co., Fulton Sylphon Division.)

require more than one expansion joint, must be fitted with anchors so that expansion cannot be accumulative from one section to another.

In low-pressure steam heating and in hot-water heating the change of pipe length may be based on the coefficients of Table 1.4 without allowances for the effects of pressure within the pipe.

10.17. Plastic Pipes. Materials such as polyethelene, polyvinyl and other synthetic resins are available for the construction of pipes and tubes instead of using metal and metal alloys. These nonmetallic conduits for fluid flow are employed in a variety of ways in industries and on farms. Special fittings are available for the connection of lengths of plastic pipe and the fastening of the assemblage to a fluid-handling system constructed of metal pipes. Available lengths of plastic pipes may be joined by means of a process known as chemical welding in which the edges are softened by the application of a solvent before they are brought together.

Plastic pipes resist ordinary corrosion and are not subject to the electrolytic action which sometimes destroys those made of metal. When made of carefully selected and prepared materials, plastic pipes may have a long life when used to convey chemicals which would cause the disintegration of metal pipes after a few days of operation. The friction losses within a plastic pipe are generally less than those in a steel pipe of the same size which handles the same volume of the same fluid. Most pipes of the smaller sizes are flexible so that long-radius bends in a system can be made without the use of either fittings or forming tools.

The uses of plastic pipes are increasing in number, and there is a likelihood that in the future new and as yet untried applications of them will be made. Individuals who have not had experience with the installation of plastic pipes under conditions identical with those to be involved in a new application should seek the advice of the manufacturer of the materials to be used or that of an authority in the matter.

REFERENCES

1. "Wrought-Iron and Wrought-Steel Pipe," *ASA Standard B36.10*, American Standards Association, New York (1950).
2. "Welded and Seamless Steel Pipe," *ASTM Specification A53-52T*, American Society for Testing Materials, Philadelphia (1952).
3. "Welded Wrought-Iron Pipe," *ASTM Specification A72*, American Society for Testing Materials, Philadelphia (1945).
4. "American Standard Code for Pressure Piping," *ASA Standard B31.1*, American Standards Association, New York (1942).
5. "API Specification 5L for Line Piping," American Petroleum Institute, New York, 15th ed. (Mar. 1956).
6. *Piping Handbook*, Sabin Crocker, McGraw-Hill Book Co., New York (1945).
7. "Steel Butt-Welding Fittings," *ASA Standard B16.9*, American Standards Association, New York (1951).
8. "Socket-Welding Fittings," *ASA Standard B16.11*, American Standards Association, New York (1946).
9. "Standard Specifications for Copper Pipe, Standard Sizes," *ASTM Specification B42-49*, American Society for Testing Materials, Philadelphia (1949).
10. *Heat Transmission*, W. H. McAdams, McGraw-Hill Book Co., New York (1954).
11. "Rapid Method of Determining the Economical Thickness of Pipe Insulation," U. W. Smith, *Heating, Piping, Air Conditioning*, 19, No. 10, 118 (1947).
12. *District Heating Handbook*, National District Heating Association, Pittsburgh, 3rd ed. (1951).
13. "Insulation Handbook," P. W. Swain, *Power*, 94, Nos. 2, 3, 5, 7, and 9 (1950).
14. "Heat Transfer Through Thick Insulation on Cylindrical Enclosures," T. S. Nickerson and G. W. Dusinberre, *ASME Trans.*, 70, No. 8, 903 (1948).

BIBLIOGRAPHY

1. *Design of Piping Systems*, Engineers of the M. W. Kellog Company, John Wiley & Sons, New York, 2nd ed. (1956).
2. *Flex-Anal Charts for Design of Piping for Flexibility*, E. A. Wert and S. Smith, Blaw-Knox Company, Pittsburgh (1941).
3. "The Significance of and Suggested Limits for the Stress in Pipe Lines Due to the Combined Effects of Pressure and Expansion," D. B. Rossheim and A. R. C. Markl, *ASME Trans.*, **62**, 443 (1940).
4. *Piping Stress Calculations Simplified*, S. W. Spielvogel (author and publisher), Lake Success, N. Y. (1955).
5. "Formula for Pipe Thickness," W. J. Buxton and W. P. Burrowes, *ASME Trans.*, **73**, 575 (1951).
6. "Why Ferrous Pipe Corrodes," F. E. Kulman, *Gas Age*, **115**, No. 10, 26 (May 19, 1955).
7. "How to Reduce Piping Costs with Welding Fittings and Flanges," J. Tolliver, *Power*, **99**, No. 4, 75 (Apr. 1955).
8. "Facts About Plastic Pipe," W. E. Jacobson, *Air Conditioning, Heating, Ventilating*, **52**, No. 1, 87 (Jan. 1955).
9. "Methods of Joining Plastic Pipe," W. E. Jacobson, *Air Conditioning, Heating, Ventilating*, **52**, No. 9, 75 (Sept., 1955).
10. "Plastic Materials for Piping," J. Bedford, *Ind. Heating Eng.*, **17**, Nos. 117, 118, 119, and 120 (1955).

PROBLEMS

1. (a) A horizontal 8-in. uncovered standard steel pipe 1000 linear ft in length carries saturated steam at a temperature of 297.7 F. The temperature of the air surrounding the pipe is 70 F. Find the total heat losses from the pipe in Btuh. (b) If the pipe of part a were filled with hot water at a temperature of 210 F and the surrounding air temperature remained at 70 F, what would be its hourly heat losses? (c) If 1000 ft of 8-in. tarnished copper tubing were used to convey hot water at 210 F when the air temperature about the tube was 70 F, what would be the hourly heat losses from its exposed surface?

2. Find the hourly heat losses for each condition of Problem 1 if each pipe or tube is covered with 85 per cent magnesia insulation 3 in. thick when the temperature of the air surrounding the covering is 60 F.

3. A length of bare pipe when uncovered loses 20,000 Btuh but when insulated the loss is reduced to 2500 Btuh. Find the efficiency of the pipe covering.

4. Calculate the hourly heat losses per linear foot of 4-in. commercial steel pipe, covered with 1 in. of 8 ply per in. corrugated and laminated asbestos paper, which conveys 180 F hot water through a space where the temperature of the air at the pipe level is assumed to be 70 F.

5. Find the linear expansion, in inches, of 600 ft of copper tubing when its temperature is raised from 75 F to 220 F.

11

Heating with Steam

11.1. Classification of Steam-Heating Systems. Steam-heating systems may be classified with respect to the piping arrangement at the heat disseminators as either one-pipe or two-pipe plants. Steam systems may also be classified as either mechanical or gravity types, depending upon whether or not a pump is used to return the condensate to the boiler. A third method of classification refers to the steam pressures within the heat disseminators; these may vary from 15 to 150 psig for high-pressure installations and from 0 to 15 psig for low-pressure plants. Low-pressure two-pipe systems may be further classified as vapor systems and vacuum systems. The latter may be subdivided into either return-line or differential-vacuum systems.

11.2. One-Pipe Systems. These systems having only a single-pipe connection to a heat disseminator require air vents for the supply mains and for each heat-emitting unit. The vents are automatic valves which discharge the air. Unless the disseminators and supply mains are kept free from air, they will not be completely filled with steam and the system will not function properly. A typical sketch of a one-pipe steam system with gravity flow is shown in Fig. 11.1. The illustration is of an upfeed system. One-pipe systems may also be installed with steam mains located in the top of the building. One-pipe steam-heating systems when properly designed and installed should not require a boiler steam pressure of more than 2 psig.

In the upfeed arrangement, Fig. 11.1, the supply main is taken from the boiler header and rises close to the basement ceiling. From this point the main is pitched downward in the direction of steam flow with a grade of not less than $\frac{1}{4}$ in. in 10 ft. The main may be carried in a

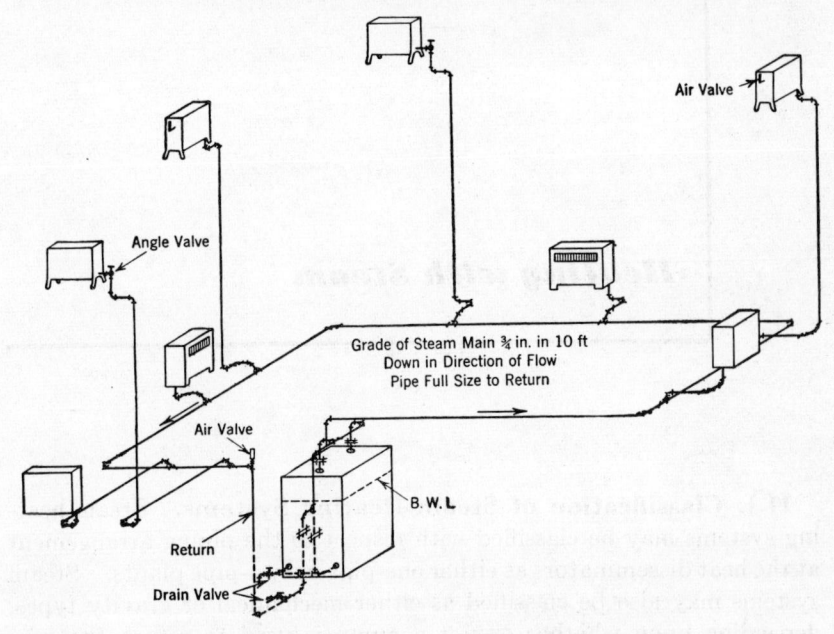

Fig. 11.1. One-pipe air-vent steam-heating system.

single loop about the basement, or it may be divided into two or more parts. All ends of mains should be maintained at least 18 in. above the normal boiler water line in order to allow for the head of water in the drops at those points. It often becomes necessary to place the boiler in a pit in order to lower the boiler water line with respect to the ends of the main or mains. When a main is long or the basement head room is limited, relaying the main, Fig. 11.2, helps to solve the problem. A drain connection at the bottom of the rise is always required; otherwise the pocket will fill with water.

Return mains are designated as either wet or dry. A wet-return main is below the boiler water level and is filled with water returning to the boiler. A dry-return main carries both condensate and air and sometimes may be filled with steam, although this is not a desirable condition. A dry-return main is above the boiler water level. In the

HEATING WITH STEAM

system of Fig. 11.1 a wet return is used but it is very short. The automatic air valve shown at the end of the steam supply main eliminates air and other gases collecting at this point. Air cannot pass through a wet return. The piping details of the return connection to the boiler, Fig. 11.1, involve what is commonly known either as a *Hartford*, or an *Underwriters*, or a *safety loop*. This protective device should be installed in all steam systems where the condensate is returned by gravity action. It is often used in systems employing a pump (see Fig. 11.8). The end of the return main of Fig. 11.1 is reduced in size below the normal boiler water-line level and the pipe is carried to near the boiler room floor. The pipe then rises vertically and is connected to the equalizer pipe which is placed in a vertical position between the supply main and the return connections near the bottom of the boiler. The connection to the equalizer pipe is through a short pipe nipple placed in a horizontal position slightly below the normal water level of the boiler. The purpose of the arrangement is to prevent back flow of water from the boiler into the return main. When

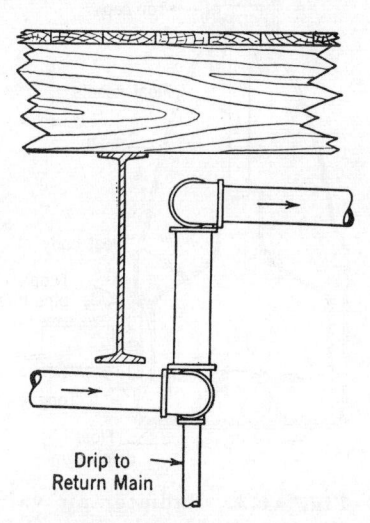

Fig. 11.2. Steam main relay with drip connection.

the boiler water level falls below the top of the short horizontal pipe connection to the equalizer pipe, this section of pipe fills with steam which is of low density and the effect is to increase the gravity head of condensate in the return piping and thereby facilitate its return to the boiler.

The gravity-return system of Fig. 11.1 may be converted to a mechanical system by the addition of a condensation pump. A pump is necessary when radiators, convectors, and condensing coils must be placed below the boiler water line or at an insufficient distance above it. Building limitations often prohibit a boiler pit so that a pump becomes necessary. A pump is also required when the boiler steam pressure is greater than that of the supply mains. The addition of a pump to a one-pipe system does not eliminate the counterflow of steam and water in the risers and runouts.

Each disseminator of a one-pipe steam-heating system must be fitted with an automatic valve. The type of air valve shown by Fig. 11.3

vents air when the system is filling with steam, prevents the loss of steam, but does not hinder the return of air when the fire in the boiler furnace is inactive. When this type of valve is used the pressure in the heat-emitting units is always equal to or greater than that of the atmosphere. The use of automatic air valves, which include a device which prevents the return of air, creates a system in which the steam pressure is less than atmospheric much of the time. The performance of a one-pipe system is improved by the addition of nonreturn air valves. However, experience has shown that these accessories do not retain their effectiveness unless careful attention is given to their maintenance.

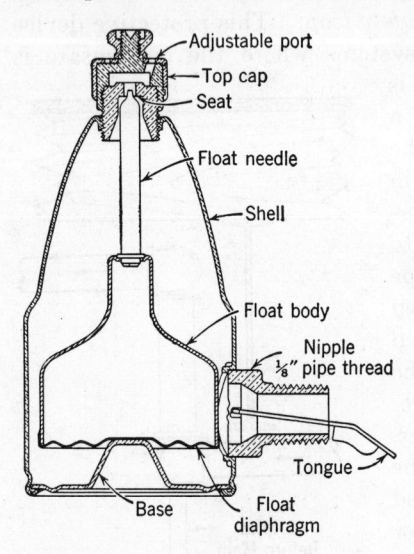

Fig. 11.3. Radiator air valve. (Courtesy Hoffman Speciality Co.)

One-pipe steam-heating systems are simpler and cheaper to install than any of the two-pipe designs. When properly laid out and installed they give satisfactory operating results. However, it is impossible to modulate the flow of steam into the heating units by hand operation of valves. Thermostatic control of the heating units must operate with the steam supply either completely "on or off."

The most suitable application of a one-pipe steam-heating system is in a private residence where control of the fire in the boiler furnace is obtained by use of a single thermostat. This operation eliminates the necessity of individual control of the disseminators. However, for satisfactory performance it is very important that each air valve used have an adjustable vent port. Proper adjustment of the air-valve vents insures simultaneous heating of all radiators and convectors.

11.3. Two-Pipe Systems. The distinguishing feature of this type of steam-heating system is the two-pipe connections to each installed heat disseminator as shown in Fig. 11.4. The condensate leaves the heat-dispersing unit through a separate pipe and it is therefore possible to throttle the flow of supply steam to any desired extent. The steam control valves may be placed at the tops of radiators where they are more convenient when operated by hand. Special valves having graduated ports are available for facilitating hand control.

HEATING WITH STEAM

Two-pipe high-pressure systems are used only in large industrial-type buildings where high-pressure steam is required for purposes other than space heating. The heat disseminators are usually unit heaters. Two-pipe low-pressure systems are either vapor or vacuum systems according to the manner in which the reduced pressure is produced.

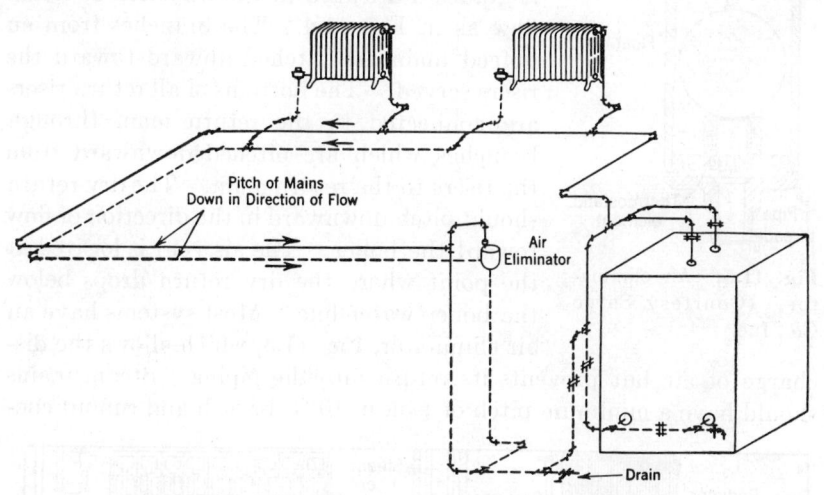

Fig. 11.4. Elementary vapor steam-heating system.

11.4. Vapor Systems. The designation "vapor system" is applied to layouts, Fig. 11.4, which function as follows:

1. The boiler steam pressure ranges from a few ounces above atmospheric as a maximum to pressures less than atmospheric. The subatmospheric pressures are obtained by condensation of steam in the heat disseminators when the boiler furnace fire is declining in activity.

2. The condensation is generally returned to the boiler by gravity. With certain vapor systems a pump may feed water to the boiler, but the pump does nothing to create a vacuum in the system.

3. All air vented from the radiators and the piping is collected at the end of a dry-return main and is discharged at that location by the action of an air eliminator.

Some form of graduated-control valve is placed at the inlet of each heat disseminator. The steam flow in a radiator is through the top row of nipples and downward through the radiator sections. The radiator may be only partially filled with steam except when the demand for heat is a maximum. The outlet connection to the return piping may be made at either end of the radiator and is always at a bottom tapping. Generally steam is prevented from flowing through into the return piping by a trap which will pass both water and air.

216　　AIR CONDITIONING AND REFRIGERATION

Vapor systems are installed as both upfeed and downfeed plants. A dry-return main is necessary as the air from radiators and supply piping is discharged into this main from which it is finally vented. The supply main is taken to a high point above the boiler from which it grades downward in the direction of steam flow as in Fig. 11.4. The branches from an upfeed main are pitched upward toward the risers served. The bottoms of all return risers are connected to the return main through branches which are pitched downward from the risers to the return main. The dry return should pitch downward in the direction of flow toward the boiler. The air vent is located at the point where the dry return drops below the boiler water line. Most systems have an air eliminator, Fig. 11.5, which allows the discharge of air but prevents its return into the piping. Steam mains should have a minimum pitch of $\frac{1}{4}$ in. in 10 ft, branch and runout con-

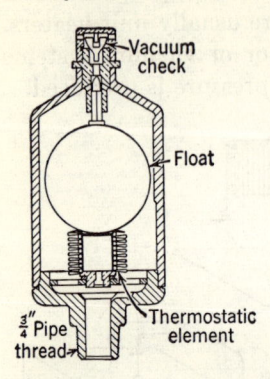

Fig. 11.5. Air eliminator. (Courtesy Sarco Co., Inc.)

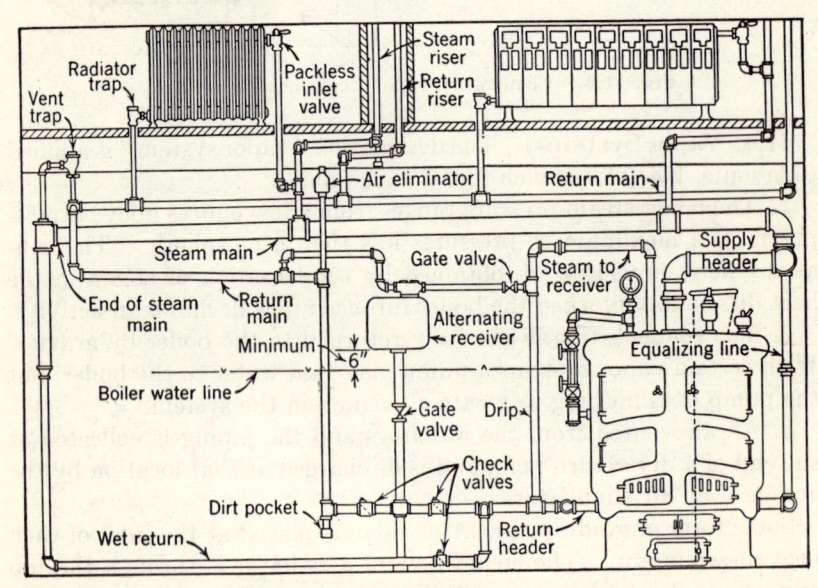

Fig. 11.6. Vapor heating system with alternating receiver. (Courtesy Sarco Co., Inc.)

nections $\frac{1}{2}$ in. per ft, dry-return mains 1 in. per 20 ft, and wet-return mains 1 in. per 40 ft. All supply piping must be large enough to limit the steam-pressure losses to 1 oz or less per 100 ft.

HEATING WITH STEAM

11.5. Alternating Receivers or Return Traps. In order to secure positive return of condensate to the boiler at all times a device known as an alternating receiver or return trap is often used in vapor systems. Figure 11.6 gives some of the details of a vapor system which has an alternating receiver.

Figure 11.7 gives a sectional view of a return trap or receiver. The unit has a float chamber in which a hollow float rises as water accumulates in it. The float is attached to a weight and lever mechanism connected to a balanced steam valve which is in a pipe leading from the boiler steam outlet. The float when lifted brings the mechanism to a point where it trips. The steam valve is quickly opened and at the same time the trap top vent to the return main and the air eliminator is closed. This action equalizes the trap pressure with that of the boiler, and the collected water in the trap flows to the boiler by gravity. When the trap has been emptied of water, the float causes the mechanism to open the air vent and to close the steam valve so that more water may accumulate. Traps of this kind are installed with the bottom of the trap from 6 to 18 in. above the normal boiler water line.

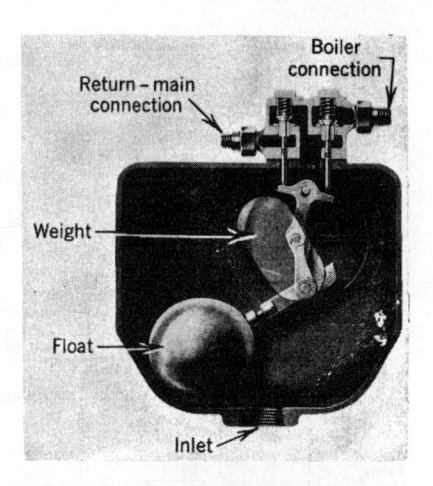

Fig. 11.7. Alternating receiver. (Courtesy Sarco Co., Inc.)

11.6. Vacuum Systems. These systems are distinguished from two-pipe vapor steam systems in that a pressure less than atmospheric in the return piping is mechanically produced by means of a vacuum pump, Fig. 11.8. Each radiator of the system is fitted, at its return outlet, with a thermostatic trap which will allow both air and water to pass but which closes against the flow of steam. All drips into return mains, from mains and supply risers, must pass through traps which prevent the loss of steam but pass both air and water.

The heat disseminator inlet valves should be of a packless type with graduated control when the units are to be regulated by hand. The details of the supply and return piping are similar to those of vapor systems, and they may be installed with either upfeed or downfeed supply mains.

For best results the vacuum pump should be located so that the return piping grades uniformly downward toward the receiver of the unit. This allows gravity flow of the condensate to the pump. Sometimes *lift fittings* are permissible in the return piping when it is necessary to raise the return piping from one level to another, but lift fittings are to be avoided whenever possible. The function of the vacuum pump is to withdraw air which is vented from the supply piping and radiators into the return mains. The vacuum pump also

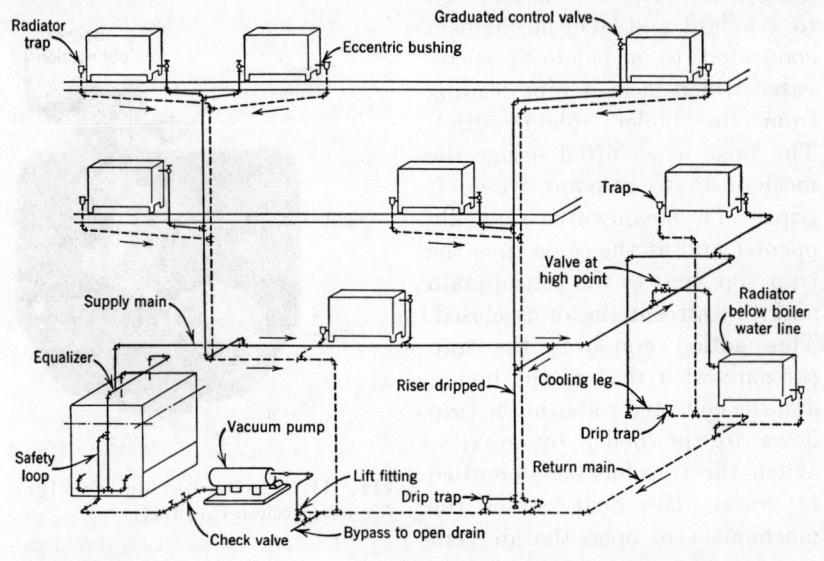

Fig. 11.8. Typical upfeed steam-heating system with a vacuum pump.

withdraws the condensate from the receiver and discharges it into low-pressure heating boilers or into the feed-water heater when high-pressure boilers are used. The air withdrawn from the system by the pump is generally separated from the condensate at the vacuum pump and discharged to the atmosphere. The piping details at the boiler inlet are usually similar to those shown in Fig. 11.8.

Vacuum systems are adapted to tall buildings and also those covering considerable ground area. The action of the vacuum pump is to produce free circulation of steam by the maintenance of a definite pressure differential between the supply and the return piping. This action enables the more rapid venting of air to take place, as steam pressure alone is not necessary to force the air out. Positive return of the condensate is also effected, and radiators may be located below the level of the boiler water line.

Vacuum steam-heating systems may be separated into two general

HEATING WITH STEAM

groups: (1) return-line systems and (2) differential or subatmospheric systems.

11.7. Vacuum Return-Line Systems. Return-line installations generally have a vacuum maintained in the return piping without any attempt to produce pressure less than atmospheric in the radiators and the supply piping. Figure 11.8 illustrates a typical upfeed system.

The steam mains should be pitched in the direction of flow not less than 1 in. in 20 ft, and the dry-return mains should have a pitch of at least 1 in. in 40 ft. Steam mains and other pipes should be designed with the maximum allowable pressure loss not to exceed 1 oz per 100 ft of pipe. The pressure differential between the two sides of a radiator trap must be at least equivalent to 8 oz. The displacement of the vacuum pump must be sufficient to handle not only the condensate and air discharged into the return piping but also the vapor produced by the re-evaporation (due to reduced pressure) of some of the condensate passed by the traps.

11.8. Differential-Vacuum Heating Systems. These systems are operated to vary the heat output from the radiators by changing both the temperature and the quantity of the steam supplied. The steam is supplied continuously but with varying rates of flow and temperature, which is controlled by adjusting its pressure in the supply piping. Depending upon weather conditions, the steam in heat-dispersal units may exist with temperatures corresponding to pressure conditions ranging downward from 2 psig to 25 in. of mercury vacuum.

The details of subatmospheric heating plants differ from return-line systems principally in the matter of control equipment. There is also a slight difference in the details of radiator inlet valves and traps. Each steam-supply valve at the heat-dispersal units has either a fixed or an externally adjustable orifice so that the steam distribution between the various units may be properly adjusted. The radiator traps have a pressure range of from 15 psig to 25 in. mercury, which is greater than that necessary in a return-line vacuum system.

Important parts of the control equipment, Fig. 11.9, are: (1) the steam-control valve, (2) the control panel, (3) the heat balancer, (4) the selector, (5) either a room thermostat, or a room resistance thermometer, (6) the differential controller, and (7) the vacuum pump.

The steam-control valve operates to vary the flow of steam to the supply piping and functions under the action of the control panel which transmits the combined actions of the heat balancer, the selector, and the room thermometer. The heat balancer is a form of convector having a pair of resistance thermometers, one of which is located above its heating element and one below it. The combination of the two thermometers measures the air-temperature differential existing on the two sides of the heating element, and the heat balancer thereby gives an indication of the heat supply to the system. The heating demand is measured by the selector, attached to the inside surface of

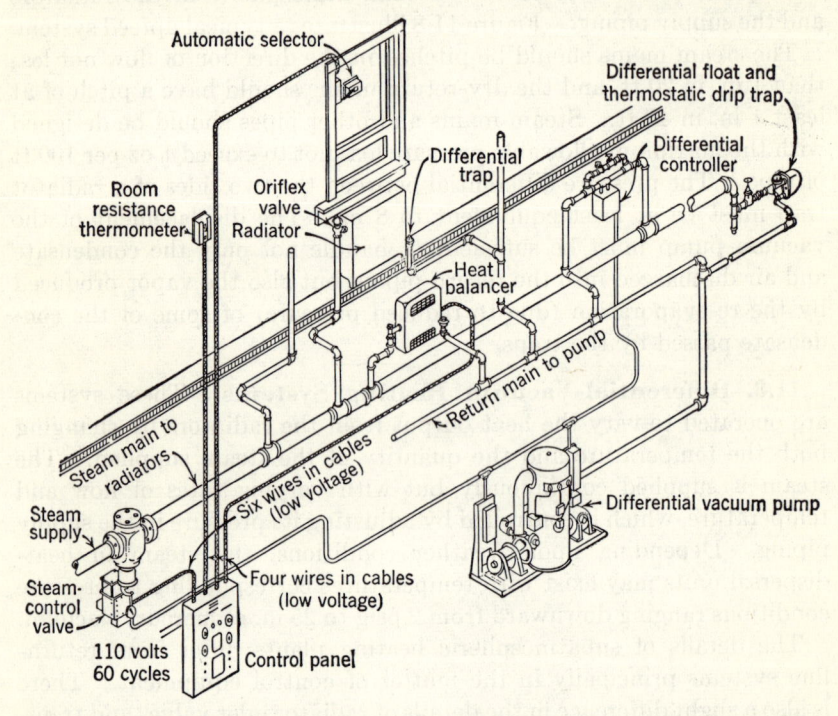

Fig. 11.9. Control equipment of Vari-Vac differential heating system. (Courtesy C. A. Dunham Co.)

a window, which is influenced by the effects of outside-air temperature and wind velocity. In connection with the control panel the room thermometer functions to compensate the effects of the selector and the heat balancer. The differential controller connected to the supply and the return piping is actuated by the pressure differential existing between the two lines and functions to start the pump when the difference is small and to stop the pump when conditions are satisfactory.

11.9. Vacuum Pumps. Air may be removed, and condensate from the returns of vacuum systems may be handled, by various types of vacuum pumps. Vacuum pumps may be high-pressure steam-driven reciprocating units, motor-driven rotary units, or rotary units

HEATING WITH STEAM

which embody both a motor drive and a low-pressure steam-turbine drive. In the last type the drive alternates according to operating conditions.

Piston vacuum pumps, either steam- or motor-driven units, should have a volumetric displacement at least six times that of the volume of condensate handled. Rotary pumps should be able to handle not less than 0.75 lb of condensate per hr per sq ft of attached radiation and 0.6 cu ft of air per min per sq ft of radiation.

11.10. Radiator Valves. The material of construction of valve bodies is usually brass, and they are made in angle, straightaway, and right- and left-hand corner patterns for both steam and hot-water service. The valve action in both opening and closing may be quick or slow. Valves are also made to give graduated or fractional opening with the amount of port opening indicated by a pointer attached to the valve stem or handle. Graduated-control valves cannot be used with one-pipe systems and are not generally used for hot water.

A packless-type angle radiator valve is illustrated in Fig. 11.10. Part A is a collar machined on the valve stem against which stainless-steel wafers B press to form a tight annular metal-to-metal seal against the leakage of steam, water, or air.

Fig. 11.10. Packless radiator valve. (Courtesy J. P. Marsh Co.)

11.11. Radiator Traps. These devices function to allow the escape of condensate and air from radiators and are used in two-pipe steam-heating systems. Present-day radiator traps are generally thermostatic devices built in various commercial forms, one of which is illustrated in Fig. 11.11. Thermostatic traps are comprised of a body, having an inlet opening and an outlet port with a seat, together with a cover to which may be attached a thermostatic element containing a volatile fluid. When the thermostatic element is cool it is contracted and the outlet port is open.

Steam entering a cool radiator by reason of its pressure forces the air of the radiator through the open port of the trap to the return piping.

Steam condensation also occurs, and the water thus produced is discharged by the trap until the radiator is filled with steam and the thermostatic element is heated. Heat from the steam then vaporizes enough of the volatile fluid to create a pressure within the element. As soon as the element has a sufficient pressure within it, the port is closed and the escape of steam prevented.

Thermostatic traps are made in angle, straightaway, and corner patterns for both high- and low-pressure duty. High-duty traps

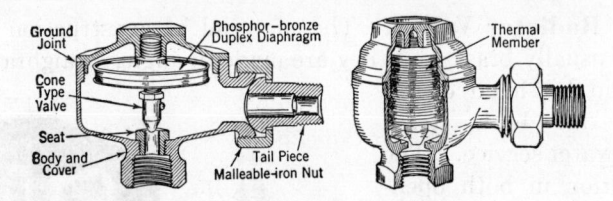

Fig. 11.11. Thermostatic radiator trap.

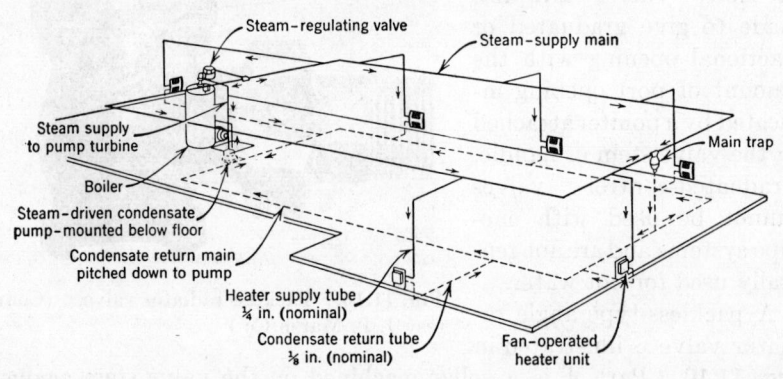

Fig. 11.12. One version of a selective-temperature steam-heating system. (Courtesy Iron Fireman Corp.)

operate with pressures ranging from 15 psig to 25 in. of mercury vacuum. Low-duty traps function with pressures ranging from 15 psig to 15 in. of mercury vacuum.

11.12. Selective-Temperature Steam Heating. A new and radically different type of steam-heating system is shown in Fig. 11.12, which represents a plant installed within a residence built upon a concrete slab. The layout is characterized by the features which follow.

1. A modulating thermostatically controlled forced convector, fitted with a small steam-turbine driven fan, located within each room. At present the convectors are made in three sizes with output capacities of 6000, 12,000, and 18,000 Btuh.

HEATING WITH STEAM

2. All of the steam piping from the supply main to the convectors is of ¼-in. nominal size and the condensate piping from the convectors to the return main is of ⅛-in. nominal size.

3. The condensate is returned to the boiler by a steam-driven pump.

4. A steam-regulating valve controls either an oil or a gas burner within the boiler furnace and maintains a constant pressure in the steam-supply main. The regulating valve also functions to deliver steam to the condensate-pump turbine before steam enters the convectors when the system begins to function.

The principal advantages offered by the system are: (1) individual room-temperature control, (2) generous circulation of filtered air within each room, and (3) a mechanical steam-heating system which is independent of electric service, providing that the fuel burner can be operated without it.

11.13. Flow of Steam in Pipes. Frictional resistance, under any condition, offered to the flow of steam in pipes may be computed by the use of equation 6.3 which is applicable to all fluids. The friction-loss chart of Fig. 11.13 was prepared from data pertaining to dry saturated steam existing in the pressure range usually employed in heating systems. Although condensate is always present in the pipes of heating systems, its volume is small compared with that of the steam and Fig. 11.13 can be applied. Figure 11.13 is useful for the estimation of friction-pressure losses and for the selection of the proper size of pipe required to handle a specified weight of steam with a limitation on the allowable loss of pressure due to friction.

11.14. Allowances for Pipe Fittings. In Art. 6.9 it was stated that allowances for friction-pressure losses in pipe fittings may be made by adding to the actual length of the pipe used an equivalent length for each fitting involved. Table 11.1 indicates the equivalent lengths of elbows, tees, and valves which are commonly used in steam-heating systems.

Example 1. Find the friction-pressure losses in a 2-in. steam pipe line which has one globe valve, six 90-deg elbows, and 140 ft of straight pipe through which 5 lb of dry saturated steam flows per minute and in which the average steam pressure is 15.5 psia.

Solution. The equivalent length of the globe valve (Table 11.1) is 46 ft. The equivalent length of each elbow is 4.3 ft. For the system the total equivalent length is 140 + 46 + (6 × 4.3) = 211.8 ft. The steam pressure falls within the range of Fig. 11.13 so that it may be used. The pressure drop in pounds per square inch per 100 ft is shown as 0.43. The friction loss for the system is 0.43 × 211.8/100 = 0.91 psi.

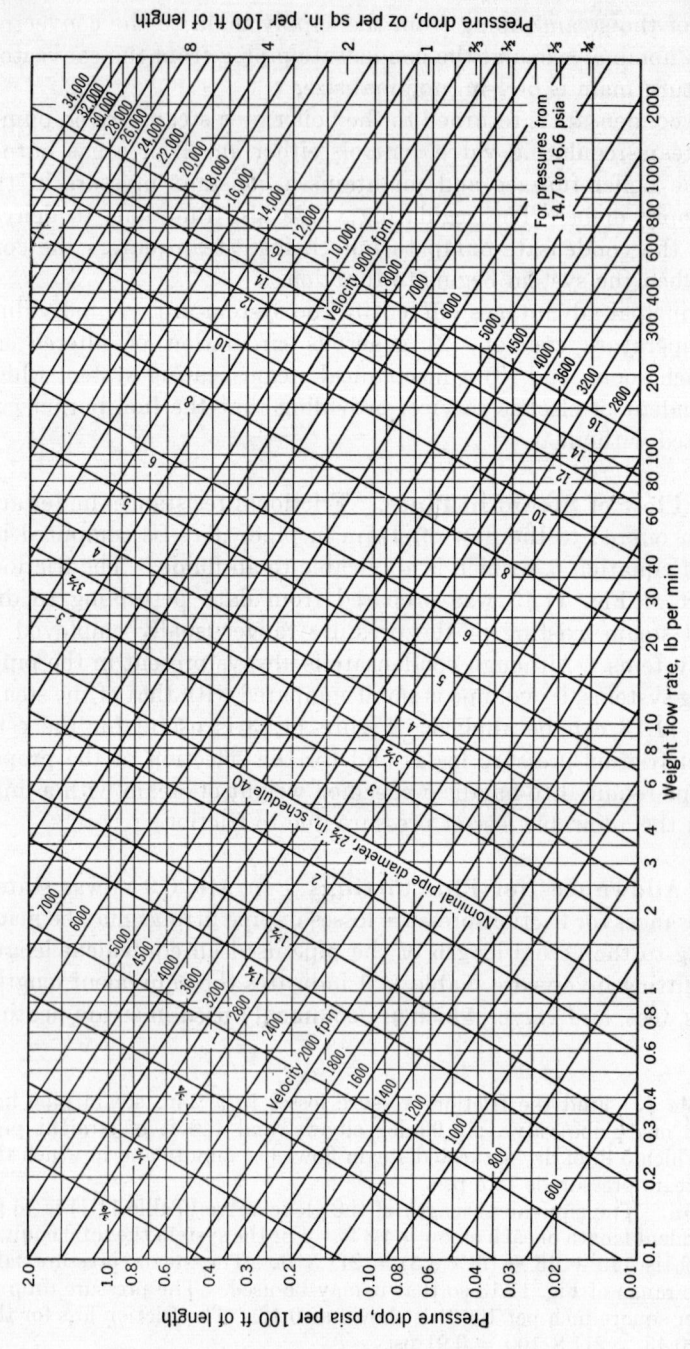

Fig. 11.13. Working chart for saturated steam flowing in schedule 40 wrought-iron or steel pipe. [Data from *Heating and Ventilating Magazine* (Aug. 1951). Used by permission.]

HEATING WITH STEAM

Example 2. Find by use of Fig. 11.13 the nominal size of a pipe needed to convey 300 lb of dry saturated steam per hour. The friction-pressure loss is not to exceed 1 oz per 100 ft of equivalent length. The steam pressure in the pipe line is to average 16 psia.

Solution. The flow rate is 300 ÷ 60 = 5 lb per min. Figure 11.13 indicates that the correct nominal pipe size is 3 in.

TABLE 11.1. Length in Feet of Pipe to be Added to Actual Length of Run—Owing to Fittings—to Obtain Equivalent Length*

Size of Pipe, In.	90-Deg Elbow	Side Outlet Tee	Gate Valve†	Globe Valve†	Angle Valve†
$\frac{1}{2}$	1.3	3	0.3	14	7
$\frac{3}{4}$	1.8	4	0.4	18	10
1	2.2	5	0.5	23	12
$1\frac{1}{4}$	3.0	6	0.6	29	15
$1\frac{1}{2}$	3.5	7	0.8	34	18
2	4.3	8	1.0	46	22
$2\frac{1}{2}$	5.0	11	1.1	54	27
3	6.5	13	1.4	66	34
$3\frac{1}{2}$	8	15	1.6	80	40
4	9	18	1.9	92	45
5	11	22	2.2	112	56
6	13	27	2.8	136	67
8	17	35	3.7	180	92
10	21	45	4.6	230	112
12	27	53	5.5	270	132
14	30	63	6.4	310	152

* From "Heating Ventilating Air-Conditioning Guide 1957."
† Valve fully open.

11.15. The Design of One-Pipe Steam-Heating Systems. The pipes of steam-heating systems are obliged to carry three media which are steam, air, and water. Careful consideration must be given to the velocity of steam flow in the risers (vertical pipes to the heat disseminators), in the runouts (horizontal connecting pipes), and in radiator valves of one-pipe systems where condensate and steam flows are in opposite directions. The pipes and the valve passageways must be large enough to hold the velocity of the flowing steam below that at which condensate will be entrained and carried with it. The critical velocity is reached when counterflow of steam and water ceases and the condensate travels with the flowing steam. The critical velocity is numerically less in small pipes than it is in large-sized pipes. For pipes of a given size the critical velocity is less in horizontally placed ones than in those which stand in a vertical position. Table 11.2 indicates steam velocities which should *not* be exceeded in the risers and runouts of one-pipe upfeed steam-heating systems of the type illustrated by Fig. 11.1. Steam velocities in the mains of one-pipe systems

TABLE 11.2. Suggested Maximum Velocities* in Risers and Runouts of One-Pipe Steam-Heating Systems

Pipe Size, in.	Riser Velocity, fpm	Runout Velocity, fpm
$\frac{3}{4}$	680	340
1	770	490
$1\frac{1}{4}$	810	650
$1\frac{1}{2}$	1130	680
2	1300	760
$2\frac{1}{2}$	1470	820
3	1640	970
$3\frac{1}{2}$	1770	1140
4	1810	1320

* Computed for a steam pressure of 15.7 psia.

having counterflow of steam and condensate should be held to those shown in Table 11.2 for runouts.

Example. A one-pipe steam-heating system has a total load of 600 sq ft of edr and is laid out in such a way that the condensate flows toward the boiler in all parts of the main (counterflow). The boiler pressure under design conditions is 1 psig. Determine the nominal pipe size of the steam main.

Solution. The condensation may be taken as 0.25 lb per hr per sq ft of the radiation rating and amounts to 150 lb per hr or 2.5 lb per min. Figure 11.13 indicates that the velocity of steam flow is 1220 fpm in 3-in. pipe and 900 fpm in $3\frac{1}{2}$-in. For similar conditions Table 11.2 indicates that for horizontal runouts the steam velocity in a 3-in. pipe should not exceed 970 fpm, whereas in a $3\frac{1}{2}$-in. pipe the velocity may be as great as 1140 fpm. Therefore, a $3\frac{1}{2}$-in. pipe is one of a suitable size.

For mains which are pitched downward in the direction of both steam and condensate flow, the principal consideration is the friction-pressure losses. When condensate is to be returned to a boiler by gravity action, the sum of the friction losses in the steam main between the boiler and its end and the friction losses in the return main, both expressed in feet of water, should not exceed the difference in feet of the elevation of the steam-main end and the normal water level in the boiler. In Fig. 11.1 the distance amounts to the difference in feet of elevation of the main at the location of its air valve and the normal boiler-water level (bwl) when both are measured above the same datum plane. A design procedure for the system of Fig. 11.1 is to compute the condensation per hour in each radiator and then by use of Table 11.2 and Fig. 11.13 select the proper pipe sizes for both the risers and the runouts. The different sections of the steam-supply main may be sized by the use of Fig. 11.13. Inasmuch as the wet-return main handles only water, it might be sized by the use of Fig. 12.5. Most designers prefer to use the data of Table 11.3. The $I = B = R^1$ recommends the use of a 2-in. steam main of uniform size throughout its length in all one-pipe systems having a total load less than 386 sq ft edr.

In the case of a larger system the usual procedure is to compute the equivalent length of the main and size the pipe so that the friction-pressure losses in its entire length will not exceed one-half of the steam pressure (in pounds per square inch gage) that will be maintained within the boiler under design conditions.

11.16. Design of Two-Pipe Systems. Usually steam and condensate flow in the same direction in the mains of two-pipe systems. A small amount of condensate in the steam mains, risers, and runouts results from heat losses from them; that produced in the heat disseminators is carried to the boiler through a separate system of return piping. Generally the pipe sizes to handle a given weight of steam are smaller in the two-pipe design than in a one-pipe installation. Figure 11.13 may be used for sizing the supply mains. The allowable friction loss per 100 ft is taken as one-half of the boiler pressure in pounds per square inch gage divided by the equivalent length of the longest main in hundreds of feet. Return mains of two-pipe systems cannot be sized with friction charts applying to pipes carrying water because they must handle water, air, and some steam. Return mains of the proper size are selected by use of the data of Table 11.3. The table column used for sizing a return main has the same pressure-drop heading as was used in selecting the proper pipe size for the steam-supply main. However, this does not mean that the friction loss in the return main will be the same as that which exists in the steam-supply main. But experience has shown that a two-pipe system will operate satisfactorily when the pipe sizing is made by the foregoing procedure. The data of Table 11.3 are based on research investigations[2,3] sponsored by the ASHAE; the reader is referred to the current ASHAE Guide for more tabular data and further suggestions pertaining to the design of the various steam-heating systems.

TABLE 11.3. Return-Pipe Capacities*

Capacity of Return Mains and Risers†

Mains

Pipe Size, In.	1/16-Lb or 1/2-Oz Drop per 100 Ft Wet	Dry	Vacuum	1/24-Lb or 2/3-Oz Drop per 100 Ft Wet	Dry	Vacuum	1/16-Lb or 1-Oz Drop per 100 Ft Wet	Dry	Vacuum	1/8-Lb or 2-Oz Drop per 100 Ft Wet	Dry	Vacuum	1/4-Lb or 4-Oz Drop per 100 Ft Wet	Dry	Vacuum	1/2-Lb or 8-Oz Drop per 100 Ft Wet	Dry	Vacuum
M	N	O	P	Q	R	S	T	U	V	W	X	Y	Z	AA	BB	CC	DD	EE
3/4	....	....	....	....	....	326	....	....	400	....	....	568	....	....	800	....	....	1,130
1	500	248	....	580	285	570	700	320	700	1,000	412	994	1,400	460	1,400	....	....	1,980
1 1/4	850	520	....	990	595	976	1,200	670	1,200	1,700	868	1,700	2,400	962	2,400	....	....	3,390
1 1/2	1,350	822	....	1,570	943	1,550	1,900	1,060	1,900	2,700	1,360	2,700	3,800	1,510	3,800	....	....	5,370
2	2,700	1,880	....	3,240	2,140	3,256	4,000	2,300	4,000	5,600	2,960	5,680	8,000	3,300	8,000	....	....	11,300
2 1/2	4,700	3,040	....	5,300	3,470	5,450	6,700	3,500	6,700	9,400	4,900	9,510	13,400	5,450	13,400	....	....	18,900
3	7,500	5,840	....	8,500	6,250	8,710	10,700	7,000	10,700	15,200	9,000	15,200	21,400	10,000	21,400	....	....	30,200
3 1/2	11,000	7,880	....	13,200	8,800	13,020	16,000	10,000	16,000	22,000	12,900	22,700	32,000	14,300	32,000	....	....	45,200
4	15,500	11,700	....	18,300	13,400	18,000	22,000	15,000	22,000	31,000	19,300	31,200	44,000	21,500	44,000	....	....	62,200
5	....	....	....	....	....	31,500	....	....	38,700	....	....	54,900	....	....	77,400	....	....	109,000
6	....	....	....	....	....	50,450	....	....	62,000	....	....	88,000	....	....	124,000	....	....	175,000

Risers

3/4	....	....	....	....	....	570	....	....	700	....	....	994	....	....	1,400	....	....	1,980
1	....	190	....	....	190	976	....	190	1,200	....	190	1,700	....	190	2,400	....	....	3,390
1 1/4	....	450	....	....	450	1,550	....	450	1,900	....	450	2,700	....	450	3,800	....	....	5,370
1 1/2	....	990	....	....	990	3,260	....	990	4,000	....	990	5,680	....	990	8,000	....	....	11,300
2	....	1500	....	....	1500	5,450	....	1500	6,700	....	1500	9,510	....	1500	13,400	....	....	18,900
2 1/2	....	3000	....	....	3000	8,710	....	3000	10,700	....	3000	15,200	....	3000	21,400	....	....	30,200
3	....	....	....	....	....	13,400	....	....	16,000	....	....	22,700	....	....	32,000	....	....	45,200
3 1/2	....	....	....	....	....	17,900	....	....	22,000	....	....	31,200	....	....	44,000	....	....	62,200
4	....	....	....	....	....	31,500	....	....	38,700	....	....	54,900	....	....	77,400	....	....	109,000
5	....	....	....	....	....	50,500	....	....	62,000	....	....	88,000	....	....	124,000	....	....	175,000

* This table is based on pipe size data developed through the research investigations of the American Society of Heating and Ventilating Engineers. Used by permission.
† Capacity expressed in square feet of equivalent direct radiation.

HEATING WITH STEAM

11.17. Heating Plans. Working drawings and specifications of proposed heating installations are essential and should give with clarity all necessary information. Symbols for pipes, fittings, and other pieces of equipment are given in Table A.2 of the Appendix. Use of these symbols saves time in the preparation of drawings and gives them clarity without a multiplicity of detail lines.

The basement piping layout for a small steam heating system may follow the general scheme of that shown for a hot-water system in Fig. 12.9. A plan of each floor in the building should indicate the intended location of all disseminators and the risers needed to serve them.

11.18. Orifices for Steam Radiators and Convectors. The distribution of steam to the various radiators and convectors of a heating system is of prime importance. Some form of a radiator orifice may be used to balance a two-pipe system.

Orifices available are those having adjustable openings and those with fixed ports. An adjustable orifice may be part of the radiator control valve or it may be a separate fitting. Although orifices can be placed at various locations in the steam piping of a heating system, the usual placement is at the steam inlet of each heat disseminator.

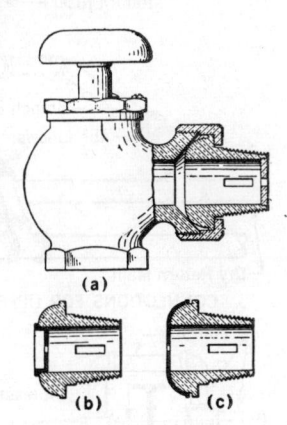

Fig. 11.14. Metering orifice.

Figure 11.14 shows an angle-type radiator valve with a union and a tail piece or nipple which is screwed into the radiator tapping bushing. A fixed-orifice plate may be placed within the nipple as shown by *b* or it may be located in the union at *c* in Fig. 11.14. As a result of investigations by Sandford and Sprenger,[4] sizes of orifices are recommended for different amounts of radiation. The data are included in tabular form in the ASHAE Guide.

11.19. Piping Details. Figure 11.15 shows some of the more common arrangements for both one-pipe and two-pipe steam systems. These details are worked out with the idea of providing for expansion, the avoidance of water pockets which cannot be drained, the drainage of the system, and the removal of air from the supply mains and radiators.

11.20. District Heating with Steam. Frequently in cities and about institutions a single plant to furnish heat to several buildings is

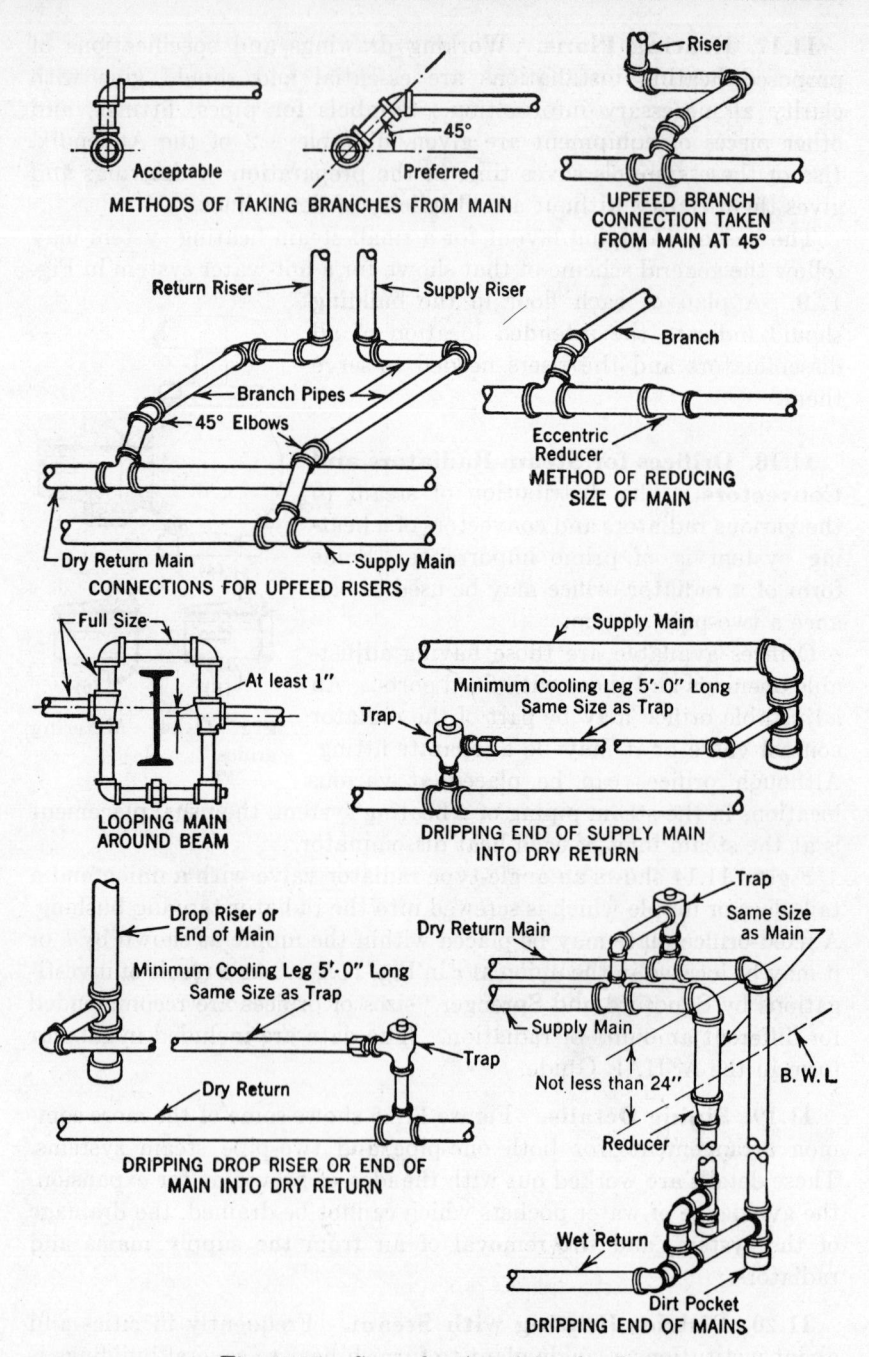

Fig. 11.15. Steam-heating piping details.

HEATING WITH STEAM

more economical than having boiler equipment located in each building. The vehicle used to carry the heat to the buildings may be either steam or water. Steam may be generated in large modern boiler plants with an efficiency of fuel utilization ranging from 70 to 88 per cent. Losses occur in the distributing mains, and the final over-all efficiency of fuel-heat utilization ranges from 60 to 65 per cent in district heating systems.

The steam delivered by the boiler plant may arrive at the place of consumption at either high or low pressure. Often the steam pressure has to be reduced at the building where it is used. This can be done by pressure-reducing valves, or, where it is desired and the steam pressure is great enough, the pressure reduction can be effected by means of steam turbines. The heating system used within the building may be one of the types previously described.

11.21. Steam Conduits. Steam distributing pipes must be protected from mechanical damage, corrosion, moisture, and the loss of heat from them; also pipe drainage and provision for expansion must be given attention.

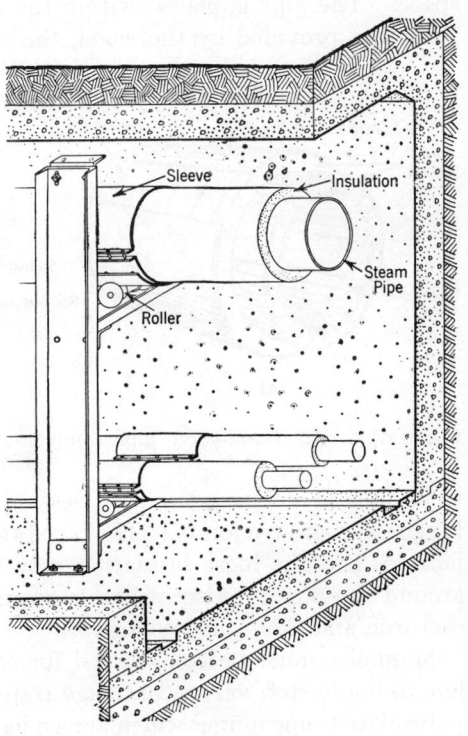

Fig. 11.16. Tunnel for underground steam mains at the University of Illinois.

Underground steam pipes are placed in tunnels or pipe conduits; the conduits have many forms. Tunnels, Fig. 11.16, are very desirable as the pipes are more accessible and the pipe insulation can be kept dry. The pipes are supported upon rollers with pipe insulation between the rollers and the mains. The insulation between the pipe and its roller support carries the weight of the pipe. The outer surface of the 85 per cent magnesia covering of the pipes is protected from abrasive action from the pipe roller by sheet-metal sleeves placed about the covering at the roller locations.

Two forms of pipe conduit which are buried in the ground are illustrated in Fig. 11.17. Conduit *a*, made in sections with tongue-and-groove joints at the ends, has wood staves bound together with wire hoops and is coated on the outside with an asphalt material as a protection against moisture from the earth. Inside the conduit is a tin-plated sheet-metal lining separated from the wooden walls by an air space. The pipe is placed within the lining without insulation other than that provided by the wood, the metal lining, and the air space.

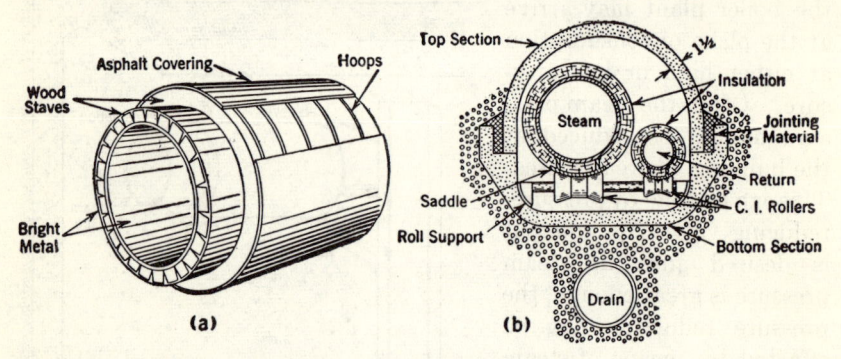

Fig. 11.17. (*a*) Wood-stave pipe conduit. (*b*) Precast concrete pipe conduit.

Conduit *b* is comprised of split sections of tile cemented together at tongue-and-groove joints. The pipes are supported on rollers properly located, and the loose insulation may be packed within the conduit around the pipes. Also available are underground conduits made of cast iron and preformed concrete.

Manholes must be constructed for conduits at intervals along the line to house stop valves, drainage traps, and expansion joints.

Insulated-pipe units,[5] which are asphalt-coated and tension-wrapped with either copper or aluminum foil, manufactured for installations of distributing mains and returns placed overhead find considerable usage.

Whenever possible the condensation from the distributing lines and that from the buildings served should be returned to the boilers to effect a saving of both heat and water. These return lines may be carried in the tunnels and conduits.

REFERENCES

1. "One-Pipe Steam-Heating Systems," *I = B = R Installation Guide No. 2*, Institute of Boiler and Radiator Manufacturers, New York (1950).

2. "Flow of Condensate and Air in Steam Heating Returns," F. C. Houghten and Carl Gutberlet, *ASHVE Trans.*, **39**, 179 (1933).
3. "Condensate and Air Return in Steam Heating Systems," F. C. Houghten and J. L. Blackshaw, *ASHVE Trans.*, **39**, 199 (1933).
4. "Flow of Steam Through Orifices into Radiators," S. S. Sandford and C. B. Sprenger, *ASHVE Trans.*, **37**, 371 (1931).
5. *Insulated Piping Systems, Underground and Overhead*, The Ric-Wil Company, Cleveland.

PROBLEMS

1. Find the friction-pressure losses in pounds per square inch in a 3-in. steam pipe line which has one gate valve fully open, five 90-deg elbows, and 130 ft of straight pipe through which 900 lb per hr of dry saturated steam flows with an average pressure of 16 psia.

2. Find the nominal size of steel pipe needed to convey 540 lb per hr of dry saturated steam under an average line pressure of 16.5 psia. The friction-pressure losses are not to exceed $1\frac{1}{2}$ oz per 100 equivalent ft of length.

3. Counterflow of steam and condensate takes place in a runout pipe which supplies 15 lb per hr of dry saturated steam entering it at 1.3 psig. Find the required nominal size of pipe. Barometer 29.92 in. of mercury at 32 F.

4. A one-pipe steam-heating system has 1152 sq ft of edr which operates with steam at 215 F in small-tube direct radiators. The steam and its condensate flow together in the supply main. Find the nominal size of steel pipe required at the entering end if the friction is not to exceed 1 oz per sq in. per 100 ft of equivalent length.

5. Assume that the one-pipe system of Problem 4 is laid out in the general form of Fig. 11.1. Specify the size of the "wet" portion of the return main if the supply main is sized to give a pressure drop of $\frac{1}{16}$ psi per 100 ft.

6. A vacuum return-line system is to serve radiators and convectors which have a total rated capacity of 7000 sq ft edr. The equivalent length of the supply main is 800 ft and the boiler steam pressure is to be 1 psig. Specify the nominal size of the steam-supply main as it leaves the boiler header and the nominal size of the dry return as it approaches the vacuum pump. The pipe material is wrought iron. Assume that the barometric pressure is standard.

12

Heating with Hot Water

12.1. Applications and Classifications of Systems. Flow of water in a system may be produced as the result of either density differences or by the action of a pump. On these bases, system classifications are gravity flow and forced circulation. Open systems have atmospheric pressure exerted upon the water of an expansion tank. Pressure systems may or may not have an expansion tank and may function with much higher water pressures and temperatures than are possible with open systems.

Open and pressure systems may have either upfeed or downfeed arrangements of the supply risers. Upfeed systems are further separated into (1) one-main installations which have a single pipe to serve as both a supply and return main, and (2) two-main plants having separate supply and return mains. Overhead downfeed systems have supply mains above the level of the highest radiators served and return mains below the lowest radiators. All overhead systems may have either single or double downfeed risers. Examples of the foregoing forms of hot-water systems are illustrated in Fig. 12.1.

Each of the system arrangements shown in Fig. 12.1 has an open expansion tank. The types of plants shown by cases *a*, *b*, and *c* may be installed and operated with closed tanks. Systems which have an

HEATING WITH HOT WATER

overhead supply main must have an open expansion tank, as in cases *d* and *e*, if air and other gases in the system are to be vented through it. All of the illustrations of Fig. 12.1 indicate gravity-flow circulation; however, each arrangement may be converted to forced circulation by the addition of a pump.

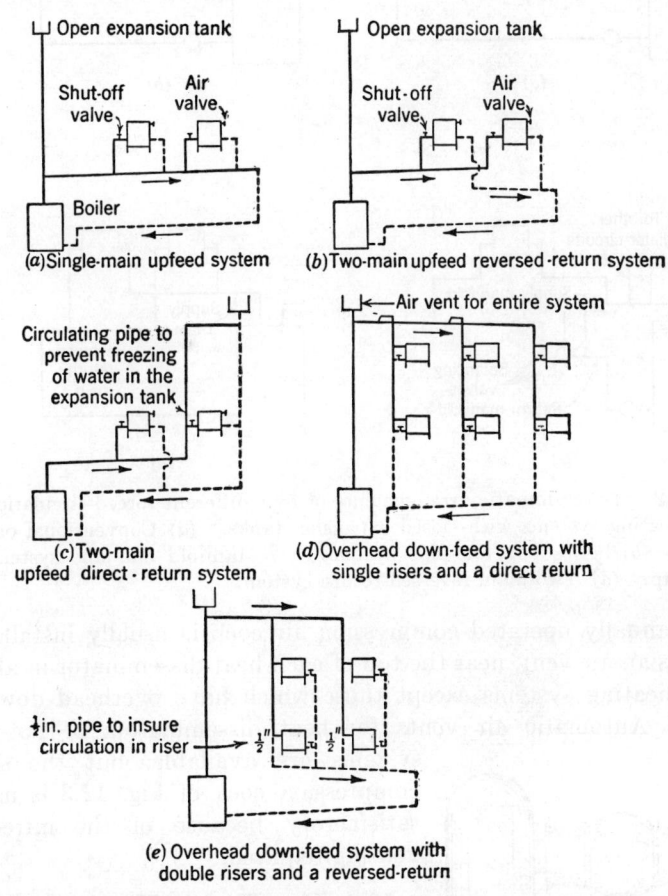

Fig. 12.1. Diagrammatic arrangements of five different gravity-flow hot-water heating systems with open expansion tanks.

Although use is made of the designs shown in Fig. 12.1, the majority of present-day systems of hot-water heating are one of the four types illustrated in Fig. 12.2. Use of a closed expansion tank allows greater flexibility of the temperature of the heated water. Action of a pump provides more positive water circulation and permits the use of smaller pipes throughout the system.

236   AIR CONDITIONING AND REFRIGERATION

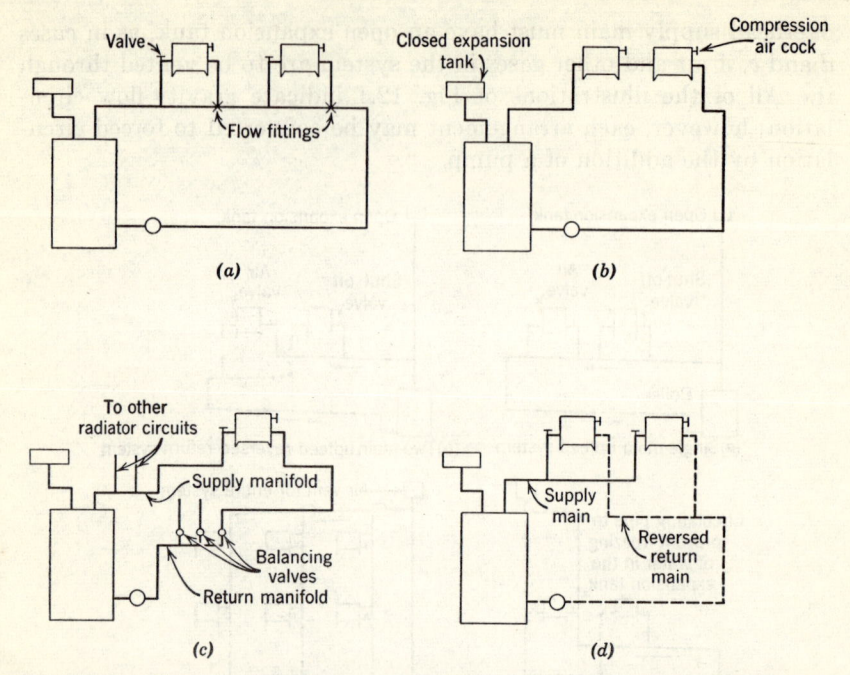

Fig. 12.2. Diagrammatic arrangements of four different forced-circulation hot-water heating systems with closed expansion tanks. (a) Conventional one-pipe system. (b) Series-loop one-pipe system. (c) Manifold one-pipe system with one pump. (d) Two-main reversed return system.

A manually operated compression air cock is usually installed, to serve as an air vent, near the top of each heat disseminator in all hot-water heating systems except those which have overhead downfeed mains. Automatic air vents for heat disseminators of hot-water systems are available, but the simple compression cock of Fig. 12.3 is usually satisfactory because of the infrequent venting required.

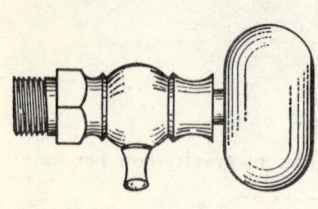

Fig. 12.3. Compression air cock.

12.2. Hot-Water Heating System Piping Details. Air will collect at all high points in the piping and radiators unless vented. All piping should have uniform grades. Mains grading downward in the direction of flow cannot be relayed or lifted in elevation unless an air vent is provided at the top of the relay. Provision must be made by properly installing and grading the pipes for complete drainage of the system at the boiler. Wherever reductions in pipe size are made

HEATING WITH HOT WATER

in horizontal lines eccentric reducing couplings are necessary, and these must be installed so that the tops of all sections of pipe joined are in alignment, thereby preventing air pockets.

12.3. Sizes of Open Expansion Tanks. All systems require some arrangement to take care of the increase of water volume when it is heated. Open systems universally employ an expansion tank to absorb the increase of volume when the water is heated. With an open-tank system the bottom of the tank is located at least 3 ft above the top of the highest radiator.

The volume of the expansion tank required depends upon the amount of water in the system and its temperature range. The maximum range of water temperature ever to be considered in an open-tank system is from 35 to 210 F. The densities of water at these two temperatures are 62.42 and 59.88 lb per cu ft respectively. Hence, when 1 cu ft of water is heated from 35 to 210 F, its final volume is $62.42 \div 59.88 = 1.0425$ times the original volume, which is an increase of 4.25 per cent. On this basis the volume of the expansion tank should be 4.25 per cent of the volume of the system. The expansion volume based on minimum and maximum temperatures other than 35 and 210 F can be determined by multiplying the system volume by $(d_{wc}/d_{wh} - 1)$, where d_{wc} and d_{wh} are the densities, pounds per cubic foot, of water at its assumed lowest and highest temperatures respectively. Extra capacity is advisable in an open expansion tank to allow for the loss of water by evaporation. In small systems the tank usually has a capacity at least double that of the smallest that could be used.

Boiler manufacturers seldom include data in their catalogues relative to the internal volumes of the units they build. The same situation also applies in the published literature for heat disseminators. Some data are usually to be found in the ASHAE Guide and average water-volume data for both heating boilers and disseminators are given in a paper[1] based on an extensive survey. The curves of Fig. 12.4 were plotted from the sources just mentioned. Information relative to the internal volumes of steel and wrought-iron pipes is given in Table 10.1. Similar data pertinent to copper tubing are included in Table 10.2. The following example illustrates the method of sizing an open expansion tank.

Example. A hot-water heating system has the following components: 210 sq ft of edr in the form of small-tube radiators, 120 sq ft of rated convector capacity, a boiler with a net rating of 66,000 Btuh, 144 linear ft of 1-in. nominal-size steel pipe, and 141 linear ft of $\frac{3}{4}$-in. pipe of the same material.

The temperatures of the water in the system are assumed to never be less than 50 F nor more than 200 F. Specify the expansion tank size in gallons.

Solution. Figure 12.4 shows that the volume of water in 210 sq ft of edr of small-tube radiators is 11.6 gal. For 120 sq ft of convector surface an internal volume of 3.12 gal is obtained as one-fourth of that indicated in Fig. 12.4 for 480 sq ft. The boiler volume, obtained from Fig. 12.4, is 15 gal, Table 10.1 gives a linear length of 166.6 ft of 1-in. steel pipe as having 1 cu ft

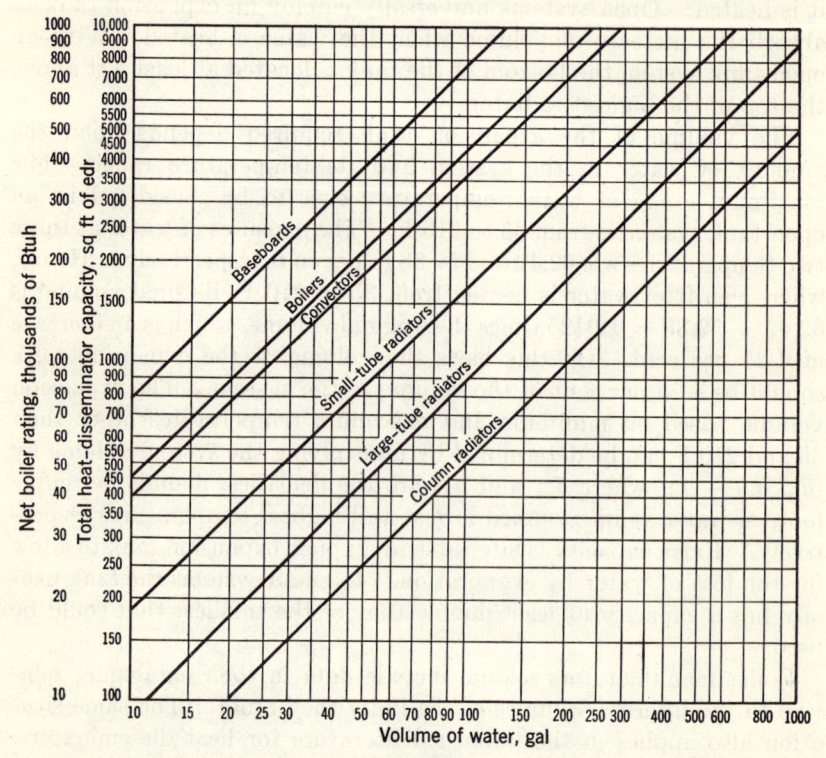

Fig. 12.4. Water capacities of boilers and heat disseminators of heating systems.

of internal volume. Therefore, the 144 ft of 1-in. pipe contains 144/166.6 or 0.864 cu ft, which is 0.864 × 7.48 = 6.46 gal. Similarly, 141 ft of $\frac{3}{4}$-in. pipe is found to have a volume of 3.91 gal. The total volume of the system is the sum of the component quantities and is 11.6 + 3.12 + 15 + 6.46 + 3.91 = 40.09 gal. Water density at 50 F is 62.38 and at 200 F 60.13 lb per cu ft. The expansion volume is found to be [(62.38/60.13) − 1] × 40.09 = 1.48 gal. Inasmuch as a small system is involved, the capacity of the tank should be twice the expansion volume or 3.0 gal.

12.4. The Sizing of Closed Expansion Tanks. The function of a closed expansion tank in a hot-water system is to maintain sufficient pressure to keep the highest radiators filled when the system is cold

HEATING WITH HOT WATER

and prevent the creation of dangerously high pressures when the system is hot. This regulation of the pressure in a closed system is provided through the expansion and contraction of an adequate volume of air which is trapped in the upper portion of the expansion tank. The greater the volume of air trapped in the tank the less will be the variation in pressure as the temperature of the water in the system fluctuates through its operating range. The procedure to be used in calculating the required size of tank for any given system can best be illustrated by an example.

Example. Specify the size of a closed expansion tank for the system of the previous article. The tops of the highest radiators in the system are located 4 ft above the water level in the expansion tank when the water temperature is 50 F. A minimum pressure of 5 psig is to be maintained at these locations. The maximum and minimum temperatures will be taken as in the preceding example and the assumption will be made that the maximum allowable pressure in the expansion tank is 25 psig when the barometric pressure is 14.7 psi.

Solution. The expansion volume is 1.48 gal as computed in the prior example. Three different pressures together with the corresponding air volumes in the tank must be considered. Before the system is filled with water the volume of the air V_1 to be considered is that of the expansion tank in cubic feet. The corresponding atmospheric pressure p_1 is that of the atmosphere or 14.7 psia. When the system is filled with cool water, at 50 F, some of it will flow into the expansion tank and compress the contained air until its pressure is equal to that in the system risers at the same horizontal level. This air pressure will be designated as p_2 which is equal to 14.7 + 5 + (4 × 62.38)/144 = 21.435 psia. The corresponding air volume at 21.435 psia is V_2. In this case when the water of the system is heated a quantity equal to 1.48 gal is forced into the expansion tank to compress its contained air to a volume of V_3 cu ft and a pressure of p_3 psia. In this case p_3 is arbitrarily taken as 25 psig or 39.7 psia. The temperature of the water in the expansion tank is assumed to remain constant as there is no fluid circulation in this part of the system.

The following relationships apply (Art. 1.12): $P_2V_2 = P_3V_3$ and $P_1V_1 = P_2V_2$. (Generally the volumes involved are expressed in cubic feet but the same results may be obtained if the quantities are consistently used in terms of gallons. Similarly, the equations of Art. 1.12 require that the pressure be in pounds per square foot, but since a pressure ratio is involved in each calculation here the units may be in pounds per square inch absolute.) Volume V_3 is equal to V_2 minus the expansion volume and is $V_2 - 1.48$ in this example; P_2 and P_3 have been previously determined; $21.435V_2 = 39.7(V_2 - 1.48)$; $V_2 = 3.22$ gal; $14.7V_1 = 21.435 \times 3.22$; and $V_1 = 4.71$ gal.

The foregoing procedure indicates the minimum required capacity of the expansion tank necessary for specified conditions. No allowance is included for loss of air from the tank. In a small system the usual practice is to install a closed tank with a volume equal to at least two times the calculated minimum volume. Therefore, in this case the recommended tank capacity is 10 gal. The discussions of this and the prior article indicate that a closed expansion tank must be much larger than an open tank for a system of the same internal volume.

12.5. Advantages and Operating Characteristics of Closed-Tank Systems. It is possible to carry water at higher temperatures without danger of boiling in closed-tank or pressure systems than in open expansion tanks. When a closed expansion tank is properly sized for the system which it serves, there is no loss of water from the relief valve so long as the expansion tank contains the necessary amount of air.

A relief valve which functions when a new system first goes into operation gives an indication that the expansion tank is too small. Water discharge by the valve after several weeks or months of operation of the plant indicates that air has been lost from the expansion tank (usually by transfer to the disseminators in the warming and cooling of the water in the system) to an extent that the volume remaining is insufficient. Tanks of inadequate capacity should be replaced by larger ones. When air transfer has occurred the expansion tank should be drained and the original volume of air restored. Special fittings which are now available may be applied at the water outlet from the boiler to reduce air transfer from the expansion tank. Frequent discharge of water by the relief valve when the system is hot necessitates its replacement by cool fresh water. Repeated feeding of raw water containing mineral materials in solution leads to the deposition of scale in a hot-water boiler. A few manufacturers design the boiler and outlet piping of their units with a top air space which is adequate to take care of the water expansion of an entire system.

12.6. Pressure Systems Without Expansion Tanks. Such systems are equipped with pressure-relief valves and automatic water feeders and may be used when good make-up water is available. When the pressure within the system becomes greater than that for which the relief valve is set, the safety device discharges water. When the pressure is insufficient the pressure-reducing valve allows water from a service line to enter the system. Many systems of this type are in operation but in some cases frequent addition of water to the system is inevitable.

12.7. Resistances to Flow in Hot-Water Heating Systems. Friction losses in the pipes of hot-water heating systems may be com-

HEATING WITH HOT WATER

puted by the use of equation 6.3. The friction-loss chart of Fig. 12.5 is a result of research by Dr. F. E. Giesecke and applies to clean commercial steel and wrought-iron pipes. Although friction-pressure losses vary somewhat with changes of water density and viscosity, the data of Fig. 12.5 may be used, without appreciable error, in sizing the pipes and in computing pump heads for all hot-water heating

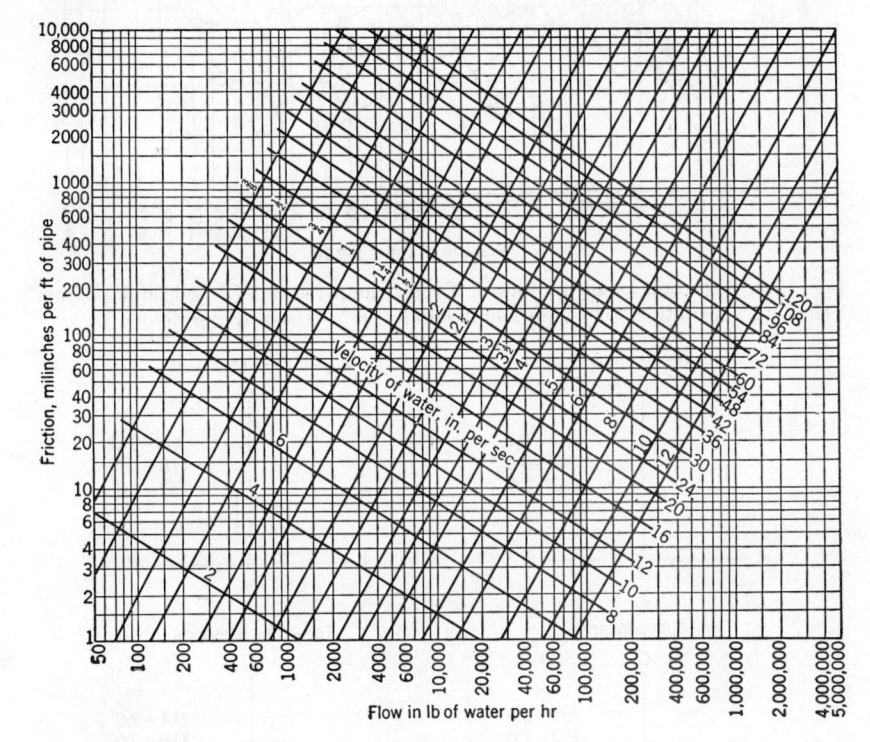

Fig. 12.5. Friction head per foot of black-iron pipe in hot-water heating systems. (Based on data from "Heating Ventilating Air-Conditioning Guide 1957." Used by permission.)

systems constructed with either steel or wrought-iron pipes. Figure 12.6 gives data for copper tubing.

Most designers of hot-water heating systems make use of the elbow equivalents of pipe fittings instead of their equivalent lengths. Elbow equivalents for both iron and copper fittings are given in Table 12.1. The elbow equivalent of a tee is greatly affected by the conditions of flow at the point of its location. Data for elbow equivalents of tees are given in Figs. 6.12 and 6.13. The diagrams of Fig. 12.7, which

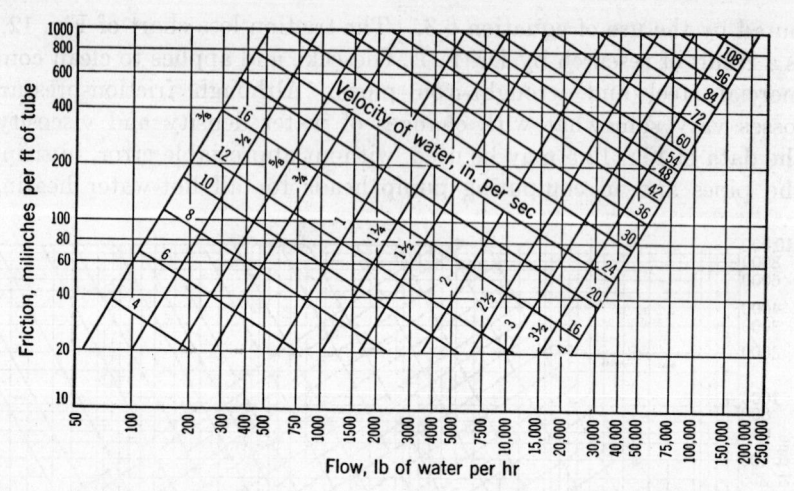

Fig. 12.6. Friction losses per foot of Type-L copper tube. (Based on data from "Heating Ventilating Air-Conditioning Guide 1957." Used by permission.)

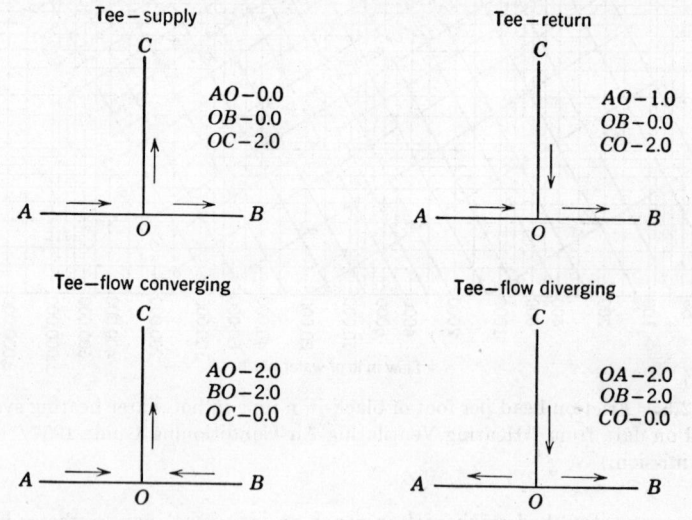

Tee – supply

AO – 0.0
OB – 0.0
OC – 2.0

Tee – return

AO – 1.0
OB – 0.0
CO – 2.0

Tee – flow converging

AO – 2.0
BO – 2.0
OC – 0.0

Tee – flow diverging

OA – 2.0
OB – 2.0
CO – 0.0

Fig. 12.7. Approximate elbow equivalents of tees. (Abstracted from $I = B = R$ *Installation Guide No. 4.*)

were taken from reference 2, are often employed to find the elbow equivalents of tees in hot-water heating systems when time is not available to make more exact determinations.

Fittings at lettered points (see Fig. 12.8) in the supply main should be included in the downstream section. Fittings at lettered points in the return main should be included in the upstream section.

HEATING WITH HOT WATER

After the sum of the elbow equivalents of fittings has been found for the portion of the circuit which is under consideration, it may be converted to equivalent length of straight pipe or tubing in feet by

$$L_e = \left(\frac{25 \times d \times E_e}{12}\right) \qquad (12.1)$$

where d is the nominal diameter of the pipe or tube, inches. The total equivalent length of the section being considered is then found

TABLE 12.1. Iron and Copper Elbow Equivalents

Fitting	Iron Pipe	Copper Tubing	Fitting	Iron Pipe	Copper Tubing
Elbow, 90-deg	1.0	1.0	Open-globe valve	12.0	17.0
Elbow, 45-deg	0.7	0.7	Angle radiator valve	2.0	3.0
Elbow, 90-deg long-turn	0.5	0.5	Radiator or convector	3.0	4.0
			Boiler or heater	3.0	3.0
Elbow, welded 90-deg	0.5	0.5	Tee, when used in place		
Reduced coupling	0.4	0.4	of an elbow	1.8	1.2
Open return bend	1.0	1.0	Flow-control valve	21	21
Open gate valve	0.5	0.7	Wide-open cock	1	1

by adding the equivalent length of the fittings and the actual length of the pipe or tubing.

12.8. Systems as Affected by Pipe Sizes. The desirable condition is that the friction losses in all pipe circuits be the same and just equal to the total head available. In gravity systems, if all pipes are too small the flow of water is retarded and its temperature drop is increased, which in turn tends to increase the operating head but also reduces the amount of heat given off by the radiators. Gravity systems are self-compensating in many respects so that systems laid out with inexact pipe sizes often, but not always, adjust themselves to a fairly satisfactory operating condition. Where the pipe sizes are all too small, changes of operation may be effected by means of a circulator. The selected pipe sizes are then those of the minimum size and cost of installation which will not unduly increase the required head to be maintained by the pump and the expense of operation.

The water may flow through a forced-circulation system so that it will cool from 10 to 30 F or any other reasonable number of degrees in the radiators. Generally a water temperature drop of 20 F in the system gives good results without excessive power requirements for circulating the water.

Heat losses from insulated mains and pipes can usually be ignored. However, if it seems desirable, the actual amount of surface in the radiators can be decreased to compensate for the heat given off by the mains, etc., after the pipe sizes have been determined. When forced-circulation systems are used in residences, the mains and all other pipes are so small that the heat loss from them is negligible, even when they are not insulated.

12.9. Sizing of Pipes for Hot-Water Heating Systems. In general the procedures for sizing the pipes of gravity-flow and forced-circulation hot-water heating systems are much the same. With gravity-flow systems the available head in a radiator circuit is dependent on the heights of two water columns and their differences in density. The design conditions thereby limit the available head in gravity-flow installations and great care must be exercised in order that the friction losses for the conditions of the design do not exceed the head available in any water circuit. With forced circulation the available head is not limited and the problem becomes one of selecting pipes of such sizes that the required amount of water can be handled with a reasonable expenditure for power.

In either system the design should be made so that the resistances to flow in all water circuits are such that some circuits are not favored over others, thereby causing one portion of the building to heat more readily than other portions. With commercial pipes of uniform sizes in the different sections it is practically impossible to balance the circuits so that all function to give exactly the desired results. Therefore it may become necessary to introduce additional resistance in some water circuits to balance their friction losses with those of other circuits. This can be done by inserting in those circuits where additional resistance is required either sections of pipe of a smaller size, balancing elbows and valves, or orifices which create an obstruction to flow. Table 12.2, based on the work of Dr. Giesecke, gives frictional data for central-opening circular-diaphragm orifices when placed in commercial pipes carrying water with different velocities of flow. Two parallel circuits can easily be balanced if a suitable cock is included in each one. Since relatively few gravity systems are currently being installed, detailed instructions for the design of this type will not be given here.

12.10. Piping Design for Two-Main Forced-Circulation Hot-Water Systems. There is a point where the velocity of flow and the required pressure head in a forced-circulation systems result in the minimum cost of installation and operation. Under ordinary con-

Table 12.2 Friction Heads (in Milinches) of Central Circular-Diaphragm Orifices in Unions*

Diameter of Orifices, In.	Velocity of Water in Pipe, In. per Sec									
	2	3	4	6	8	10	12	18	24	36
¾-in. Pipe											
0.25	1300	2900	5000	11,300	20,800	32,000	45,000			
0.30	650	1450	2500	5,700	10,400	16,000	23,000	57,000		
0.35	330	740	1300	2,900	5,200	8,000	12,000	26,000	47,000	
0.40	170	380	660	1,500	2,600	4,000	6,800	13,000	24,000	53,000
0.45	...	185	330	740	1,300	2,000	2,900	6,500	12,000	27,000
0.50	...	...	155	350	620	970	1,400	3,200	5,700	13,000
0.55	...	...	75	170	300	480	700	1,600	2,800	6,400
1-In. Pipe											
0.35	900	2000	3500	7,800	14,000	22,000	32,000			
0.40	460	1000	1800	4,000	7,200	12,000	17,000	37,000	65,000	
0.45	270	570	1000	2,300	4,100	6,400	9,300	21,000	37,000	
0.50	160	330	580	1,400	2,300	3,700	5,400	12,000	22,000	50,000
0.55	...	190	330	750	1,300	2,200	3,000	7,000	13,000	28,000
0.60	...	...	200	440	800	1,300	1,800	4,200	7,400	17,000
0.65	...	...	120	260	460	720	1,100	2,400	4,300	10,000
1¼-In. Pipe											
0.45	1000	2250	4000	8,900	16,000	25,000	36,000			
0.50	660	1450	2600	5,800	10,400	16,400	23,000	53,000		
0.55	430	950	1700	3,800	6,800	10,500	15,000	34,000	60,000	
0.60	280	630	1100	2,500	4,400	6,900	10,000	22,000	40,000	
0.65	190	420	750	1,700	3,000	4,700	6,700	15,000	27,000	60,000
0.70	...	285	510	1,150	2,000	3,100	4,500	10,000	18,000	40,000
0.75	...	190	330	750	1,300	2,100	3,000	6,700	12,000	26,000
1½-In. Pipe†											
0.55	850	1900	3300	7,400	13,000	21,000	30,000			
0.60	600	1300	2300	5,400	8,600	16,800	21,000	50,000		
0.65	400	850	1500	3,600	7,200	10,400	14,000	30,000	53,000	
0.70	260	600	1100	2,600	4,400	7,000	10,000	21,000	39,000	
0.75	180	400	760	1,800	3,000	5,000	7,000	14,000	28,000	
0.80	...	300	540	1,200	2,200	3,200	5,000	10,200	19,000	45,000
0.85	...	200	380	860	1,600	2,300	3,000	7,800	13,000	30,000
2-In. Pipe†											
0.70	890	1850	3500	7,400	14,000	22,300	33,000			
0.80	470	975	1800	3,900	7,400	11,700	17,000	37,000		
0.90	255	560	1000	2,200	4,200	6,500	9,500	20,500	38,000	
1.00	160	340	610	1,320	2,520	4,000	5,800	12,500	23,000	49,000
1.10	...	214	375	850	1,600	2,500	3,700	7,900	14,000	30,000
1.20	...	...	195	460	950	1,360	1,910	4,200	8,100	16,800
1.30	...	...	...	275	525	980	1,375	3,100	4,400	8,850

* From "Heating Ventilating Air-Conditioning Guide 1956." Used by permission.

† The losses of head for the orifices in the 1½-in. and 2-in. pipe were calculated from those in the smaller pipes, the calculations being based on the assumption that, for any given velocity, the loss of head is a function of the ratio of the diameter of the pipe to that of the orifice. This had been found to be practically true in the tests to determine the losses of head in orifices in ¾-in., 1-in., and 1¼-in. pipe, conducted by the Texas Engineering Experiment Station and also in the tests to determine the losses of head in orifices in 4-in., 6-in., and 12-in. pipe, conducted by the Engineering Experiment Station of the University of Illinois (*Bulletin 109*, Table 6, p. 38, Davis and Jordan).

ditions the economical friction-pressure losses will amount to 140 to 400 milinches of water per ft of run.

The following example illustrates the procedure to be followed in designing two-main forced-circulation reversed-return hot-water heating systems.

Example. Assume a system having five radiators, one boiler, and a pump as indicated in Fig. 12.8. The lengths of straight pipe and the elbows of both

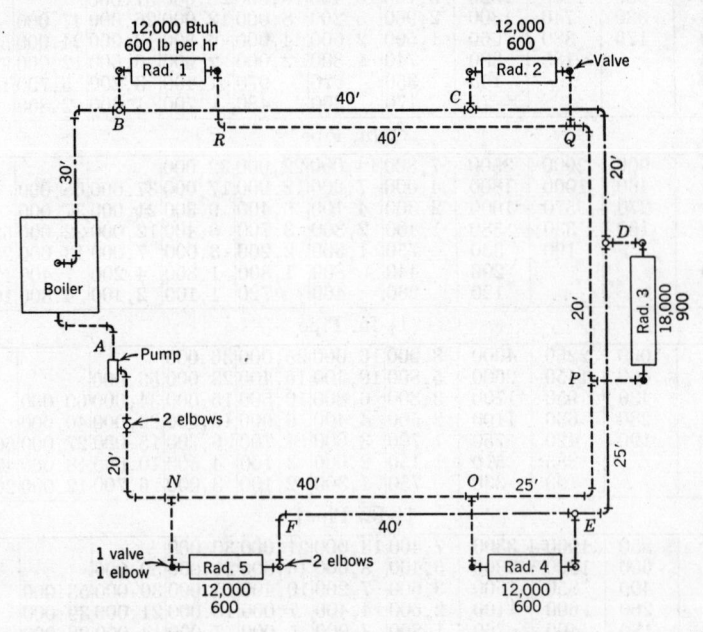

Fig. 12.8. Forced-circulation reversed-return hot-water heating system. Expansion-tank connections not shown.

the supply and return mains are shown, together with the heat outputs of the radiators and the weights of water to be handled by the various portions of the system. No allowances will be made for the loss of water temperature in the supply mains and pipes, and the decrease in the water temperature at all radiators will be taken as 20 F. Each radiator has 3 ft of pipe in the supply connection from the main and 4 ft of pipe in the outlet connection to the return main. All radiators have two elbows in the inlet connection and radiators 1, 2, 3, and 4 also have in addition a connection through a tee to the supply main. Each radiator has at its outlet an angle valve, one 90-deg elbow, and all radiators with the exception of number 1 have a connection through a tee to the return main. Determine the pipe sizes, assuming that iron pipe will be used throughout.

Solution. An arrangement such as Table 12.3 is a convenience in tabulating the calculated values. The headings of the table are self-explanatory. Elbow equivalents of tees are given in Fig. 12.7.

HEATING WITH HOT WATER

TABLE 12.3. Pipe-Sizing Data for a Reversed-Return Forced-Circulation Hot-Water Heating System

Section	Water Handled lb per hr	Pipe Size, in.	Water Velocity, in. per sec	Length of Straight Pipe, ft	Elbow Equivalents of Fittings	Equivalent Lengths of Fittings, ft	Total Equivalent Length of Section, ft	Friction Loss per Foot, milinches	Friction Loss in Section, milinches	Accumulative Total Friction Losses, milinches
Supply main										
AB	3300	1	30	30	8	16.7	46.7	400	18,700	18,700
BC	2700	1	24	40	0	0	40.0	280	11,200	29,900
CD	2100	1	18	20	1	2.1	22.1	190	4,200	34,100
DE	1200	3/4	16	25	1	1.6	26.6	180	4,790	38,890
EF	600	3/4	9	40	0	0	40.0	57	2,280	41,170
Return main										
AN	3300	1	30	20	4	8.3	28.3	400	11,300	11,300
NO	2700	1	24	40	1	2.1	42.1	280	11,800	23,100
OP	2100	1	18	25	2	4.2	29.2	190	5,550	28,650
PQ	1200	3/4	16	20	2	3.1	23.1	180	4,150	32,800
QR	600	3/4	9	40	1	1.6	41.6	57	2,380	35,180
Radiator 1	600	3/4	9	7	11	17.2	24.2	57	1,380	
Radiator 2	600	3/4	9	7	12	18.7	25.7	57	1,460	
Radiator 3	900	3/4	13	7	12	18.7	25.7	125	3,210	
Radiator 4	600	3/4	9	7	12	18.7	25.7	57	1,460	
Radiator 5	600	3/4	9	7	11	17.2	24.2	57	1,380	

The values which appear in the table for section AB in the supply main were derived as follows. The water handled was obtained by adding the weights required in each of the five radiators of the system. The nominal pipe size was chosen with the aid of Fig. 12.5. A 1-in. pipe was selected, as the friction loss with this size falls in the recommended range of 140 to 400 milinches per ft. The velocity was obtained from Fig. 12.5, the length of straight pipe from Fig. 12.8. The elbow equivalents of the fittings were computed as follows: two elbows between the pump and the boiler plus three, which is the elbow equivalent of the boiler, plus three elbows between the boiler and point B. The elbow equivalent of the fittings in section AB is therefore $2 + 3 + 3 = 8$. The equivalent length of the fittings in this section was calculated with equation 12.1 as $L = (25 \times 1 \times 8)/12 = 16.7$ ft. The total equivalent length of the straight pipe and fittings is $30 + 16.7 = 46.7$ ft. The friction loss per foot, obtained from Fig. 12.5, is 400 milinches and the friction losses for the section are $46.7 \times 400 = 18,700$ milinches. Similar data for the other sections of the supply main can be obtained in an identical manner. (The friction losses within the supply main from tees are negligible.)

Friction-loss data applicable to the various sections of the return main were determined as in the preceding paragraph with the exception that tees were included in the computations of the equivalents of the pipe fittings. The elbow equivalent of the fitting in section NO is that of the tee at N and from Fig. 12.7 is 1.0.

Data for each of the five radiator circuits were computed as previously outlined. No pipe sizes less than $\frac{3}{4}$ in. were included so that orifices may be used to balance circuits when necessary. The fittings considered in the circuits of radiators 2, 3, and 4 are the tee in the supply main, for which the elbow equivalent (Fig. 12.7) is 2.0; two elbows between the supply main and the radiator; three elbow equivalents for the radiator; two elbow equivalents for the radiator angle valve (Table 12.1); one elbow between the valve and the return main; and the tee in the return main with an elbow equivalent of 2.0 as indicated in Fig. 12.7. The circuits of radiators 1 and 5 have one elbow and one tee. The completion of the example involves the calculations of the friction losses in each complete circuit of all five radiators and the selection of orifices needed for balancing. The necessary data are tabulated in Table 12.4.

The complete circuit, which includes radiator 3, has the greatest friction losses and it therefore becomes the control radiator. The resistance of each of the other circuits must be made to equal that of the control circuit if the correct weight of water per hour is to be circulated through it. The simplest method of balancing the system is the inclusion of an orifice of the proper diameter (Table 12.2) in each circuit except that of the control radiator. The velocity of water flow required in the selection of an orifice from Table 12.2 is included in Table 12.3. When fixed orifices are not available, the resistance of any circuit other than that of the control one may be increased either by changes of pipe sizes in the branch circuits or by the employment of adjustable resistance fittings.

The head to be overcome by the action of a pump is equal to the resistance of the control circuit. In this case the pump head is $65,960/(12 \times 1000) = 5.5$ feet of water.

HEATING WITH HOT WATER

TABLE 12.4. Diameters of Orifices Required to Balance the Radiator Circuits of the System of Fig. 12.8

Radiator Number	Friction Loss in Supply Main, Pump to Tee, milinches	Friction Loss in Radiator Circuit, Supply Main to Return Main, milinches	Friction Loss in Return Main, Tee to Pump, milinches	Total Friction Loss in Complete Circuit, milinches	Additional Resistance Required, milinches	Diameter of Required Orifice, in.
1	18,700	1380	35,180	55,260	10,700	0.30
2	29,900	1460	32,800	64,160	1,800	0.45
3	34,100	3210	28,650	65,960	0	—
4	38,890	1460	23,100	63,450	2,510	0.40
5	41,170	1380	11,300	53,850	12,110	0.30

The same general procedure which was used in the preceding example can be employed for sizing the pipes in a forced-circulation direct-return system. In a hot-water system of this type the circuits through the different radiators are of different lengths so that it is necessary to choose the radiator farthest from the boiler as the control radiator.

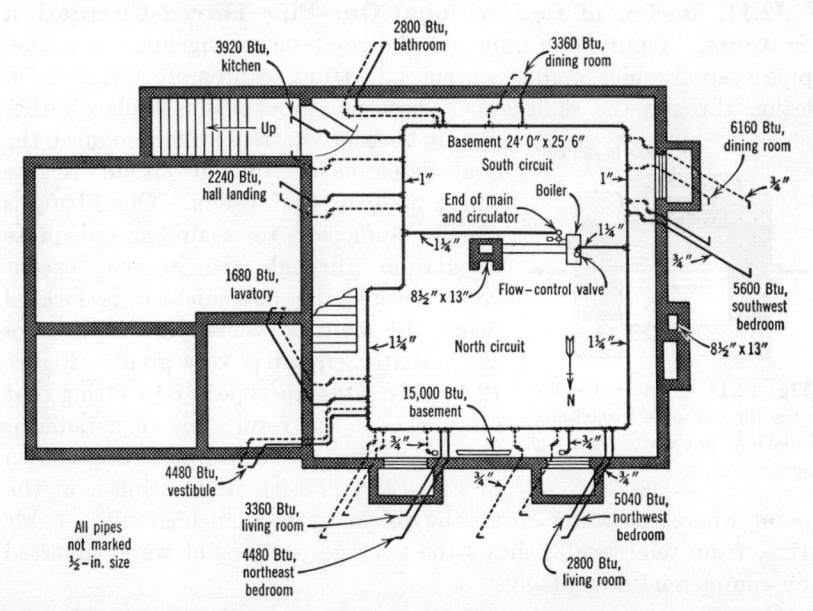

Fig. 12.9. Layout of basement piping for a one-pipe forced-circulation hot-water heating system.

When the heat lost from the piping is neglected, the mean temperature of the water in all of the heat disseminators of any two-pipe hot-water heating system, designed on the basis of an over-all system-temperature drop of 20 F, is 10 F less than the temperature of the water in the boiler.

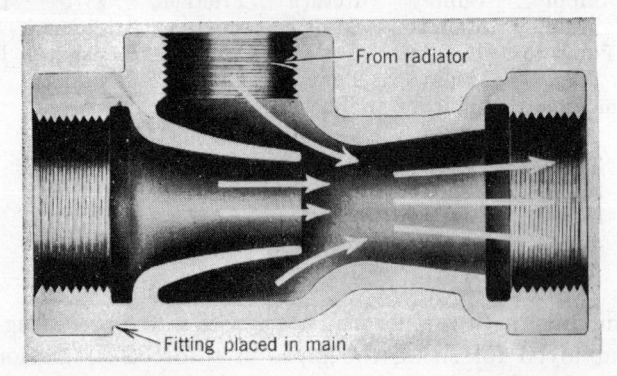

Fig. 12.10. Induced-flow fitting for single-main forced-circulation hot-water heating system. (Courtesy Trane Co.)

12.11. Design of Conventional One-Pipe Forced-Circulation Systems. Figure 12.9 indicates the common arrangement of a one-pipe system which employs a special fitting to promote water circulation through the radiators. These one-pipe fittings replace either one or both of the tees which connect the heat-disseminator branch circuit to the single main of the system. One fitting is usually sufficient to secure an adequate circulation through the branch, except where either the disseminator is located below the main or where the resistance of the radiator circuit is very great. Figure 12.10 illustrates one type of flow fitting that may replace the return tee of a radiator circuit. When the style of fitting shown in Fig. 12.11 is used, its location is at the point where the water enters the disseminator-branch circuit. Table 12.5, from reference 3, shows the average portions of water diverted by commercial flow fittings.

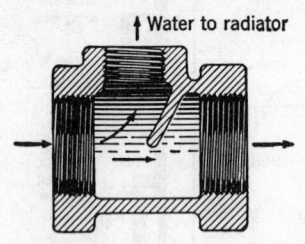

Fig. 12.11. Water distributor for one-pipe forced-circulation hot-water heating system.

Conventional one-pipe systems may be laid out with one or more flow circuits. The system of Fig. 12.9 consists of a trunk which divides into two circuits.

HEATING WITH HOT WATER

The mean temperature of the water in the various disseminators of a one-pipe system is not the same, as is the case in a two-pipe arrangement. The procedure to be used in determining the actual mean temperatures which can be expected can best be illustrated by an example.

TABLE 12.5. Per Cent of Water Flowing in a Main Diverted Through a Heat Disseminator by a One-Pipe Fitting

Nominal Size of One-Pipe Fitting, in.	Location Above Main, per cent	Below Main, per cent
$\frac{3}{4} \times \frac{1}{2}$	34.0	29.0
$1 \times \frac{1}{2}$	15.0	12.8
$1 \times \frac{3}{4}$	25.0	21.3
$1\frac{1}{4} \times \frac{1}{2}$	10.0	8.5
$1\frac{1}{4} \times \frac{3}{4}$	16.0	13.6
$1\frac{1}{2} \times \frac{1}{2}$	8.0	6.8
$1\frac{1}{2} \times \frac{3}{4}$	11.0	9.4

Example. When a constant boiler-water temperature of 200 F is maintained, find the mean water temperature in each heat disseminator served by the south circuit, Fig. 12.9, comprised of 1-in. nominal-size pipe. Design for an over-all system-temperature drop of 20 F and arrange the computed data in table form.

Solution. The computations of the mean temperature of the water in the heat disseminator of the southwest bedroom follow (see Table 12.6). The disseminator load is 5600 Btuh and the total circuit load is 24,080 Btuh. The circuit water-temperature drop caused by the disseminator is $(5600 \times 20)/24{,}080$ or 4.66 F. This quantity, 4.66 F, subtracted from the upstream circuit temperature, 200 F, gives 195.34 F as the temperature of the water entering the next disseminator. Table 12.5 indicates that 25 per cent of the circuit water passes through the disseminator when it is located above the 1-in. main and has a $\frac{3}{4}$-in. branch connection. The disseminator temperature drop is $(4.66 \times 100)/25$ which is 18.65 F, and the heat-disseminator mean water temperature is $200 - (18.65/2) = 190.68$ F. Other computed data shown in Table 12.6 for all disseminators served by the same main circuit were obtained in the same manner. When more than one heat-emitting device is placed in a space, as in the dining room, the treatment is as though two or more separate rooms are involved and their total heat losses per hour are equal to those estimated for the single room. When the mean water temperatures are known the selection of radiators, convectors, or other devices may be made by the methods of Chapter 8.

The problem of determining the friction to be overcome in a conventional one-pipe hot-water heating system is simpler than when a two-pipe system is involved, as it is not necessary to calculate the resistances of the radiator-branch circuits. Resistances introduced into circuits by the installation of flow fittings must be considered. Figure 12.12 gives data for the resistances of one make of flow fittings. The lower portion of the graph shows the resistance when only one flow fitting is used to divert water through a heat disseminator. The upper portion of the figure gives the resistance added when both the

TABLE 12.6. Data Pertinent to Determination of Heat-Disseminator Mean Water Temperatures in a Specified Circuit of a One-Pipe Hot-Water Heating System

Room Served	Room Number	Heat-Disseminator Load, Btuh	Up-stream-Circuit Water Temperature, deg F	Circuit Water Temperature Drop, deg F	Per Cent of Water Diverted	Heat-Disseminator Temperature Drop, deg F	Mean Temperature of Water in heat Disseminator, deg F
Southwest bedroom	1	5,600	200.00	4.66	25	18.65	190.68
Dining room 1	2	6,160	200.00	5.12	25	20.48	189.76
Dining room 2	3	3,360	190.22	2.79	15	18.60	180.92
Bathroom	4	2,800	187.43	2.32	15	15.47	179.70
Kitchen	5	3,920	184.11	3.26	15	21.70	173.26
Hall landing	6	2,240	181.85	1.85	15	12.32	175.69
Total load		24,080					

upstream and the downstream tees are replaced by flow fittings. A method of calculation of the head to be overcome by a pump used to circulate water in a conventional form of one-pipe hot-water heating system is indicated by the following example.

Example. Compute the total resistance of the two-circuit one-pipe hot-water heating plant of Fig. 12.9.

Solution. The computations and data are presented in the form of a table (see Table 12.7). The necessary pipe lengths are obtainable from measurements of a scale drawing of the plant and the fittings involved may be ascertained by a study of the piping arrangements of Fig. 12.9. The weight of water flowing in the south circuit (1204 lb per hr) is the total heat load from Table 12.6 divided by 20. The weight handled by the north circuit is obtained in the same manner from the data given in Fig. 12.9. The weight passing through the trunk is the sum of the amounts handled by the circuits. The total resistance of the system is equal to 9000 + 10,752 = 19,752 milinches.

The preceding computations involved elbow equivalents of the fittings which were obtained from Table 12.1 and the equivalent lengths which were computed by use of equation 12.1. The friction loss produced by each single-flow fitting in the north circuit, i.e., the one which is entirely made of $1\frac{1}{4}$-in. pipe and which handles 1842 lb per hr of water, is obtained from the lower

HEATING WITH HOT WATER

portion of Fig. 12.12. The loss amounts to 1000 milinches per fitting and the total loss is 6000 milinches. The resistance of the double-flow fitting (two fittings are used for the heat disseminator serving the basement because it is located below the main which feeds it) is shown by the upper portion of Fig. 12.12 and is 2000 milinches.

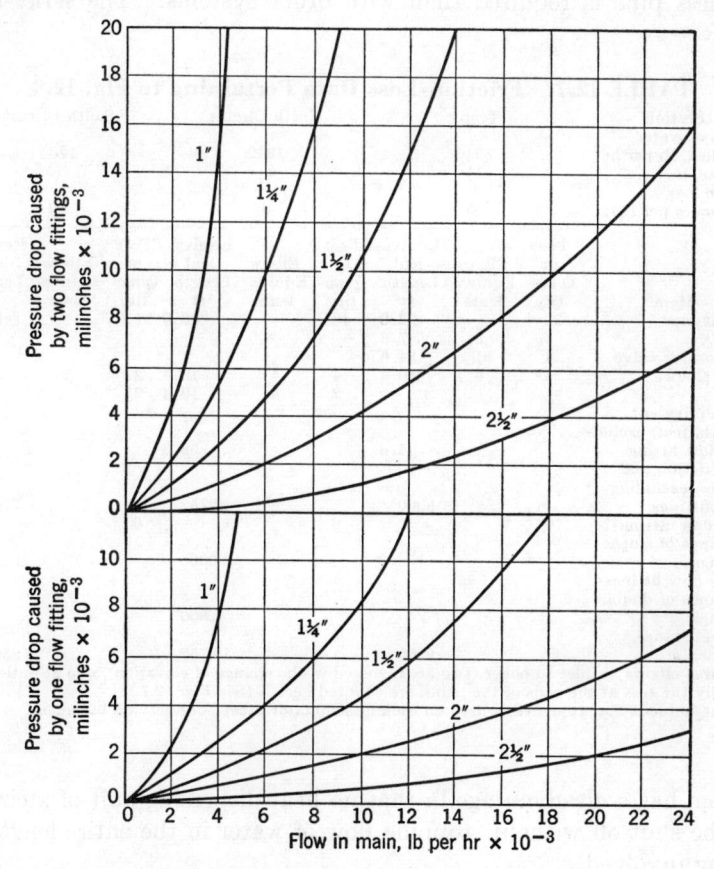

Fig. 12.12. Resistance of "monoflo" fittings. (Replotted from Bell and Gossett *Handbook*.[4] Used by permission.)

In order to have exactly the desired weight of water flowing in each of the two main circuits, it is necessary to install a cock in the south circuit so that 10,752 − 9303 or 1449 milinches of resistance may be added by partially closing it.

12.12. Design of a One-Pipe Series-Loop Hot-Water Heating System. One-pipe series-loop systems, Fig. 12.2b, are especially advantageous in buildings constructed upon a concrete slab. Piping from the boiler room to the beginning of each circuit and from the end of each circuit to the boiler room may be laid before concrete is poured about it. The circuit pipes continue from heat disseminator to heat disseminator without a parallel main. The series loop is often used in buildings which have basements because flow fittings are not needed and less pipe is required than with other systems. The series-loop

TABLE 12.7. Friction-Loss Data Pertaining to Fig. 12.9

Section	Trunk			North Circuit			South Circuit		
Weight of water handled, lb per hr	3046			1842			1204		
Pipe size, in.	1¼			1¼			1		
Friction loss, milinches per foot	95			40			62		
Item	Feet or Quantity	Elbow Equivalents	Equivalent Length, ft	Feet or Quantity	Elbow Equivalents	Equivalent Length, ft	Feet or Quantity	Elbow Equivalents	Equivalent Length, ft
Straight pipe	22		22.0	48		48.0	31.0		31.0
Boiler	1	3	7.8						
Flow-control valve	1	21	54.6						
90-deg elbows	4*	4	10.4	4	4	10.4	2	2	4.2
Tees†				2	4	10.4	2	4	8.4
Total equivalent length, feet, excluding flow fittings			94.8			68.8			43.6
Total friction, milinches, excluding flow fittings			9000			2752			2703
Single-flow fittings‡				6			6		
Resistance of single fittings						6000			6600
Double-flow fittings				1					
Resistance of double fittings						2000			
Total resistance, milinches			9000			10,752			9303

* Three elbows at inlet to boiler (two are involved in the change of elevation), one at outlet.
† Only the tees at the ends of the trunk are counted here. (See Fig. 12.7.)
‡ One induced-flow-type fitting used on each disseminator except one serving basement.

system has a disadvantage in that no heat-dispersing unit of a circuit can be shut off without stopping flow of water in the entire length of piping involved.

The general procedure of computing the mean water temperatures within heat-dispersal units of the series-loop arrangement is the same as that given for ordinary one-pipe systems except that the circuit water-temperature drops caused by heat disseminators and the temperature drops within them are the same since all of the water flowing in the circuit passes through each unit. The method of determination of the total system resistance is also unchanged except that all of the pipe and all of the fittings of the circuits must be included. Elbow equivalents of radiators and convectors are given in Table 12.1. Equivalent lengths of cast-iron baseboards are generally given in

HEATING WITH HOT WATER

catalogues pertaining to them. The friction of finned pipe is the same as that of plain pipe having the same diameter.

12.13. Design of Manifold One-Pipe Systems. The manifold one-pipe system shown in Fig. 12.2c has a separate circuit to each room served. Balancing of the several parallel circuits is accomplished by the use of special valves which are generally placed at the end of each circuit as it joins the return manifold. The arrangement of piping of any circuit serving only one disseminator is series loop. The

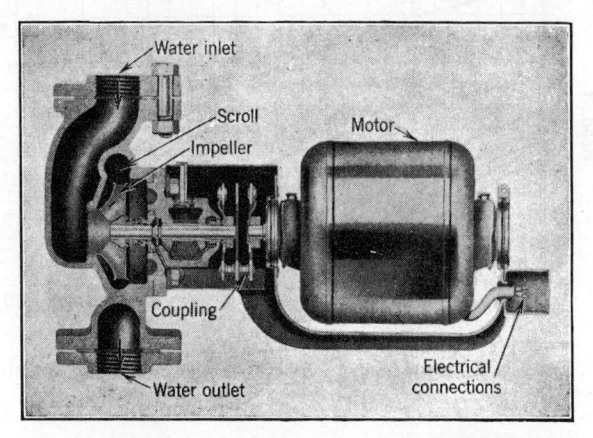

Fig. 12.13. Water circulator. (Courtesy Trane Co.)

mean temperature of the water in any disseminator depends upon that of the boiler water and the adjustment of the balancing valve. In case a circuit serves only a single disseminator, its mean water temperature may be assumed as only 10 F less than that of the boiler water when the system temperature drop is 20 deg. The total resistance of a manifold system is equal to that from the return manifold to the supply manifold plus that of the circuit which has the greatest resistance.

12.14. Circulator Selection. Figure 12.13 shows a sectional view of a small single-stage centrifugal pump suitable for service as a circulator in a hot-water heating system. Figure 12.14 shows performance curves for one manufacturer's line of circulators. Preliminary steps in the selection of the proper size are the calculation of the amount of water which must be circulated and the estimation of the friction head against which the unit must operate. In general, the sizing of pipes throughout the system should be such that the total head required to overcome friction is between 8 and 11 ft of water, except for small systems for residences. When the required volume in gallons per minute and the total head in feet of water are known, the unit in a particular manufacturer's line of circulators which is best suited to the system may be selected by referring to performance curves such as those appearing in Fig. 12.14. The circulator selected should be the one whose curve is nearest the intersection of a vertical line representing the required gallons per minute and a horizontal line

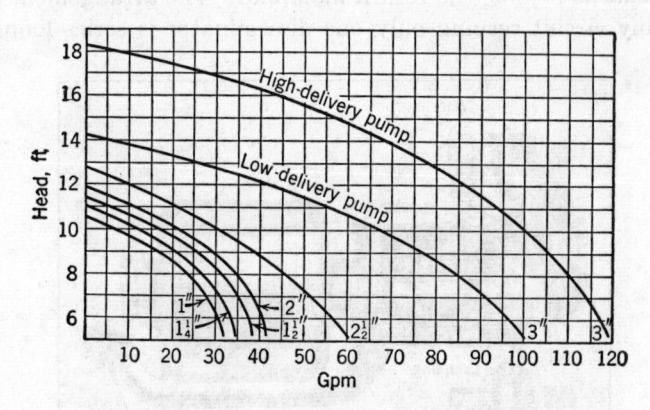

Fig. 12.14. Performance curves of Trane water circulators with various operating heads.

representing the total friction head of the system. If the choice is in doubt due to the point of intersection falling between the curves for two different pumps, the larger of the two pumps should be chosen.

12.15. Power for Pumping. It is necessary to estimate power required to circulate the water through a hot-water heating system in order that the proper size of the driving motor may be specified when it is not furnished by the pump manufacturer. The power requirement in terms of horsepower is given by the following equation:

$$\text{hp} = \frac{P \times q}{33000 \times E_m} \tag{12.2}$$

where P = pressure head, pounds per square foot.

12.16. Boiler-Piping Connections and Methods of Temperature Control in Hot-Water Heating Systems. The details of the water piping in the vicinity of a boiler are determined, in the case of home-heating installations, by the method of providing hot water for use in the kitchen and bathroom. When a separate domestic water heater is to be installed so that the boiler serves the heating system

HEATING WITH HOT WATER

alone, the arrangement of Fig. 12.15a may be used. With such an installation the room thermostat controls both the fuel burner and the water circulator, and the boiler-water temperature is allowed to fall when there is no demand for heat. When the boiler must furnish service water in addition to that sent through heat-dispersal units, one

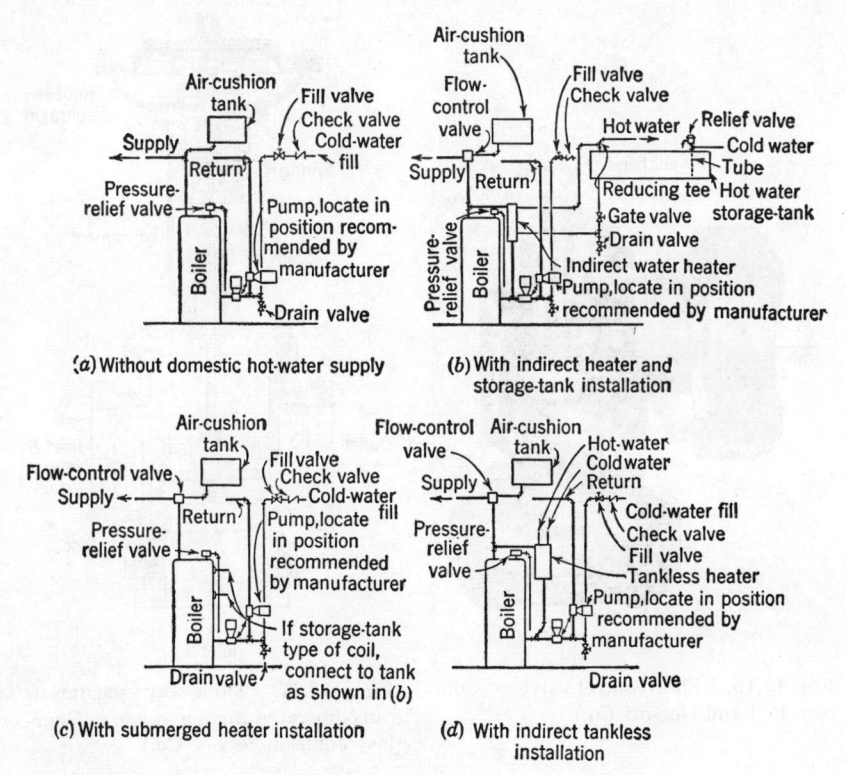

Fig. 12.15. Typical boiler-piping connections in single-main forced circulation hot-water heating systems. (From $I=B=R$ *Installation Guide No. 100*.[3] Used by permission.)

of the arrangements, either *b*, *c*, or *d*, of Fig. 12.15 must be used. These installations require control of the burner by a separate temperature-sensitive device (*aquastat*) which is located in the water of the boiler. The aquastat has both a high and a low setting and functions to maintain the boiler water, at all times, between certain limits such as between 190 and 210 F.

In addition to an aquastat, a hot-water heating system which supplies service hot water must include a *flow-control valve*, Fig. 12.16. The function of the flow-control valve is to prevent gravity circulation of hot water from the boiler to the heat disseminators when the water circulator is not in operation.

Some hot-water heating installations employ a bypass around the boiler and a mixing valve, Fig. 12.17. With this arrangement the temperature of the water supplied to the heat disseminators may be at

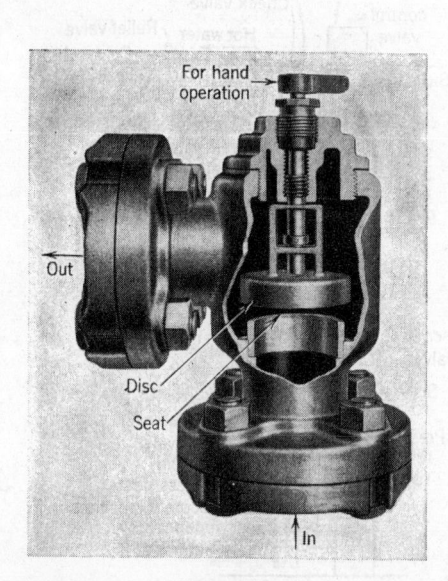

Fig. 12.16. Flow-control valve. (Courtesy Bell and Gossett Co.)

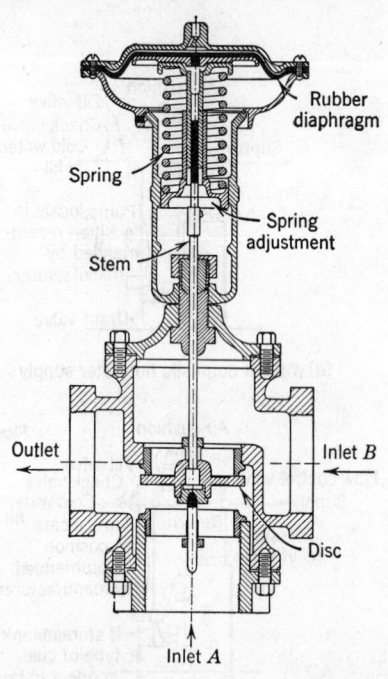

Fig. 12.17. Three-way pneumatically operated mixing valve. (Courtesy Johnson Service Co.)

any desired level between that maintained in the boiler and that of the room air. If desired the mixing valve can be controlled by an arrangement employing an outdoor thermostat which will gradually raise the temperature of the mixture as the outdoor temperature falls. This scheme of operation is particularly advantageous in hot-water panel heating.

12.17. High-Temperature Water Heating Systems. Some hot-water heating systems serving large buildings employ water at temperatures up to 400 F. The use of high-temperature water rather than high-pressure steam has the advantages that traps are not required and the temperature of the heating medium may be more easily modulated. The use of water at high temperature increases the heat output of the heat-dispersal units over that possible when the water temperatures are within the usual ranges; savings in costs may result from the use of smaller sized units. High-temperature water systems may be designed to operate with a greater temperature drop than the 20 deg commonly used in ordinary systems.

The friction charts of Figs. 12.5 and 12.6 are not applicable to high-temperature systems; the head to be overcome by a pump must be obtained either by calculations which employ the fundamental equations of Chapter 6 or by use of a special friction chart which is applicable to the conditions involved. The output rates of the special heat disseminators necessary should be obtained from the manufacturers of such equipment. The general design procedure, with the exceptions mentioned, are the same as those previously discussed. (See Bibliography.)

12.18. Use of High-Temperature Media Other than Water. When media temperature greater than 215 F are needed in heating systems or in process work, consideration should be given to the use of fluids other than water. Certain chemical compounds can be heated to high temperatures while in a liquid state without the necessity of maintaining a high pressure as is the case when water is used. (See Bibliography.)

REFERENCES

1. "Compression Tank Selection for Hot-Water Heating Systems," H. A. Lockhart and G. F. Carlson, *Heating, Piping, Air Conditioning*, **25**, No. 4, 132 (Apr. 1953).
2. "Two Pipe Reverse Return Gravity Hot-Water Heating System," *I =B =R Installation Guide No. 4*, Institute of Boiler and Radiator Manufacturers, New York (1947).
3. "One Pipe Forced Circulation Hot Water Heating Systems," *I =B =R Installation Guide No. 100*, Institute of Boiler and Radiator Manufacturers, New York (1950).
4. *Handbook*, Bell and Gossett Co., Morton Grove, Ill. (1949).

BIBLIOGRAPHY

High-Temperature High-Pressure Hot-Water Heating

1. "High-Pressure Hot-Water Heating Systems Adapted to Existing Steam Plant," *Steam Engr.*, **7**, No. 7, 275 (Apr. 1953).

2. "Swiss District Heating System Uses High-Temperature Water," *Heating and Ventilating*, **34**, No. 12, 39 (Dec. 1937).
3. "High-Pressure Hot-Water System Heats Molds at Plastics Plant," A. Westbrook, *Heating, Piping, Air Conditioning*, **10**, No. 6, 381 (June 1936).
4. "Conversion of Factory Process Heating System From Steam to High-Pressure Hot Water," M. Freund, *Ind. Heating Eng.*, **5**, No. 19, 49 (July 1943).
5. "High-Pressure Hot-Water Boilers for Heating and Process Requirements," *Ind. Heating Engr.*, **5**, No. 19, 49 (July 1943).
6. "High-Temperature Heating for Finishing Plants," P. L. Geiringer, *Textile World*, **94**, No. 6, 104 (June 1944).
7. "Design of High-Pressure Hot-Water Heating Installations," J. Porges, *Air Treatment Engr.*, **10**, No. 2, 37 (Feb. 1947).
8. "High-Pressure Hot Water," J. R. Kell, *Heating, Piping, Air Conditioning*, **20**, Nos. 4, 6, 8, and 10 (Apr., June, Aug., and Oct. 1948).
9. "High-Pressure Hot-Water Heating," P. Reschke, *Heating, Piping, Air Conditioning*, **21**, No. 6, 98 (June 1949).
10. "Boiler-Plant for High-Temperature Hot Water," J. B. Pinkerton, *Combustion Eng.*, **5**, No. 3, 95 (Mar. 1951).
11. "High-Temperature Hot-Water Heating in a S.A. Plastics Plant," *Ind. Heating Engr.*, **13**, No. 63, 25 (Jan. 1951).
12. "H-P, H-T, Hot Water Saves Fuel, Cuts Industrial Heating Maintenance," O. S. Lieberg, *Power*, **96**, Nos. 5 and 6 (May and June 1952).
13. "Pressurizing HPHW Systems with Nitrogen and Other Methods," J. R. Kell, *Heating, Piping, Air Conditioning*, **25**, No. 2, 84 (Feb. 1953).
14. "High-Temperature Hot Water," S. A. Kuil, *Heating, Piping, Air Conditioning*, **26**, Nos. 1 and 2 (Jan. and Feb., 1954).
15. "High-Temperature Liquid-Processing Piping," W. F. Moore, *Heating, Piping, Air Conditioning*, **23**, No. 5, 100 (May 1951).

Use of High-Temperature Media Other than Water

1. "Liquid-Dowtherm Heating and Cooling Units," T. C. Stack and J. E. Friden, *Chem. Eng. Progr.*, **48**, No. 8, 409 (Aug. 1952).
2. "Design and Operation of Large Heat-Transfer Installation," N. Brearly, *Inst. Mech. Eng. (London) Proc.*, **166**, 350 (1952).

PROBLEMS

1. An open-tank hot-water heating system is of a design for use with water at a minimum temperature of 55 F and a maximum one of 185 F. The plant consists of 150 linear feet of $1\frac{1}{4}$-in., 100 ft of $\frac{1}{2}$-in. nominal-size steel pipe and a boiler with a net rating of 160,000 Btuh. The radiation installed includes 400 sq ft of baseboard units and 500 sq ft of small-tube radiators. Find the required (minimum) capacity, in gallons, of the expansion tank. Specify the capacity of the tank to be installed.

2. Find the minimum expansion-tank capacity in gallons for a closed hot-water heating system which has a boiler net rating of 180,000 Btuh, 800 sq ft of convector surface, 150 sq ft of small-tube radiation, 100 linear ft of $1\frac{1}{4}$-in. and 50 ft of $\frac{3}{4}$-in. nominal-size steel pipe. The water temperature ranges

HEATING WITH HOT WATER

from 50 to 218 F, and a minimum pressure of 5 psig is to be maintained at the topmost radiator the upper portion of which is 15 ft above the water level in the tank. The barometric pressure is 14.2 psi. The maximum pressure in the tank is not to exceed 30 psig. Specify the tank size in gallons which you would install.

3. A small-tube direct radiator, rated at 105 sq ft edr, handles water at 200 F inlet and 170 F outlet temperatures when the air of the room in which it stands is at a temperature of 70 F. The radiator supply and return pipes are each 1-in. nominal size. Find the weight of water handled in pounds per hour. Compute the volume of the water in gallons per minute, at 200 F, and its velocity of flow in feet per second in the supply pipe.

4. A closed circuit of 2-in. copper tubing includes 50 ft of straight pipe, two standard 90-deg elbows, one open-globe valve, three 90-deg long-turn elbows, one heater, one flow-control valve, and one radiator. Find the equivalent length of tubing in feet.

5. A water-flow circuit is composed of one radiator, 80 ft of straight 4-in. steel pipe, seven 90-deg welded elbows, one tee used in place of an elbow, one open gate valve, one boiler, and one flow-control valve. Find the equivalent length of the circuit in feet.

6. The hot-water heating plant of Fig. 12.8 is to be installed in a locality where the heat output of each radiator must be 25 per cent greater than indicated. The pipe fittings and lengths of steel pipe shown remain unchanged. Determine applicable data for all of the items enumerated by Tables 12.3 and 12.4; water temperature drop is 20 F.

7. When a constant boiler-water temperature of 200 F is maintained, find the mean water temperature in each radiator served by the north main of Fig. 12.9, comprised of $1\frac{1}{4}$-in. nominal-size steel pipe. Design for an over-all system-temperature drop of 20 F with the computed data arranged as in Table 12.6. Assume two flow fittings are used to divert flow to the unit in the basement.

8. Refer to Fig. 12.9 and Table 12.7. Compute the total system resistance in milinches that would be encountered if the pipe in the north circuit were changed from $1\frac{1}{4}$-in. to 1-in. nominal size. What resistance in milinches would be required from the balancing cock in the south circuit to secure the desired weight distribution between the two circuits with this arrangement?

9. A pump is required to force 9250 lb per hr of water at 165 F through a heating system against a total resistance of 82,300 milinches of water. If the mechanical efficiency of the pump is 65 per cent, find the required horsepower input to it.

13

Panel Heating, Snow Melting, and Panel Cooling

13.1. Definition of Panel Heating. The heating of rooms intended for human occupancy by means of large warmed panels is referred to as panel heating or radiant heating. The term panel heating is the one preferred by the authors because the other one implies that all the heat output of such a system is in the form of radiant heat, which is not the case. A heating system may be classified as a panel type when the heating medium, which may be electricity, water, or air, is circulated through either wires, pipes, or ducts incorporated in large panels forming sections of the floor, ceiling, or walls. Heat is transferred by radiation from each warmed panel to all objects within the room and to all room surfaces except the one in which it is located. The air in the room is warmed by convection currents passing over the surface of the panel or over the surfaces of objects receiving radiant heat from the panel.

13.2. Fundamentals of Panel Heating. It has been pointed out in Chap. 3 that the human body loses heat to its environment by convection, radiation, and evaporation. In the effective-temperature range which produces bodily comfort, the loss of heat by evaporation is comparatively small and nearly constant over a considerable range

262

PANEL HEATING

of dry-bulb temperatures and relative humidities. Therefore under comfort conditions the heat loss from the body is chiefly related to the combined effect of convection and radiation. The rate of heat loss by convection depends upon the average temperature difference between the surfaces of the exposed skin and outer clothing and the surrounding air, the area of such surfaces, and the velocity of the air movement over them. The rate of heat loss by radiation depends upon the area of the exposed skin or clothing surfaces and upon the difference between the average temperature of the surfaces and the mean radiant temperature of the surrounding walls, floors, ceilings, or objects. The mean radiant temperature may be defined as a uniform temperature of all the surrounding surfaces which would result in the same loss (or gain) of heat by radiation as that to (or from) the same objects and surfaces at their actual surface temperatures. The same feeling of comfort can therefore be produced with a comparatively high mean radiant temperature and a comparatively low air temperature, or vice versa. However, in the operation of a panel-heating system, it is not possible to control the mean radiant temperature and the air temperature independently.

Under comfortable still-air conditions during the heating season the mean of the skin and outer clothing temperatures of persons normally clothed is approximately 81 F. Since there are considerable portions of the body such as the insides of the arms and legs which radiate their heat to other body surfaces, the effective surface exposed to loss by radiation is less than the effective surface exposed to loss by convection. The average adult person has an effective surface exposed to radiation amounting to 15.5 sq ft and a total surface exposed to the loss of heat by convection amounting to 19.5 sq ft. The normal rate of heat production in an average-sized sedentary individual is about 400 Btuh of which approximately 100 Btuh are rejected in the form of vaporization of insensible perspiration on external skin areas and evaporation into the air that is taken into the body through the process of respiration. The remainder, or approximately 300 Btuh, must be absorbed by the combination of loss by radiation and loss by convection. In an environment at 70 F the loss by convection from the clothed body amounts to approximately 140 Btuh, and the remaining 160 Btuh must be dissipated by radiation to all surfaces to which the body is exposed. If the mean radiant temperature of the surrounding envelope were raised to 81 F, which is the average surface temperature of a fully clothed man, the loss of heat by radiation would be reduced to zero, and the air temperature would have to be reduced to the point where loss of heat by convection alone would amount to

300 Btuh. The still-air temperature that would be required to produce comfort for the individual under these conditions is approximately 60 F.

In the actual application of panel heating, a certain portion of the envelope surrounding a room, such as either the floor or the ceiling, is heated to whatever temperature is required to maintain the desired effect. The temperature of the remainder of the envelope and the temperature of the air in the room will vary with changing weather conditions. Early designers thought it possible to raise the mean radiant temperature of the envelope of a room, by means of heated panels, to a value considerably above that which would prevail when other means of heating are used, thereby producing comfort conditions for people with a considerably decreased room-air temperature. However, experience has shown that convection currents over the surfaces of the heated panels result in the heating of the air in the room to the point where it approaches the temperature maintained by other systems of heating. Consequently, the relationships of convection and radiation in a room heated by large warmed panels are not greatly different from those found in rooms heated by either warm air or warm water in conventional heating plants.

13.3. Advantages of Panel Heating. Heating panels which are incorporated in the floor, ceiling, or walls of a heated room appear to have the following advantages over the ordinary steam, hot-water, and warm-air heating systems.

1. The heat-disseminating elements are completely out of sight and do not interfere with any decorative plan.
2. The heating panels do not interfere in any way with any desired placement of rugs and furniture.
3. Since the panels are incorporated in the structural frame of a building, the heating plant may be started as soon as this portion of the construction has been completed, thus facilitating the completion of interior work when construction is carried on during cold weather.
4. When the heated panels are in either the floor or ceiling, this type of heating system does not require consideration in the placing of interior partitions and does not interfere with later moving of such partitions. This advantage is of greatest value in office buildings where it may be desirable to rearrange interior partitions from time to time.
5. Warmer floors are produced by heating panels than by heat disseminators such as radiators, convectors, or warm-air registers.
6. Heating panels are arranged so that it is not possible for small

PANEL HEATING

children to throw the system out or adjustment by tampering with valves or dampers.

7. Air currents within a room heated by panels have smaller velocities than those in rooms which have other forms of heat disseminators; one result is that dust particles, which may carry disease-producing organisms, can settle readily.

8. Heating panels may also be used for radiant cooling in climates where warm-weather relative humidities are consistently low. Also they may be used to supplement other arrangements for summer air conditioning in all climates.

13.4. Disadvantages of Panel Heating. The principal disadvantages of panel heating are:

1. Because of the heat storage in the large panels, which are used as heat disseminators, the problem of maintaining uniform room comfort conditions through rapid changes in the outdoor air temperature is usually more difficult than with other heating systems.

2. Heating panels are usually constructed on the job by the building or heating contractors and are not as readily adaptable to standardization as other types of heat-dispersing units.

3. Panel-heating systems do not offer the possibility of supplying outside air for ventilation to the heated rooms as do certain other types.

4. Heating panels are incorporated as an integral part of the building structure, and the repair of a leak which may develop after the system is placed in service can be very expensive.

13.5. Floor Versus Wall Versus Ceiling Panels. There is little advantage in one location over either of the other two from the standpoint of temperature gradient from the floor to the ceiling. A radiant panel in any of the three possible room surfaces produces more uniform temperatures throughout the height of a room than the usual steam and hot-water radiators. As indicated in Table 13.1, the room surface chosen for the location of the panel has an important bearing on the operative heat-dissipating capacity of the unit in Btuh per square foot of exposed surface. If a floor location is strongly favored for other reasons, a panel in this location may be supplemented with one or more smaller panels installed either in a wall or in the ceiling if necessary in order to obtain the required capacity. However, from an economic standpoint, it is desirable to choose a location which will permit a panel of sufficient capacity to be installed in a single room surface.

Floor panels are more easily installed, particularly in basementless buildings where the network of pipes for circulating the heating

TABLE 13.1. Approximate Heat Output of Radiant Panels Under Design Conditions

Room-air temperature assumed at 70 F; mean radiant temperature of other five room surfaces assumed at 65 F

Location of Panel	Maximum Allowable Surface Temperature, Deg F	Maximum Total Heat Output, Btu per Sq Ft per Hr	Maximum Heat Output by Radiation, Btu per Sq Ft per Hr	Maximum Heat Output by Convection, Btu per Sq Ft per Hr	Per Cent of Heat Output by Radiation	Per Cent of Heat Output by Convection
Floor	90	43	24*	19	55*	45*
Wall	110	65	42*	23	65*	35*
Ceiling	120	83	58*	25	70*	30*

* Data taken with permission from *Hot Water Heating and Radiant Heating and Radiant Cooling*, by F. E. Giesecke.

medium is supported the proper distance above a fill of suitable material after which a concrete floor slab may be poured around the pipes.

13.6. Steel Pipes Versus Copper Tubing.

Heated water is the medium most generally used in panel systems, and the pipes conducting it through the panels may be made from wrought iron, steel, or copper. Wrought iron is a special grade of steel so that the physical

TABLE 13.2. Comparison of Advantages of Wrought-Iron or Steel Pipe and Copper Tubing when Used as Conductors of Hot Water in Panel Heating

Wrought-Iron or Steel Pipe	Copper Tubing
1. Lower cost for piping	1. Tubes are easier to bend on the job
2. Has practically the same coefficient of expansion as concrete	2. Copper is completely resistant to corrosion by any chemical compound that is likely to be found in the circulating water
3. Pipe walls are strong enough to resist damage in handling, in fabricating, and while waiting for slab to be poured	3. Tubing walls and joints are smoother, thus reducing friction losses from those existing in the same diameter of wrought-iron pipe handling the same volume of water
4. Welded joints in wrought-iron pipe are stronger than soldered joints used for connecting lengths of copper tubing	4. Soldered joints in copper tubing are easier to make than welded joints in wrought-iron pipe
5. Larger pipe diameters can be used in floor panels for the same expenditure, thus reducing pumping costs due to lower friction losses	5. Practicable to use smaller diameter tubing in ceiling panels, thus reducing the thickness of plaster required and consequently the thermal storage of the panel

properties of these two materials are almost identical. Table 13.2 gives a comparison of the merits of either wrought iron or steel with those of copper, obtained from an impartial survey of claims made by each of the two interested industries.

Greater thermal conductivity is sometimes claimed as an advantage in favor of the use of copper, and greater emissivity of the bare pipe is likewise claimed as an advantage in favor of wrought-iron or steel, but, since the tubes or pipes are usually embedded in concrete or plaster, the emissivity of the bare pipe is not worthy of consideration, and the resistance of the pipe wall is a negligible factor in the transfer

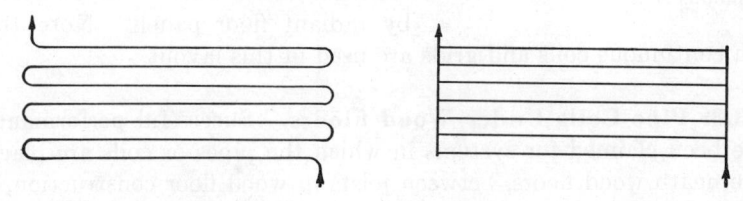

Fig. 13.1.   Continuous coil.   Fig. 13.2.   Grid-type coil.

of heat from the water to the air of the room regardless of which of the two metals is used. Proponents of the use of copper point out that the material is perfectly capable of withstanding the compressive forces set up in a tube because its coefficient of expansion is greater than that of concrete. They also point out that the bond between the copper tubing is strong enough to withstand the strain on it and that the stress in the tube and the tendency of the force created to break the bond is not accumulative and therefore is not greater in long embedded tubes than in short ones. Likewise, the proponents of wrought-iron or steel pipe explain that the chances that conduits made of this material and used in a radiant panel will fail because of corrosion are remote, if they are protected from corrosive materials such as cinders.

13.7. Coil Design. Pipe elements for heating panels are usually made in one of two basic patterns, namely, continuous coil as shown in Fig. 13.1 or grid as shown in Fig. 13.2. The continuous coil is usually less expensive to fabricate especially when copper tubing is used. However, the grid pattern has one important advantage in that the velocity of the water in any section of the unit is less, and consequently the friction losses are less when the water flows through adjacent pipes in parallel instead of in series as is the case when the pipe is laid in a continuous coil.

Figure 13.3 shows a pattern which may be used to combine several continuous coils of moderate length in a semigrid pattern. This layout produces a network of pipes which is less expensive to fabricate than a grid pattern and one in which the frictional resistance to flow of the required amount of water is considerably less than that in a continuous coil providing the same total length of pipe.

Figure 13.4 shows the complete layout of piping for a one-story basementless house which is to be heated by radiant floor panels. Note that both continuous coils and grids are used in this layout.

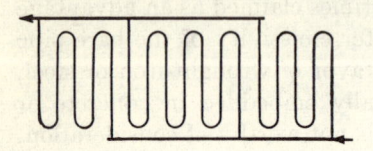

Fig. 13.3. Combination grid and continuous coil.

13.8. Pipe Coils Under Wood Floors. Successful performances have been claimed for systems in which the pipes or coils are placed underneath wood floors, between joists in wood floor construction, or between the sleepers in wood floors placed on top of a concrete slab. When this type of panel construction is used, the wood floor becomes

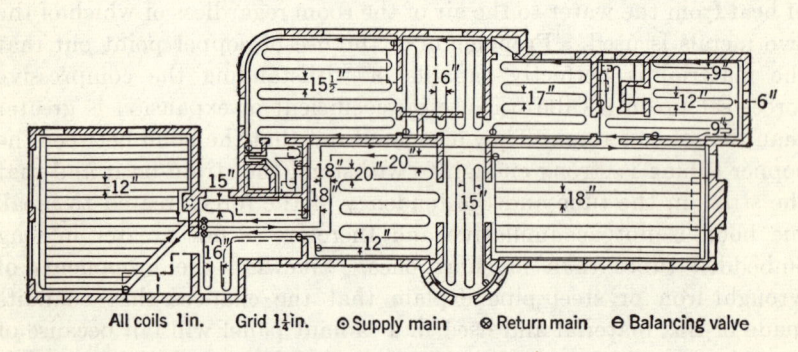

Fig. 13.4. Typical radiant-heating piping. (Courtesy A. U. Byers Co.)

the radiant panel, heat being transferred from the pipes to it by radiation and convection through the air space between the joists or sleepers.

Experiences with several installations of this type indicate that wood flooring will not be damaged by the heat from the hot pipes, provided that the wood of which the floors are made has been thoroughly dried. It has been found that the heat transferred from the water to the air of the room is approximately 1.75 Btuh per sq ft of pipe surface per degree temperature difference when plain pipes are used. Extended surfaces when properly applied facilitate the transfer of heat from the

PANEL HEATING

pipes to the wood floor and produce an over-all rate greater than the amount mentioned in the preceding sentence.

13.9. Air Vents and Expansion Tanks for Heating Panels. Air vents should be provided at all high points in a panel-heating system, if the circulation is to be by gravity flow. However, when a pump is used, the flowing water can be depended upon to carry the air to other points which may be more convenient for the locations of vents. Coils, pipes, and mains should be arranged to reduce to a minimum the number of points requiring air vents.

The procedure to be used in sizing the expansion tank for a water-heated panel plant is the same as for any other hot-water system.

13.10. Control of Panel Heating. One of the most difficult problems in connection with the design and installation of heating panels is the selection of the proper system for regulating their heat output. Improper selection of controls is the real reason for failure in the majority of cases where the systems have not been able to produce the degree of comfort which had been expected. Insufficient pipe surface in a panel can, to a considerable extent, be compensated by raising the temperature of the circulating water above that on which the design is based, but phenomena such as "flashing" and "hunting" due to an improper control are in some instances most difficult to handle. Heating panels in which pipes are embedded in concrete or plaster are certain to possess considerable thermal inertia and also considerable resistance to the flow of heat from the water to the surface of the panel. Building structures vary greatly in regard to thermal inertia because some have walls made of either one or more thin sheets of metal or wood products while others have walls including several inches of solid masonry. Building structures also vary greatly in regard to thermal resistance as some walls may contain several inches of insulation while others have none. Buildings having a large proportion of their wall areas taken up by windows or doors have very little thermal resistance.

Large thermal capacity and thermal resistance in heating panels cause slow responses to temperature controls, whereas large capacity and resistance in the structure produce delays in the reflection of external load changes within the building. Usually such structural lag is advantageous from a control standpoint as it permits the use of a control element that is actuated by changes in the outdoor air temperatures for anticipation of the load changes.

The factor of greatest importance from the standpoint of control of a panel-heated building is the relationship between the thermal inertia of the panel and that of the structure. If the time required for an externally imposed load to be reproduced within the building is equal to the time required for a change in circulating-water temperature to produce a change in the surface temperature of the panel, no serious control problem exists and any properly designed control system using an outside temperature bulb will maintain comfort conditions within the structure even through a rapid change in the outdoor weather. However, when a building having very little thermal inertia, such as a house provided with an unusual amount of glass area, is heated with a panel having large thermal inertia, such as a concrete floor panel, maintenance of comfort conditions within the house through a sudden change in outdoor temperature is impossible with any control equipment that is available at the time of this writing. On the other hand, a concrete-floor panel in a house having concrete walls supplemented with some insulation to produce thermal resistance as well as thermal inertia could be controlled in a satisfactory manner. The best results are possible when the thermal inertia of the panel and that of the structure are equal. Satisfactory results can be obtained when the thermal inertia of the panel is less than the thermal inertia of the structure. However, the control of the panel is certain to be unsatisfactory when the thermal inertia of the panel is appreciably greater than that of the structure.

13.11. Design of Panel-Heating Systems Using Heated Water. Many different procedures have been proposed for the design of panel-heating systems, but there is at present no one method which has been generally accepted as standard. Many factors affect the performance of a heating panel, such as temperature of the panel and that of the unheated surfaces, emissivity of the various surfaces, air temperature, size of pipes, spacing of pipes, conductivity of covering material, and depth of cover. A completely accurate design procedure taking into consideration all of the factors is exceedingly involved and laborious. In all the methods which have been proposed for practical application, certain assumptions are made in regard to some of the less important factors in order to simplify the solution. Some of the proposed methods include a mathematical analysis of each problem while others make use of a rather extensive series of charts. Due to space limitations it is not possible to present a specific design procedure here. Complete instructions for the design of water-heated floor and ceiling panels may be found in the ASHAE Guide. Simplified design procedures with supporting tables and charts may be obtained from some of the large manufacturers of steel pipe and also from certain companies which produce copper tubing.

PANEL HEATING

13.12. Snow Melting. Either steel pipes or copper tubes may be embedded in concrete sidewalks and driveways for the purpose of melting falling snow and thereby preventing its accumulation on surfaces used for foot and vehicular traffic. The heated liquid which is circulated through the embedded coils is usually a freezeproof solution of water and ethylene glycol or some other suitable compound.

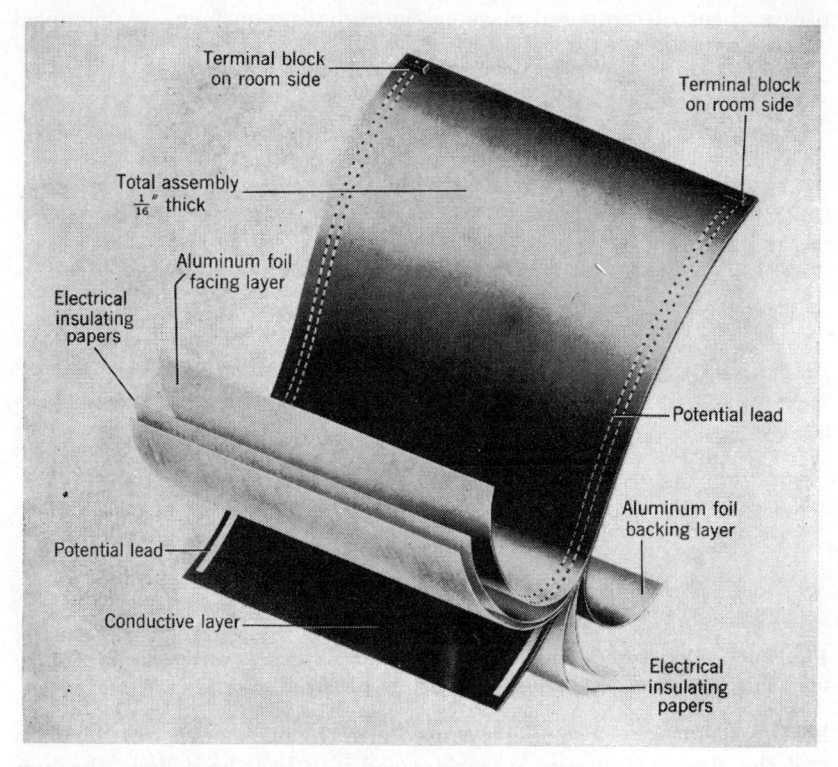

Fig. 13.5. Electric heating panel employing a layer of conductive rubber. (Courtesy United States Rubber Co.)

The circulating fluid may be heated by either a separate boiler or by a heat exchanger attached to a heating boiler. (See Bibliography.)

13.13. Use of Electricity as a Heating Medium in Panel Heating. Due to their low initial cost and the elimination of all combustion devices, various types of heating panels employing electrical energy as a source of heat are becoming increasingly popular. Electric heating panels may be constructed by incorporating special resistance wires within the plastering of a ceiling. Commercial electric heating panels ready for installation are available in several designs.

Figure 13.5 shows the details of a panel which employs a conductive layer of rubber as the means of converting electrical energy into useful heat. The panel is $\frac{1}{16}$ in. thick and is attached to a wall or a ceiling by means of an adhesive. The standard sizes of the panels are 3 ft by 4 ft, 4 ft by 4 ft, and 6 ft by 4 ft. The capacity of output is 75 Btuh per sq ft. Panels are available to operate with either 115 or 230 volts.

Figure 13.6 shows the parts of an electrically heated panel which uses a sheet of tempered glass as the radiating surface. A chemical

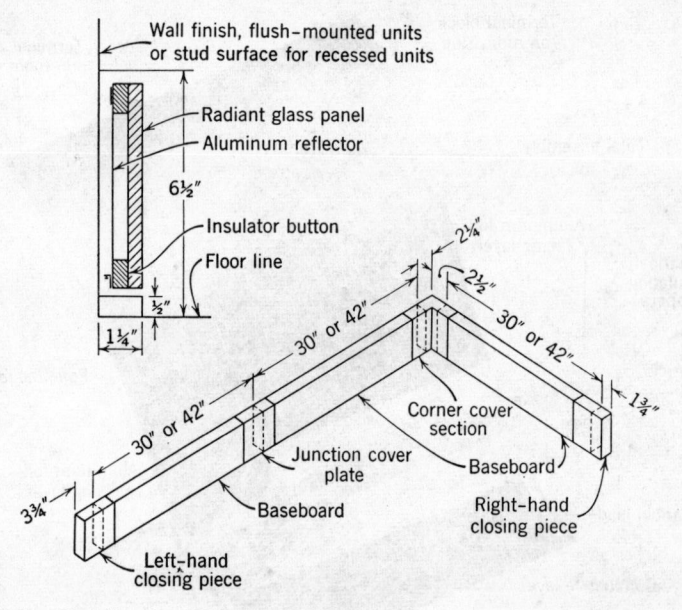

Fig. 13.6. Details of a glass heating panel and a typical arrangement of three units in a baseboard installation. (Courtesy Electriglas Corp.)

heating element is permanently fused into the back surface of the glass and the electrical energy is transformed into heat as it flows through the element. The heat generated is absorbed by the glass and then transmitted to the room air by radiation and convection. An aluminum reflector at the back of the panel assembly reduces heat losses from the reverse side. Panels of this type are made in a variety of shapes, sizes, and capacities. Long narrow panels may be placed along exposed walls, as shown in the lower part of Fig. 13.6. This arrangement produces a heating effect similar to that of radiant baseboards. Glass heating panels are also available for installation in ceilings or in walls at strategic positions to produce special desired effects.

13.14. Use of Air as the Heating Medium in Panel Systems. Panel systems using air as the heating medium can in general be

PANEL HEATING

installed at lower initial cost than that of systems employing hot water and they are free from the danger that a leak may develop and cause serious damage. Inasmuch as it is not necessary to embed pipes or tubes in the substance used for constructing them, ceiling or wall panels which have to support only their own weight may be much thinner when air is used as the heating medium instead of water. A thinner panel results in less thermal inertia and less resistance to heat conduction from the circulating medium to the surface which is exposed to the room that is being heated. Another advantage in using heated air instead of heated water is that local hot spots in the panel are eliminated and there is less danger that the plaster will crack. The only serious disadvantages in using air instead of water are the lower density and the lower specific heat of this medium, necessitating the circulation of a greater weight and a much greater volume for the same heat output.

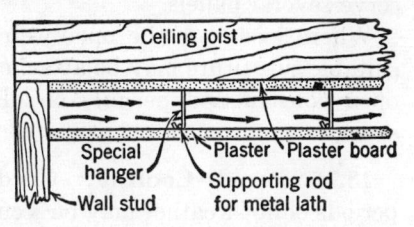

Fig. 13.7. Suspended warm-air panel.

Heated air can be circulated through hollow tile under concrete floors or between the joists of wood floors, but the ceiling panel offers the possibility of achieving light weight with conventional lath and plaster or its equivalent. If wood joists are used in the ceiling construction it is necessary to provide a circulating space below the bottom of the joists. This may be accomplished as shown in Fig. 13.7.

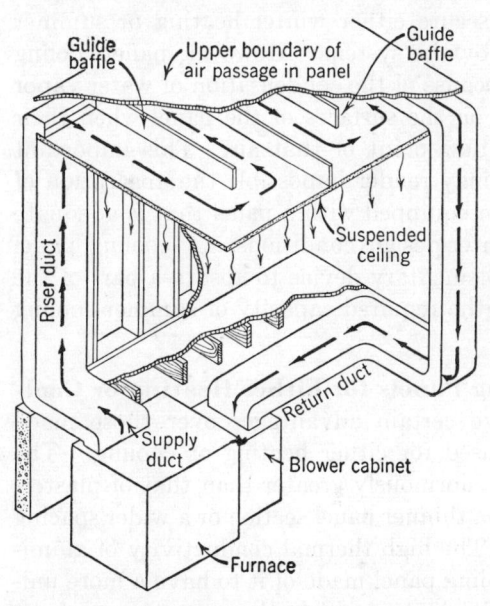

Fig. 13.8. Schematic diagram of a warm-air panel-heating system.

One method of supplying heated air, directing its flow through the panel and returning it to the furnace, is illustrated in Fig. 13.8. The furnace is the same type as is used in conventional forced-circulation warm-air heating systems. The fan used may also be the conventional type. Figure 13.8 shows a furnace serving only one panel, but by means of suitable supply and return trunk ducts a single furnace can serve several panels.

Where joists are the open-web steel type, the lath and plaster or suitable substitute may be attached to the bottoms of the joists, as the open construction permits circulation of the air through the joist spaces.

13.15. Panel Cooling. Conditions of human comfort during periods of hot weather may be secured by the use of some form of panel cooling in which either water or air may be circulated in one of the respective heating systems previously described. A heat exchanger designed to cool the circulating medium replaces either the furnace or the boiler of a panel-heating plant. By providing a suitable arrangement for the direction of the circulating medium through either the heater or chiller as necessary, use may be made of the same coils or baffling in the panels to secure either winter heating or summer cooling of the spaces served by the system. However, panel cooling is limited in its application because of the condensation of water vapor held by the air of the space on the surfaces of the panel when their temperatures are below the dew point of that air. This important limitation of the equipment may render impossible the absorption of the entire heat gain of a room equipped with a panel that is adequate to heat the space under winter-design conditions. A cooling panel can always be used as a supplementary device to absorb a part of the heat gain and thereby lessen the required capacity of summer-cooling equipment.

13.16. Aluminum Ceiling Panels for Either Heating or Cooling. Aluminum panels have certain advantages over those made with plaster surfaces when used for either heating or cooling. The conductivity of aluminum is enormously greater than that of plaster; thus the metal permits either a thinner panel section or a wider spacing of the water tubes or both. The high thermal conductivity of aluminum permits a heating or cooling panel made of it to have a more uniform face temperature (that is nearly equal to the temperature of the circulating water) than is possible with other materials. These advantages are especially important in panel cooling because it is easier to secure a mean panel-surface temperature low enough to absorb the sensible-heat gain of a space without having a temperature in localized areas which is below the dew point of the air-vapor mixture in the room. Contact between aluminum panels and water tubing of the same material may be made by brazing or by a mechanical contact maintained by a snap-on arrangement. Aluminum ceiling panels may be perforated to provide acoustical treatment without an appreciable loss of effectiveness for either heating or cooling the space beneath them.

13.17. Warm-Air Systems Combining Panel and Convection Heating or Cooling. The advantage of warmer floors, characteristic of panel-heated rooms, can be obtained together with air circulation, in the spaces served, and air humidification by employing a combination panel and convection system. In the combination system the warm air first passes through passages in the panel, thus increasing the temperature of its surface, after which it is caused to pass into the room through suitable diffusers. Several different arrangements for such a system are possible, but the ones most commonly used are the perimeter-loop and perimeter-radial systems which were described in Arts. 7.6 and 7.7.

Combination warm-air panel and convection heating systems can be adapted to summer air conditioning if the weight of air circulated is great enough to permit the necessary absorption of the total heat gain without the delivery of air to the panel at a temperature below that of the room-air dew point.

BIBLIOGRAPHY

Panel Heating

1. "Comparative Performances of a Warm-Air Ceiling Panel System and a Convection System," R. W. Roose, M. E. Childs, G. H. Green, and S. Konzo, Eng. Exp. Sta., *Bull No. 401, University of Illinois* (1952).
2. "Effects of Room Size and Non-Uniformity of Panel Temperature on Panel Performance," L. F. Schutrum and J. D. Vouris, *Heating, Piping, Air Conditioning*, **26,** No. 9, 133 (Sept. 1954).
3. "Heat Flow Characteristics of Hot Water Floor Panels," E. L. Sartain and W. S. Harris, *Heating, Piping, Air Conditioning*, **20,** No. 1, 183 (Jan. 1954).
4. "Effects of Non-Uniformity and Furnishings on Panel Heating Performance," L. F. Schutrum and C. M. Humphreys, *Heating, Piping, Air Conditioning*, **26,** No. 2, 131 (Feb. 1954).
5. "Field Studies of Floor Panel Control Systems," A. B. Algren, E. F. Snyder, Jr., and R. R. Head, *Heating, Piping, Air Conditioning*, **26,** No. 4, 117 (Apr. 1954).
6. "Heat Exchanges in Ceiling Panel Heated Room," L. F. Schutrum, G. V. Parmelee, and C. M. Humphreys, *ASHVE Trans.*, **59,** 197 (1953).

7. "Heat Exchanges in Floor Panel Heated Room," L. F. Schutrum, G. V. Parmelee, and C. M. Humphreys, *ASHVE Trans.*, **59**, 495 (1953).
8. "Laboratory Studies of the Thermal Characteristics of Plaster Panels," C. M. Humphreys, C. V. Franks, and L. F. Schutrum, *ASHVE Trans.*, **57**, 363 (1951).
9. "Field Studies of Heat Losses from Concrete Floor Panels," C. M. Humphreys, C. V. Franks, and L. F. Schutrum, *ASHVE Trans.*, **57**, 221 (1951).
10. "Effect of Panel Location on Skin and Clothing Surface Temperature," L. P. Herrington and R. J. Lorenzi, *Heating, Piping, Air Conditioning*, **21**, No. 10, 107 (Oct., 1949).
11. "Ground Temperature Distribution with a Floor Panel Heating System," A. B. Algren, *ASHVE Trans.*, **54**, 321 (1948).
12. "Space Heating by Electric Radiant Panels and by Reverse Cycle," Louis Slegel, *Eng. Exp. Sta., Bull. No. 24, Oregon State College* (1948).
13. "Surface and Water Temperatures for Radiant Heating Systems," L. J. LaTart, N. W. Smith, and L. D. Mills, *Heating and Ventilating*, **51**, No. 2, 101 (Feb. 1954).
14. "The Case for Designing Radiant Heating Panels to an Assumed Panel Water Temperature," W. P. Chapman, *Heating and Ventilating*, **50**, No. 10, 83 (Oct. 1953).
15. "Radiant Heating with Warm-Air Split System," E. O. Thatcher, *Heating and Ventilating*, **51**, No. 7, 93 (July, 1954).
16. "Radiant Glass Heating Panels," P. R. Achenbach, *Heating and Ventilating*, **50**, No. 1, 88 (Jan. 1953).
17. "Radiant Heated Hardwood Floors," J. S. Mathewson, *Heating and Ventilating*, **49**, No. 1, 72 (Jan. 1952).
18. "Radiation and Convection from Surfaces in Various Positions," G. B. Wilkes and C. M. F. Peterson, *ASHVE Trans.*, **44**, 513 (1938).
19. "Air Temperature Gradients in a Panel Heated Room," J. M. Ayres and B. W. Levy, *ASHVE Trans.*, **54**, 131 (1948).
20. "An Analytical Solution of Heat Flow vs. Wire Temperature for Electric Cables Buried in Plaster," J. E. Gott, *Trans. Am. Inst. Elec. Engrs.*, **72**, Part 2, 160 (July 1953).
21. "Floor Panel Heating, Some Design Data," N. S. Dillington, *J. Inst. Heating Ventilating Engrs.* (*London*), **21**, No. 218, 256 (Oct. 1953).
22. *Radiant Heating*, T. Napier Adlam, Industrial Press, New York (1947).
23. *Radiant Heating*, R. W. Shoemaker, McGraw-Hill Book Co., New York (1948).

BIBLIOGRAPHY

Snow Melting

1. "Operating Experience and Data from a Sidewalk Snow Melting System," L. A. Stevens and G. D. Wimans, *ASHVE Trans.*, **57**, 51 (1951).
2. "Avoiding Thermal Stresses in Snow Melting Systems," W. P. Chapman, *Heating, Piping, Air Conditioning*, **27**, No. 8, 104 (Aug. 1955).
3. "Snow Melting, Questions and Answers," *Heating, Piping, Air Conditioning*, **23**, No. 10, 111 (Oct. 1951).

4. "Design Conditioning for Snow Melting," W. P. Chapman, *Heating and Ventilating*, **49**, No. 11, 88 (Nov. 1952).
5. "Design of Snow Melting Systems," W. P. Chapman, *Heating and Ventilating*, **49**, No. 4, 96 (Apr. 1952).
6. "Electrically Heated Highway Snow Melting System," *Heating and Ventilating*, **48**, No. 1, 64 (Jan. 1951).
7. "Snow Melting Design for a Shipping Center," J. E. Ziegler, *Heating and Ventilating*, **49**, No. 5, 83 (May 1952).
8. "Data on Snow Melting Systems," T. W. Reynolds, *Heating and Ventilating*, **49**, No. 9, 84 (Sept. 1952).
9. *Snow Melting*, T. Napier Adlam, Industrial Press, New York (1950).

Panel Cooling

1. "Panel Cooling for a Residence," R. R. Irwin, *Heating, Piping, Air Conditioning*, **27**, No. 5, 175 (May 1955).
2. "Effect and Temperature Requirements for Cooling Panels Removing Internal Radiation," Merl Baker, *Heating, Piping, Air Conditioning*, **24**, No. 6, 118 (June 1952).
3. "Design Factors in Panel and Air Cooling Systems," C. S. Leopold, *ASHVE Trans.*, **57**, 61 (1951).
4. "Cooling Panels, Conditioned Air, Used at New Glass Plant," *Heating, Piping, Air Conditioning*, **24**, No. 4, 102 (Apr. 1952).
5. "Ceiling Panels Cool Office Building," *Eng. News-Record*, **148**, No. 3, 40 (May 1951).

Panel Heating and Cooling

1. "Radiant Heating and Cooling, Part I," C. O. Mackey, L. T. Wright, Jr., R. E. Clark, and N. R. Gay, *Bull. No. 32, Eng. Exp. Sta.*, Cornell University (1943).
2. "Aluminum Panels for Heating, Cooling," J. J. Mann, *Power*, **96**, No. 11, 71 (Nov. 1952).
3. "Radiant Heating, Cooling," Gunnar Frenger and D. W. Day, *Heating, Piping, Air Conditioning*, **24**, No. 9, 88 (Sept. 1952).
4. "New Alcoa Building Uses Ceiling Heating-Cooling Panels," B. R. Small, *Heating, Piping, Air Conditioning*, **22**, No. 9, 93 (Sept. 1950).
5. "Heating and Cooling with Air in a Radiant Tile Floor," L. B. Nye, Jr., *Heating and Ventilating*, **48**, No. 7, 72 (July 1951).
6. "Heat Flow Analysis in Panel Heating or Cooling Sections, Case I, Uniformly Spaced Pipes Buried Within a Solid Slab," L. E. Hulbert, H. B. Nottage, and C. V. Franks, *ASHVE Trans.*, **56**, 189 (1950).
7. "Heat Flow Analysis in Panel Heating or Cooling Sections, Case II, Floor Slab on Earth with Uniformly Spaced Pipes or Tubes at the Slab Earth Interface," H. B. Nottage, D. V. Franks, L. E. Hulbert, and L. F. Schutrum, *ASHVE Trans.*, **59**, 527 (1953).
8. "Aluminum Ceiling Panels for Heating and Cooling," E. S. Howarth, S. C. Huddleston, and R. M. Kock, *ASHVE Trans.*, **57**, 343 (1951).
9. *Panel Heating and Cooling Analysis*, B. F. Raber and F. W. Hutchinson, John Wiley & Sons, Inc., New York (1947).
10. *Hot Water Heating and Radiant Heating and Radiant Cooling*, F. E. Giesecke, Technical Book Co., Austin, Tex. (1947).

Combination Panel and Convection Systems Adapted to Either Heating or Cooling

1. "Heat Emission Characteristics of Warm-Air Perimeter Heating Ducts," J. R. Jamieson, R. W. Roose, H. T. Gilkey, and S. Konzo, *Bull. No. 412, Eng. Exp. Sta.*, University of Illinois (1953).
2. "Comparative Performances of Two Warm-Air Perimeter Systems and Three Convection Systems," M. E. Childs, R. W. Roose, H. T. Gilkey, and S. Konzo, *Bull. No. 403, Eng. Exp. Sta.*, University of Illinois (1952).
3. "Cooling a Small Residence Using a Perimeter-Loop Duct System," D. R. Bahnfleth, C. F. Chen, and H. T. Gilkey, *Heating, Piping, Air Conditioning*, **26**, No. 2, 111 (Feb. 1954).
4. "Performance of Warm-Air Perimeter-Loop and Perimeter-Radial Systems in a Residence," H. T. Gilkey, R. W. Roose, and M. E. Childs, *ASHVE Trans.*, **59**, 473 (1953).
5. "Air Panel Heating-Cooling for Floor or Ceiling Use," P. R. Goemann, *Heating, Piping, Air Conditioning*, **24**, No. 3, 98 (Mar. 1952).
6. "Warm Air Forced Through Cellular Steel Floors in an Exhibition Home," J. S. Bobbio, *Heating and Ventilating*, **48**, No. 1, 55 (Jan. 1951).
7. "Warm Air Perimeter Heating—Heat Emitted from Floor Surface," J. R. Jamieson, R. W. Roose, and S. Konzo, *ASHVE Trans.*, **57**, 437 (1951).

14

Air Conveying and Distribution, Fans, Duct Design, and Diffusion

14.1. The Determination of Velocities of Air Flow. Instruments employed for measuring air velocities are Pitot tubes and anemometers. Pitot tubes (discussed in Art. 6.3) may be used to determine the velocity of any fluid; in general these instruments serve to measure a pressure from which a velocity of flow may be computed. Anemometers function by either the rotation of a component part, the impact of the fluid stream upon a member, or by the change of temperature of a wire of an electric circuit.

14.2. Anemometers. Figure 14.1 illustrates a type of instrument used for the determination of air velocities ranging from 150 to 1500 fpm. This form of anemometer is a delicate instrument which requires frequent calibration and careful handling. The air flow causes a vaned wheel to rotate and operate the indicating mechanism.

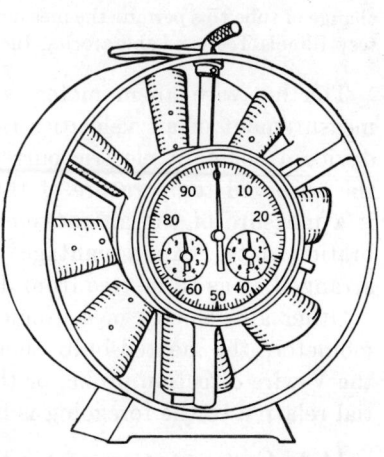

Fig. 14.1. Anemometer.

279

An instrument known as a *vane anemometer* or *velometer* is valuable because it gives an instantaneous reading without the necessity of either timing, calculations, or reference to charts. This device is used ordinarily with air and is of the form shown in Fig. 14.2; it consists of a delicately balanced and magnetically stabilized vane within a case through which the air flows when the instrument is pointed into the air stream to receive its impact. Velocities as low as 50 fpm may be measured, and attachments may be provided to adapt the instrument to the measurement of the flow of air in ducts.

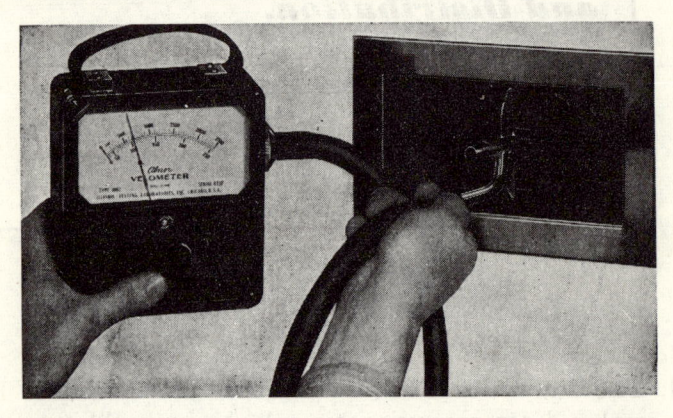

Fig. 14.2. Determination of return-grille-face air velocity with a velometer. A change of tube tips permits the measurement of supply-grille air velocities. (Courtesy Illinois Testing Laboratories, Inc.)

The hot-wire anemometer is another instrument adapted to the measurement of air velocities ranging from 20 to 2000 fpm. In this device a constant electric current is maintained through a wire and the observed temperature of the wire, as indicated by its resistance, is a measure of the air movement when referred to the proper calibration data. An advantage of the instrument is that it can be arranged so as to be free from directional effect.

Other air-velocity measurements involve the use of the kata thermometer, the heated-bulb thermometer, the heated thermocouple, the V-wire direction probe, or the smoke-filament techniques. Material relative to the foregoing is included in the Bibliography.

14.3. Computation of Air Velocity from Its Velocity Pressure. The simple expression $v = \sqrt{2gP_v}$ expresses velocity of flow in feet per second, when the velocity pressure is measured in feet head of the fluid involved. However, the velocity pressures in air ducts are usually measured and recorded in terms of inches of water (or in. WG).

AIR CONVEYING AND DISTRIBUTION

The following equation includes factors which convert velocity pressure measured in. WG to feet head of air.

$$v = \sqrt{2g \frac{p_v d_w}{12 d_a}} \qquad (14.1)$$

The average velocity of flow in a duct can only be accurately determined by a traverse (Art. 6.3).

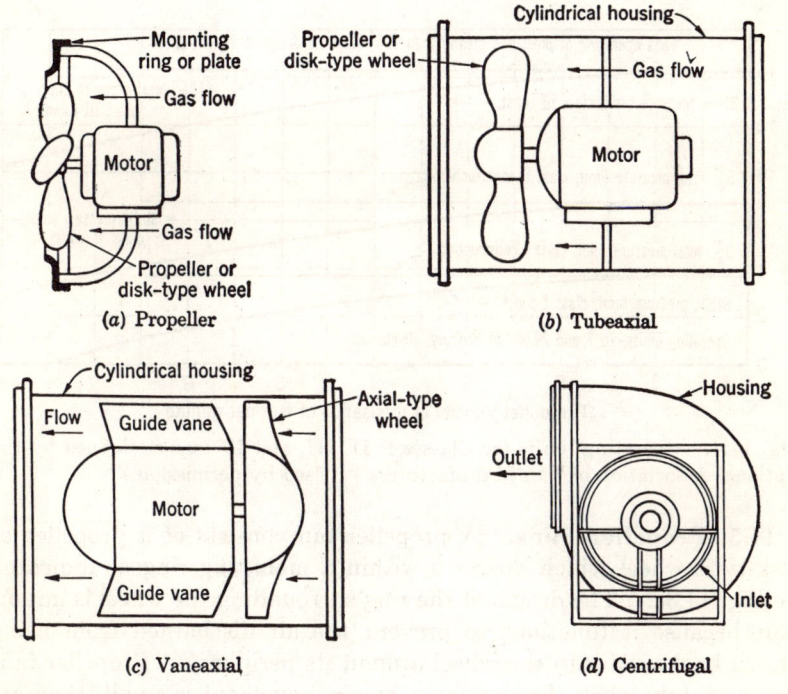

Fig. 14.3. Fan types. (Used by permission of the National Association of Fan Manufacturers.)

Whenever a velocity pressure is required, equation 14.1 is arranged as

$$p_v = \frac{12 v^2 d_a}{2 g d_w} \qquad (14.2)$$

14.4. Fans. Air movements through heating, ventilating, and air-conditioning apparatus are usually produced by the action of some form of a fan. The National Association of Fan Manufacturers[1] has named and defined types of fans as illustrated in Fig. 14.3. Essentially a fan consists of a rotating wheel which is surrounded by a stationary member designated as a housing. Energy is transmitted to the air handled by the power-driven wheel and a pressure difference is created to produce flow of the material handled. Irrespective of their type of construction, fans may function as either blowers or exhausters. Blowers discharge air against a pressure at their outlet. Exhausters remove gases from a space by suction. Fans are separated into classes according to their ability to move gases against resistance, as shown graphically in Fig. 14.4.

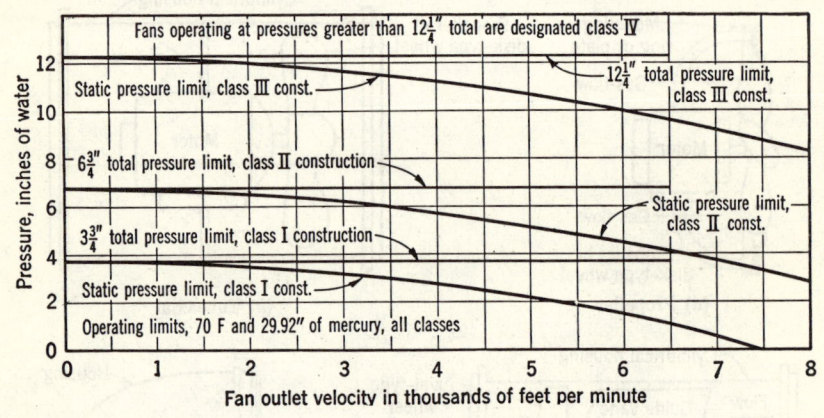

Fig. 14.4. Operating limits for Classes I, II, III, and IV fans as defined by the National Association of Fan Manufacturers. (Used by permission.)

14.5. Propeller Fans. A propeller fan consists of a propeller or disk-type wheel which operates within a mounting ring as indicated by Fig. 14.3a. The design of the ring surrounding the wheel is important because it functions to prevent the air discharged from being drawn backward into the wheel around its periphery. Propeller fans are used only when the resistance to air movement is small; they are useful for the ventilation of attic spaces, the removal of cooking odors from kitchens, the ventilation of lavatories and bathrooms, and many other applications where little or no ductwork is involved.

14.6. Tubeaxial Fan. This form of fan consists of a propeller wheel supported by bearings within a cylinder. The wheel drive may be either from an electric motor within the cylinder directly connected to its shaft, or it may be through a belt arrangement from a motor mounted outside the encircling housing. These fan units are designed to handle a wide range of air volumes against medium discharge pressures. The wheels are constructed to minimize re-entry losses about their perimeters. A large-diameter hub, Fig. 14.3b, is used and die-

AIR CONVEYING AND DISTRIBUTION

formed blades are designed to produce uniform air velocity from hub to tips. The air discharge from a tubeaxial fan follows a spiral path as it leaves the cylindrical housing.

14.7. Vaneaxial Fan. This form of fan unit combines the wheel, the driving mechanism, and the housing construction of the tubeaxial fan with stationary directional guide vanes, Fig. 14.3c. Spiral flow

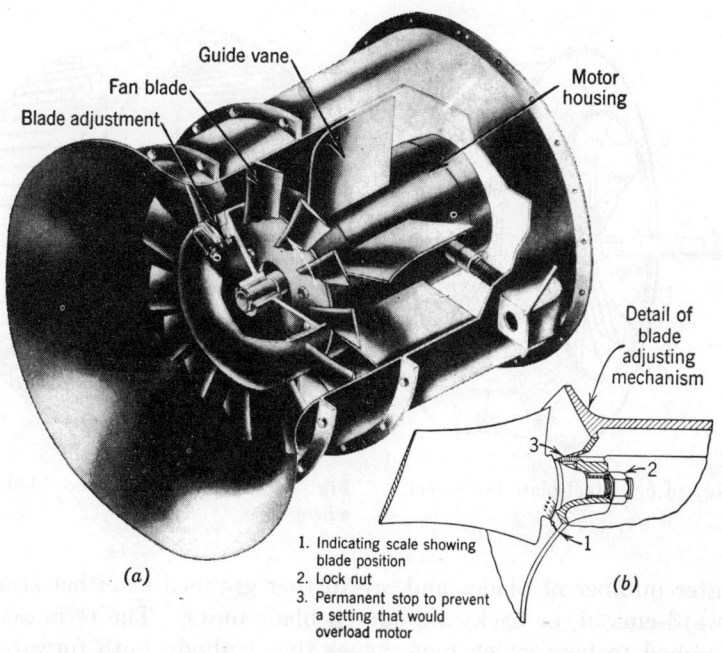

Fig. 14.5. Axial-flow fan with adjustable blades. (Courtesy Joy Manufacturing Co.)

of the discharge air is eliminated, turbulence of flow is reduced, and the efficiency of operation and the pressure characteristics are better than those of the tubeaxial form of construction. Also the resulting straight-line flow leaving the fan assures quieter operation. Figure 14.5 shows a vaneaxial fan with adjustable blades which may be reset, if necessary, after installation to achieve the best possible results of operation. Because the inlet and discharge openings are in alignment in both tubeaxial and vaneaxial fans they have an advantage of simplified installation in a duct system over that possible with centrifugal fans.

14.8. Centrifugal Fans. All centrifugal fans, an example of which is shown in Fig. 14.3d, have a wheel which is rotated within a scroll-shaped housing. Centrifugal-fan wheels are built in many forms to give suitable performance characteristics so that a proper fan may be selected for use with any air-handling problem. Centrifugal fans, as a class, are adaptable to a greater range of operating pressures than those previously discussed. Two general classifications of centrifugal fans are *steel-plate* and *multiblade* units. The latter have a relatively

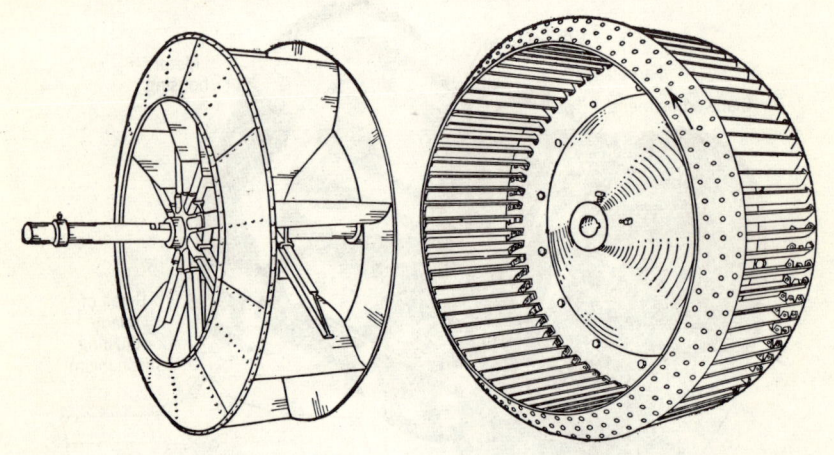

Fig. 14.6. Steel-plate fan wheel.

Fig. 14.7. Forward-curved blade fan wheel.

greater number of blades and are further grouped as either straight, forward-curved, or backward-curved blade units. The term *conoidal* is applied to fans which have vanes that embody both forward and backward curvature of each of its blades.

Steel-plate fan wheels are useful in handling either air or other gases which contain dust, cinders, or other solids. As shown in Fig. 14.6, they usually consist of two spiders each fitted with from 6 to 12 arms. Each pair of arms carries a flat steel float or blade of some radial length. The wheel floats may be either straight or curved forward or backward, depending upon the desired operating characteristic. Wheel constructions vary; a blade may be attached to a single arm only in some units, in others the two spiders are replaced by a single one and a hub plate. Special designs intended for the handling of stringy materials such as cotton and wool have blades that extend from the hub and are welded to it in such a way that there are no edges upon which the material may lodge. Units that have blades made

AIR CONVEYING AND DISTRIBUTION

from heavy steel plates can be used for all classes of services and are sometimes referred to as *industrial fans*.

The multiblade fan wheels of Figs. 14.7, 14.8, and 14.9 have blades that are shallow in depth when compared with those of the steel-plate type. Practically all commercial multiblade fans are equipped with wheel blades that are curved either forward or backward with respect to the direction of rotation.

A large percentage of the centrifugal fans installed in air-conditioning systems are equipped with wheels that have narrow blades curved

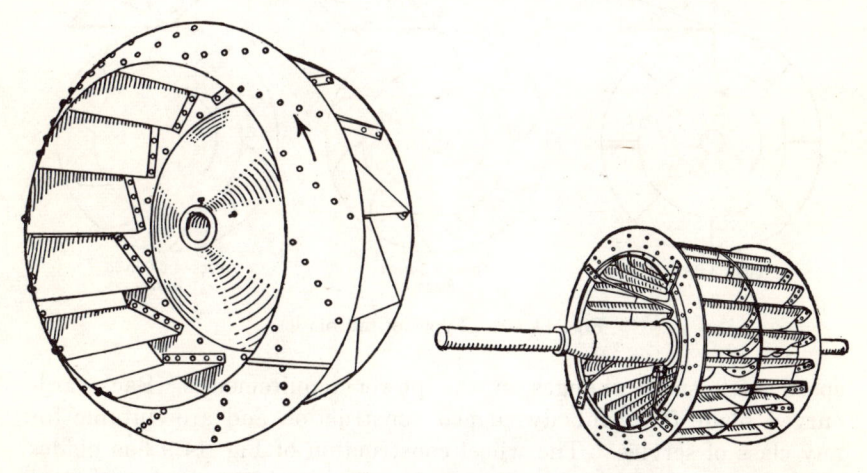

Fig. 14.8. Backward-curved blade fan wheel.

Fig. 14.9. Conoidal double-inlet, double-width fan wheel.

forward in the direction of rotation as shown by Fig. 14.7. Inasmuch as the blades are very shallow in depth, the diameter of the housing air-inlet opening more nearly approaches that of the wheel than in any other type. The ample inlet opening, together with the streamlined hub of the wheel, promotes a smooth flow of air into the rotating blades and thereby increases the efficiency of the fan and reduces its noise. Forward-curved fan blades are more capable of overcoming the attached-duct system resistance when their operation is at low speeds than is the case with other types. This desirable feature is due to the high resultant velocity of the air leaving the blade tips as indicated in Fig. 14.10. Fans of the type just mentioned are made for Class I and Class II service only.

The fan wheel of Fig. 14.8 is equipped with blades which are much deeper than those shown in Fig. 14.7; also they are curved backward with respect to the direction of rotation. Because of the low resultant velocity shown in Fig. 14.10, backward-curved blades must be operated at a much higher speed of rotation than forward-curved ones if the same static pressure is to be produced in each case. In some cases the higher required speed may be an advantage because of a possible direct connection to the driving motor. Fan wheels that have properly designed backward-curved blades operate at high efficiency and have non-overloading power characteristics as discussed in Art. 14.14; they also offer the advantage of wide ranges of capacity at constant

Straight Blade   **Forward Curved Blade**   **Backward Curved Blade**

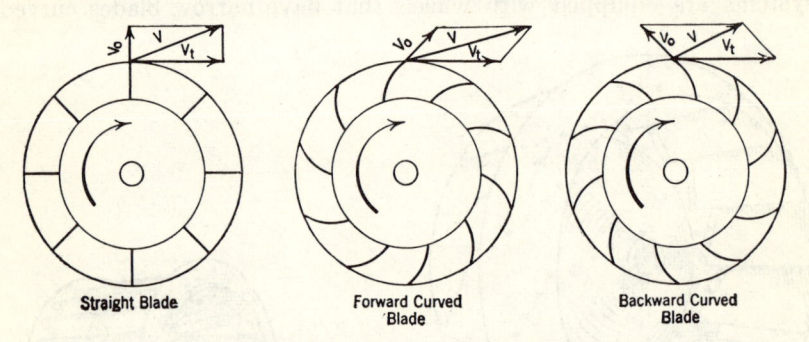

Fig. 14.10. Types of fan blades.

speed with small changes in the power requirements. Backward-curved blade fans embody rugged construction and are suitable for any class of service. The wheel construction of Fig. 14.9 has blades which have more than one curvature. The term conoidal is applied to blades of this type as they are in the form of segments of surfaces from two tangent cones. This design gives each blade a forward curve at its inner edge and a marked backward curvature at the periphery or exit edge. A blade producing desirable operating characteristics is secured which also has sufficient rigidity without the use of stiffening rings. Fans equipped with conoidal blades have non-overloading power characteristics similar to those of fans which have blades which are tipped backward throughout their entire lengths. Such fans are usually made for either Class I or Class II service conditions.

14.9. Effects of Fan-Blade Shapes on Air Velocities. The effects of blade shapes upon the resultant velocity of air are indicated in Fig. 14.10. The vector diagrams indicate velocities by the following symbols: blade-tip velocity V_t, velocity of the air flowing along the blade face V_o, and the final resultant velocity of the air V. When all the wheels shown are operated with the same blade-tip speed, the forward-curved blades give the highest and the backward-curved blades the smallest resultant air velocity. The resultant velocity

# AIR CONVEYING AND DISTRIBUTION

of the air as it leaves the fan blades is of importance from the standpoints of operating speed required to produce a given static pressure and the noise which may be originated by the fan.

14.10. Fan Nomenclature. Fans of the steel-plate and multi-blade types are designated according to (1) the number of inlets, single or double; (2) the width of the wheel, single or double; (3) the discharge, top, bottom, vertical, horizontal, or angular; (4) the housing, full, seven-eighths, or three-quarters; and (5) clockwise and counter-clockwise rotation as viewed from the drive side.

A full-housed fan is one in which the fan scroll is completely above the base upon which the fan rests, as illustrated in Fig. 14.11. Seven-eighths or three-quarters full-housed fans have the scroll extending below the top of the supporting base. The standardized designations of the hand of a centrifugal fan and drive arrangements are shown in Fig. 14.12.

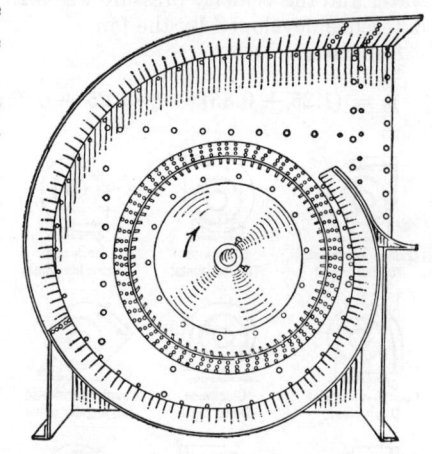

Fig. 14.11. Fan wheel and scroll.

14.11. Total-Pressure Difference Developed by a Fan. Fans operate with the gases handled arriving at the inlet to the fan housing under atmospheric pressure when no inlet duct is used and at pressures less than atmospheric when exhausting through an inlet duct. Certain types of blowers such as gas boosters have pressures greater than atmospheric in the inlet duct.

If the fan has no suction duct the entry losses to the fan housing are considered as part of the fan losses and are reflected in the mechanical efficiency of the fan. If the fan has no discharge duct leading away from its outlet the discharge static pressure is zero and the total pressure at this point is equal to the average velocity pressure. In any event, the average total-pressure difference (total pressure) created by a fan is the average total pressure at the fan outlet minus the average total pressure at the fan inlet. Static pressures less than atmospheric are considered negative. Velocity pressure is always positive.

Example 1. A fan having no inlet duct discharged air at 1.00 in. of water static pressure and an average velocity pressure of 0.15 in. of water. Find the total pressure difference developed by the fan.

Solution.

$$p_t = (1.00 + 0.15) - 0 = 1.15 - 0 = 1.15 \text{ in. of water}$$

Example 2. A fan maintained at its outlet an average static pressure of 1.25 in. of water with an average outlet velocity pressure of 0.35 in. of water. In the suction duct near the fan inlet the static pressure was −1.25 in. of water and the velocity pressure was 0.25 in. of water. Find the total-pressure difference developed by the fan.

Solution.

$$p_t = (1.25 + 0.35) - (-1.25 + 0.25) = 1.60 + 1.0 = 2.60 \text{ in. of water}$$

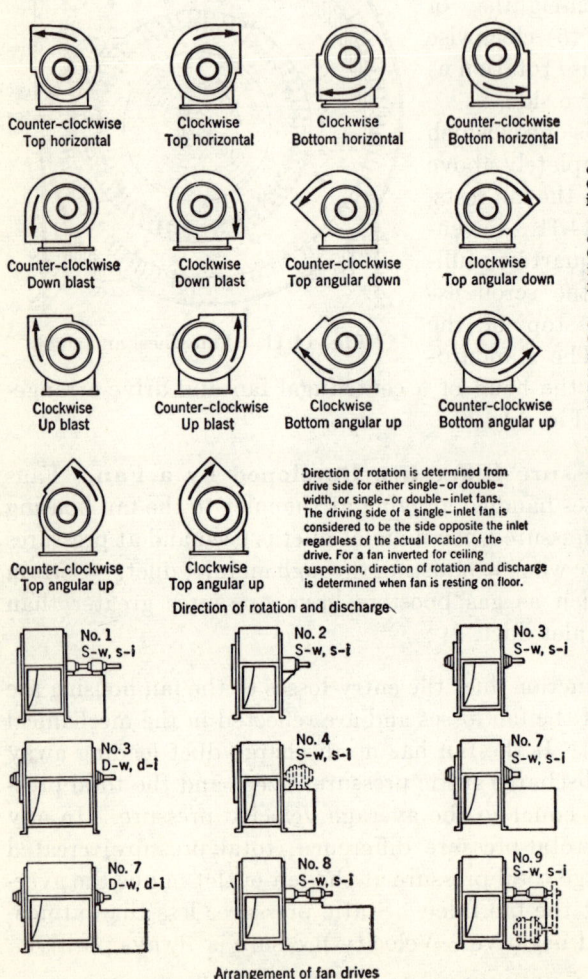

DRIVE ARRANGEMENTS

1. S-w, s-i wheel for belt drive or direct connection. Overhung wheel. Two bearings on base.

2. S-w, s-i wheel for belt drive or direct connection. Overhung wheel. Bearings in brackets supported by fan housing.

3. S-w, s-i wheel for belt drive or direct connection. One bearing on each side and supported by fan housing. Not recommended in sizes 27-in. wheel and smaller.

3. D-w, d-i wheel for belt drive or direct connection. One bearing on each side and supported by fan housing.

4. S-w, s-i wheel for direct drive. Overhung wheel on prime mover shaft. No bearings on fan. Base mounted or equivalent for the prime mover.

7. S-w, s-i wheel for belt drive or direct connection. No. 3 plus base for prime mover. Not recommended in sizes 27-in. diameter and smaller.

7. D-w, d-i wheel for belt drive or direct connection. No. 3 plus base for prime mover.

8. S-w, s-i wheel for belt drive or direct connection. No. 1 plus base for prime mover.

9. S-w, s-i wheel for belt drive. No. 1 designed for mounting prime mover on side of base.

Fig. 14.12. Fan discharges and arrangements of drive.

AIR CONVEYING AND DISTRIBUTION

The total pressure at the fan outlet in an actual system should be computed as the velocity pressure at the farthest point of discharge plus all pressure losses in the path taken by the air to reach its destination.

The total pressure at the fan inlet in a system which includes a suction duct and apparatus such as filters and coils within it is always equal to the total frictional resistance in that part of the system. The total pressure at this point in a system of that type is always negative and numerically *less* than the static pressure at the same location. In Example 2 the static pressure at the fan inlet is -1.25 and the total pressure is -1.0.

In selecting a fan from catalogue data it is usually necessary to know the volume of air which it must handle and the static pressure against which it must operate. The static pressure *as used for this purpose* is the sum of the static pressure at the fan outlet minus the negative *total* pressure at the fan inlet, $1.25 - (-1.0) = 2.25$ in the case of Example 2.

The static pressure for the purpose of fan selection is often taken as the total system resistance (inlet and discharge). This procedure, although sufficiently accurate for many practical purposes, is technically correct only when the discharge system consists of one duct which has the same cross-sectional area as the fan outlet and no side branches. (See the first example of Art. 14.22.)

14.12. Fan-Air Horsepower. The power output of a fan is expressed in terms of air horsepower (air hp) and represents work done by the fan.

$$\text{Air hp} = \frac{m_a \times P_t}{33{,}000} \qquad (14.3)$$

In terms of the volume of air handled in cubic feet per minute and the total pressure difference created by the fan in inches of water

$$\text{Air hp} = \frac{q \times p_t \times d_w}{12 \times 33{,}000} \qquad (14.4)$$

14.13. Fan Efficiencies. The ratio of the air hp output to the driving power [brake horsepower (bhp)] required at the fan shaft expresses the mechanical efficiency of the fan.

$$\text{Fan mechanical efficiency in per cent, } E_m = \frac{\text{Air hp}}{\text{Bhp}} \times 100 \qquad (14.5)$$

Air hp is a function of the total pressure difference created by a fan which is the sum of the static pressure difference and the velocity pressure at the fan outlet. The velocity pressure is usually small in comparison with the static pressure that a fan must develop and generally tables giving performance data for fans do not include it. Consequently, it is usually more convenient to calculate the static efficiency of a fan rather than its mechanical efficiency. Static efficiency is expressed as

$$E_s = \frac{\text{Air hp, based on static pressure only}}{\text{Bhp}} \times 100 \quad (14.6)$$

The air hp, based on static pressure only, may be computed using the static pressure in place of the total pressure in either equation 14.3 or 14.4. The static efficiency is always less than the mechanical efficiency, which is sometimes referred to as the *total efficiency*. Static efficiency is not a true performance characteristic; it does provide a convenient basis for the comparison of two fans that are under consideration for a given installation. If the mechanical efficiency is known, the static efficiency E_s equals $E_m \times p_{st}/p_t$.

The term *brake horsepower* is used in common practice in place of *horsepower input*, which is more descriptive of the quantity involved.

The mechanical efficiencies of steel-plate fans range from 40 to 60 per cent; those of multiblade-centrifugal or axial-flow fans range from 50 to 90 per cent. Disk- and propeller-type fans operate at efficiencies approaching 100 per cent when the resistance to air flow is small, but the efficiency decreases rapidly as the resistance is increased.

14.14. Fan Performance. Fans should be tested according to the test codes adopted by both the National Association of Fan Manufacturers and ASHAE.[1] The conditions under which the capacities are to be given involve standard air weighing 0.075 lb per cu ft.

Centrifugal and axial fans are arranged for blowing tests by setting them up without a duct connection at the inlet and with a discharge duct having a uniform cross-sectional area equal to that of the fan discharge outlet and not less than ten duct diameters in length. When a transformation piece is used at the fan outlet to change the shape of the section from rectangular to round, the greatest angle between the longitudinal axis of the duct and any element in the sides of the adapter is not to be more than 7 deg.

Pressure measurements are made at a point not less than three-fourths of the duct length away from the fan outlet. The duct outlet is arranged so that there can be different amounts of opening ranging from 0 to 100 per cent. Fans under test are first operated at con-

AIR CONVEYING AND DISTRIBUTION

stant speed and data are taken with the percentages of the duct opening ranging from 0 to 100 per cent. The taking of such data permits the development of characteristic curves similar to those of Figs. 14.13, 14.14, and 14.15. The maximum efficiency of a fan occurs when

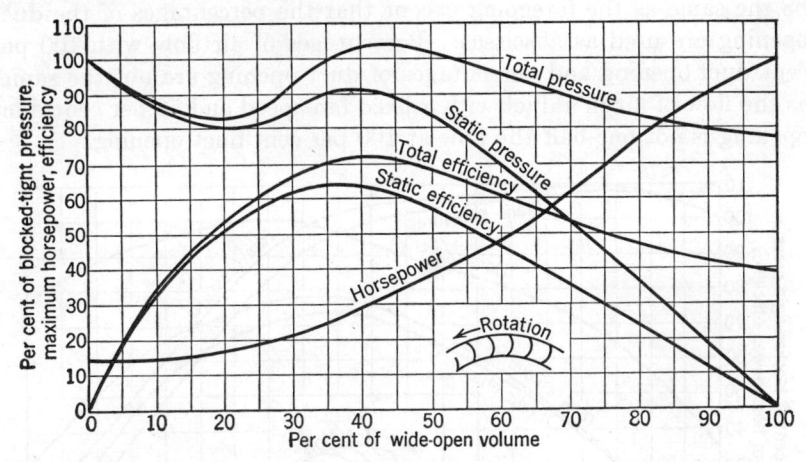

Fig. 14.13. Operating characteristics of a fan with blades curved forward. (From "Heating Ventilating Air-Conditioning Guide 1947." Used by permission.)

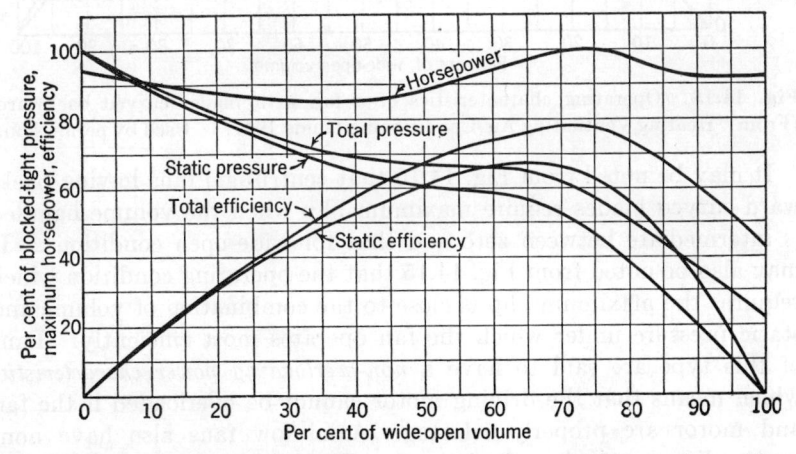

Fig. 14.14. Operating characteristics of axial-flow airfoil-type fan. (From "Heating Ventilating Air-Conditioning Guide 1947." Used by permission.)

operated at constant speed at some duct opening between 0 and 100 per cent.

The several manufacturers of fans have different methods of plotting the characteristic curves and care must be taken to ascertain how the material is presented. Thus the characteristic curves may be plotted, as shown in the above-mentioned figures, with the percentage of wide-open volume as abscissa and the percentages of blocked-tight pressures, maximum bhp, and efficiencies as ordinates. Again, the curves may be the same as the foregoing except that the percentages of the duct opening are used as abscissas. Percentages of air flow with 100 per cent duct opening and percentages of duct opening are not the same, as the flow of air in a duct with a fixed fan speed and 50 per cent duct opening is not one-half the flow at 100 per cent duct opening.

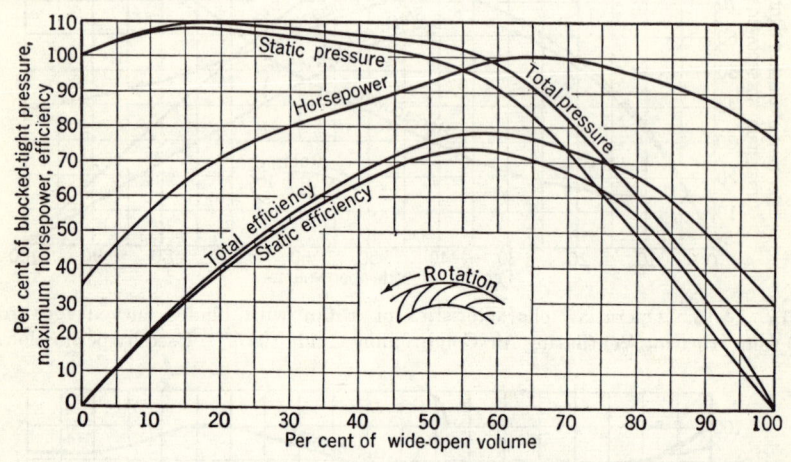

Fig. 14.15. Operating characteristics of a fan with blades curved backward. (From "Heating Ventilating Air-Conditioning Guide 1947." Used by permission.)

It may be noted from Fig. 14.15 that centrifugal fans having backward-curved blades require maximum bhp when the volume handled is intermediate between zero and that for wide-open conditions. It may also be noted from Fig. 14.15 that the operating condition which requires the maximum bhp is close to the combination of volume and static pressure under which the fan operates most efficiently. Fans of this type are said to have a *non-overloading power characteristic*, which means that the driving motor cannot be overloaded if the fan and motor are properly selected. Axial-flow fans also have non-overloading power characteristics (see Fig. 14.14), but centrifugal fans whose blades are tipped forward require an ever-increasing amount of power as the volume is increased from zero to that flowing under wide-open conditions (see Fig. 14.13). However, this type of centrifugal fan provides greater static pressure for a given blade-tip velocity than the other types which have been discussed and is commonly used in air-conditioning systems in spite of this disadvantage. Changes in a

AIR CONVEYING AND DISTRIBUTION

system which might reduce its over-all resistance to flow should be made with caution when a fan of this type is included, as such a change may result in a serious overload on the driving motor.

Sound intensity, in decibels, is an added performance characteristic which is supplied by some fan manufacturers.

14.15. Effect of System Resistance on Volume Delivered by a Fan. When a fan is in actual operation as part of an air-handling system the static pressure it develops must be the same as the resistance of the attached system. The resistance of any fixed system varies nearly as the square of the flow and may be represented as a curve in which the static pressure necessary to overcome it is plotted against the volume handled. Curve A of Fig. 14.16 may be assumed to show the resistance of a certain system for varying amounts of flow. Such a curve may be referred to as a system characteristic. Curve 1 may be assumed to show the static pressure which a certain fan is capable of developing with the same variation in air volume when operated at a certain speed.

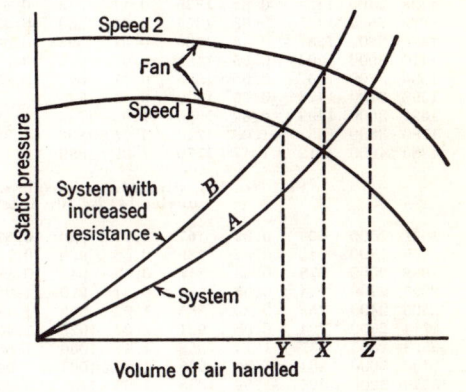

Fig. 14.16. Operating characteristics of a fan and duct system.

It may be assumed that the fan whose static pressure characteristic at a certain speed is curve 1 becomes part of a system whose resistance characteristic is curve A. When the fan is put into operation the resistance of the system will increase with the increase in air volume, along curve A, until this curve intersects curve 1. At this point the resistance of the system is equal to the static pressure which the fan is capable of producing, and the flow becomes stabilized at the amount X on the volume scale directly below the intersection. If the resistance of the system should be increased by the partial clogging of a filter, the system characteristic might be changed from curve A to curve B, in which case the volume delivered would be reduced from amount X to amount Y. It would be possible to maintain volume X against the increased resistance of the system represented by curve B by increasing the fan speed so as to change the fan static pressure characteristic from curve 1 to curve 2, provided that the motor were capable of delivering the increased power requirement. At the

TABLE 14.1. Performance Characteristics of Six Sirocco Fans*
Air density 0.075 lb per cu ft

Volume, cfm	Outlet Velocity, fpm	0.5 In. S.P. rpm	bhp	1.0 In. S.P. rpm	bhp	1.5 In. S.P. rpm	bhp	2.0 In. S.P. rpm	bhp	2.5 In. S.P. rpm	bhp	3.0 In. S.P. rpm	bhp
\multicolumn{14}{c}{Fan No. 90, single inlet, single width, wheel dia. 9 in., outlet $9\frac{9}{16}$ by $7\frac{1}{16}$, inlet 9 in. dia.}
752	1600	**919**	**0.13**	1180	**0.20**	1421	0.29						
846	1800	962	0.17	1205	0.24	1421	0.33						
940	2000	1011	0.21	1240	0.28	1452	**0.37**	1643	0.48				
1034	2200	1068	0.26	1280	0.34	1476	0.44	**1661**	**0.54**	1837	**0.67**		
1128	2400	1124	0.32	1325	0.41	1513	0.51	1690	0.60	1850	0.75	**2013**	**0.88**
1222	2600	1181	0.38	1375	0.49	1550	0.59	1718	0.71	1873	0.83	2026	0.99
1316	2800	1237	0.45	1430	0.57	1593	0.66	1753	0.82	1905	0.91	2044	1.09
1410	3000	1294	0.55	1485	0.67	1642	0.77	1796	0.93	1937	0.99	2070	1.20
1504	3200	1357	0.65	1535	0.77	1691	0.88	1838	1.05	1963	1.15	2104	1.30
1598	3400	1428	0.75	1590	0.89	1740	1.03	1881	1.16	2008	1.30	2139	1.40
1692	3600	1499	0.89	1650	1.02	1789	1.18	1923	1.33	2055	1.46	2174	1.61
1786	3800	1555	1.03	1710	1.16	1838	1.32	1980	1.50	2103	1.62	2217	1.82
1880	4000	1612	1.17	1770	1.32	1899	1.49	2036	1.67	2150	1.78	2278	2.03
\multicolumn{14}{c}{Fan No. 135, single inlet, single width, wheel dia. $13\frac{1}{2}$ in., outlet $14\frac{3}{8}$ by $10\frac{3}{4}$, inlet $14\frac{3}{8}$ in. dia.}
1671	1600	**609**	**0.32**	767	**0.47**	920	**0.67**	1055	**0.88**	1179	1.10		
1879	1800	642	0.42	789	0.56	928	0.77	1060	1.01	1181	1.25	1294	1.49
2088	2000	678	0.53	817	0.68	944	0.88	1067	1.15	**1183**	**1.40**	1297	1.66
2297	2200	714	0.65	850	0.84	970	1.02	1082	1.29	1188	1.58	**1300**	**1.84**
2505	2400	753	0.80	885	1.02	998	1.19	1103	1.45	1194	1.77	1305	2.03
2714	2600	794	0.96	921	1.22	1028	1.40	1129	1.66	1215	1.98	1315	2.24
2923	2800	837	1.14	958	1.44	1060	1.66	1159	1.91	1239	2.21	1331	2.48
3132	3000	881	1.35	998	1.68	1094	1.94	1192	2.21	1267	2.49	1354	2.74
3340	3200	925	1.59	1038	1.95	1131	2.26	1228	2.56	1298	2.82	1381	3.04
3549	3400	972	1.87	1078	2.23	1170	2.61	1264	2.95	1331	3.24	1409	3.40
3759	3600	1020	2.17	1121	2.54	1210	2.99	1302	3.36	1364	3.70		
3968	3800	1058	2.55	1165	2.87	1252	3.38	1341	3.81	1399	4.16		
4177	4000	1103	2.95	1212	3.26	1295	3.80	1381	4.29	1434	4.64		
\multicolumn{14}{c}{Fan No. 200, single inlet, single width, wheel dia. 20 in., outlet $21\frac{5}{16}$ by $15\frac{13}{16}$, inlet $21\frac{1}{4}$ in. dia.}
3675	1600	**384**	**0.65**	492	**1.00**								
4135	1800	403	0.82	504	1.20	597	**1.60**	687	2.15				
4594	2000	422	1.03	520	1.45	606	1.90	**689**	**2.37**	760	**2.87**		
5053	2200	443	1.27	537	1.73	619	2.22	694	2.64	766	3.25	838	4.24
5513	2400	466	1.54	554	2.06	633	2.58	705	3.05	774	3.65	**842**	**4.51**
5972	2600	492	1.86	573	2.43	651	3.00	718	3.53	783	4.13	847	5.10
6432	2800	520	2.26	593	2.82	666	3.45	733	4.03	795	4.66	855	5.28
6891	3000	548	2.71	614	3.25	684	3.95	750	4.60	810	5.25	867	5.91
7350	3200	577	3.25	636	3.75	703	4.50	767	5.20	826	5.88	880	6.58
7810	3400	606	3.63	659	4.35	722	5.10	785	5.85	843	6.57	895	7.35
8269	3600	640	4.52	684	5.02	742	5.72	802	6.60	860	7.34		
8729	3800	670	5.29	709	5.78	762	6.43	820	7.32	878	8.19		
9188	4000	700	6.19	735	6.56	786	7.30	840	8.17	896	9.10		

increased speed this fan would handle the volume represented by point Z if the filter were cleaned or replaced so as to return the system resistance to that represented by curve A.

14.16. Fan Tables. The usual tabular data relative to the performance of a given size of a centrifugal fan are similar to those of Table 14.1 which is a condensation of the data of six tables of a manufacturer's catalogue. The complete tables include more extensive

AIR CONVEYING AND DISTRIBUTION

TABLE 14.1. (Continued)

Volume cfm	Outlet Velocity fpm	0.5 In. S.P. rpm	bhp	1.0 In. S.P. rpm	bhp	1.5 In. S.P. rpm	bhp	2.0 In. S.P. rpm	bhp	2.5 In. S.P. rpm	bhp	3.0 In. S.P. rpm	bhp

Fan No. 300, single inlet, single width, wheel dia. 30 in., outlet 31⅞ by 23⅝, inlet 31¼ in. dia.

8,240	1600	272	1.43	349	2.14	423	2.98	487	3.91				
9,270	1800	286	1.85	356	2.63	425	3.49	489	4.43	538	5.43		
10,300	2000	302	2.30	367	3.20	431	4.11	491	5.05	540	6.11		
11,330	2200	317	2.86	381	3.83	439	4.81	496	5.79	543	6.91	597	8.09
12,360	2400	332	3.51	394	4.56	449	5.65	502	6.67	547	7.79	599	9.01
13,390	2600	348	4.23	408	5.36	462	6.56	510	7.67	553	8.82	602	10.02
14,420	2800	365	5.04	423	6.32	475	7.62	520	8.84	560	9.94	608	11.23
15,450	3000	382	6.01	438	7.41	489	8.82	532	10.15	570	11.26	616	12.65
16,480	3200	399	7.09	455	8.62	502	10.12	544	11.53	581	12.72	625	14.20
17,510	3400	417	8.30	469	9.88	517	11.53	557	13.06	594	14.35	635	15.93
18,540	3600	433	9.63	486	11.27	532	13.09	572	14.72	607	16.11	647	17.75
19,570	3800	448	11.05	502	12.84	547	14.75	586	16.54	621	18.03	660	19.75
20,600	4000	466	12.59	519	14.58	563	16.52	601	18.52	635	20.14	672	21.85

Fan No. 440, single inlet, single width, wheel dia. 44 in., outlet 48¼ by 34⅞, inlet 50 in. dia.

18,240	1600	177	2.85	232	4.56	280	6.36						
20,520	1800	183	3.69	236	5.44	282	7.38	323	9.56				
22,800	2000	191	4.60	242	6.47	285	8.64	325	10.83	360	13.22		
25,080	2200	201	5.77	249	7.69	290	9.95	327	12.34	362	14.94	394	17.57
27,360	2400	213	7.25	254	9.19	295	11.35	331	14.12	366	16.63	397	19.54
29,640	2600	225	8.89	260	10.91	301	13.13	336	16.00	369	18.79	399	21.69
31,920	2800	236	10.59	269	12.67	308	15.09	343	18.02	374	21.17	402	24.17
34,200	3000	248	12.55	279	14.87	313	17.60	348	20.32	379	23.56	407	26.67
36,480	3200	259	14.91	290	17.48	320	20.02	354	22.86	385	26.15	412	29.55
38,760	3400	271	17.33	302	20.54	327	22.68	359	25.70	391	29.13	419	32.64
41,040	3600	279	19.87	313	23.68	338	26.03	366	29.28	397	32.32	425	36.15
43,320	3800	289	23.01	326	27.04	348	29.61	373	32.87	402	36.45	431	39.96
45,600	4000	299	26.17	337	30.63	358	33.17	382	36.75	409	40.64	437	44.14

Fan No. 657, single inlet, single width, wheel dia. 65¾ in., outlet 71¼ by 52, inlet 72⅞ in. dia.

40,320	1600	118	6.31	155	10.07	187	14.07						
45,360	1800	122	8.16	158	12.02	189	16.32	215	21.14				
50,400	2000	127	10.17	162	14.30	191	19.09	217	23.95	240	29.23		
55,440	2200	134	12.75	166	17.01	193	22.00	219	27.28	242	33.03	263	38.84
60,480	2400	143	16.02	169	20.31	197	25.10	221	31.21	244	36.76	265	43.20
65,520	2600	150	19.65	174	24.11	201	29.03	224	25.38	246	41.55	267	47.96
70,560	2800	158	23.42	180	28.01	205	33.36	229	39.83	250	46.80	269	53.44
75,600	3000	165	27.75	186	32.87	209	38.91	233	44.92	253	52.09	272	58.96
80,640	3200	173	32.96	194	38.65	213	44.26	236	50.54	257	57.80	276	65.33
85,680	3400	181	38.31	201	45.42	218	50.14	240	56.81	261	64.41	280	72.17
90,720	3600	186	43.93	209	52.35	225	57.54	244	64.74	265	71.44	284	79.93
95,760	3800	193	50.87	217	59.78	233	65.47	249	72.67	269	80.59	288	88.36

* American Blower Co., used by permission.

data and cover a wider range of static pressures; data for many other sizes of fans are also given. The fan manufacturer builds other types of fans for which similar data are given in bulletins pertinent to them. The objective of the compilation of Table 14.1 is to provide a minimum amount of data which can be used for the solution of problems involving the selection of fans as illustrated in the first example of Art. 14.22.

Many fan tables, as in the case of Table 14.1, indicate by either boldface or italic type the most efficient conditions of operation at each static pressure. Whenever possible a fan should be selected that will operate under conditions near to those for its maximum efficiency. With the limited amount of data of Table 14.1 this may not be possible but when all of the manufacturer's data are available a satisfactory selection may be made. Fan selection technique is the same from actual catalogue data as from Table 14.1.

14.17. Fan Laws. Generally basic fan laws are concerned with (1) a change of fan speed when the unit is attached to a system which has a fixed characteristic; (2) variations of the density of the gas handled; (3) an increase or decrease of the fan size, based on the maintenance of a strict proportionality of its parts, when a fixed point of rating is maintained; and (4) system resistance or static head. Other pertinent laws may be built up from considerations of the basic laws. The following statements are applicable in determining fan performances under the conditions stated.

1. Change of speed with a definite fan size, duct system, and gas density. (a) The capacity changes directly as the speed ratio, (b) the static pressure varies as the square of the speed ratio, and (c) the hp required to drive the fan varies as the cube of the speed ratio. Thus when the speed of a fan is doubled the speed ratio is 2, the capacity is doubled, the static pressure is four times as large, and the required driving power (bhp) is eight times that necessary before the speed was doubled.

2. Gas-density change with a fixed-size fan operated with unchanged speed to give a constant volume discharge into a fixed or constant system. Both the power required and the static pressure vary directly as the gas-density ratio.

3. Gas-density changes with a fixed size of fan operated at variable speeds to discharge a constant weight of gas into a fixed or constant system. (a) The volume, the static pressure, and the fan speed (rpm) vary inversely as the density ratio, and (b) the required driving power (bhp) varies inversely as the square of the density ratio.

4. Gas-density variation with a fixed size of fan operated at variable speed to produce a constant discharge pressure against the resistance offered by a fixed or constant duct system. The volume discharged, the fan speed (rpm), and the required power (bhp) vary inversely as the square root of the ratio of the gas densities.

5. Fan-size change with the fan operated at constant speed (rpm), constant air density, uniform fan proportions, and a fixed point of rating. (a) The discharge volume varies as the cube of the ratio of wheel diameters, (b) the static pressure varies as the square of the wheel diameters' ratio, (c) the blade-tip speed varies as the ratio of the wheel

AIR CONVEYING AND DISTRIBUTION

diameters, and (*d*) the bhp varies as the fifth power of the ratio of the wheel diameters.

6. *Fan-size change with the fan operated with constant blade-tip speed, unchanged air density, uniform fan proportions, and a fixed point of rating.* (*a*) The discharge volume and the required power (bhp) vary as the square of the wheel diameters' ratio, (*b*) the discharge pressure remains constant, and (*c*) the speed (rpm) varies inversely as the ratio of the wheel diameters.

14.18. General Procedure in Fan Selection. The volume of air which must be handled and the system resistance (static pressure) must be known before a fan can be selected. Resistances offered to air flow by commercial air filters, washers, coils, and other equipment can be furnished by the builders. Pressure losses in ducts, elbows, fittings of various types, and straight lengths of ducts are discussed later in this chapter. When the density of the air handled is not 0.075 lb per cu ft (standard condition), application of fan law 4 must be made as illustrated in the following example because all tabular data represent performance when handling air at standard density.

Example. From Table 14.1 select a fan to handle 1150 cfm of air having a density of 0.0607 lb per cu ft against a static pressure of 0.5 in. WG. Find the required operating speed and the bhp that will be needed.

Solution. Fan law 4 is applied to find the volume which a suitable fan would handle against the same static pressure when the air density is standard. (The volume to look for in the fan table.) $\sqrt{0.0607/0.075} = \sqrt{0.81} = 0.90$. $1150 \times 0.90 = 1034$ cfm. In Table 14.1 it is found that fan No. 90 handles 1034 cfm of standard air against 0.5 in. static pressure when operated at 1068 rpm and the bhp required is 0.26. Fan law 4 is now used again to convert the volume, speed, and bhp found in the table (based on an air density of 0.075) to data applicable to the actual operating density. Actual volume is $1034 \div 0.90 = 1150$ cfm, the actual speed is $1068 \div 0.90 = 1188$ rpm, and the actual bhp is $0.26 \div 0.9 = 0.289$ bhp.

In the foregoing example it was possible to find in Table 14.1 the exact volume and the exact static pressure that were sought. If this were not the case, exact results would require interpolation in the table for both the exact volume and the exact static pressure as illustrated in the first example of Art. 14.22.

14.19. Duct Friction. The method of Art. 6.6 which is applicable to all fluids may be applied to air. Figure 14.17 indicates the friction losses per 100 ft of clean, circular, galvanized-metal ducts, which have approximately 40 joints in the stated length of straight duct, when the air handled has a density of 0.075 lb per cu ft. *When air of greater or less density is handled the chart is used as though the density were*

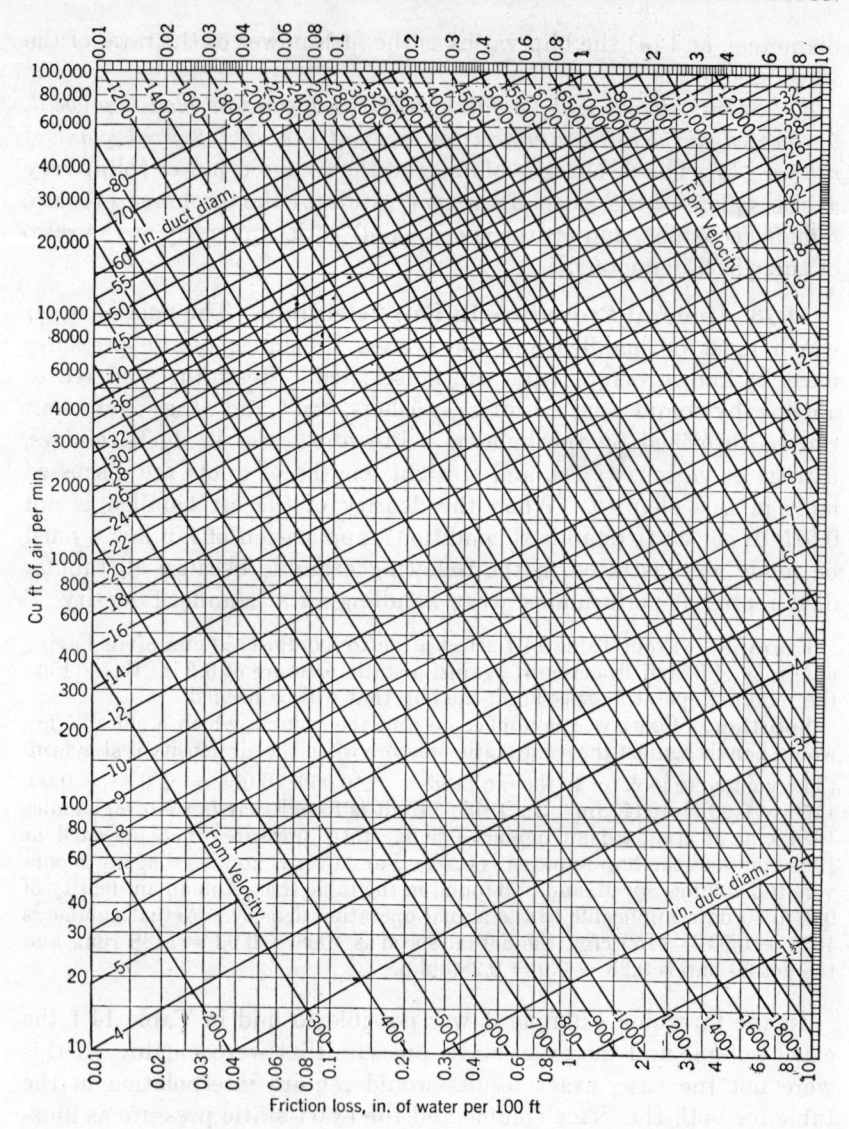

Fig. 14.17. Friction-pressure losses of air in circular ducts. Based on standard air (density 0.075 lb per cu ft) flowing through average, clean, round, galvanized metal ducts having approximately 40 joints per 100 ft. Friction losses are expressed in inches of water at 62 F (density 62.3 lb per cu ft). (From "Heating Ventilating and Air-Conditioning Guide 1948." Used by permission.)

AIR CONVEYING AND DISTRIBUTION

standard, and the friction loss as given by the chart is corrected by multiplying it by the ratio of the actual density of the air handled to 0.075. For the average application, values obtained from the chart are sufficiently accurate without correction for any air temperature between 50 F and 90 F, for any relative humidity, and, except for cities at a high elevation above sea level, for any normal variation in barometric pressure. Figure 14.17 should not be used to obtain values below those shown, by extrapolation, because critical flow would occur in this region, and the data would be unreliable. For unusually rough pipes, the method of Art. 6.6 should be used, as the friction loss may be more than double that obtained by use of the chart. Additional charts giving roughness correction data may be found in reference 2.

The friction losses of a rectangular duct may be calculated by the method of Art. 6.7. When the data of Fig. 14.17 are to be used in connection with a rectangular duct, it is first necessary to determine the *circular equivalent* which is the diameter, in inches, of a round duct that has the same friction losses per foot of length when handling the same volume of air having the same density. Equation 14.7, taken from the 1957 ASHAE Guide, may be used for a satisfactory solution.

$$d_c = 1.30 \frac{(ab)^{0.625}}{(a+b)^{0.25}} \quad (14.7)$$

where d_c = the circular equivalent of the duct, for equal friction and capacity, diameter in inches.
a = the greater dimension of a rectangular duct, inches
b = the lesser dimension of a rectangular duct, inches

Equation 14.7 was derived by equating equations for friction losses in circular and rectangular ducts in which equal volumes of air flow at the same density.

Example. Find the circular equivalent of a rectangular duct which has the dimensions 22 in. by 16 in.
Solution. $a = 22$, $b = 16$, $ab = 352$, and $(a+b) = 38$. $ab^{0.625} = 352^{0.625} = 39.05$, $(a+b)^{0.25} = 38^{0.25} = 2.483$. Hence, $d_c = 1.3 \times 39.05/2.483 = 20.4$ in.

A table of circular equivalents which covers a range of rectangular dimensions from 3 in. to 96 in. is given in reference 2.

14.20. Shock Losses. The additional friction losses caused by fittings have been discussed in Art. 6.9. Losses from causes other than pipe friction are called *shock losses* or *dynamic losses*. It was mentioned in Art. 6.9 that such losses may be expressed either as an applicable constant times the velocity pressure or as an equivalent

TABLE 14.2. Pressure Losses due to Elbows

Additional equivalent losses in excess of friction to intersection of center lines

Type	Conditions	C^*	L/D	L/W
N, deg	Rectangular or round; with or without vanes	$\dfrac{N}{90}$ times value for similar 90-deg elbow [4]		
90-deg round section	Miter	1.30†	65 [3]	
	$R/D = 0.5$	0.90		
	0.75	0.45	23	
	1.0	0.33	17	
	1.5	0.24	12	
	2.0	0.19	10	
	H/W R/W			
90-deg rectangular section	0.25 { Miter	1.25†		25 [3]
	0.5	1.25		25
	0.75	0.60		12
	1.0	0.37		7
	1.5	0.19		4
	0.5 { Miter	1.47		49
	0.5	1.10		40
	0.75	0.50		16
	1.0	0.28		9
	1.5	0.13		4
	1.0 { Miter	1.50		75
	0.5	1.00		50
	0.75	0.41		21
	1.0	0.22		11
	1.5	0.09		4.5
	4.0 { Miter	1.38		110
	0.5	0.96		65
	0.75	0.37		43
	1.0	0.19		17
	1.5	0.07		6
	R/W R_1/W R_2/W			
90-deg square section with splitter vanes	Miter 0.5			28 [5]
	0.5 0.4	0.70 [4]		19
	0.7 0.6			12
	1.0 1.0	0.13		7.2
	1.5	0.12		
	Miter 0.3 0.5			22 [5]
	0.5 0.2 0.4	0.45		16
	0.75 0.4 0.7	0.12		
	1.0 0.7 1.0	0.10		
	1.5 1.3 1.6	0.15		
Miter with turning vanes	Plate vanes	0.35 [4]		
	Formed vanes	0.10 [4]		
Miter tee with vanes	Consider equal to a similar elbow. Base loss on entering velocity.			
Radius tee				

* Values based on f (friction factor) values of approximately 0.02.
† Values calculated from L/D and L/W values of reference 3 for $f = 0.02$.
Note: Superscript numbers refer to references at end of chapter.
(From "Heating Ventilating Air-Conditioning Guide 1957." Used by permission.)

AIR CONVEYING AND DISTRIBUTION

TABLE 14.3. Pressure Losses due to Area Changes

Type	Illustration	Conditions A_1/A_2	Loss Coefficient C_1	Loss Coefficient C_2
Abrupt expansion		0.1	0.81	81
		0.2	0.64	16
		0.3	0.49	5
		0.4	0.36	2.25
		0.5	0.25	1.00
		0.6	0.16	0.45
		0.7	0.09	0.18
		0.8	0.04	0.06
		0.9	0.01	0.01
		θ	C_r^*	
Gradual expansion		5°	0.17[4]	
		7°	0.22	
		10°	0.28	
		20°	0.45	
		30°	0.59	
		40°	0.73	
Abrupt exit	$(A_2 = \infty)$	$A_1/A_2 = 0.0$	1.00	
Square edge orifice exit		A_0/A_1	C_0	
		0.0	2.50[6]	
		0.2	2.44	
		0.4	2.26	
		0.6	1.96	
		0.8	1.54	
		1.0	1.00	
Bar across duct		E/D	C	
		0.10	0.7[4]	
		0.25	1.4	
		0.50	4.0	
Pipe across duct		E/D	C	
		0.10	0.20[4]	
		0.25	0.55	
		0.50	2.0	
Streamlined strut across duct		E/D	C	
		0.10	0.07[4]	
		0.25	0.23	
		0.50	0.90	

Type	Illustration	Conditions A_2/A_1	Loss Coefficient C_2
Abrupt contraction, square edge		0.0	0.34[12]
		0.2	0.32
		0.4	0.25
		0.6	0.16
		0.8	0.06
Gradual contraction		θ	
		30°	0.02[4]
		45°	0.04
		60°	0.07
Equal area transformation		$A_1 = A_2$	C
		$\theta \leq 14°$	0.15[4]
Flanged entrance		$A = \infty$	C
			0.34[6]
Duct entrance		$A = \infty$	C
			0.85[4]
Formed entrance		$A = \infty$	C
			0.03[4]
Square edge orifice entrance		A_0/A_2	C_0
		0.0	2.50[6]
		0.2	1.90
		0.4	1.39
		0.6	0.96
		0.8	0.61
		1.0	0.34
Square edge orifice in duct	$A_1 = A_2$	A_0/A	C_0
		0.0	2.50[6]
		0.2	1.86
		0.4	1.21
		0.6	0.64
		0.8	0.20
		1.0	0.0

* C_r is to be applied with the proper C_1 or C_2 value given in the section above for abrupt expansion.

Note 1: Subscript on C indicates cross section at which velocity is calculated, i.e., 1, 2, or orifice.

Note 2: Superscript numbers refer to references at end of chapter.

(From "Heating Ventilating Air-Conditioning Guide 1957." Used by permission.)

length of straight pipe or duct. The following equation may be used to compute the shock loss due to any entrance, turn, change of section, or other incongruity for which data are given in either Table 14.2 or 14.3.

$$p_{sh} = Cp_v \tag{14.8}$$

302 **AIR CONDITIONING AND REFRIGERATION**

where C is a constant obtained from either Table 14.2 or 14.3. Table 14.2, applying to various types of bends, also includes data on equivalent lengths (given as L/D or L/W).

Before using Table 14.2 in the case of elbows in rectangular ducts it is necessary to determine the applicable R/W and H/W ratios. R is the radius of the center line of the elbow, W is the distance across the

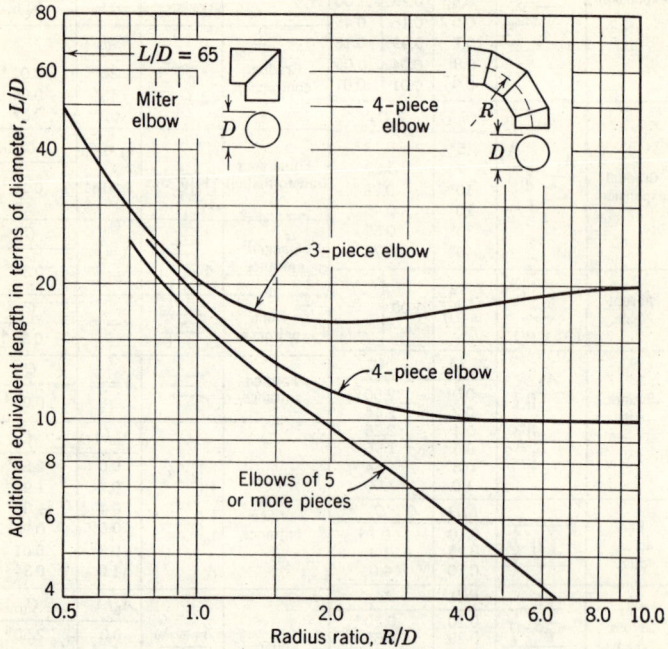

Note: Equivalent lengths given are to be added to the actual lengths measured to intersection of center lines at the elbow. (See Fig. 14.20.)

Fig. 14.18. Loss in 90-deg elbows of circular cross section.

duct in the plane in which the bend occurs, and H is the other dimension of the rectangle. The ratio H/W is often called the *aspect ratio*.

Example 1. Find the shock loss in inches of water that would be expected in a 90-deg rectangular elbow in a sheet metal duct 12 in. high and 24 in. wide handling 4000 cfm of essentially dry air at a temperature of 200 F. The room temperature is 70 F. The turn will be in the vertical plane and the center-line radius is 1 ft. Assume standard barometric pressure.

Solution. Equation 14.2 will be used to find the velocity pressure. $v = 4000/2 \times 60 = 33.3$ fps, $d_a = 0.06015$, and $d_w = 62.27$ lb per cu ft. $p_v = (12 \times \overline{33.3}^2 \times 0.06015)/(64.4 \times 62.27) = 0.200$ in. of water. The distance across the duct in the plane of the bend (W) is 1.0 ft and the other dimension (H) is 2.0 ft. The ratio R/W is $1.0/1.0 = 1.0$ and H/W is $2.0/1.0 = 2.0$;

AIR CONVEYING AND DISTRIBUTION

C by interpolation in Table 14.2 is $0.22 - \frac{1}{3}(0.22 - 0.19) = 0.21$; and from equation 14.8 p_{sh} equals $0.21 \times 0.20 = 0.042$ in. of water.

Example 2. Find the equivalent length of straight duct of the same size and shape that would have the same friction loss as the elbow of Example 1.

Solution. As in Example 1, $H/W = 2.0$ and $R/W = 1.0$. From Table 14.2 the ratio L/W is $11 + \frac{1}{3}(17 - 11) = 13$. W is 1.0 ft in this case and L equals $13 \times 1.0 = 13.0$ ft (the equivalent length of the elbow).

Figure 14.18, taken from reference 3, may be used to find the equivalent length of round-section sheet-metal elbows made from two or more angle-cut pieces of straight pipe.

14.21. Types of Duct Systems and Friction Losses at Branch Take-offs. Air from a single fan may be distributed from several

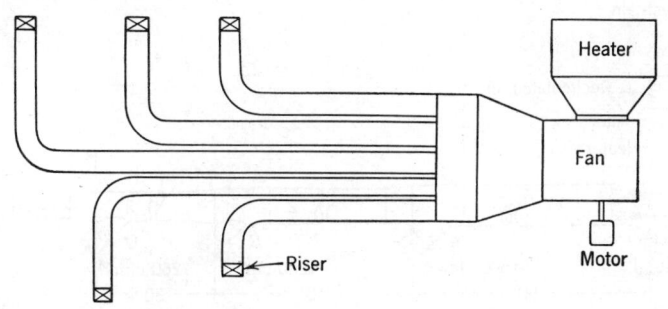

Fig. 14.19. Individual duct system.

duct outlets, located at various distances from it, in many different ways. Commonly used types of systems are discussed in subsequent paragraphs; space limitations in this work prevent the inclusion of material relative to the design of all types. Additional information is contained in the references at the end of this chapter.

The individual duct system of Fig. 14.19 has a separate branch from the plenum to each discharge grille. When the branches are proportioned in size in accordance with the volume to be handled at a velocity common to all, the friction losses per unit of equivalent length will be the same in each and it is necessary to compute only the resistance offered by the longest individual duct. In order to properly proportion the flow among the branches it is necessary to have a control damper in each. With the duct arrangement shown in Fig. 14.19 and with an air velocity common to all branches, the loss of head between the plenum chamber and the branches may be neglected. The shock loss between the fan outlet and the plenum is due to a change of air velocity; it should be computed by equation 14.8, using the proper value of C from Table 14.3.

Example. A fan discharges air into a plenum chamber with an arrangement like that of Fig. 14.19. The air velocity at the fan outlet is 2000 fpm and that within the plenum is 1000 fpm. The total included angle between the sides of the transition piece is 40 deg. The air density is 0.075 lb per cu ft and the temperature of the gage fluid is 70 F. Find the head lost in inches of water.

Solution. Equation 14.8 will be used. The velocity entering the transition is 2000 fpm or 33.3 fps. The velocity pressure is $(12 \times 33.3^2 \times 0.075)/(64.4 \times 62.27) = 0.249$. A_1/A_2 equals 0.5 and C_1 for abrupt expansion from Table 14.3 is 0.25. C_r ($e = 40$) from Table 14.3 is 0.73 and p_{sh} equals $0.25 \times 0.73 \times 0.249 = 0.0454$ in. of water.

In a similar manner the friction loss caused by practically any type of duct formation which may be encountered can be computed with

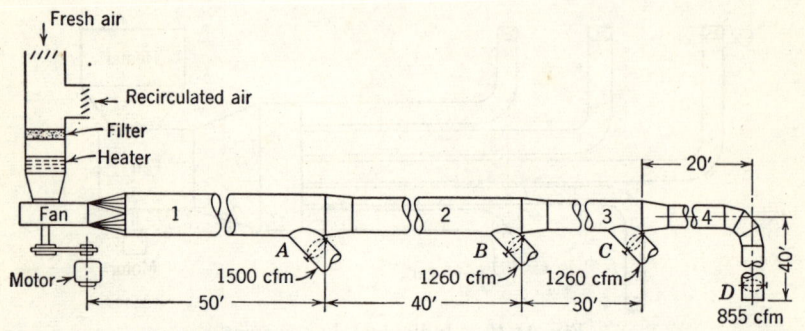

Fig. 14.20. Fan-heating system with a circular graduated-size main trunk duct.

equations 14.2 and 14.8, using the proper C value from Table 14.3. Where more than one C value is given in Table 14.3, note 1 of the footnotes should be observed. The footnote pertaining to C_r for gradual expansion should also be noted.

Information about an individual duct system which employs a box plenum is given in reference 7. The extended plenum type of individual duct system is described in reference 8. The trunk and branch ducts of Fig. 14.20 are circular in section and the trunk is reduced in size as air is removed from the main stream in order to maintain a constant frictional resistance per unit of length (see the first example of Art. 14.22).

When a branch duct is to be designed so that a desired air volume can be made to flow through it without the use of a control damper, the loss of air pressure at the point where it leaves the trunk and enters the branch must be ascertained; Fig. 14.21 supplies recent data for such losses.[9] The curve for a 90-deg angle was drawn from data computed by use of an equation, given in reference 9, which is based on a study of several laboratory investigations and an analysis of

AIR CONVEYING AND DISTRIBUTION

the flow conditions. The curves for the 60-deg and the 45-deg angles also came from reference 9 and are based on results obtained in two different investigations made in Germany. The loss coefficient λ,

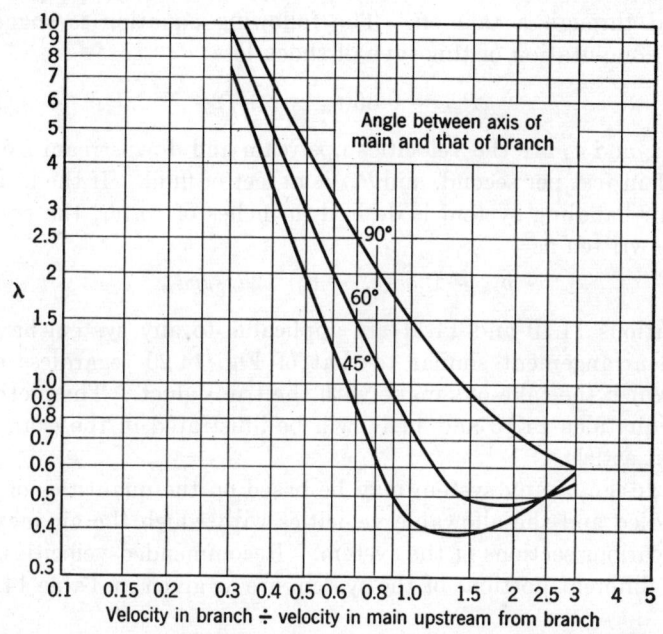

Fig. 14.21. Value of loss coefficient λ for branch take-offs from trunk ducts. (λ is the ratio of the total head lost in the take-off to the velocity pressure in the branch.)

given by Fig. 14.21, is the ratio of the head lost in the take-off to the velocity pressure in the branch. The total head lost is

$$p_{sh} = \lambda p_{vb} \qquad (14.9)$$

where λ is the loss coefficient from Fig. 14.21 and p_{vb} is the velocity pressure in the branch in inches of water. The method of making use of the data of Fig. 14.21 is illustrated by the following example.

Example. Air flows into a branch duct which has a take-off at an angle of 60 deg with the trunk duct. The air velocity in the trunk is 2000 fpm and in the branch it is 1000 fpm. The density of the air handled is 0.06 lb per cu ft. Find the loss of total head in the take-off expressed in inches of water at 70 F.

Solution. The ratio of the velocity of the air in the branch to that in the trunk is 0.5 and λ from the center curve of Fig. 14.21 is 3.0. The velocity pressure in the branch, when calculated by use of equation 14.2, is $[12(1000/60)^2 \times 0.06]/(64.4 \times 62.27) = 0.05$ in. of water at 70 F. The head lost, as determined by use of equation 14.8, is $p_{sh} = 3.0 \times 0.05 = 0.150$ in. of water.

A loss of head also occurs in the main stream when a portion is diverted through a take-off. The following equation is suggested[9] for the computation of this type of shock loss:

$$P_{sh} = 0.35(v_u - v_d)^2/2g \qquad (14.10)$$

where v_u and v_d are the velocities upstream and downstream from the take-off in feet per second, and P_{sh} is in feet of fluid. If the head lost in an air-handling system is desired in inches of water, the equation may be written as

$$p_{sh} = 0.35(v_u - v_d)^2 12d_a/2gd_w \qquad (14.11)$$

Equations 14.10 and 14.11 are applicable to any system having a general arrangement similar to that of Fig. 14.20 regardless of the angle which the take-offs make with the trunk duct. The method of using equations 14.10 and 14.11 will be illustrated in the example of the next article.

The design of any system may be based on the quantities of air to be handled and the allowable velocities with which the air may flow in the various sections of the system. Recommended velocities of air flow in different portions of the system are as given in Table 14.4.

TABLE 14.4. Recommended and Maximum Duct Velocities

	Recommended Velocities, fpm			Maximum Velocities, fpm		
Designation	Residences	Schools, Theaters, Public Buildings	Industrial Buildings	Residences	Schools, Theaters, Public Buildings	Industrial Buildings
Outside air intakes*	500	500	500	800	900	1200
Filters*	250	300	350	300	350	350
Heating coils*	450	500	600	500	600	700
Air washers	500	500	500	500	500	500
Suction connections	700	800	1000	900	1000	1400
Fan outlets	1000–1600	1300–2000	1600–2400	1700	1500–2200	1700–2800
Main ducts	700–900	1000–1300	1200–1800	800–1200	1100–1600	1300–2200
Branch ducts	600	600–900	800–1000	700–1000	800–1300	1000–1800
Branch risers	500	600–700	800	650–800	800–1200	1000–1600

* These velocities are for total face area, *not* the net free area; other velocities are for net free area.

(From "Heating Ventilating Air-Conditioning Guide 1957." Used by permission.)

This scheme of design makes the calculation of pressure losses in the separate parts of the system laborious because the friction losses are not constant when the ducts are proportioned on the basis of allowable velocities of flow. The easier method of sizing the ducts of a system, such as the one shown in Fig. 14.20, is to determine the size of the last outlet on the basis of the allowable outlet velocity and the quantity of

AIR CONVEYING AND DISTRIBUTION

air to be handled. From the size so determined and the quantity of air discharged, the friction losses per 100 ft of duct may be ascertained and the system designed using this value as the constant friction-pressure loss. When the design is so made, the velocities of air flow are not uniform but are greatest near the fan outlet.

With any arrangement of ducts the fan must create enough pressure to discharge the air with the required quantity and velocity at the outlet which has the greatest resistance to flow between it and the fan. This usually, but not always, means the longest run of duct. The other ducts are either so proportioned that the available head is used up within them or they are dampered to give additional resistance so that the proper quantity of air will pass through them.

14.22. Trunk-Duct Design Procedure and Fan Selection. The system should be laid out to deliver the air with the least expenditure of power, materials, and space. After the sizing of the ducts is completed, the static pressure against which the fan must operate should be calculated. The following will serve to illustrate the layout of a duct system by the constant friction-pressure-loss method.

Example. A fan discharges air through a system of circular ducts, as shown in Fig. 14.20, at an average temperature of 130 F at the outlets when the barometric pressure is 29.92 in. of mercury (density 0.06729 lb per cu ft). The quantities of air delivered at the outlets are as indicated by Fig. 14.20. The outlet air velocities are approximately 800 fpm. All outlets are to be dampered to control the amount of air flowing. The ratio of the three-piece-elbow radius to the duct diameter is 1.5. The actual friction losses at the heater and the fan inlet are 0.624 in. of water. Find the sizes of the duct, select a fan from Table 14.1, determine the speed at which it must operate, and the power input (bhp) to drive it.

Solution. The sizes of the duct outlets are:

$$A: (1500 \times 144) \div 800 = 270 \text{ sq in. or } 18.5 \text{ in. diameter}$$
$$B \text{ and } C: (1260 \times 144) \div 800 = 226.8 \text{ sq in. or } 17 \text{ in. diameter}$$
$$D: (855 \times 144) \div 800 = 154 \text{ sq in. or } 14 \text{ in. diameter}$$

The friction-pressure-loss chart, Fig. 14.17, shows that a duct 14 in. in diameter handling 855 cfm, with a density of 0.075 lb per cu ft, will have friction-pressure losses amounting to 0.067 in. of water per 100 ft of run. Considering this value a constant friction loss, the sizes of the other portions of the main duct are determined to be as indicated in Table 14.5.

The diameter of section 4 and the friction losses within that portion of the system were determined by locating the intersection of a horizontal line designating the specified air volume of 855 cfm and an inclined one which represents the specified velocity of 800 fpm. The duct diameter and the air velocity for each of the other sections in each case was found by locating the intersection of a horizontal line which represented the volume handled and a vertical line indicating the friction loss, 0.067 in. of water per 100 ft, that existed in section 4.

The equivalent length of the three-piece elbow ($R/D = 1.5$) in section 4 is found, by use of Fig. 14.18, to be 17 diameters, or $17 \times 14/12 = 19.8$ ft. The equivalent length of the longest run of duct is therefore $180 + 19.8 = 199.8$ ft. The friction losses when corrected for the actual air density are $(0.06729/0.075) \times (0.067 \times 199.8)/100 = 0.12$ in. of water.

In addition to the pressure loss due to friction in the straight pipe and the elbow of the discharge system, there will be shock losses in the transition at the fan outlet and at each of the take-offs A, B, and C. The head lost at the transition cannot be computed until the fan has been selected, so that will be temporarily neglected. The head lost in the trunk due to the take-off at A, using equation 14.11 and data from Table 14.5, is $0.35(1225/60 - 1120/60)^2 \times 12 \times 0.06729/(64.4 \times 62.27) = 0.00023$ in. of water at 70 F. Similarly, the head lost in the trunk at take-offs B and C is found to be 0.00044 in. and

TABLE 14.5

Section	Cubic Feet per Minute	Duct Diameter, in.	Velocity of Air Flow, fpm	Length of Duct Section, ft
4	855	14	800	60
3	2115	20	970	30
2	3375	23.5	1120	40
1	4875	27	1225	50
				Total 180 ft

0.00056 in. respectively. The velocity pressure in section 4 is $[12(800/60)^2 \times 0.06729]/(64.4 \times 62.27) = 0.0358$ in. of water. The total pressure at the beginning of section 1 (see Art. 14.11) is $0.0358 + 0.12 + 0.00023 + 0.00044 + 0.00056 = 0.15703$ in. of water. The velocity pressure in section 1 equals $[12(1225/60)^2 \times 0.06729]/(64.4 \times 62.27) = 0.0842$ in. and the static pressure at this point is $0.15703 - 0.0842 = 0.07283$ in. The tentative static pressure which will be used in selecting the fan is $0.07283 - (-0.624) = 0.69683$ in. (This is based on the assumption that the outlet area of the fan will be equal to that of section 1.)

The volume of the air handled under actual conditions is 4875 cfm at 130 F and an absolute pressure of 14.696 psia. Under rating conditions a suitable fan would handle $4875 \sqrt{0.06729/0.075} = 4875 \times 0.945 = 4600$ cfm against the specified static pressure according to fan law 4.

An effort should be made to locate in a fan performance table the combination of a discharge volume of 4600 cfm of standard air against a static pressure of 0.69683 in. of water.

Table 14.1 indicates that fan No. 200 will deliver 4594 cfm of standard air against a static pressure of 0.5 in. of water when its speed is 422 rpm and the power requirement is 1.03 bhp. This fan will be selected and the friction loss in the transition at the fan outlet can now be estimated with equation 14.8. The velocity in the outlet of fan No. 200 when delivering 4594 cfm is 2000 fpm. Under actual operating conditions the velocity will be $2000 \times 4875/4594 = 2120$ fpm. The velocity pressure in the fan outlet will be $[12(2120/60)^2 \times 0.06729]/(64.4 \times 62.27) = 0.251$ in. of water at 70 F. The ratio of areas A_1/A_2 will be $1225/2120 = 0.578$, and the constant C_1 for an abrupt expansion, from Table 14.3 by interpolation, equals 0.178. Assuming that the maximum

AIR CONVEYING AND DISTRIBUTION

angle between sides of the expanding tapered transition will be 20 deg, C_r from Table 14.3 is 0.45. The loss in total head in the transition at the fan outlet is $0.45 \times 0.178 \times 0.251 = 0.020$ in. of water. The total pressure required at the fan outlet is $0.15703 + 0.02 = 0.17703$ in., which will be taken as 0.177, and the static pressure at the fan outlet will be $0.177 - 0.251 = -0.074$. (There will actually be a slight suction at the fan outlet in this particular system.) The static pressure against which the fan will actually operate is $-0.074 - (-0.624) = 0.55$ in. Interpolation will now be made in Table 14.1 for the exact volume under standard conditions (4600 cfm) and the exact static pressure (0.55 in.) as given in Table 14.6.

TABLE 14.6

	0.5 In. Static Pressure			1.0 In. Static Pressure	
Volume	rpm	bhp		rpm	bhp
4594	422	1.03		520	1.45
5053	443	1.27		537	1.73
4600	422.3	1.033		520.2	1.454

At 0.55 in. static pressure the rpm is $422.3 + 0.05/0.5(520.2 - 422.3) = 432.1$ rpm, and the bhp is $1.033 + 0.05/0.5(1.454 - 1.033) = 1.075$. The desired data obtained from Table 14.1 by interpolation must now be converted to operating conditions by applying fan law 4 again. Volume equals $4600 \div 0.945 = 4875$ cfm, speed equals $432.1 \div 0.945 = 458$, and bhp equals $1.075 \div 0.945 = 1.14$.

From the standpoint of the cost of operation it is important that the best available fan be selected for each installation. A careful study should be made of more than one manufacturer's unabridged performance data for the entire range of sizes of the type of fan to be used. The type and size of the fan selected should be such that it will deliver the required volume of air against the computed resistance of the system with the least bhp.

The method used in the foregoing example is exact in the light of present knowledge. The loss of head in the trunk duct due to the effect of the transition at the fan outlet and the branch take-offs was considered, as well as the static pressure regain due to the decreasing velocity as the air flows toward the outlet of section 4.

If the static pressure had been taken as equal to the system resistance, neglecting the small resistance at the transition and take-offs *and the static regain*, the result would have been $0.12 - (-0.624) = 0.744$ in. of water and both the speed and bhp indicated would have been considerably higher than necessary in this case. Usually the static pressure regain in a system similar to that of Fig. 14.20 is greater than the shock losses and the use of the simplified procedure mentioned at the end of Art. 14.11 results in a quantity of air delivery that is slightly more than required. If after installation the fan speed is adjusted to give exactly the required volume, the hp requirement will usually be less than that which was estimated by the simplified method.

Some designers assume that the fan static pressure will be the total system resistance minus one-half of the theoretical static pressure regain. The theoretical static pressure regain is the velocity pressure at the fan outlet minus the velocity pressure at the end of the system. By this method the fan static pressure of the preceding example would be $0.744 - (0.251 - 0.0358)/2 = 0.6364$ in. of water.

In the prior example of this article the air velocity in each of the branch ducts is the same as that in section 4 of the trunk, i.e., 800 fpm. The calculation of resistances in branch ducts is not necessary when dampers are used to obtain the proper air volumes. When a branch is to be designed to give the required delivery of air and to use all of the available head without the aid of a branch damper, the air velocity in each damperless branch must be greater than that in the longest run. Under such conditions the proper size of the branch duct must be determined by a trial-and-error solution.

Example. Assume that branch A of Fig. 14.20 has an angle of 45 deg with the trunk and has an equivalent length of 75 ft not including that of the take-off. Find the proper size of the branch duct if it is to deliver 1500 cfm without damper controls.

Solution. The total pressure in the trunk at take-off A will be the velocity pressure at the end of section 4 plus all losses between location A and that point. The length of trunk duct involved will be taken as 132 ft and the friction loss, using data available from the previous example, is $152/200 \times 0.12 = 0.091$ in. The total head just upstream from take-off A is $0.0358 + 0.091 + 0.00023 + 0.00044 + 0.00056 = 0.12803$ in. It will be necessary to select by trial and error a duct size such that the take-off loss and the friction and shock losses in the branch will use up the difference between this total pressure and the velocity pressure corresponding to the desired velocity of exit (800 fpm). This velocity pressure from the preceding example is 0.0358 in. and the head to be used up is $0.12803 - 0.0358 = 0.09223$ in. of water. The loss at the take-off and that of the transition which will now be required at the end of branch A will be tentatively neglected and on this basis the friction loss in this branch must be $0.09223/0.75 = 0.123$ in. of water per 100 ft. The friction chart, Fig. 14.17, indicates that a duct approximately 15.5 in. in diameter will handle 1500 cfm of standard air with a friction loss of 0.12 in. of water per 100 ft. A duct 16 in. in diameter will be tried. The velocity will be $1500 \times 144/201.06 = 1075$ fpm and the actual friction loss per 100 ft will be $0.10 \times 0.06729/0.075 = 0.09$ in. of water. The take-off loss using equation 14.9 and Fig. 14.21 (velocity in branch/velocity in main = 0.88) together with available data from the previous example is $[0.5 \times 12 \times (1075/60)^2 \times 0.06729]/(64.4 \times 62.27) = 0.0323$ in. The total head lost excluding the tapered transition at the end is $0.0323 + 0.09 \times 0.75 = 0.0998$ in. This is approximately 0.01 in. more than the head available, and it does

AIR CONVEYING AND DISTRIBUTION

not include the loss in the tapered transition at the end which, if it had a 20-deg total included angle, would amount to $0.09 \times 0.45 \times 12 \times (1075/60)^2 \times 0.06729/(64.4 \times 62.27) = 0.00262$ in. If a 16-in. diameter duct were used the volume delivered would be less than 1500 cfm, but only slightly less as the friction and shock losses decrease rapidly as the velocity is reduced. The solution could be reworked with a slightly larger trial size but in general the foregoing selection would be close enough for practical purposes.

It is seldom necessary to size branch ducts to use up the head available instead of using dampers. Use of dampers in branches of a system such as Fig. 14.20 does not increase the fan head or the fan bhp.

14.23. Air-Duct Construction. All duct construction should be as smooth as possible where the air passes over its inner surfaces.

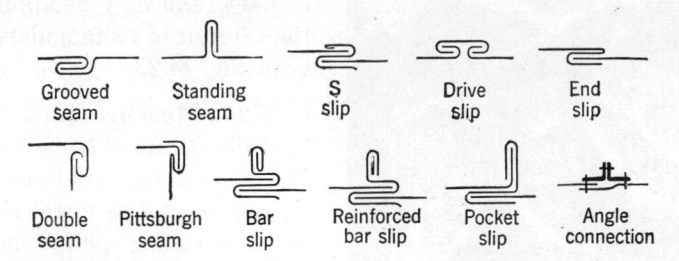

Fig. 14.22. Forms of seams and joints in sheet-metal duct construction. (From "Heating Ventilating Air-Conditioning Guide 1948." Used by permission.)

The ducts should be airtight and rigid to prevent vibration of the sheets. Recommended sheet-metal gages for rectangular duct construction are given in reference 2. Figure 14.22 shows several methods of joining two edges, in duct construction.

14.24. Heat Loss or Heat Gain Through Duct Walls, Duct Insulation. When no insulation is applied to sheet-metal ducts the principal resistances to heat transmission are the inside- and the outside-air films, as the resistance of the metal is negligible. An average value for the conductance of the outside film is 1.5 Btuh per sq ft per deg F temperature difference, and this value can be applied to practically all installations. The conductance of the inside film depends principally on the velocity of the air and the diameter of the duct. Calculations of sufficient commercial accuracy for the inside-surface conductance may be made by use of

$$f_i = 0.32 v^{0.8} \div D^{0.25} \qquad (14.12)$$

The over-all coefficient U may be computed as in Art. 4.3.

Any insulating material in suitable form for application to ducts may be used and for large ducts may be applied to either the inside or the outside surface. However, any insulating material applied to the inside surface of ducts should be noncombustible. Noncombustible insulating material applied to the inside of ducts is helpful in reducing noise, but great care must be used in its application to secure ,the smoothest possible surface, and the character of the final interior surface must be taken into consideration in estimating the friction loss of the system. Insulating material in blanket form may be applied to the exterior of rectangular ducts as in Fig. 14.23.

Fig. 14.23. Exterior insulation applied to sheet-metal ducts. (Courtesy Grant-Wilson Co., Inc.)

14.25. Measurement and Specifications of Noise Levels. Many contracts made to cover equipment and its installation in air-conditioning plants include specifications relative to noise produced by machinery and air flow in the system that may be transmitted to occupied spaces.

Noise intensities are usually expressed in terms of decibels which are units that indicate the relation between either two amounts of power or two sound-pressure levels.

$$\text{db} = 20 \log_{10} (P/0.0002) \tag{14.13}$$

where P is the sound pressure in dynes per square centimeter or

$$\text{db} = 10 \log_{10} I/10^{-16} \tag{14.14}$$

in which I is the sound intensity in watts per square centimeter.

A stated sound level in decibels is related to a threshold of sound of either 0.0002 dyne or 10^{-16} watt per sq cm. The relation of the decibel scale to sound pressure and sound intensity is given in Table 14.7. One db is about the smallest change in sound level that can be perceived by the human ear.

It is most satisfactory to measure noise with a sound-level meter called an acoustimeter which is now available for this purpose. An acoustimeter usually consists of a microphone, a high-gain audio-

AIR CONVEYING AND DISTRIBUTION

amplifier, and a rectifying milliammeter. The instrument is usually designed to indicate sound intensities directly in decibels.

Since a sound intensity is ten times the logarithm of a ratio of two powers and not a definite power, the intensity of a combination of two different sounds superimposed one on the other cannot be predicted by adding the two separate sound levels. In general, it has been found that combining two separate sounds of equal powers

TABLE 14.7. Decibel Scale Versus Sound Pressures and Sound Intensities*

Decibel Level	Pressure, dynes per sq cm	Intensity, watts per sq cm	Decibel Level	Pressure, dynes per sq cm	Intensity, watts per sq cm
0	0.000200	1.000×10^{-16}	40	0.0200	1.000×10^{-12}
1	0.000224	1.259×10^{-16}	50	0.0631	1.000×10^{-11}
2	0.000252	1.585×10^{-16}	60	0.200	1.000×10^{-10}
3	0.000282	2.000×10^{-16}	70	0.631	1.000×10^{-9}
4	0.000317	2.520×10^{-16}	80	2.00	1.000×10^{-8}
6	0.000399	4.000×10^{-16}	90	6.31	1.000×10^{-7}
8	0.000503	6.310×10^{-16}	100	20.0	1.000×10^{-6}
10	0.000631	1.000×10^{-15}	110	63.1	1.000×10^{-5}
20	0.00200	1.000×10^{-14}	120	200.0	1.000×10^{-4}
30	0.00631	1.000×10^{-13}			

* From "Heating Ventilating and Air-Conditioning Guide 1956." Used by permission.

results in a combined intensity that is 3 db higher than that of the individual noises regardless of the actual levels.

It is necessary in specifying maximum noise levels from machines to agree on the exact location of the microphone of the acoustimeter when the acceptance test is made. Noise levels in occupied spaces vary from a range of 10–20 db in radio broadcasting studios to 75–90 in vehicular tunnels. Typical noise levels for many other situations are given in reference 2.

In general, air-conditioning apparatus is regarded as acceptable from a noise standpoint if its operation does not increase the original noise level of the conditioned spaces by more than 3 db. This means that the equipment noise alone must not be greater than the room noise alone.

Noise levels above 120 db cause actual pain in the human ear and hearing may be affected by continued exposure to levels exceeding 90 db and in some cases less. People differ greatly in their susceptibility to hearing damage from exposure to high noise levels. The medical departments of some industrial organizations now give to new employees working in noisy departments frequent tests to determine their ability to continue the same work without damage to their sense of hearing.

14.26. Control of Noise Level. A thorough discussion of the various treatments which may be used to reduce noise is beyond the scope of this book. Transmission of mechanical noises from the fan through the duct system can be prevented by use of either a canvas or an asbestos cloth to connect the fan inlet and the fan outlet to the respective parts of the system. Transmission of vibration noises through the building itself can be prevented by mounting the apparatus on pads of resilient material such as rubber. In general, rotational or vortex noises from a fan will be at acceptable levels if a fan is selected such that it will operate at or near its maximum mechanical efficiency. Noises due to the flow of air through the system can be kept down to acceptable levels without the use of sound insulation by proportioning all ducts so that the velocity will not exceed 900 to 1200 fpm in main ducts or 600 to 1000 fpm in branches. Air velocity leaving registers or grilles should not exceed 200 to 400 fpm where quiet operation is required.

14.27. The Problem of Air Distribution. Regardless of how well the rest of the equipment may be designed, the performance of heating, ventilating, or air-conditioning systems will be unsatisfactory unless suitable provision is made for the uniform distribution of the heated air, the fresh air, or the conditioned air throughout the space that is being treated. The problem of securing proper distribution in large rooms without producing objectionable drafts in local areas is sometimes a difficult one. Air movements must be limited to 30 fpm in all spaces that may be occupied by people who are seated if ideal conditions are to be maintained. A somewhat higher velocity may not be noticed if the temperature and relative humidity of the air both fall within the upper limits of the comfort zone. The average person, as usually clothed for a public gathering, is especially sensitive to air movements over the neck or the ankles. In air-conditioned spaces where occupants will always be moving about, velocities up to 120 fpm may be tolerated without a feeling of discomfort.

Perfect distribution would be achieved by a system in which the air is introduced at room temperature through one entire room surface, uniformly perforated, and removed through the opposite room surface perforated in a similar manner. However, practical considera-

AIR CONVEYING AND DISTRIBUTION

tions require the use of air diffusers such as registers or grilles which occupy only a small portion of the room surface in which they are placed. Because of the limited area of the distribution outlets the outlet velocity required for the delivery of the required amount of air is usually several times the velocity which can be tolerated by occupants without discomfort. Fortunately, the movement of air withdrawn from a room through grille openings is not noticeable to occupants seated near them even when the free-area velocity is as high as 750 fpm. It is therefore possible to project the conditioned air into the conditioned space at velocities which are high enough to secure satisfactory distribution through outlets located either in the ceiling or in the upper part of the side walls and then remove it through grilles which are placed in the floor or in the side walls close to floor level. Definite schemes are described in detail in Arts. 15.18 through 15.21.

14.28. Behavior of a High-Velocity Jet of Air. A stream of air introduced into a room with considerable velocity of flow entrains room air adjacent to it. This action causes the static pressure of the air near the jet to be lessened, and, as a result, room air moves to produce a secondary circulation as indicated by the arrows of Fig. 14.24. This secondary circulation is sometimes advantageous when the temperature of the jet of air is above or below the room temperature, as it produces a mixing process that is desirable. The same "dragging effect" which causes the room air adjacent to the jet to be carried along with it also causes the outer edge of the jet to be pulled off to the side, thus causing a gradual "decay" of the projected stream. The *throw* of an air jet may be defined as the distance between the outlet and a position at which air motion has been reduced to a maximum velocity of 50 fpm. The throw depends upon the initial velocity and initial size and shape of the diffuser. In large rooms with high ceilings such as auditoriums the diffusers may be few and large; in low-ceilinged rooms such as railroad cars a perforated type of diffuser may be required in order that the projected streams may "decay" completely before reaching the heads of occupants only a few feet away.

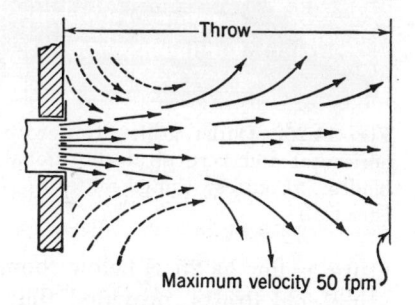

Fig. 14.24. Secondary air circulation produced by jet action.

14.29. Diffusion Devices. Several different arrangements have been developed for the purpose of providing in one device an archi tecturally acceptable mask for the end of the supply duct and a control of the stream of conditioned air which issues from it. A few of the more commonly used types will be discussed in the following paragraphs.

Figure 14.25 shows a grille with two sets of directional vanes. The vertical vanes may be turned by use of a special tool to adjust the spread of the jet in a horizontal direction and a second set back of the first is used to regulate the angle of the jet in a vertical plane.

A special type of grille shown in Fig. 14.26 is designed to create maximum secondary circulation and is very useful where it is necessary to introduce air at temperatures considerably below the comfort level. With this type of grille properly placed, incoming air may have a temperature as low as 25 F below room temperature without danger of creating cold drafts, provided that the grilles are placed at a sufficient distance from the occupants of the room.

Fig. 14.25. Outlet grille fitted with horizontal and vertical air-directional blades. (Courtesy United States Register Co.)

Figure 14.27 is an illustration of the use of a diffuser which consists of several specially formed concentric rings. The diffuser shown is designed for installation in ceilings and arranged in such a way that all of the air discharged is directed downward at some definite angle with the horizontal. This type of diffuser may be obtained in a variety of vane arrangements so as to provide "blows" in one, two, three, or four directions. Suitable arrangements may be placed in a side wall instead of the ceiling. In regard to any outlet grille or register, the ratio of the length to the width of its core is called its *aspect ratio*.

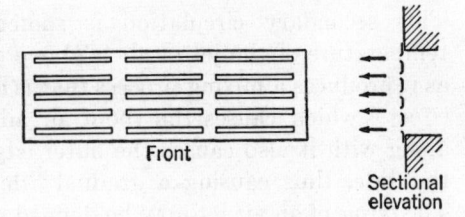

Fig. 14.26. Injection-type air-flow grille.

One type of diffuser consists of a perforated sheet-metal panel. It is well adapted to use in rooms where satisfactory results could not be achieved with any of the diffusers which have been previously discussed because of low ceilings. Continuous slots arranged to discharge the conditioned air in a horizontal direction near the ceiling have also been

AIR CONVEYING AND DISTRIBUTION

used with good results in low-ceilinged rooms which are not more than 40 ft wide. An advantage of the perforated type of diffuser is that it is not readily discernible when incorporated in acoustical ceilings.

14.30. Automatic Static Pressure Control in Duct Systems. In a system where several rooms are served by a single duct, occupants of certain rooms may alter damper settings in such a way that the static pressure in other portions of the duct may be changed, causing the delivery of air to those rooms to be increased or decreased to the point where discomfort to the occupants results. Unsatisfactory delivery of conditioned air to specific rooms of a system due to varying static pressure can be prevented by the installation of an automatic static pressure controller working on the principle illustrated in Fig. 14.28.

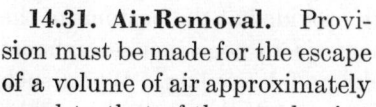

Fig. 14.27. Concentric-ring air diffuser. (Courtesy Air Devices, Inc.)

14.31. Air Removal. Provision must be made for the escape of a volume of air approximately equal to that of the supply air to avoid building up a noticeable pressure in a conditioned space which would interfere with the closing of doors. In some types of public buildings where exterior doors are in

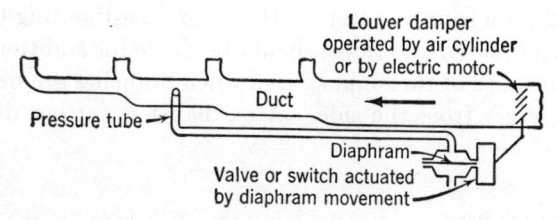

Fig. 14.28. Flow control by static pressure tube in duct system.

almost constant use it may not be necessary to install either grilles or ducts for exhausting air from the room. However, in most air-conditioning systems it is necessary to provide outlet grilles and ducts for conducting the exhausted air either to the outside of the building or to the air-conditioning apparatus for recirculation to the space served.

The location of exhaust grilles is far less critical than that of supply outlets because the air flows toward them from areas which are much larger than the openings, and there is no concentrated high-velocity stream at any appreciable distance away (see Fig. 14.29). Exhaust grids are usually located in a side wall with the bottom of the opening at or near floor level. However, ceiling locations are recommended for bars, kitchens, dining rooms, etc., where warm air will gravitate to the ceiling level. Some ceiling outlets of special design combine the supply and return openings in a single unit. Low side-wall locations should not be placed closer than 5 ft to an area normally occupied by people who are seated. Exhaust grilles set in the floor will perform satisfactorily but are objectionable from the standpoint of dirt collection. Allowable velocities through exhaust grilles are largely dictated by pressure-loss and noise-level considerations. Gross-area velocities between 400 and 800 fpm are usually satisfactory from every standpoint.

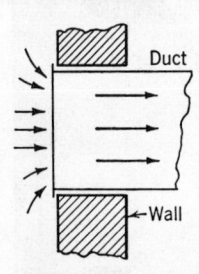

Fig. 14.29. Air entering an exhaust duct.

It is sometimes economical to use corridors as exhaust ducts. Air from the individual rooms passes into the corridors through grids set in the partitions or through an opening provided at the bottoms of the doors. Another frequently used scheme is to construct an exhaust duct at the top of a corridor by providing a false ceiling. Where such a duct is constructed without the use of metal, the plastering of the ceiling must be very tight to prevent dirt streaks caused by air entering the duct from the corridor through small cracks. The same procedure may be used for designing an exhaust system as was explained in Art. 14.22 for a supply system. However, in estimating the friction loss of such a system, allowance should be made for additional friction due to disturbance of the main stream when branches are brought into an exhaust trunk from the side. Air exhaust is further discussed in Art. 15.22.

REFERENCES

1. *Standards, Definitions, Terms, and Test Codes for Centrifugal, Axial and Propeller Fans, Bull. No. 110*, National Association of Fan Manufacturers, Detroit, 2nd ed. (1952).
2. "Heating Ventilating Air-Conditioning Guide," ASHAE, New York (1957).

3. "Energy Losses in 90-Degree Duct Elbows," D. W. Locklin, *ASHVE Trans.* **56,** 479 (1950).
4. *Modern Air Conditioning, Heating and Ventilating,* W. H. Carrier, R. E. Cherne, and W. A. Grant, Pitman Publishing Co., New York, 2nd ed., 248 (1950).
5. "Effect of Vanes in Reducing Loss in Seven-Inch Square Ventilating Duct," M. C. Stuart, C. F. Warner, and W. C. Roberts, *ASHVE Trans.*, **48,** 409 (1942).
6. "Pressure Losses Due to Bends and Area Changes in Mine Airways," G. E. McElroy, *Information Circular I.C. 6663,* U. S. Bureau of Mines (1932).
7. "Investigation of the Pressure Characteristics and Air Distribution in Box-Type Plenums for Air Conditioning Systems," S. F. Gilman, R. J. Martin and S. Konzo, *Bull. No. 393, Eng. Exp. Sta.*, University of Illinois (1951).
8. "Investigation of the Pressure Losses of Take-Offs for Extended-Plenum Type Air Conditioning Duct Systems," S. Konzo, S. F. Gilman, J. W. Holl, and R. J. Martin, *Bull. No. 415, Eng. Exp. Sta.,* University of Illinois (1953).
9. "Pressure Losses of Divided Flow Fittings," S. F. Gilman, *Heating, Piping, Air Conditioning,* **27,** No. 4, 141 (Apr. 1955).

BIBLIOGRAPHY

Fans

1. *Flow and Fans,* C. H. Berry, Industrial Press, New York (1954).
2. *Fan Engineering,* Buffalo Forge Co., Buffalo, 5th ed. (1948)
3. *Centrifugal Pumps and Blowers,* A. H. Church, John Wiley & Sons, New York (1944).
4. *Fans,* Theodore Baumeister, McGraw-Hill Book Co., New York (1935).
5. "How to Pick Centrifugal Fans," R. G. Lubinsky, *Industry and Power,* **67,** No. 1, 64 (July 1954).
6. "Power's Handbook on Fans," T. Hicks, *Power,* **95,** No. 10, 87 (Oct. 1951).
7. "Axial Flow Fans—Design for Production," *J. Inst. of Heating Ventilating Engrs.* (*London*), **18,** No. 180, 178 (July 1950).
8. "Vaneaxial Fan Fundamentals," R. Mancha, *ASHVE Trans.*, **56,** 277 (1950).
9. "High Temperature Fans," H. Chase, *Machine Design,* **22,** No. 2, 119 (Feb. 1950).
10. "Relation of Fan Characteristics to Common Ventilating Problems," H. M. Chapman, *Heating and Ventilating,* **48,** No. 6, 77 (June 1951).
11. "Contra-Rotating Axial Flow Fan," R. H. Young, *J. Inst. Heating Ventilating Engrs.* (*London*), **18,** No. 187, 448 (Mar. 1951).
12. "Design and Performance of a High Pressure Axial Flow Fan," L. S. Marks and Thomas Flint, *ASME Trans.,* **57,** 383 (1935).
13. "Fan Noise Level, What Does It Mean?" R. Getlitz, *Heating, Piping, Air Conditioning,* **23,** No. 6, 93 (June 1951).
14. "Volume Control of Mechanical Draft Fans by Adjustable Speed Fluid Drives," R. G. Olson, *Combustion,* **21,** No. 2, 43 (Aug. 1949).

System Design

1. "Pipes, Ducts and Fittings for Warm Air Heating and Air Conditioning," *Simplified Practice Recommendation R207-49,* U. S. Government Printing Office (1949).

2. "Branch Fitting Performance at High Velocity," C. M. Ashley, S. F. Gilman, and R. A. Church, *Heating, Piping, Air Conditioning*, **27**, No. 12, 117 (Dec. 1955).
3. "Fitting Losses for Extended Plenum Forced Circulation Forced Air Systems," H. H. Korst, N. A. Buckley, S. Konzo, and R. W. Roose, *ASHVE Trans.*, **56**, 259 (1950).
4. "Heating Ventilating Air-Conditioning Guide," ASHAE, New York (1956).

Air Velocity, Air Volume Measurement, and Air Distribution

1. "Temperature, Humidity, and Air Motion Effects in Ventilating," O. W. Armspach and Margaret Ingels, *ASHVE Trans.*, **28**, 103 (1922). (Kata thermometer is described.)
2. "The Heated Thermometer Anemometer," C. P. Yaglou, *J. Ind. Hyg. Toxic.*, **20**, No. 8 (Oct. 1938).
3. "A Simple Heated Thermocouple Anemometer," H. B. Nottage, *ASHVE Trans.*, **56**, 431 (1950).
4. "A V-Wire Direction Probe," H. B. Nottage, J. G. Slaby, and W. P. Gojsza, *ASHVE Trans.*, **58**, 79 (1952).
5. "Smoke Filament Technique for Experimental Research in Room Air Distribution," H. B. Nottage, J. G. Slaby, and W. P. Gojsza, *ASHVE Trans.*, **58**, 399 (1952).
6. "Air Flow Measurements from Intake and Discharge Openings," G. L. Tuve and D. K. Wright, Jr., *ASHVE Trans.*, **46**, 313 (1940).
7. "The Measurement of the Flow of Air Through Registers and Grilles," L. E. Davies, *ASHVE Trans.*, **36**, 201 (1930).
8. "Wide Area, Low Velocity Air Diffusion," E. J. Kurek, *Heating and Ventilating*, **51**, 85 (Oct. 1954).
9. "Room Air Distribution for Year 'Round Air Conditioning, Part I," S. F. Gilman, H. E. Straub, A. E. Hersey, and R. B. Engdahl, *ASHVE Trans.*, **59**, 151 (1953). Part II, H. E. Straub and S. F. Gilman, *ASHVE Trans.*, **60**, 249 (1954).
10. "Maximum Downward Travel of Heated Jets from Standard Long Radius ASME Nozzles," Linn Helander, Shee-Mang Yen, and R. E. Crank, *ASHVE Trans.*, **59**, 241 (1953).
11. "Air Velocities in Ventilating Jets," G. L. Tuve, *ASHVE Trans.*, **59**, 261 (1953).
12. "Isothermal Ventilation Jet Fundamentals," H. B. Nottage, J. G. Slaby, and W. P. Gojsza, *ASHVE Trans.*, **58**, 107 (1952).
13. "The Control of Air Streams from a Long Slot," Alfred Koestel and Chia-Yung Young, *ASHVE Trans.*, **57**, 407 (1951).
14. "Comparative Study of Ventilating Jets from Various Types of Outlets," Alfred Koestel, Phillip Herman, and G. L. Tuve, *ASHVE Trans.*, **56**, 459 (1950).
15. "Air Streams from Perforated Panels," Alfred Koestel, Phillip Herman, and G. L. Tuve, *ASHVE Trans.*, **55**, 283 (1949).
16. "Computation Charts and Theory for Rectangular and Circular Jets," H. G. Elrod, Jr., *Heating, Piping, Air Conditioning*, **26**, No. 3, 149 (Mar. 1954).
17. "Performance of Side Outlets on Horizontal Ducts," D. W. Nelson and G. E. Smedberg, *ASHVE Trans.*, **49**, 58 (1943).

Noise Measurement and Sound Control

1. *Handbook of Noise Measurement,* A. P. G. Peterson and L. L. Beranek, General Radio Co., Cambridge, Mass. (1953).
2. *Acoustics in Modern Building Practice,* Fritz Ingerslev, Architectural Press, London (1952).
3. *Acoustic Measurements,* L. L. Beranek, John Wiley & Sons, New York (1949).
4. "Introduction to Noise," J. O. Kraehenbuehl, *Am. Foundryman,* **25,** No. 6, 52 (June 1954).
5. "Noise in Ventilating Systems," H. J. Purkis, *J. Inst. Heating Ventilating Engrs. (London),* **20,** No. 206, 255 (Oct. 1952).
6. "Measuring and Evaluating Noise," A. F. Fiske, Jr., *Gen. Elec. Rev.,* **53,** No. 11, 29 (Nov. 1950).
7. "Reducing Product Noise with Acoustic Materials," *Product Eng.,* **23,** No. 9, 162 (Sept. 1952).
8. "Combating the Effect of Noise," W. F. Scholtz, *The Foundry,* **82,** No. 4, 112 (Apr. 1954).
9. "Some Fundamental Principles of Noise Control," A. C. Pietrasanta, *Refrig. Eng.,* **62,** No. 4, 37 (Apr. 1954).
10. "Experimental Approaches to the Study of Noise and Noise Transmission in Piping Systems," W. L. Rogers, *ASHVE Trans.,* **59,** 347 (1953).
11. "Acoustic Power Level Determination for Machinery in Hard Rooms," W. J. Burch, *Heating, Piping, Air Conditioning,* **29,** No. 6, 161 (June 1957).
12. "Criteria for Room Noise from Air Conditioning," C. M. Ashley, *Heating, Piping, Air Conditioning,* **29,** No. 7, 145 (July 1957).
13. "Noise Control Problems in Air Conditioning Equipment," R. J. Wells, *Heating, Piping, Air Conditioning,* **29,** No. 8, 138 (Aug. 1957).
14. "Sound Standards for Testing and Rating," H. C. Hardy, *Heating, Piping, Air Conditioning,* **29,** No. 9, 148 (Sept. 1957).
15. "Application of Sound Standards to Equipment," R. E. Parker, *Heating, Piping, Air Conditioning,* **29,** No. 9, 154 (Sept. 1957).
16. "Estimating Octave Band Levels of Noise Generated by Air Conditioning Systems," F. B. Holgate and Sidney Baken, *Heating, Piping, Air Conditioning,* **29,** No. 9, 160 (Sept. 1957).

PROBLEMS

1. Velocity-pressure measurements of 0.04, 0.25, 0.81, 1.0, and 2.25 in. WG were made by use of a Pitot tube placed in a duct in which dry air flowed at a temperature of 100 F and an absolute pressure of 29.0 in. of mercury at 32 F. (*a*) Find the rate of air flow, in feet per second, represented by each velocity pressure, when measured with the gage fluid at 120 F. (*b*) Calculate the velocities of flow, in feet per second, when the gage-fluid temperature is 40 F.

2. A fan has an inlet duct with a diameter of 25 in. and an outlet 24 in. by 19 in. The fan discharges 9500 cfm of dry air measured at 75 F and an absolute pressure of 29.47 in. of mercury at 32 F. The operating static pressures are −0.75 in. near the inlet and 1.5 in. WG at the outlet. Find the total-pressure difference which the fan creates. Assume the gage-fluid temperature is 60 F.

3. A fan operates with an inlet-duct entrance loss of 0.05 in. WG, a friction loss in the inlet duct of 0.03 in. WG, an average air velocity in the fan inlet of 25 fps, a fan-outlet air velocity of 50 fps, and a discharge static pressure of 0.75 in. WG at the fan outlet. The air handled is dry and is at 80 F and under a barometric pressure of 29.5 in. of mercury at 32 F. The density of the gage fluid is 62.3 lb per cu ft. Find the total pressure difference, in inches of water, created by the fan.

4. A fan with a free inlet discharges 15,300 cfm of standard air with a velocity pressure of 0.23 in. WG and a static pressure of 1.15 in. WG when the gage fluid weighs 62.3 lb per cu ft. The power input to the fan is 4.3 bhp. Find the static and the mechanical efficiencies.

5. A fan delivers 2297 cfm of standard air when operated at a speed of 1300 rpm and a power input of 1.84 bhp when its static pressure is 3 in. WG. Determine by calculation its performance when the operating speed is changed to 958 rpm with no change of duct system, air density, or water density.

6. The fan of Problem 5 is operated to deliver to a fixed and constant system the same weight of air as under the first condition when the air density is 0.062 lb per cu ft. Determine its conditions of operation.

7. A fan handling air having a density of 0.05 lb per cu ft operates at 400 rpm and delivers 8000 cfm against a static pressure of 1.5 in. WG. Find the speed that would be required to maintain the same static pressure in the same system if the air density were standard. What volume of standard air would be delivered?

8. A fan handles 1500 cfm of dry air at 65 F against a static pressure of 0.20 in. WG and requires 0.10 bhp. Find the static pressure, and the bhp required to deliver the same volume when the air temperature is increased to 165 F. Assume standard atmospheric pressure and that the density of the gage fluid remains constant.

9. From Table 14.1 select a fan to handle 11,000 cfm of dry air at a temperature of 120 F when the barometric pressure is 28.85 in. of mercury at 32 F and the discharge pressure is 1.00 in. of water. State the required operating speed and hp.

10. A fan must handle 50,000 cfm of air at a temperature of 150 F, when the barometric pressure is 29.56 in. of mercury at 32 F and discharge it against a total static pressure of 1.5 in. WG. Determine the size of the fan, its required speed of operation, and the necessary input of power to drive it.

11. A rectangular duct has dimensions of 12 in. by 30 in. and is to handle 2000 cfm of standard air. (*a*) Find the circular equivalent of the duct in inches. (*b*) Find the friction-pressure losses in a length of 150 ft of straight duct by use of Fig. 14.17. (*c*) If the air volume remains unchanged but its density becomes 0.085 lb per cu ft, what friction-pressure loss would exist within the duct?

12. A fan discharges dry air at a velocity of 1800 fpm into a transition piece which has a total included angle of 40 deg. The air velocity in the plenum is to be 900 fpm when the air exists at a temperature of 85 F and a barometric pressure of 29.92 in. of mercury at 32 F. The gage-fluid temperature is 70 F. Find the head loss due to shock in inches WG.

13. A fan discharges air with a density of 0.065 lb per cu ft with an outlet velocity of 1600 fpm. The transition piece between the fan outlet and a plenum chamber contracts at an angle of 60 deg to give the air a velocity of

AIR CONVEYING AND DISTRIBUTION

2000 fpm as it enters the distributing section. The gage-fluid temperature is 70 F. Find the shock loss due to contraction of the section in inches WG.

14. How much would the shock loss be decreased if in Problem 12 the total included angle were reduced from 40 to 20 deg?

15. Find the shock loss, in inches WG that would be expected in a 90-deg rectangular elbow in a sheet-metal duct 12 in. high and 24 in. wide handling 3000 cfm of dry air at a temperature of 240 F. The turn will be in the horizontal plane and the center-line radius is 2.0 ft. Assume standard barometric pressure and a gage fluid temperature of 70 F. What is the additional equivalent length of the elbow?

16. A three-piece, 90-deg elbow (similar to that in Fig. 14.20) exists in a 16-in. diameter round duct. The center-line radius of the elbow is 24 in. Find the additional equivalent length of the elbow due to shock loss, in feet.

17. Air which weighs 0.07 lb per cu ft flows in a trunk duct with a velocity of 1500 fpm and at 900 fpm in a branch which is taken off at an angle of 45 deg. Find the loss of total head in the take-off expressed in inches of water at 70 F.

18. A duct has a diameter of 36 in. and a length of 90 ft in addition to three five-piece 90-deg elbows which have center-line radii of 1.25 times the duct diameter. Air is handled at 1200 fpm with a temperature of 140 F and a pressure of 14.696 psia. Find the actual friction-pressure loss in inches of water. Length of straight pipe given has been measured to intersection of center lines as in Fig. 14.20. Assume dry air. The effect of the static pressure on the density of the air may be neglected.

19. Assume that branch B of Fig. 14.20 has an angle of 45 deg with the trunk and an equivalent length of 40 ft not including that of the take-off. Find the proper size of the branch duct if it is to deliver 1260 cfm without damper controls. All conditions are as in the first example of Art. 14.22.

20. Calculate the inside-surface conductance for a duct which has a diameter of 24 in. and in which air flows with a velocity of 50 fps.

21. A sound intensity in a certain room is 50 db. Find the sound pressure in dynes per square centimeter and the sound intensity in watts per square centimeter.

22. If an air conditioner which creates a sound intensity of 50 db is installed in the room of Problem 21, what sound intensity in decibels can be expected in the space when the unit is operating?

15

Ventilation and Air Purification

15.1. Necessity for Ventilation. Air changes in spaces may be required for any or all the following reasons: (1) heat and moisture given off by the occupants, (2) vitiation of the air by respiratory processes and emanations from the skin, (3) chemical product fumes and vapors, (4) oxygen depletion as the result of the combustion of fuels and other materials, (5) excessive heat from all sources, (6) moisture liberated in manufacturing and other operations, (7) toxic gases, (8) odors, (9) bacteria, (10) smoke and fog productions, and (11) dirt, dust, lint, etc., liberated within a space to float in its air.

15.2. Air Vitiation. The respiratory processes of men reduce the oxygen and increase the carbon dioxide and moisture contents of air surrounding them and pollute it with organic materials from the nose, mouth, throat, and lungs. Skin emanations are another source of air pollution. Air odors come from many sources and they may or may not be indicative of harmful materials in the air. From a psychological standpoint they are objectionable and often lead the room occupant to feel that the ventilation of a space is poorer than it actually is.

Fresh air from the open country may have from 3 to 4 parts of carbon dioxide, CO_2, per 10,000 parts of air. Air as exhaled by an adult usually has about 400 parts. Concentrations of from 7 to 8 parts of CO_2 per 10,000 parts of air are not excessive. For the most part the usual concentrations of CO_2 as found in occupied spaces are not considered to be of great importance.

Carbon monoxide, even in very small quantities, is a deadly gas as far as human life is concerned. Moreover, it is not noticeable to the individual inhaling it. Carbon monoxide is absorbed by portions of the blood so that proper assimilation of oxygen cannot be made. In the ventilation of garages and in any other spaces where quantities of CO are given off, the amounts of ventilation air must be such that the concentration will not exceed 1 part in 10,000 parts of air in the space (0.01 per cent) where the exposure is continuous. Fuel-burning automotive vehicles are the sources of CO most commonly encountered by the public. A code[1] covering the ventilation of garages has been formulated. Many other toxic gases and vapors are possible air contaminants, some of which are explosive in their nature. Air impurities other than gases and vapors are often called *aerosols*.

15.3. Quantities of Ventilating Air Required. When the source of air contamination within a space is its human occupant,

TABLE 15.1. Outdoor Air Required, per Person, for Summer Air-Conditioning Operations*

Type of Service	Amount per Person, cfm
Auditoriums, churches, and theaters (no smoking)	5 to 7.5
Barber shops, beauty parlors, funeral homes, open spaces in banks, and retail shops	7.5 to 10
Apartments, drugstores with lunch counters, hospital rooms, hotel rooms, general offices, and restaurants	10 to 15
Brokers' board rooms, directors' rooms, night clubs, private offices, and taverns (heavy smoking)	20 to 30

* Data from The Trane Company *Air-Conditioning Manual*.[4]

factors which seem to affect the minimum quantity of ventilating air to be brought in from outdoors are: objectionable body odors, the room-air temperature and relative humidity, tobacco smoke, the volume of the room per occupant, and the odor-absorbing capacity of the air-conditioning processes. Data obtained from studies made by the Harvard School of Public Health[2] pertaining primarily to ventilation requirements during the heating season are given in the ASHAE Guide. Residential spaces may have sufficient fresh air admitted to them by leakage through window and door cracks. For ventilation requirements of commercial and industrial buildings see reference 3.

Because of the loads which may be imposed on air-conditioning plants in hot weather by the use of outside air, there is a tendency to reduce the amounts to a minimum when the air is to be cooled and dehumidified (see Table 15.1).

Ventilation specifications are usually given in terms of cubic feet per minute of standard air (density 0.075 lb per cu ft) regardless of the actual design conditions in regard to temperatures and pressures. When interpreted in this manner the weight of fresh air involved is the same regardless of the design conditions which are assumed. *In general, regardless of other considerations, the ventilation rate should cause at least one air change per hour.*

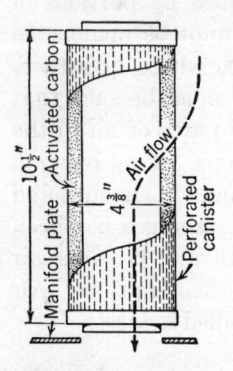

Fig. 15.1. Canister with activated carbon for odor adsorption. (Courtesy W. B. Connor Engineering Corp.)

15.4. Odor Suppression. Individuals when exposed to odors for some period of time become less conscious of them because of the dulling of their olfactory sense. In appraising odor conditions within a space attention should be given to the impressions of individuals entering the rooms as well as to those of the room occupants.

Some odor-bearing materials and vapors carried by air are successfully adsorbed by the use of activated carbon (charcoal made from coconut shells) placed in perforated canisters, Fig. 15.1, mounted on manifold plates. These assemblages, Fig. 15.2, are installed in duct sections through which the air to be deodorized flows. Whenever the activated carbon becomes ineffective as an adsorbent, it may be regenerated by heating it to a temperature of approximately 1000 F. Each canister has an air capacity ranging from 25 to 35 cfm with air resistances of from 0.15 to 0.20 in. of water pressure. Special cells with face dimensions approximately 24 in. by 24 in. and air-handling capacities up to 1000 cfm are available. Activated-carbon requirements vary from 25 lb per 1000 cfm of air in residential applications to 35 lb per 1000 cfm

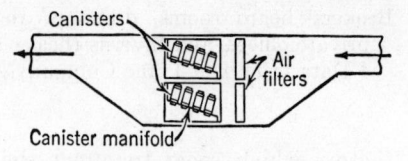

Fig. 15.2. Installation of odor-adsorbing equipment in an air duct. (Courtesy W. B. Connor Engineering Corp.)

VENTILATION AND AIR PURIFICATION

in commercial installations. Filtration of the air entering the odor-absorbing units is advisable as their effectiveness is reduced by accumulations of dust on the activated carbon.

Ozone, O_3, has been employed to mask odors as it has a pungent smell resembling that of weak chlorine. Because of its likelihood of causing injury to humans its concentrations are limited to 0.01 to 0.05 ppm (parts per million) of air. With these low concentrations ozone is not effective either as a masking or a sterilizing agent.

15.5. Air Sterilization. Bacteria from human and other sources exist in occupied spaces. These germs may be harmless or otherwise. Dust, lint, and other materials floating in air afford a method of conveyance for bacteria, and the removal of such particles in air washers, filters, electrostatic cleaners, and other devices provides one method of at least partially freeing air from harmful germs.

Special ultraviolet-ray lamps which give radiations in the neighborhood of 2600 *angstrom units* (an angstrom unit equals 1/250 millionth of an inch) have been successful in sterilizing air.[5] Possible locations for ultraviolet-ray lamps are in the upper portions of rooms and in the air-supply ducts just outside of the spaces. In any event lamps used for air sterilization must be shielded so that the eyes and the skin of the occupants of the space will not be exposed to them.

Progress has been made[6-8] in the use of small quantities of finely divided droplets or mists of either propylene or triethylene glycol sprayed into air to kill bacteria. Effective air sterilization depends upon either the radiations of lamps or the mists of germicides being at all times capable of reaching the organism to be killed.

Investigations[9] have shown that the life span of certain air-borne organisms, sprayed into the atmosphere from a liquid suspension, is dependent upon the relative humidity; the mortality rate was very high when the relative humidity was maintained near 50 per cent. The mortality rate decreased when the relative humidity was either higher or lower than 50 per cent.

15.6. Classification of Aerosols. Drinker and Hatch[10] divide aerosols into four classifications: dust, fumes, smoke, and mists or fogs.

Dust is formed by reducing earthy materials to small size without the involvement of chemical reactions. Common examples are the dusts of coal, cement, and wheat. *Lint* is a special string-like form of dust resulting from the handling of materials such as cotton and wool.

Fumes are solid particles generated by condensation and solidification of a material from its gaseous state; their production is often accompanied by a chemical reaction such as oxidation. A common example is the zinc-oxide fume produced from a zinc vapor.

Smoke is a term generally applied to a visible collection of aerosols; its formation is usually in connection with the combustion of organic materials such as tobacco, wood, oil, or coal.

Mists or fogs are the result of the condensation of water vapor on small-sized solid particles of materials already suspended in the air; they can purposely be produced by the atomization of liquid water. The foregoing aerosols are usually man-made; the following ones result from processes of nature.

Pollens from weeds, trees, flowers, and other forms of vegetation may become suspended in the outdoor air and be brought into the interiors of structures by air infiltration through the cracks about window and door openings and through the fresh-air intakes of air-conditioning systems. Some pollens are harmless; others may be very injurious to individuals susceptible to their effects.

Bacteria are minute living organisms which range in their effect on human beings from harmless to very dangerous.

Dirt is the conglomeration of particles which characteristically collect on filters, porches, walks, floors, and furniture.

The sizes of aerosols, as discussed in technical literature, are invariably measured in terms of microns. A *micron* is one millionth of a meter which is approximately 1/25,000th of an inch. The size ranges in each of the classifications are generally as follows: dust, 0.25 to 20; fumes, 0.01 to 0.75; smoke, less than 0.5; mists and fogs, greater than 5.0; pollens, causing hay fever, 10 to 50; and bacteria, which vary greatly in size and shape. Some bacteria will pass through a circular opening as small as 0.02 micron in diameter.

15.7. Cleaning Efficiency. Much work has been done by various investigators in the matter of testing and rating cleaners. Instructions are given in an ASHVE code.[11]

The nature and the size of the dust particles have much to do with their removal from air by cleaners. Coarse particles are much more easily removed from air than those of fine size. Results of cleaner tests can best be compared when made with standard dust.

Cleaning efficiency or dust arrestment when determined by the weight method is expressed as

$$E = \frac{(m_1 - m_2)100}{m_1} \qquad (15.1)$$

where m_1 and m_2 are weights of dust per unit volume of air.

VENTILATION AND AIR PURIFICATION

Cleaner efficiencies may be evaluated by means of (1) particle counts, (2) weights of dust introduced into air and the extraction of dust from cleaned air by passing it through a porous crucible, and (3) the dust-spot or blackness test as devised by the National Bureau of Standards.

15.8. Air Cleaner Classifications. The nature of the impurities to be removed and the thoroughness of air cleaning necessary fix to a considerable degree the particular type of equipment to be used for given conditions. In general, cleaners may be separated into the following groups according to the principle of their operation: (1) filters with viscous-coated surfaces upon which dirt particles are impinged, (2) dry filters which consist of cellulose or cloth materials which screen out dirt particles, (3) devices which operate with a change of direction of the air flow so that dirt particles are thrown out of the air stream by the action of either inertial or centrifugal forces, (4) electrical precipitation in which dirt particles are ionized and given a positive charge of electricity which allows them to be deposited upon surfaces or plates having a negative charge, and (5) air washers which by means of water sprays and the impingement of dirt materials upon wetted surfaces accomplish air cleaning.

Air cleaners may also be classified as *atmospheric*, which ordinarily are used to remove objectionable-material particles from outside air, and *industrial air and gas cleaners*, which in general deal with aerosols or gases dispersed into the atmosphere as the result of either manufacturing processes or the handling of materials of a dusty nature. Atmospheric air cleaners are used in heating and air-conditioning systems and seldom handle dust loadings greater than 4 grains per 1000 cu ft of air. Industrial air and gas cleaners must deal with wide ranges in particle size and with dust loadings varying up to 20,000 grains per 1000 cu ft of air.

Dirt and other materials collected by air cleaners must be removed from them, and the schemes employed are classed as (1) automatic and (2) nonautomatic. Automatic rejuvenation of air cleaners will be discussed in connection with particular types of equipment illustrated. Nonautomatic cleaners, when dirty, may be rehabilitated by (1) discarding (throwaway) dirty elements and replacing them with clean ones; (2) manually cleaning the dirty parts without removing them from place; and (3) removing those parts in need of attention, renovating them, and replacing them in the apparatus.

Applications of air cleaners are in the following categories: (1) general uses which include central systems of air-conditioning and ventilation, unit ventilators, window installations, and warm-air furnaces; (2) removal of objectionable particles from smoke and stack gases; (3)

the collection of dust in exhaust systems; and (4) the collection of dust from a local process in which an excessive amount is produced. Nonautomatic cleaners, according to the maximum resistance they offer to air flow in inches of water, are: low, 0.18; medium, 0.50; and high, 1.0 or more. The pressure losses through any particular air filter depend upon the velocity of air flow and the state of cleanliness of the unit.

Principles and modifications and some performance data are covered in ensuing articles.

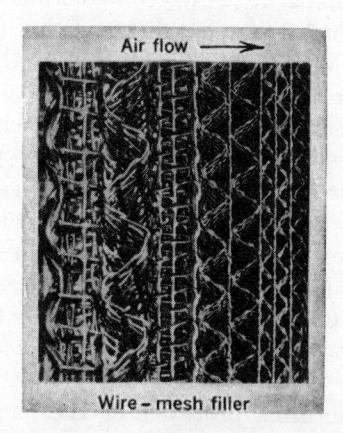

Fig. 15.3. Viscous-impingement air-filter panel, holding frame, and wire-screen mesh. (Courtesy Air-Maze Corp.)

15.9. Viscous-Impingement Air Filters. Units of this type are built in several forms, one of which is the panel or cellular form illustrated in Fig. 15.3. The frame of the cell is sometimes of steel and other times of cardboard. The cardboard construction is for those units which are discarded after use has rendered them unfit for further service. The cells are filled with such materials as steel shavings, vegetable fibers, spun-glass fibers, animal hair or bristles, and in the unit shown in Fig. 15.3 the filtering surfaces are formed by the use of crimped wire-mesh screens diagonally placed within the panel. Regardless of the material used as a filling, the surfaces upon which dirt impingement takes place are coated with a suitable oil.

When the filter agents are made of different sizes of fibrous materials, the coarser elements are placed in the front of the cell where the larger

VENTILATION AND AIR PURIFICATION

particles of the dust are removed. Filters of this type continue to collect dust as long as the dirty surfaces have their dust particles oil-coated. The need to clean the surfaces and recoat them with oil or to dispose of the units occurs when either they become ineffective or the resistance to air flow becomes excessive.

Automatic viscous filters have surfaces which are either continuously or periodically moved. Figure 15.4 is the side sectional elevation of an automatic viscous unit with two endless movable filter curtains, which are mounted on roller chains driven by sprocket wheels keyed to the shafts of the curtain rollers. The curtains have a number of removable panels each constructed of a single layer of bronze screen wire to which are attached layers of woven copper mesh. Lint is caught on the bronze screen wire which is always on the air-entering side of the curtain. The inner mesh affords considerable surface for the collection of dust particles. The first curtain is the denser of the two and dips in an oil bath through which it passes for cleaning. The curtain at the exit side of the unit does not enter the oil bath but serves to prevent entrained oil from passing with the air from the cleaner. A solenoid valve operates compressed-air jets working at 1½ psig pressure to blow through the curtain panels just below where they enter the main air stream to remove the excess oil from the clean surfaces.

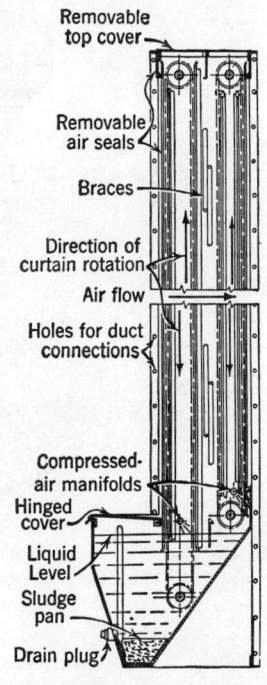

Fig. 15.4. Side sectional elevation of automatic viscous air filter. (Courtesy Dollinger Corp.)

An automatic filter, which employs a discardable viscous-coated medium in the form of a roll, is shown in Fig. 15.5. This filter offers the advantages of low initial cost, in the automatic field, and requires no maintenance between roll changes. With dust loadings which are considered normal for atmospheric type of service, a roll may be expected to give continuous service for as long as one year.

Viscous-impingement air filters are used principally in atmospheric cleaning service. A resistance of 0.5 in. of water at a face velocity of 500 fpm is typical with both automatic and nonautomatic filters of this type.

15.10. Dry Filters. The principal results of dry units are obtained oy the use of filter materials which serve as strainers. The cleaner media may be wool felts, specially woven cotton fabrics, glass cloths, cloths made from synthetic fibers, or cellulose. Neither adhesive nor viscous materials are used to coat the cleaning surfaces. The filtering

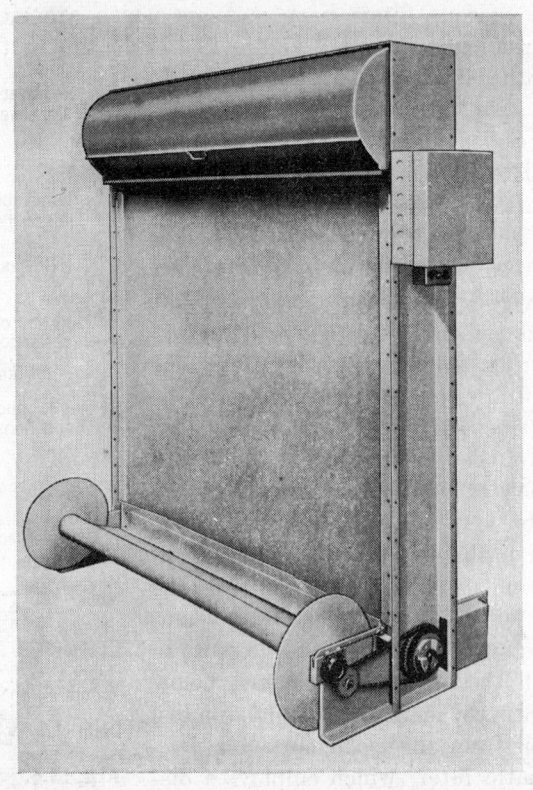

Fig. 15.5. Automatic air filter with viscuous-coated discardable roll. (Courtesy American Air Filter Co.)

materials are often arranged in pocket or bag forms to give the necessary surfaces without excessive space requirements. The principle of a small bag filter is shown in Fig. 15.6. Cleaning of the filtering surfaces of such units may involve operations of shaking, rapping, movement of a traveling ring discharging jets of compressed air, and momentary reversal of air flow through the equipment. Some dry-type filters employ a layer of cellulose spread upon expanded-metal holders arranged in pocket form. With this arrangement the filtering material is replaced when the collection of dirt causes excessive resistance.

Properly engineered and maintained dry filters are capable of collecting 99 per cent or more of dusts as small as 0.5 micron. However, it is necessary to design such units for very low velocities of air flow ranging from 0.5 to 10 fpm. When a lower efficiency may be tolerated or where an extremely high resistance can be overcome, velocities as high as 35 fpm may be used. The usual range of velocity[12] is from

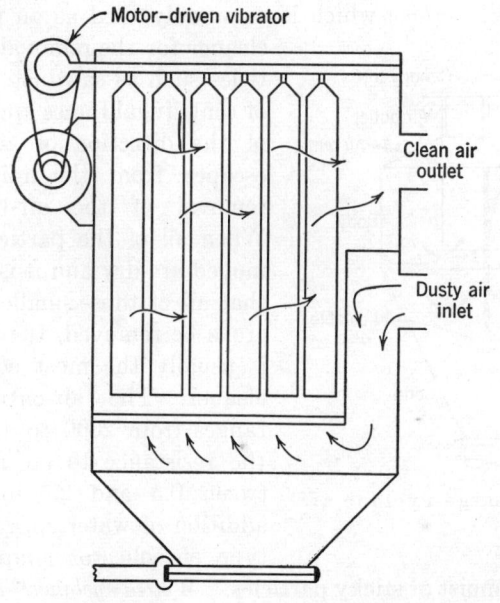

Fig. 15.6. Bag type of dry filter.

2 to 5 fpm and the resistance commonly varies from 0.2 to 0.5 in. of water when the unit is clean and from 2.0 to 4.0 in. when it is loaded with collected materials. The efficiency of dry filters improves with dust loading and where practically 100 per cent cleaning efficiency is essential, as is the case with radioactive dusts, the method used for removing the collected material must *not* be too thorough. Dry filters are not suitable for the collection of mists or any type of aerosol which is of a sticky nature.

Electrostatic effects[13] which increase filtering efficiency may be produced in dry wool felts by the addition of resin powders. Experiments[14] indicate that certain plastics, when formed into thin films and then shredded to produce a porous mass, can develop an electrostatic charge of sufficient magnitude to attract and retain aerosols. Electrostatic filters of the latter type can be cleaned, without impairing their efficiency, by rinsing them with cold water.

15.11. Change of Direction Air-Flow Cleaning Devices. Some separation of suspended particles larger than 100 microns may be accomplished in a gravity or settling chamber which is simply an enlargement of the pipe in which the air flows. The separating action may be accelerated by the incorporation of a baffle which introduces inertial forces in addition to the action of gravity. Figure 15.7 shows a *large-diameter cyclone* which is commonly used as an industrial air cleaner for the removal of large particles such as sawdust. The action of centrifugal forces and the reversal of the direction of air flow as it escapes from the unit effects the removal of the air-borne solids. When all of the particles to be removed are dry and it is not essential that all of those smaller than 10 microns be removed, the large cyclone is usually the most economical air cleaner. The air-entry velocity[12] ranges from 2000 to 4000 fpm and the resistance to air flow falls between 0.5 and 2.5 in. WG. The addition of water sprays makes this type of collector adaptable to the separation of moist or sticky particles. *Wetted cyclones*[12] are in general more efficient in the removal of all types of aerosols; however, the operating costs are higher than those of the dry type.

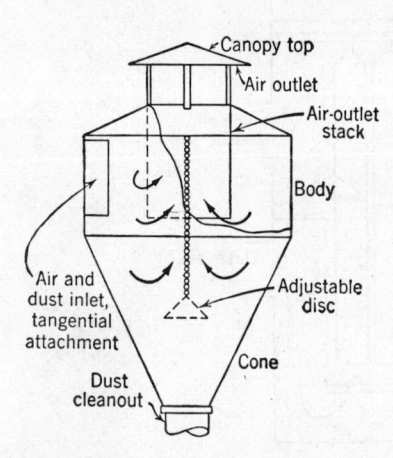

Fig. 15.7. Large-cyclone air cleaner.

Analyses[10] of the operation of cyclone separators indicate that the *separation factor* varies with the square of the gas velocity and inversely with the radius of the path of the rotating air stream. Small-diameter cyclones, several of which are usually incorporated in a single unit as in Fig. 15.8, provide more efficient air cleaning than their large-diameter counterparts. In a comparison of large- and small-diameter cyclones it appears that the entry velocities are about the same and that the resistance of the small units is in the range[12] from 2.0 to 4.5 in. WG, which is greater than that for the large-diameter type.

A piece of equipment which makes use of centrifugal separation and dynamic precipitation of dust and dirt is that shown in Fig. 15.9. This particular unit consists of a scroll-shaped casing within which a specially designed turbine-type impeller wheel, with many blades, operates at the speed of a conventional electric motor. Under centrifugal action dirt-laden air entering the wheel at its center flows outward

VENTILATION AND AIR PURIFICATION

along the wheel blades. The actions of centrifugal forces move both the light and heavy particles to the outer edge of the impeller. About the periphery of the wheel is a narrow annular space through which the separated dirt passes into a surrounding dust chamber. The blade tips extend into the dust chamber and produce a secondary-air circulation which in turn conveys the dirt to a storage hopper. The cleaned air leaving the blades and passing out into the scroll is discharged either to the outdoors or is returned to the occupied space after having been passed through a secondary-air cleaner. Entry velocities may range from 2500 to 4000 fpm.

This process of air cleaning is used in industrial plants. Equipment of a similar type is available where water sprays are used to wet dust and dirt particles and thereby increase their masses; this aids in separating them from the air.

15.12. Air Cleaning by Electrical Precipitation.

Electrostatic cleaners are effective in removing small particles that some other devices cannot separate from air.

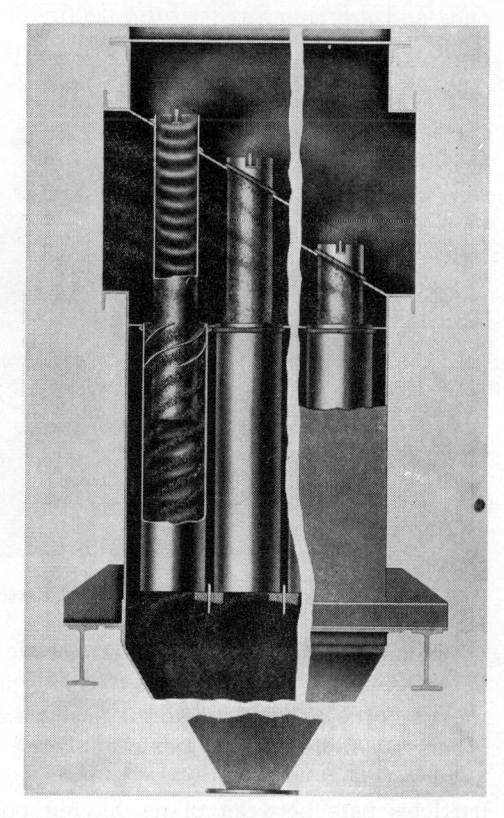

Fig. 15.8. Air-cleaning unit with small-diameter cyclones. (Courtesy American Blower Corp.)

Fundamentally, the principles of the electrical precipitator are as shown in Fig. 15.10, although the actual construction details of commercial units vary with different designs. Generally the source of electrical energy for their operation is 110 to 115 volts alternating current of either 25 or 60 cycles frequency and single phase. By means of devices in a power pack a transformation is made to direct current which is available at potentials of 12,000 and 6000 volts for atmospheric service in air conditioners.

The ionizing wire I, Fig. 15.10, is energized at 12,000 volts and thereby sets up an electrostatic field between it and the ground tubes G and G'. These tubes have grounded plates attached to them. The polarity of both the tubes and the attached plates is minus or negative. Positively charged plates are spaced between the negatively charged ones and are connected to the 6000-volt d-c terminals of the power pack. Aerosols of all types flowing with the air through the field

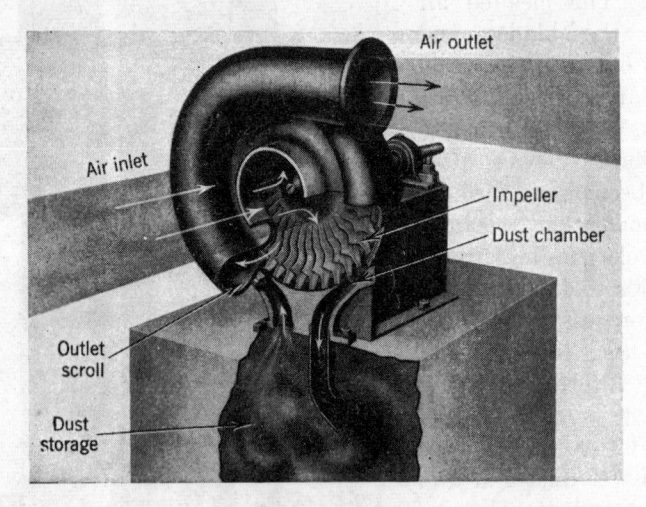

Fig. 15.9. Roto-Clone air cleaner. (Courtesy American Air Filter Co.)

between I and G and G pick up electric charges as the air becomes ionized. Some manufacturers of this type of equipment state that all the air-borne particles acquire a positive charge of electricity, whereas others say more than 80 per cent of the bodies get a positive charge and less than 20 per cent a negative charge. In any event, the charged particles pass between plates having positive and negative electric charges, and they are deposited upon a surface having a charge opposite in sign to that which they carry. The path of a positively charged particle is shown in Fig. 15.10.

Electrostatic filters may be installed as one or more removable cells or as automatic units. A special throwaway paper which is electrostatically charged while in service is also available together with the equipment needed to employ it.

High-voltage electrostatic precipitators[15] (75,000 volts) have much higher dust-holding capacities than those units previously discussed. They are used in industrial applications where the dust loadings are higher. Precipitators of this type are made most commonly in elec-

VENTILATION AND AIR PURIFICATION

trode and flat-plate units which are cleaned by vibration. Air velocities usually range from 200 to 500 fpm. Inasmuch as the required voltage is close to that at which a spark-over may occur, this type of collector is not suited to the removal of dusts which are combustible.

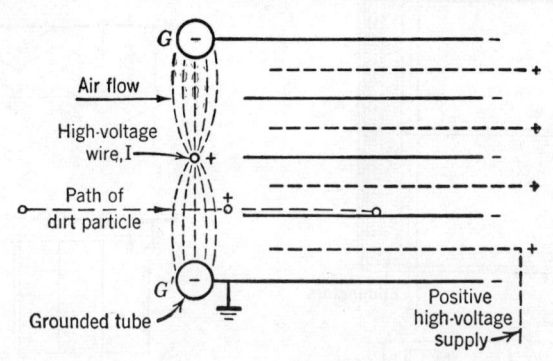

Fig. 15.10. Principle of electrostatic air cleaner.

15.13. Air Washers. Air-washing equipment serves as an air cleaner; it also functions as either an air humidifier or air dehumidifier, depending upon the conditions of operation. Air washers remove certain dusts and dirts very effectively; other impurities such as fine soot are not readily removed from air by washing. This form of equipment functions by bringing the air to be conditioned in contact with (1) water broken into fine droplets, (2) a combination of water-wetted surfaces and water discharged from spray nozzles, and (3) surfaces continuously wetted by water.

The air washer of Fig. 15.11 operates as explained in item 2. The assembled device consists of a spray chamber fitted with one bank of water nozzles and a set of water eliminators. The diffusers at the air-inlet end give more uniform distribution of the air across the spray-chamber section. The action of the washer is secured by having the air pass through finely divided water sprays produced by nozzles placed in the path of the flowing air. The water is withdrawn from the sump or tank of the spray chamber and is forced through the nozzles by a pump usually operating with a discharge pressure of around 25 psig. Water not absorbed by the air in passing through the sprays falls to the sump where a constant level is maintained by the action of a float valve in the make-up water supply line. The sump is also fitted with an overflow to permit the escape of excess water.

Washers similar to the one of Fig. 15.11 are also built with two or three banks of spray nozzles. Whenever the outside-air temperature falls below 32 F a preheater or air-tempering coil must be installed in the duct just ahead of the washer air inlet to prevent freezing of the spray water. The velocity of the air in the spray chamber should be limited to 500 to 550 fpm.

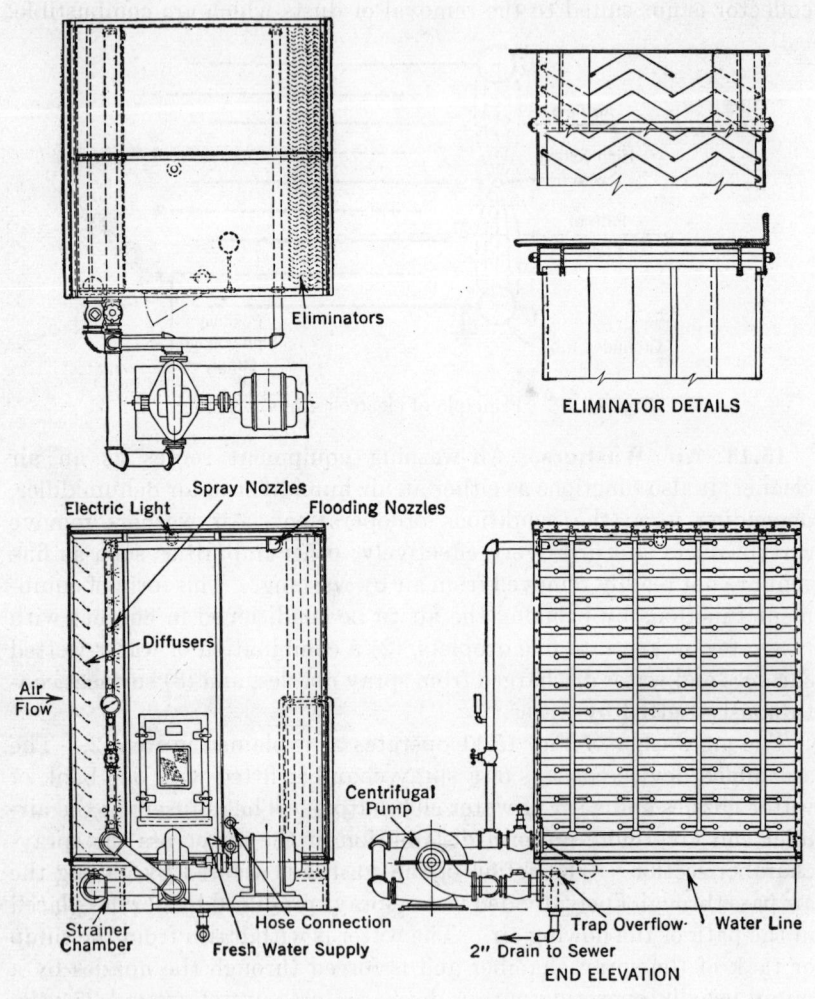

Fig. 15.11. Air washer. (Courtesy Westinghouse Electric Corp., Sturtevant Division.)

Most of the air cleaning is done as the aerosols impinge on the eliminators which serve as scrubber plates and are flooded with spray water. The scrubber surfaces are the parts of the eliminator assembly with which the exit air first comes in contact. Any water-soluble gases in the air are readily removed in a washer and some odor-causing

VENTILATION AND AIR PURIFICATION

vapors are also absorbed. Unless the spray water is changed frequently it may become saturated with odor-bearing materials and thereby be a source of odors.

The basic feature of *cellular air washers* is the employment of water-sprayed *capillary cells*, Fig. 15.12, through which the air to be processed flows. Each cell is completely filled with fine glass filaments which occupy about 1.5 per cent of its volume. A thin layer of the filaments is placed parallel to each cell face where the air enters and leaves the unit. The remainder of the filaments are placed practically parallel to the direction of air flow. Parallel arrangement c the filaments in the main b y of the cell reduces resistances to flow, permits small moving streams of air and water to make maximum contacts with excellent heat and moisture exchanges, and gives better cleaning efficiency than can be achieved in a spray chamber.

Fig. 15.12. Construction of fiber-filled air-washer cells. (Courtesy Air and Refrigeration Corp.)

Each cell unit of this particular design is capable of handling 1100 cfm of air, but a nominal rating for all conditions is 1000 cfm, which gives a face-area velocity of approximately 350 fpm. Cellular units are assembled in one or more tiers and with a sufficient number of cells per tier to give the desired air-handling capacity.

Cellular washers are used only in atmospheric service where the dust loading is light; the dirt separated from the air, as it passes through the cells, is usually carried into the sump with the water. When the glass fibers held in the cells become soiled they usually can be cleaned, without removal, by adding a suitable wetting agent to the water held in the sump.

A special type of washer known as a *packed tower*,[15] Fig. 15.13, is designed specifically for the removal of contaminants from gases. Packed towers are extensively used in the chemical industry as gas scrubbers to remove unwanted water-soluble gases and vapors from whatever may be passing through the device. The packing material is usually either irregular ceramic shapes, coke, or gravel. An average depth of the packing bed is 4 ft, the velocities[15] of the flowing media range from 100 to 300 fpm, and the resistances offered to their flow vary between 1.5 and 3.5 in. WG. Packed towers are used primarily in industrial air cleaning; they are not adapted to dust loadings greater than 1 grain per cu ft.

Fog, supplied by nozzles operated with water pressures ranging from 350 to 500 psig, is an effective cleaning agent for industrial gases.[16] A Venturi scrubber[17] can break cleaning water into very small droplets by the action of a high gas velocity instead of a high water pressure;

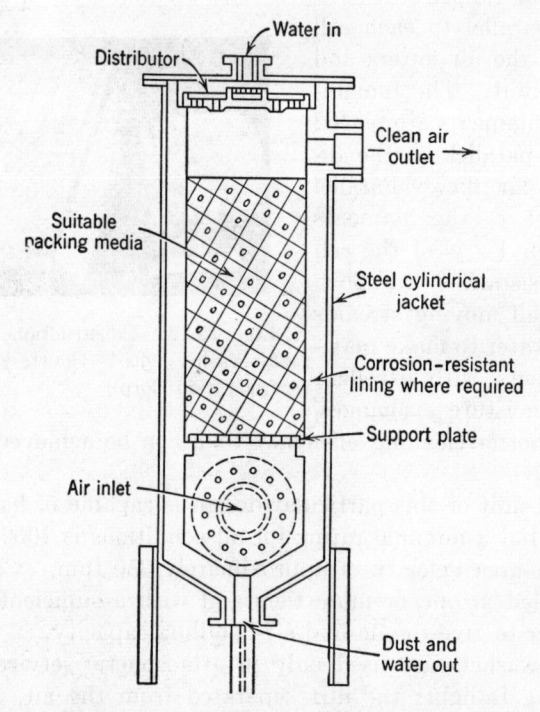

Fig. 15.13. Packed tower for air cleaning.

in operation it has produced cleaning efficiencies up to 99 per cent when used to remove contaminants such as sodium sulfate dust and sulfuric acid mists.

15.14. Systems of Ventilation. Ventilation is accomplished by natural and mechanical means. Natural systems comprise open windows, roof ventilators, and vertical ducts. Natural systems of ventilation cannot be used where air has to be passed through washers, coils, or filters. Mechanical systems embody either central plants with distributing ducts or unit ventilators, with little or no ductwork. All mechanical systems require fans.

VENTILATION AND AIR PURIFICATION

15.15. Roof Ventilators. These units are constructed in a variety of forms, four of which are shown in Fig. 15.14. Roof ventilators should prevent the entry of rain and snow and permit the outflow of air regardless of wind direction and velocity.

Continuous-ridge ventilators are used with industrial buildings where the rate of heat release from process operations, such as metal casting, is very high.

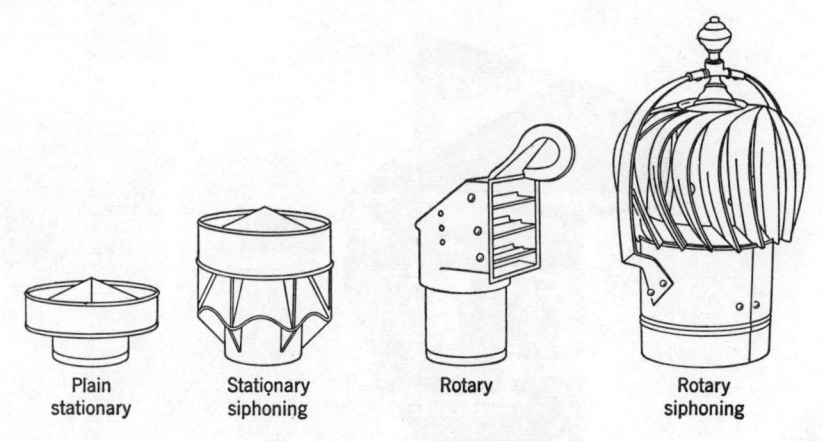

Plain stationary   Stationary siphoning   Rotary   Rotary siphoning

Fig. 15.14. Roof ventilators.

A method of computing the velocity of air movement through a roof ventilator which functions by virtue of natural draft is given in reference 18.

The roof ventilator, Fig. 15.15, includes a fan and its capacity is independent of weather conditions.

15.16. Central Ventilation Systems. These systems having one or more fans operate: (1) to blow air through ducts into spaces with the air leaving the room through vent stacks or other openings, (2) to exhaust air through a duct system with the fresh supply of air entering at inlet grilles in the walls or window openings, and (3) both to blow air into and to exhaust air from spaces to be ventilated. In the *split system* the heat losses from the building are supplied by either direct radiators or convectors and the ventilation air is handled by a fan which delivers tempered air at room temperature. The arrangement lends itself to various classes of buildings where ventilation is not required continuously. The use of direct radiators and convectors permits spaces to be brought to and maintained at the desired temperature during periods of unoccupancy. The ventilating part of the equipment can be arranged for complete air conditioning for either winter or summer conditions.

15.17. Unit Ventilators. Central systems of ventilation require considerable space in the building and its walls for the fans, heaters, and ducts. With unit systems one or more small pieces of equipment are located within each space to be ventilated. A typical unit for both

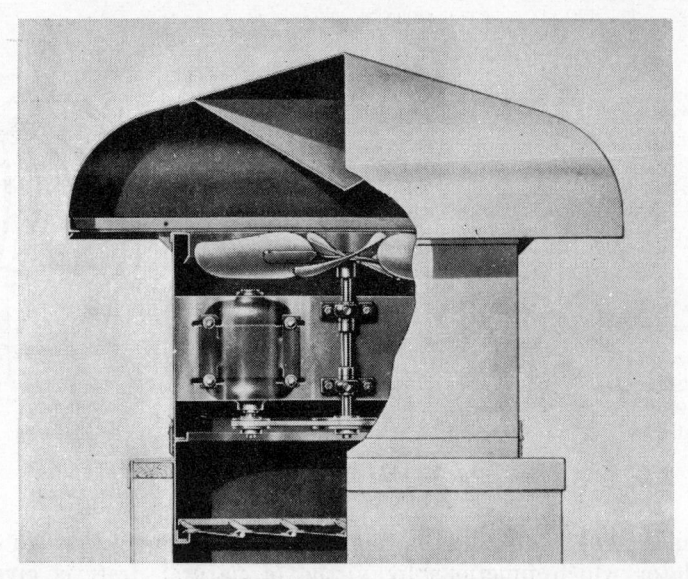

Fig. 15.15. Roof ventilator equipped with propeller fan. (Courtesy Burt Manufacturing Co.)

heating and ventilating, Fig. 15.16, consists of a housing, motor-driven fans, and a steam coil for warming the air handled. It is often desirable to have the unit ventilator merely heat the air to room temperature or slightly less and to use direct radiation and convectors for the space heating. Unit ventilators are installed to handle outside air taken through short connections made through the building outside wall along which they are placed. Arrangement for the recirculation of inside air is highly desirable as the operation of the unit with recirculated air is of great aid in warming up a cold room. The vitiated air from the room can be disposed of through a central exhaust system, through a partially opened window, or by outward leakage. Practically all unit ventilators contain an air filter. Units of this sort are also available which in reality are unit air conditioners having within a single housing the necessary features to clean, heat, humidify, and

VENTILATION AND AIR PURIFICATION

also to clean, cool, and dehumidify air. The discharge at the top outlet is adjusted to deliver the air in any desired direction. By this arrangement drafts can be avoided and turbulence created in the air above the occupied zone.

Unit ventilators are particularly useful in meeting the specific needs[19] of schoolrooms and specially designed equipment is available for this

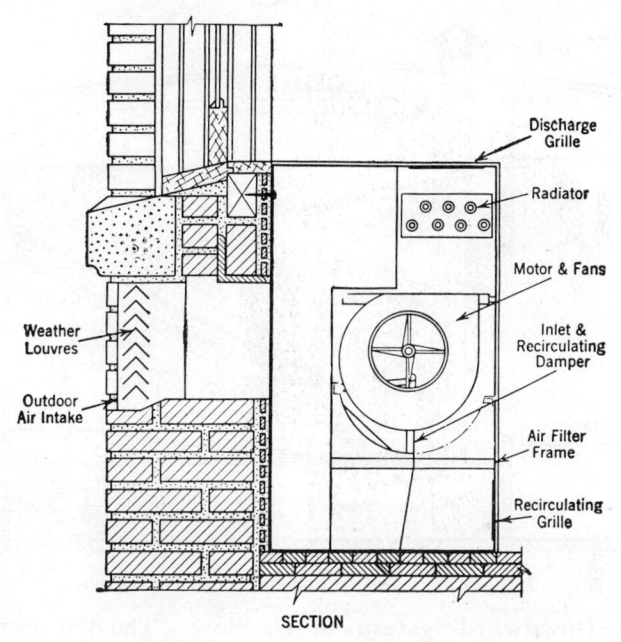

Fig. 15.16. Unit ventilator. (Courtesy Westinghouse Electric Corp., Sturtevant Division.)

class of service. Some unit ventilators function to handle all of the room heating requirements; others are designed to work in conjunction with radiators, convectors, or baseboard units.

15.18. Systems of Air Distribution. Central ventilation systems may be further classified according to the direction of air flow into the room, as upward, downward, and ejector systems.

15.19. Upward Flow Systems. The ventilating air supply in upward systems of ventilation is brought into the space through inlets near the floor line, through openings in the floor, or through the pedestals of theater chairs and seats. The foul-air outlets are located in the side walls near the ceiling or in the ceiling. Upward systems of air flow may be used in rooms where there is a marked tendency for air heated by the occupants to rise and carry with it the vitiating products from their bodies. These sytems involve difficulties in the introduction of the ventilating air so that the occupants are not subjected to drafts. Upward systems of ventilation are not satisfactory where air cooler than the room temperature is to be introduced.

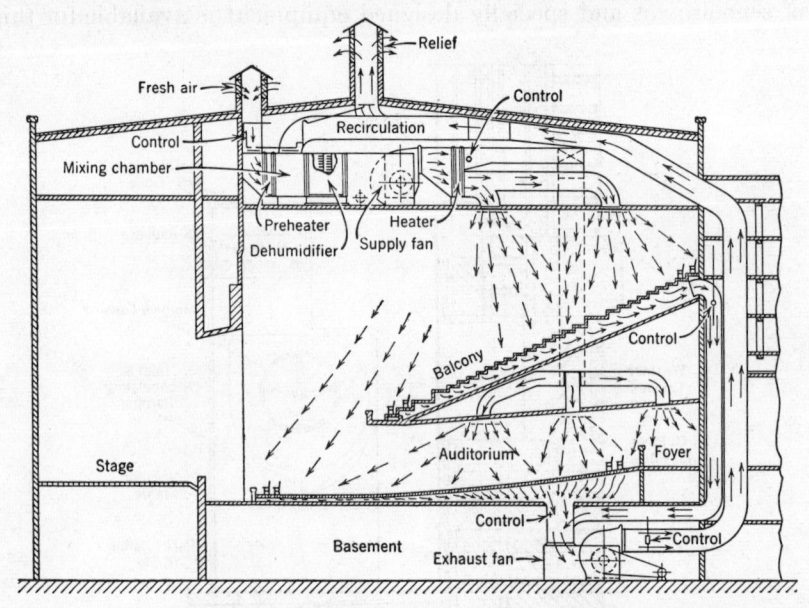

Fig. 15.17. Downward system of ventilation.

15.20. Downward Systems of Air Flow. The downward system of ventilation, Fig. 15.17, introduces the ventilating air through openings located in the ceiling and removes the vitiated air through openings either in or near the floor. Any system of downward ventilation should endeavor to spread the incoming air uniformly above the occupied zone. Downward systems of ventilation, if properly installed, are satisfactory for theaters, auditoriums, schools, and offices.

15.21. Ejector Systems. The ejector system of air distribution, which is a form of downward distribution, is feasible and satisfactory where the air may be blown through an unobstructed space for a considerable distance. The air is introduced at high velocity through properly designed grilles at the rear of the room and is discharged well above the occupied zone. The jet action of the incoming air produces considerable aspirating effects, giving a better diffusion of the fresh air within the space. The vent outlets are placed in the rear of the building, located so that the return air flows uniformly into the faces of the audience.

15.22. Air Exhaust.

With any system of fresh-air distribution there should be a positive scheme of vitiated-air removal so designed and located as to aid in the distribution of the fresh air and to prevent short-circuiting of the air between the supply and the vent outlets.

When the air is supplied to maintain a pressure within the space at a value slightly greater than atmospheric pressure, leakage outward through cracks, etc., may be sufficient to care for the disposal of vitiated air. A better method is to provide a system of vent flues or ducts through which the air can escape. These ducts may have air flow produced by either gravity action, air pressure in the room, or the action of an exhaust fan.

Where fans are used in connection with exhaust systems, allowable velocities in branches, risers, and main ducts are given in Table 14.4. Gravity flow in exhaust systems is always more or less erratic in action, and if positive air removal is desired or if the exhaust duct system is of considerable length a fan should always be used in connection with it. Rooms such as kitchens and toilets should be maintained with the air at pressures less than the other spaces to promote air inflow into the room rather than outflow.

REFERENCES

1. "Code of Minimum Requirements for Heating and Ventilating Garages," *ASHVE Trans.*, **1**, 30 (1935).
2. "Ventilation Requirements," C. P. Yaglou, E. C. Riley, and D. J. Coggins, *ASHVE Trans.*, **42**, 133 (1936).
3. "Code of Minimum Requirements for Comfort Air Conditioning," *ASHVE Trans.*, **44**, 27 (1938).
4. *Air-Conditioning Manual*, The Trane Co., LaCrosse, Wis. (1953).
5. "Ultra-Violet Light Control of Air-Borne Infections in a Naval Training Center," S. M. Wheeler, H. S. Ingraham, A. Hollaender, N. D. Lill, J. G. Cohen, and E. W. Brown, *Am. J. Public Health*, **35**, 457 (1945).
6. "Factors of Importance in the Use of Triethylene Glycol Vapor for Aerial Disinfection," Wm. Lester, Jr., Saul Kaye, O. H. Robertson, and Edwin Dunkin, *Am. J. Public Health*, **40**, 813 (1950).
7. "Triethylene Glycol Vapor Distribution for Air Sterilization," Edward Brigg, B. H. Jennings, and F. C. W. Olson, *ASHVE Trans.*, **53**, 393 (1947).
8. "Glycol Vapors for Disinfecting Purposes," editorial in *J. Am. Med. Assn.* (Mar. 8, 1947).
9. "The Lethal Effect of Relative Humidity on Air-Borne Bacteria," E. W. Dunkin and T. P. Puck, *J. Exp. Med.*, **87**, 82 (Feb. 1948).
10. *Industrial Dust*, Phillip Drinker and Theodore Hatch, McGraw-Hill Book Co., New York (1954).
11. "Standard Code for Testing Air Cleaning Devices Used in General Ventilation Work," *ASHVE Trans.*, **39**, 225 (1933).

12. "Characteristics of Commercial Air Cleaners," M. W. First and Leslie Silverman, *Heating and Ventilating*, **51**, No. 1, 78 (Jan. 1954).
13. "Electrostatic Effects in Fiber Filters for Aerosols," A. T. Rossano, Jr., and Leslie Silverman, *Heating and Ventilating*, **51**, No. 5, 102 (May 1954).
14. "Self-Charging Electrostatic Air Filters," W. T. Van Orman and H. A. Endres, *Heating, Piping, Air Conditioning*, **24**, No. 1, 157 (Jan. 1952).
15. "Operation, Application and Effectiveness of Dust Collection Equipment," J. M. Kane, *Heating and Ventilating*, **49**, No. 8, 87 (Aug. 1952).
16. "Effectiveness of the Fog Filter," F. A. Thomas, *Heating and Ventilating*, **50**, No. 2, 113 (Feb. 1953).
17. "A Venturi Scrubber for Cleaning Industrial Gases," W. P. Jones, *Heating and Ventilating*, **48**, No. 4, 73 (Apr. 1951).
18. *Plant and Process Ventilation*, W. C. L. Hemeon, Industrial Press, New York (1955).
19. "Fundamentals of Schoolroom Heating and Ventilating," Henry Wright, *Heating and Ventilating*, **48**, No. 2, 87 (Feb. 1951).

BIBLIOGRAPHY

Ventilation

1. *Plant and Process Ventilating*, W. C. L. Hemeon, Industrial Press, New York (1955).
2. *Design of Industrial Exhaust Systems*, J. L. Alden, Industrial Press, New York (1948).
3. *Exhaust Hoods*, J. M. Dalla Valle, Industrial Press, New York (1952).
4. "Cooling Hot Spots in Industry," A. B. Wason, *Heating and Ventilating*, **51**, No. 9, 88 (Sept. 1954).
5. "Ventilating of Commercial Laundries," Sidney Marlow, *Heating, Piping, Air Conditioning*, **26**, No. 11, 149 (Nov. 1954).
6. "Exhaust for Hot Processes," W. C. L. Hemeon, *Air Conditioning, Heating, Ventilating*, **51**, No. 8, 83 (Aug. 1954).
7. "New Laboratory Fume Hoods Cut Air Conditioning Load," J. F. Turner, *Heating, Piping, Air Conditioning*, **23**, No. 1, 113 (Jan. 1951).
8. "Fundamentals of Natural Ventilation of Houses," J. B. Dick, *J. Inst. Heating Ventilating Engrs. (London)*, **18**, No. 179, 123 (June 1950).
9. "Experimental Studies in Natural Ventilating of Houses," J. B. Dick, *J. Inst. Heating Ventilating Engrs. (London)*, **17**, No. 173, 420 (Dec. 1949).
10. "Measurement of Ventilation Using Tracer Gas Technique," J. B. Dick, *Heating, Piping, Air Conditioning*, **22**, No. 5, 131 (May 1950).
11. "Study of Air Flow, Ventilation and Air Movement in Small Rooms as Affected by Open Fireplaces and Ventilation Ducts," T. C. Angus, *J. Inst. Heating Ventilating Engrs. (London)*, **17**, No. 172, 379 (Nov. 1949).
12. "Minimal Replenishment Air Required for Living Spaces," W. V. Consolazio and L. J. Pecora, *ASHVE Trans.*, **53**, 127 (1947).
13. "Using Solvents Safely," A. D. Brandt, *Heating, Piping, Air Conditioning*, **24**, No. 8, 106 (Aug. 1952).
14. "Types and Applications of Power Roof Ventilators," R. H. Avery, *Air Conditioning, Heating, Ventilating*, **54**, No. 1, 57 (Jan. 1957).

15. "Ventilation of Industrial Ovens," Benjamin Feiner and Irving Kingsley, *Air Conditioning, Heating, Ventilating*, **28**, No. 12, 117 (Dec. 1956).
16. "Radioactive Process Ventilation," S. H. Glassmire, *Heating, Piping, Air Conditioning*, **28**, No. 12, 117 (Dec. 1956).
17. "Supply and Exhaust Ventilation for Metal Pickling Operations," G. M. Hama, *Air Conditioning, Heating, Ventilating*, **54**, No. 9, 61 (Sept. 1957).

Air Sterilization and Odor Control

1. "Disinfecting Air with Sterilizing Lamps," R. Nagy, G. Mouromseff, and F. H. Rixton, *Heating, Piping, Air Conditioning*, **26**, No. 4, 82 (Apr. 1954).
2. "Determining and Reducing the Concentration of Air-Borne Micro-Organisms," Mathew Luckiesh and A. H. Taylor, *ASHVE Trans.*, **53**, 191 (1947).
3. "The Bacteria and Odor Control Problem in Occupied Spaces," R. L. Kuehner, *Heating, Piping, Air Conditioning*, **24**, No. 12, 113 (Dec. 1952).
4. *Odors, Physiology and Control*, C. P. McCord and W. N. Witheridge, McGraw-Hill Book Co., New York (1949).
5. "Control of Odors," E. R. Weaver, *Circular No. 491*, U. S. Nat. Bur. Standards (Apr. 17, 1950).
6. "Odor Control with ClO_2," E. R. Woodward and E. G. Fenrich, *Chem. Eng.*, **59**, No. 4, 170 (Apr. 1952).
7. "New Device, Wider Concept, Helps to Measure Odors Quantitatively," V. E. Gex and J. P. Snyder, *Chem. Eng.*, **59**, No. 12, 200 (Dec. 1952).
8. "Pollution Control Symposium," *Chem. Eng.*, **58**, No. 5, 111 (May 1951).
9. "The Effects of Moisture Content on the Diffusion of Odors in the Air," R. L. Kuehner, *ASHVE Trans.*, **53**, 77 (1947).

Air Cleaners

1. "New Air Filter Code, Part I," A. Nutting and R. F. Logsdon, *Heating, Piping, Air Conditioning*, **25**, No. 6, 77 (June 1953).
2. "A Test Method for Air Filters," R. S. Dill, *ASHVE Trans.*, **44**, 379 (1938).
3. "Air Filter Performance as Affected by Kind of Dust, Rate of Dust Feed, and Air Velocity," F. B. Rowley and R. C. Jordan, *ASHVE Trans.*, **44**, 415 (1938).
4. "Air Filter Performance as Affected by Low Rate of Dust Feed, Various Types of Carbon and Dust Particle Size and Density," F. B. Rowley and R. C. Jordan, *ASHVE Trans.*, **45**, 339 (1939).
5. "The Effect of Lint on Air Filter Performance," F. B. Rowley and R. C. Jordan, *ASHVE Trans.*, **47**, 29 (1941).
6. "Economical Air Velocities for Mechanical Air Filtration," F. B. Rowley and R. C. Jordan, *ASHVE Trans.*, **47**, 391 (1941).
7. "Overloading of Viscous Air Filters," F. B. Rowley and R. C. Jordan, *ASHVE Trans.*, **48**, 437 (1942).
8. "Guides in Dust Collector Selections," J. M. Kane, *Air Conditioning, Heating, Ventilating*, **51**, No. 10, 77 (Oct. 1954).
9. "Sampling and Analyzing Air for Contaminants," Leslie Silverman, *Air Conditioning, Heating, Ventilating*, **52**, No. 8, 88 (Aug. 1955).
10. "Graphical Selector for Air Cleaners," J. R. Kayse, *Heating and Ventilating* **50**, No. 7, 80 (July 1953).
11. "Size Distribution and Concentration of Air-Borne Dust," K. T. Whitby,

A. B. Algren, and R. C. Jordan, *Heating, Piping, Air Conditioning*, **27**, No. 8, 121 (Aug. 1955).
12. *Industrial Health Engineering*, A. C. Brandt, John Wiley & Sons, New York (1947).
13. "Theory and Practice of the Fog Filter Air Scrubber," D. G. Hudson, *Heating and Ventilation*, **48**, No. 3, 57 (Mar. 1951).
14. "Filtration of Micro-Organisms from Air by Glass Fiber Media," H. M. Decker, J. B. Harstad, F. J. Piper, and M. E. Wilson, *Heating, Piping, Air Conditioning*, **26**, No. 5, 155 (May 1954).
15. "Removal of Bacteria and Bacteriophage from the Air by Electrostatic Precipitators and Spun Glass Filter Pads," H. M. Decker, F. A. Geile, H. E. Moorman, and C. A. Glick, *Heating, Piping, Air Conditioning*, **23**, No. 10, 125 (Oct. 1951).
16. "Performance of Inertial Collectors," Leslie Silverman, *Heating and Ventilating*, **50**, No. 2, 87 (Feb. 1953).
17. "Simple Comparison of Dry and Wet Collection Equipment for Air-Borne Contaminants," Leslie Silverman, *Heating and Ventilating*, **49**, No. 4, 109 (Apr. 1952).
18. "A Survey of Electrostatic Precipitators," E. A. Walker and J. E. Coolidge, *Heating, Piping, Air Conditioning*, **23**, No. 3, 107 (Mar. 1951).
19. "New Developments in Reverse Jet Filters," H. J. Hersey, Jr., *Heating and Ventilating*, **51**, No. 2, 109 (Feb. 1954).
20. "Efficiency Studies of Wet-Type Dust Collectors," B. D. Bloomfield, *Heating and Ventilating*, **51**, No. 4, 89 (Apr. 1954).
21. "Resistance Gradients Through Viscous Coated Air Filters," F. B. Rowley and R. C. Jordan, *ASHVE Trans.*, **56**, 237 (1950).
22. "Fibrous Filters for Fine Particle Filtration," D. J. Thomas, *J. Inst. Heating Ventilating Engrs. (London)*, **20**, No. 201, 35 (May 1952).
23. "High Efficiency Industrial Filtration," E. A. Stokes, *Ind. Heating Engr.*, **13**, No. 68, 163 (June 1951).
24. "Removal of Air-Borne Particulates and Allergens by a Portable Electrostatic Precipitator," Leslie Silverman and Richard Dennis, *Air Conditioning, Heating, Ventilating*, **53**, No. 12, 75 (Dec. 1957).

PROBLEMS

1. Automobiles passing through a certain vehicular tunnel may release as much as 10,000 cu ft of carbon monoxide per hr (measured at a pressure equal to that of the atmosphere). Calculate the volume of ventilation air needed in cubic feet per minute.

2. The sanctuary of a certain church contains 600 seats. Compute the minimum amount of standard fresh air in cubic feet per minute that should be supplied by a summer air conditioner. Find the weight of fresh air in pounds per hour.

3. An industrial air cleaner receives air from a grinding room which contains 1000 grains of foreign material per 1000 cu ft. The air leaving the cleaner has a dust loading of 100 grains per 1000 cu ft. What cleaner efficiency is represented by this operation? Calculate the weight of dust collected in an 8-hour day in pounds if the unit handles dust-laden air at the rate of 10,000 cfm.

16

Heating with Central Fan-Coil Systems and Unit Heaters

16.1. Fan-Coil Heating. For many years the scheme of heating which requires a fan, heater units, and a duct system was termed *hot blast*. This form of forced-convection indirect heating differs from mechanical warm-air furnace systems only in the equipment used for warming the air. Hot-blast heaters which have been used are assemblages of cast-iron sections, pipe coils, and copper and brass tubes with fins to give extended surfaces.

The location of the heater in a fan-coil system, with respect to the fan, is either on its discharge side, a *blow-through arrangement*, or on its suction side, which is termed a *draw-through* placement (Fig. 14.20).

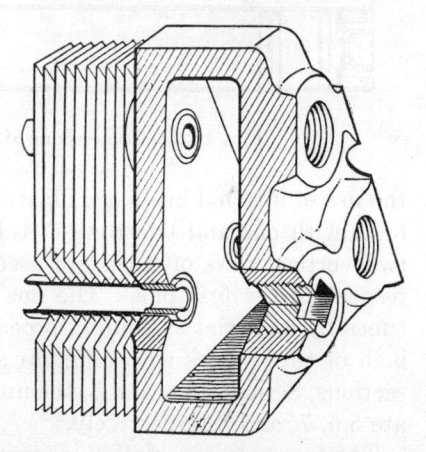

Fig. 16.1. Header arrangement of Trane blast coils.

As compared with heating systems using direct radiation, fan-coil

349

350   AIR CONDITIONING AND REFRIGERATION

apparatus has the following points in its favor: absence of direct radiators which occupy room space, requires a smaller amount of heating surface than direct radiation, and ability to supply air for ventilation.

16.2. Nonferrous Extended-Surface Blast Coils.

A representative form of extended-surface heating-coil construction is indicated by the details of Figs. 16.1 and 16.2. These sections have cast-iron headers into which either copper or brass tubes are expanded by

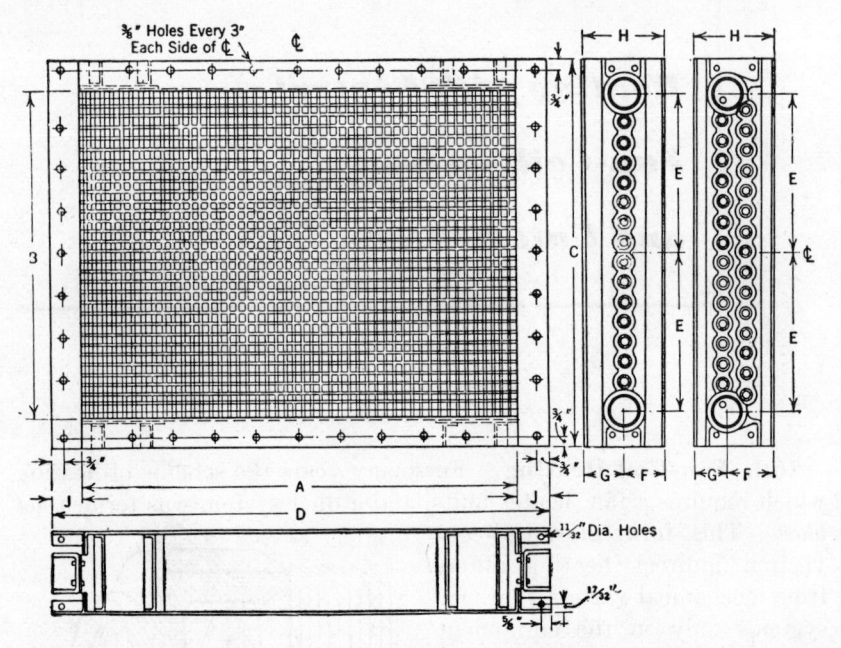

Fig. 16.2. Dimensions of Trane Type-E blast coils.

the use of internal bushings, illustrated in Fig. 16.1. The sections are built with one and two rows of tubes, as shown in Fig. 16.2. When two vertical rows of tubes are used the second row is staggered with respect to the first one. The fins are mechanically held against the tubes by a special expanding process. The number of fins per linear inch of tube varies with different sections. Thus, for Trane Type-E sections, Series 6, 8A, and 8, the number of fins per inch of tube length are 5.5, 7, and 8 respectively.

The face area of a section is given by $A \times B$, Fig. 16.2. Table 16.1 gives the face area in square feet for several combinations of coil width and length. Two or more sections may be assembled as in Fig. 16.3 to give the total required face area. The assembly should be as

FAN-COIL SYSTEMS AND UNIT HEATERS

TABLE 16.1. Square Feet Net Face Area, Not Including the Casing, Trane Type-E Heating Coils

Nominal Coil Width B, in.	Nominal or Ordering Length A, in.																			
	12	18	24	30	36	42	48	54	60	66	72	78	84	90	96	102	108	114	120
6	0.5	0.75	1.0	1.25	1.5	1.75	2.0	2.25	2.5	2.75	3.0								
9	0.75	1.13	1.5	1.88	2.25	2.63	3.0	3.38	3.75	4.13	4.5								
12	1.0	1.5	2.0	2.5	3.0	3.5	4.0	4.5	5.0	5.5	6.0	6.5	7.0	7.5	8.0	8.5	9.0	9.5	10.0
15		1.88	2.5	3.13	3.75	4.38	5.0	5.63	6.25	6.88	7.5	8.13	8.75	9.38	10.0	10.63	11.25	11.88	12.5
18		2.25	3.0	3.75	4.5	5.25	6.0	6.75	7.5	8.25	9.0	9.75	10.5	11.25	12.0	12.75	13.5	14.25	15.0
21			3.5	4.38	5.25	6.13	7.0	7.88	8.75	9.63	10.5	11.38	12.25	13.13	14.0	14.88	15.75	16.63	17.5
24			4.0	5.0	6.0	7.0	8.0	9.0	10.0	11.0	12.0	13.0	14.0	15.0	16.0	17.0	18.0	19.0	20.0
30				6.25	7.5	8.75	10.0	11.25	12.5	13.75	15.0	16.25	17.5	18.75	20.0	21.25	22.5	23.75	25.0
33					8.25	9.63	11.0	12.38	13.75	15.13	16.5	17.88	19.25	20.63	22.0	23.38	24.75	26.13	27.5

nearly square as possible. The number of rows of tubes to give the final required air temperature is secured by banking the sections, as shown in Fig. 16.3. The section castings are flanged and drilled, Fig. 16.2; when bolted together they do not require any other housing and are easily attached to the ductwork. If a duct attached to a single coil is provided with a matching flange, turned outward, its cross-sectional dimensions can be exactly the same as the nominal length and width of the coil (dimensions A and B, Fig. 16.2). When it is necessary to use more than one coil in a bank as in Fig. 16.3, the height of the attached duct will be greater than the sum of the nominal coil widths because of the width of the flange on each coil. The flange on Trane Type-E coils is 1.5 in. wide (dimension C, Fig. 16.2 = $B + 3$ in.).

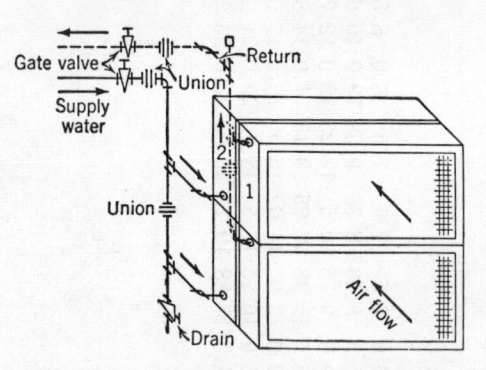

Fig. 16.3. Extended-surface heaters arranged in two tiers and two banks for hot-water counterflow operation.

Information relative to the performance of the foregoing coils, when used for heating air with saturated steam, is given in Fig. 16.4. The air friction-pressure losses for the same equipment operating with various face-area velocities, in feet per minute, are given in Table 16.2.

16.3. Required Face Area and Rows of Tubes. The selection of fan coils is made on the required amount of face area and the depth of the heater expressed in rows of tubes. Face-area requirements are determined by the quantity of air to be handled per minute and the allowable face-area velocity. Both the quantity of air handled and its allowable velocity are expressed in terms of standard air. Satisfactory face-area velocities are 600 fpm for public buildings and up to 1200 fpm in industrial plants. The number of rows of tubes required depends upon the initial-air temperature, the final desired air temperature, the average temperature of the heating medium within the coil sections, and the permissible face-area velocity. Where reliable data, based on performance tests, are available, their use is much simpler than attempts made to calculate the heater performance based on the known laws of heat transfer.[1] The data of Fig. 16.4 are based on actual tests with steam and may be used for the determination of the

FAN-COIL SYSTEMS AND UNIT HEATERS 353

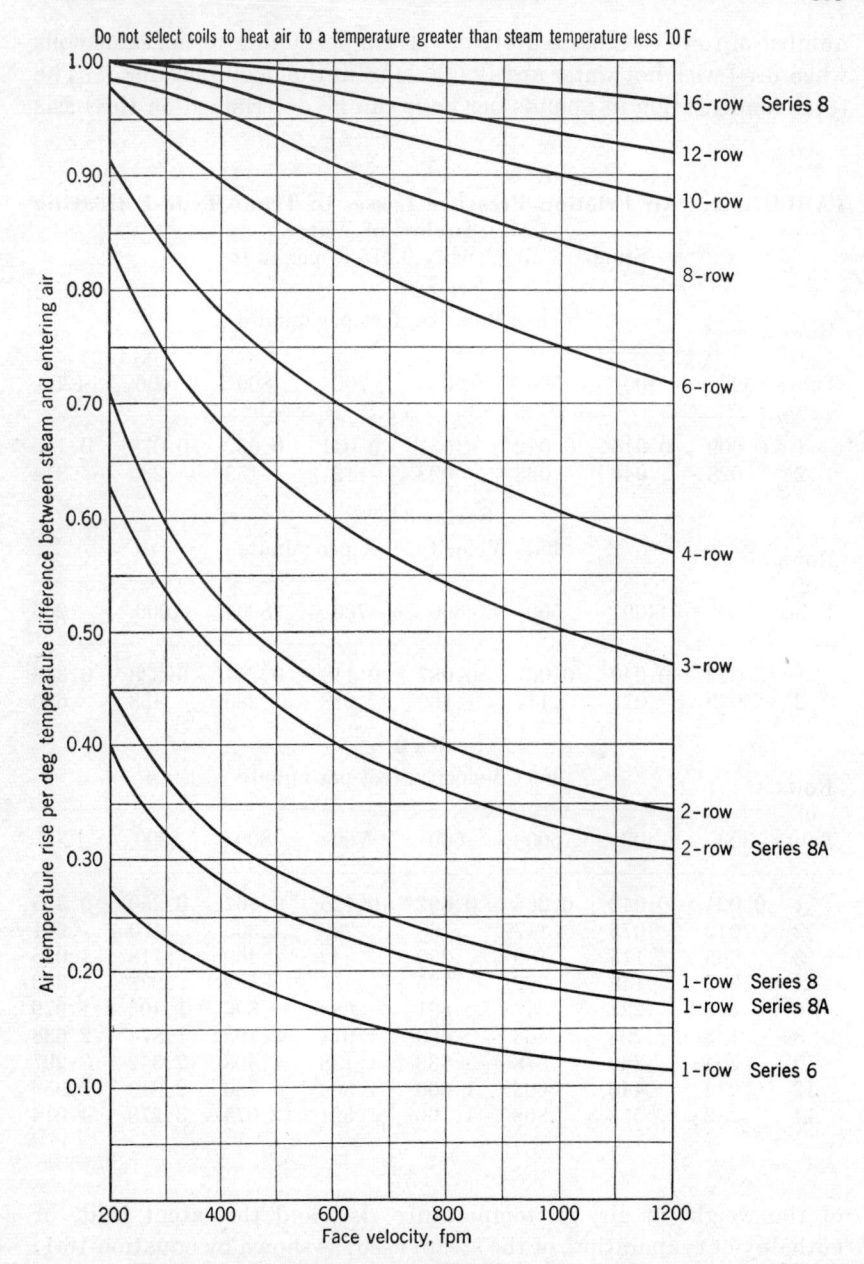

Fig. 16.4. Performance data for Trane Type-E blast coils.

number of rows of tubes required. (Data pertaining to the same coils when used with hot water are available from the manufacturer.) The total condensation in pounds per hour can be determined on the basis

TABLE 16.2. Air Friction-Pressure Losses in Trane Type-E Heating Coils, Inches of Water
Standard air, density 0.075 lb per cu ft

Series 6

Face Velocity, feet per minute

Rows of Tubes	300	400	500	600	700	800	1000	1200
1	0.009	0.015	0.022	0.031	0.041	0.052	0.078	0.110
2	.028	.046	.068	.093	.121	.155	.230	.315

Series 8A

Face Velocity, feet per minute

Rows of Tubes	300	400	500	600	700	800	1000	1200
1	0.022	0.040	0.062	0.087	0.119	0.154	0.238	0.338
2	.039	.071	.111	.160	.218	.286	.458	.630

Series 8

Face Velocity, feet per minute

Rows of Tubes	300	400	500	600	700	800	1000	1200
1	0.024	0.042	0.065	0.092	0.125	0.162	0.250	0.355
2	.042	.075	.117	.168	.229	.300	.470	.661
3	.065	.115	.180	.260	.342	.460	.718	1.015
4	.082	.148	.232	.335	.456	.598	.938	1.320
6	.122	.221	.347	.501	.683	.896	1.406	1.979
8	.163	.295	.463	.669	.911	1.195	1.874	2.638
10	.204	.369	.578	.836	1.138	1.493	2.342	3.297
12	.243	.440	.693	1.000	1.364	1.780	2.810	3.956
14	.282	.513	.808	1.166	1.591	2.078	3.278	4.614

of the weight of air, its temperature rise, and the latent heat, or enthalpy of evaporation, of the steam used, as shown by equation 16.1:

$$M_s = \frac{0.24 M_a (t_2 - t_1)}{h_{fg}} \quad (16.1)$$

FAN-COIL SYSTEMS AND UNIT HEATERS

16.4. Correction of Heater Air Friction-Pressure Losses to Actual Conditions. The air friction losses through the heater at standard conditions must be corrected to actual conditions. The mean temperature of the air as it passes through the heater is required in making the correction, as the average air density is necessary. The mean temperature of the air may be most easily determined in this case by first finding the mean temperature difference between that of the steam and that of the air passing through the coil. The entering and leaving temperatures of the air are always known. However, the variation in the temperature between the entrance to and the exit from the heat exchanger is unknown. The mean temperature difference based on a straight-line variation is called the *arithmetic mean temperature difference*. However, it is logical to suppose that the temperature of the air will increase more rapidly in the first part of its passage through the coil because of the greater temperature difference at that point and that the temperature variation will follow a curved line. If this is the case the actual mean temperature difference is less than that based on an arithmetical average of the entering and leaving temperatures. The following equation is often used for obtaining this lesser value, which is usually called the *logarithmic mean temperature difference*.

$$\text{Logarithmic mean temp. diff.} = \frac{\text{greater temp. diff.} - \text{lesser temp. diff.}}{\log_e (\text{greater temp. diff.}/\text{lesser temp. diff.})} \quad (16.2)$$

(see example of Art. 16.5.)

Equation 16.2 is applicable for finding the logarithmic mean temperature difference in a case where a medium is either heated by steam or cooled by a refrigerant at constant temperature and to either heating or cooling when the temperature of neither medium involved is constant. However, in a case where neither medium remains at constant temperature, it is necessary to assume a straight-line variation for one of them in finding the mean temperature of the other.

When the mean temperature of the air has been computed the mean density is easily obtained. The corrected air friction loss is given by

$$P_{fa} = P_{fs} \times 0.075/d_a \quad (16.3)$$

where P_{fs} is from a table such as Table 16.2 and d_a is based on the mean temperature of the air as it passes through the coil.

It may be noted that the correction of friction loss from a condition at standard air density to one at actual density is different here than in Art. 14.19. The reason for the difference is that air passing through the coil is being heated whereas in the other case its temperature remains constant as it passes along the duct.

16.5. Selection of a Heating Coil. The following example serves to illustrate the method of selecting a heating coil.

Example. A building is to be heated by use of draw-through fan-coil equipment similar to Fig. 14.20. The following air temperatures are required: at the heater outlet, approximately 132 F; entering the occupied zone, 130 F; and at the breathing-line level 5 ft above the floor, 60 F. The outside-air temperature is -20 F, the building heat losses amount to 330,800 Btuh for the stated conditions, and one-fourth of the air handled is to be drawn from the outside for ventilation purposes. The heater is to operate with a face-area velocity of 1000 fpm at standard conditions for the air.

Find the required size of the heater when it is operated with dry saturated steam at 5 psig and the barometric pressure is 14.7 psia. Calculate the weight of steam condensed in pounds per hour and determine the air friction-pressure loss in the heater in inches of water.

Solution. The weight of air handled per hour is $330{,}800/[0.24(130 - 60)]$ = 19,690 lb. The weight of the ventilating air is $19{,}690/4 = 4922$ lb per hr, and the weight of recirculated air becomes $19{,}690 - 4922 = 14{,}768$ lb per hr. The air entering the heater is assumed to be dry and its average temperature is $[(14{,}768 \times 60) + (4922 \times -20)]/19{,}690 = 40$ F. Standard air has a density of 0.075 lb per cu ft. The volume of the air entering the heater, measured under standard conditions, is $19{,}690/(60 \times 0.075) = 4375$ cfm.

The required face area of the heater is $4375/1000 = 4.38$ sq ft. Inspection of Table 16.1 indicates that a single section with a 21-in. width and a length of 30 in. has the necessary face area. The rectangular air-inlet opening has dimensions near enough to those of a square one so that the duct attached to it will be reasonably well proportioned. The required air-temperature rise per degree of temperature difference between those of the steam and the entering air must be computed before Fig. 16.4 can be used to determine the necessary number of tube rows. The assumed air-temperature rise in the heater is $132 - 40 = 92$ F. For the conditions stated the pressure of the steam is 19.7 psia and the corresponding temperature, from Table 1.6, is 227 F. The steam and entering-air temperature difference is $227 - 40 = 187$ F. The required air-temperature rise, per degree of difference of the temperatures of the steam and the entering air, is $92/187 = 0.491$ F. Figure 16.4 indicates that a three-row Series-8 coil will be satisfactory with a face velocity of 1000 fpm. The required heater depth may be secured by banking one two-row and one one-row coil section to give the necessary three rows of tubes over which air flow will occur. The weight of the required steam M_s is computed by use of equation 16.1 as $[0.24 \times 19{,}690 \times (132 - 40)] \div 960.7 = 453$ lb per hr.

The logarithmic mean temperature difference, (equation 16.2), between the steam and the air is $[(227 - 40) - (227 - 132)]/\log_e[(227 - 40)/(227 - 132)] = 135$ F and the mean temperature of the air is $227 - 135 = 92$ F. At this temperature $d_a = 0.0721$ lb per cu ft. Table 16.2 indicates that the air friction-pressure loss is 0.718 in. of water under standard conditions. The

FAN-COIL SYSTEMS AND UNIT HEATERS

air friction-pressure loss under actual conditions of air flow (equation 16.3) is $0.718 \times (0.075/0.0721) = 0.747$ in. of water.

16.6. Unit Heaters. An assemblage of a coil and a fan within an enclosing housing embodies the essential requirements of a unit heater. Generally the design is so constituted that the complete unit, which is factory assembled, can be quickly installed. For the most part no ductwork is attached to unit heaters. When such a provision is made the duct may be only a short connection through an outside wall for the purpose of bringing in ventilating air or it may take cool air from a floor when the heater is placed some distance above it.

Classification of unit heaters may be distinguished by (1) the heating medium, i.e., steam, hot water, electricity, or the products of combustion; (2) the type of fan used for air circulation, which may be either a propeller or a centrifugal unit; and (3) the arrangement of the heating surfaces with respect to fan location, either blow-through or draw-through arrangement.

Low- and high-pressure steam and mechanically circulated hot water commonly serve as heating fluids. Many installations employ gas as a fuel that is burned within the unit. Electric motors are commonly used to operate heater fans. However, steam-turbine drives are available with the exhaust steam discharged to the heater coil.

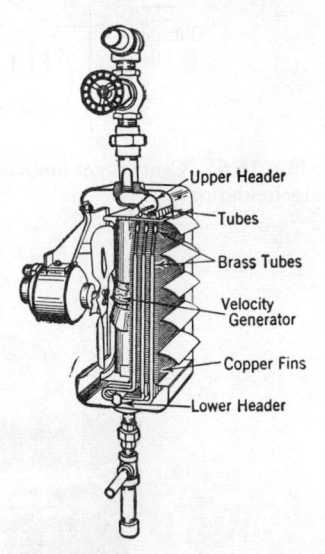

Fig. 16.5. Suspended-type unit heater. (Courtesy Modine Manufacturing Co.)

Unit heaters have wide applications for the heating of industrial buildings, machine and other workshops, garages, stores, and lobbies. The advantages of unit heaters, as compared with central fan-coil heating systems and those equipped with direct radiators, are their low first cost and the small amount of space required in the building.

16.7. Suspended Unit Heaters. The support of such a unit heater may be either from the steam or water-supply piping, Fig. 16.5, or otherwise as in Fig. 16.6; in any event the heater is placed above the zone of occupancy. The discharge outlets of the vertical suspended unit heater of Fig. 16.7 are rotated by a small electric motor An advantage of the revolving outlets is that heat is supplied to the

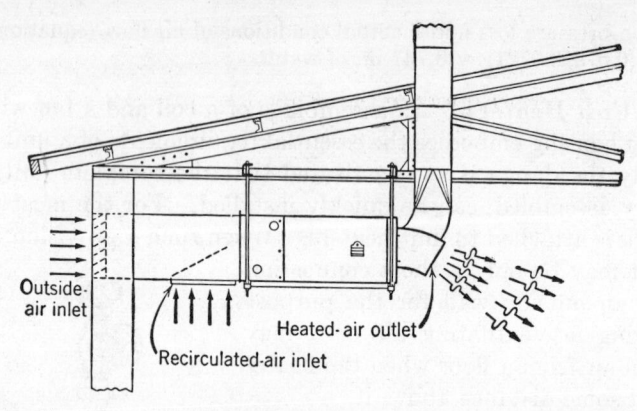

Fig. 16.6. Centrifugal-fan suspended unit heater arranged for ventilation and air recirculation.

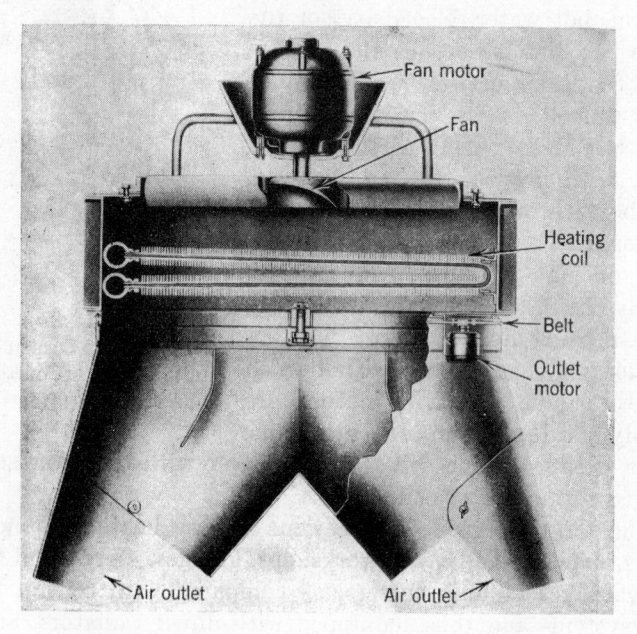

Fig. 16.7. Vertical-discharge suspended unit heater with revolving outlets. (Courtesy L. J. Wing Manufacturing Co.)

occupied zone without subjecting any individual to a continuous air movement. Figure 16.8 illustrates the essential parts of a gas-fired unit heater, which is a suspended type with a propeller fan producing horizontal discharge of the air. Similar units are available for installation in short horizontal ducts.

FAN-COIL SYSTEMS AND UNIT HEATERS 359

Suspended unit-heater installations are sometimes made with fresh-air connections from the outside together with provisions for the recirculation of room air as indicated by the arrangement of Fig. 16.6, which is fitted with a centrifugal fan. When the air dampers of the

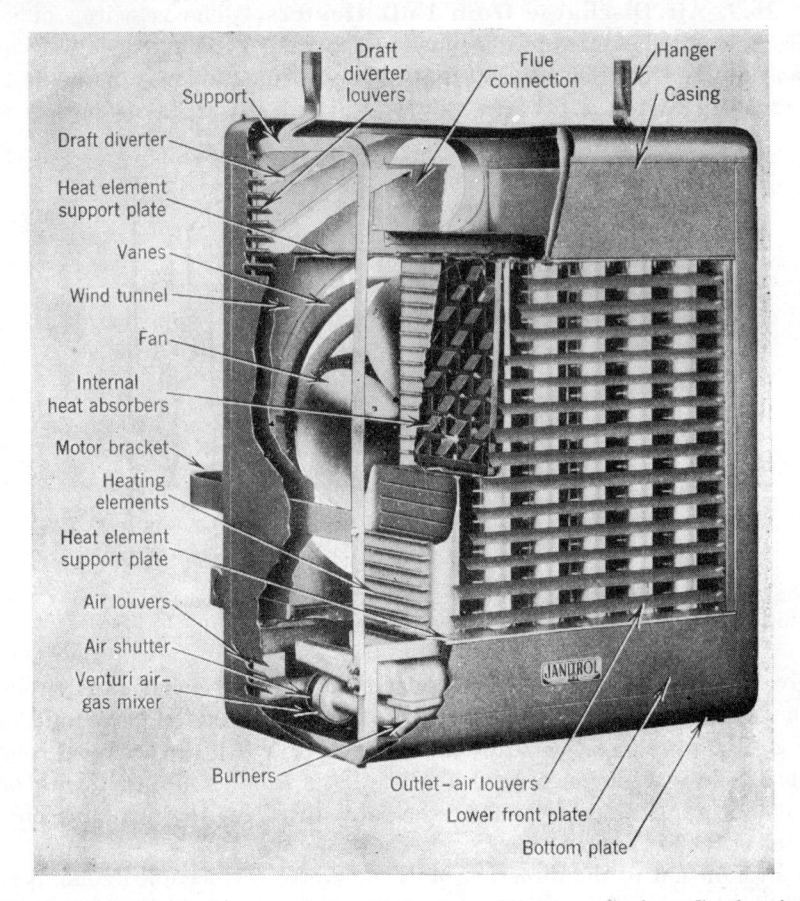

Fig. 16.8. Gas-fired suspended unit heater. (Courtesy Surface Combustion Corp.)

unit are properly adjusted and its heater coil inoperative, the equipment can be used as a ventilator to withdraw air from a space and to discharge it outside the structure.

16.8. Floor Types of Unit Heaters. A typical floor-type unit heater is shown in Fig. 16.9. The unit consists of a housing with an air inlet at the floor and air outlets at the top which may be arranged to discharge in two directions. The centrifugal fans draw the air to be heated through the heating coils placed just above the air inlet which is at the bottom. The air discharged may leave the outlets at high velocities just above the top level of the occupied zone.

16.9. Air Discharge from Unit Heaters. The velocities of air leaving unit heaters are dependent upon their type and location and may range from 300 to 2500 fpm. The warmed air may be made to travel up to 140 ft. Various devices are placed at heater outlets to

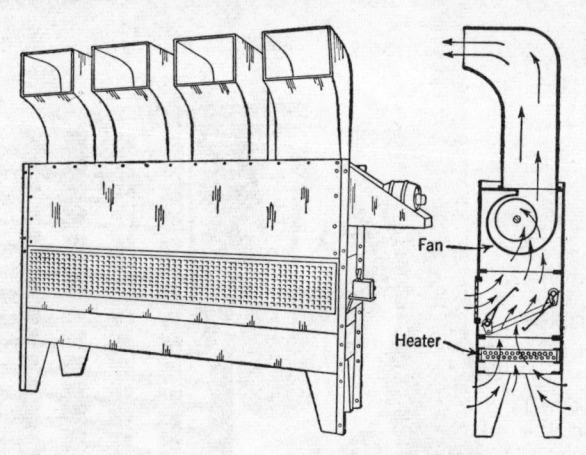

Fig. 16.9. Floor-type unit heater. (Courtesy Westinghouse Electric Corp., Sturtevant Division.)

give the desired direction and velocity of air movement as it enters the space served. The louvers of the horizontal-delivery unit of Fig. 16.5 may be adjusted to cause the air to reach the occupied zone at the desired distance from the unit. A cone-jet diffuser, with or without adjustable blades, may be substituted for the revolving outlets on the vertical-discharge unit shown in Fig. 16.7.

In addition to the direction of air flow into the occupied zone, it is advisable to arrange unit heaters to give rotational circulation of the air in a space. This is accomplished by placing the outlets to blow at an angle with the side walls of the building so that all heaters working together will produce a gentle movement of all of the air of the space. Suggested arrangements for several types of rooms are given in reference 2.

16.10. Typical Performance Data of Unit Heaters. Table 16.3 gives performance data for one manufacturer's standard line of horizontal-delivery unit heaters which are equipped with three-speed

FAN-COIL SYSTEMS AND UNIT HEATERS

TABLE 16.3. Performance Data Relative to Modine Standard Horizontal-Delivery Unit Heaters
Steam pressure, 2 psig. Entering air at 60 F and 29.92 in. of mercury barometric pressure

Model	Motor Power, bhp	Speed, rpm	Air Volume, cfm	Outlet Velocity, fpm	Sound Rating db	Heat Output, btuh	Sq Ft, edr	Condensate, lb per hr	Final Air Temp, deg F
H-75	$\frac{1}{85}$	1620	310	745	40	18,000	75	19	112
		1350	260	630	36	16,200	67		116
		1000	190	460	29	13,500	56		124
H-100	$\frac{1}{45}$	1620	370	750	44	24,000	100	25	119
		1350	310	636	40	21,500	90		123
		1000	230	463	33	18,000	75		131
H-160	$\frac{1}{33}$	1570	620	740	46	38,400	160	40	116
		1350	530	638	43	35,000	146		120
		1000	390	475	35	29,200	122		128
H-210	$\frac{1}{15}$	1580	720	866	50	50,400	210	52	123
		1350	615	748	44	45,900	192		127
		1000	455	561	36	38,300	160		136
H-280	$\frac{1}{20}$	1580	980	646	52	67,200	280	70	123
		1350	838	556	46	61,200	255		126
		1000	620	418	38	51,100	213		135
H-345	$\frac{1}{8}$	1135	1350	900	53	82,800	345	86	115
		850	1010	679	46	69,700	290		122
		570	680	465	38	54,800	228		133
H-455	$\frac{1}{8}$	1135	1920	815	55	109,200	455	113	111
		850	1440	609	48	92,000	383		118
		570	970	424	40	72,000	300		127
H-575	$\frac{1}{6}$	1135	2180	935	56	138,000	575	143	117
		850	1630	700	49	116,000	484		124
		570	1100	482	41	91,000	379		135
H-720	$\frac{1}{6}$	1135	2980	960	59	172,800	720	179	112
		850	2240	727	51	145,000	604		118
		570	1500	497	42	114,000	475		129
H-860	$\frac{1}{4}$	1135	3050	1000	62	206,500	860	214	121
		850	2290	750	55	174,000	725		129
		570	1530	519	45	136,000	566		140
H-1170	$\frac{1}{2}$	1135	4730	1160	63	280,800	1170	290	114
		850	3550	878	56	239,000	970		121
		570	2380	604	46	194,000	810		133

electric motors. Similar data are also available for other types of horizontal-delivery as well as vertical-discharge units employing steam as an air-heating medium. Nearly all builders of unit heaters are prepared to furnish performance data for their products when operated with hot water. They can also provide conversion factors for extending the data of tables such as 16.3 to steam pressures other than 2 psig and to entering air temperatures other than 60 F.

16.11. Air-Temperature Control with Unit Heaters. Manual control of the fan motor of a unit heater may be all that is required either with a single-speed motor or with capacitor motors having two or more operating speeds. The output is also controlled by the action of a room thermostat which starts and stops fan motors. Temperature is also regulated by means of a thermostatically controlled valve in the steam-supply line to vary the amount of steam admitted to the heater coils. With this method of control a thermostat placed in the return line is desirable to prevent the operation of the fan motors when steam is not being supplied to the coils. Another method of control involves the pneumatic operation of a face and bypass damper. This scheme of regulation provides low-cost modulating devices but it is applicable only to units which are equipped with centrifugal fans.

A dual-temperature control for vertical-discharge units, which involves a thermostat at the 5-ft level above the floor and another at the level of the unit heater, may be effective in reducing heating costs. The upper thermostat keeps the fan in operation at all times when the temperature at its level is above its setting although flow of the heating medium to the unit has ceased. The heated air, which tends to collect at the top of the room, is thereby returned to the floor level where it is used in the promotion of human comfort.

REFERENCES

1. *Heat Transfer*, W. H. McAdams, McGraw-Hill Book Co., New York (1954).
2. "79 Pointers for the Selection, Arrangement and Piping of Unit Heaters," C. H. Koper, *Heating and Ventilating*, **46**, No. 7, 68 (July 1949).

BIBLIOGRAPHY

Unit Heaters

1. "How to Select and Install Unit Heaters," *Mill and Factory*, **48**, No. 3, 93 (Mar. 1951).

FAN-COIL SYSTEMS AND UNIT HEATERS

2. "Piping and Controls Help Unit Heater Performance," J. H. Carpenter, *Power*, **94**, No. 11, 96 (Nov. 1950).
3. "What You Should Know About Venting Unit Heaters," H. H. Aronson, *Heating, Piping, Air Conditioning*, **26**, No. 10, 97 (Oct. 1954).
4. "The Care and Maintenance of Steam and Hot Water Unit Heaters," *Bull No. 12*, Industrial Unit Heater Association, Detroit (1952).
5. "Getting the Most Out of Unit Heaters," J. H. Carpenter, *Power*, **94**, No. 10, 82 (Oct. 1950).
6. "Practical Application of Unit Heaters," E. B. Tanner, *J. Inst. Heating and Ventilating Engrs. (London)*, **18**, No. 176, 18 (Mar. 1950).
7. "How to Care for Unit Heaters," *Heating, Piping, Air Conditioning*, **24**, No. 7, 68 (July 1949).
8. "Characteristics of Downward Jets of Heated Air from Vertical Discharge Unit Heater," Linn Helander, S. M. Yen, and L. B. Knee, *ASHVE Trans.*, **60**, 359 (1954).
9. "Spot Heating with Unit Heaters," H. E. Smith, *Heating and Ventilating*, **49**, No. 11, 152 (Nov. 1952).
10. "Piping Details for Unit Heaters," *Heating and Ventilating*, **50**, No. 5, 9 (May 1953).
11. "Standard Code for Testing and Rating Steam Unit Heaters," *Bull. No. 10*, Industrial Unit Heater Association, Detroit (1950).
12. "Standard Code for Testing and Rating Hot-Water Unit Heaters," *Bull. No. 11*, Industrial Unit Heater Association, Detroit, 2nd ed. (1953).
13. *American Standard Approval Requirements for Gas Unit Heaters*, American Gas Association, New York (1940).

PROBLEMS

1. A 30-in. by 30-in. two-row, Series-8A heater coil operates with steam at 227 F and air passing its face with a velocity of 800 fpm under standard conditions. The barometric pressure is 28.5 in. of mercury at 32 F and the air enters the heater at 30 F. Under the conditions specified find (a) the final air temperature to be expected, (b) the weight of air in pounds that the heater will warm each hour, and (c) the required weight of dry steam in pounds per hour.

2. A fan-coil heater has three rows of tubes and is made up of 21-in. by 42-in. Series-8 sections two tiers high. The conditions of heater operation are: barometric pressure, 29.0 in. of mercury at 70 F; steam pressure, 5.8 psig; temperature of air entering heater, 10 F; and face velocity of air, 1000 fpm under standard conditions. Determine (a) the final air temperature, (b) the weight of air handled each hour, and (c) the hourly weight of steam required.

3. Find the actual air friction-pressure losses for the operating conditions of Problem 1.

4. Under the actual conditions of Problem 2 what air friction-pressure losses exist?

5. Trane Type-E blast coils are to be used in a blow-through unit in a factory building. The heat losses from the structure are 1,271,000 Btuh, the temperature of the occupied space 60 F, and the outside-air temperature −10 F. The heated air is to leave the duct outlets at an average temperature of 140 F.

A loss of 3 F may be expected in the duct system between the heater outlet and the points of air discharge to the space to be heated. The heat loss represented by the 3 F temperature drop is not available for space heating. Other necessary data are: face-area air velocity, 1200 fpm at standard conditions; steam temperature, 227 F; and the weight of outside air for ventilation purposes one-third of the total amount of air circulated. (*a*) Find the required face area in square feet, specify the series to be used, and describe the exact arrangement of each bank. Calculate the dimensions of the duct attached to the heater inlet. (*b*) Determine the number of rows of tubes required and describe the complete heater arrangement. (*c*) Find the weight of dry saturated steam required per hour. Assume dry air. Assume that the condensate will leave the heater at the steam temperature.

6. The fan-coil system of Fig. 14.20 is to operate in a building with heat losses of 829,000 Btuh when the inside-air temperature is 65 F and the outside-air temperature is 0 F. Neglect heat lost from the ducts and assume that the air will be discharged to the occupied zone at an average temperature of 125 F. The face velocity is to be 800 fpm. One-fourth of the air handled is to be taken from outdoors and the remainder of the air passing through the heater is to be recirculated air. Steam is available at the heater at 9.61 psig when the barometric pressure is 29.92 in. of mercury at 32 F. Assume that the air is dry. (*a*) Select the nominal width and length of the coils to be used and state the number that will be required in each bank. (*b*) Determine the number of rows of tubes that will be required if Series-8 coils are used and the face area velocity at standard conditions is 800 fpm. (*c*) Find the total static pressure against which the fan must work if the friction losses in the system, other than that in the heater, total 0.688 in. of water. (*d*) Select a fan for this installation from Table 14.1. (*e*) Find the exact speed at which it must operate and the hp of the motor to drive the fan if it is belt-driven and the efficiency of the drive is 98 per cent.

17

Mechanical Refrigeration—
Applications to Cooling
and Heating

17.1. Applications for Mechanical Refrigeration. At least one household refrigerating unit for food preservation is now regarded as a necessity by practically every American family, and the market for refrigerating equipment is now being expanded at a phenomenal rate by the ever-increasing demand for summer air conditioning. In addition to the common applications of mechanical refrigeration in the manufacture of ice, in the preservation of foods, in the creation of skating rinks, and in the air-conditioning field, there are many other uses for its effects. The production of blast furnaces can be substantially increased by dehumidifying the air that is supplied. Many manufactured products, such as certain synthetic rubbers, optical goods, plastics, delicate instruments, and electrical goods, cannot be made in the best quality without air conditioning, which in turn employs mechanical refrigeration. Low temperatures produced by mechanical refrigeration are now used in shrink-fitting of metal parts, and the low-temperature treatment of metals is a new field which is likely to expand. Mechanical refrigeration is necessary in the creation of conditions suitable for testing engines and other types of aircraft equipment that are to fly at high altitudes. Mechanical refrigeration is now also applied in surgery, in the treatment of diseases, and in the preparation and preservation of blood plasma and other medical supplies.

17.2. Types of Mechanical Refrigeration Systems. Several different systems of mechanical refrigeration are used in present-day air-conditioning practice. The more commonly used types, which will be discussed in this chapter, are the compression system, the absorption system, and the steam-jet system. The first system mentioned is extensively used for summer cooling and is now being adapted to winter heating in an arrangement which is popularly called the "heat pump."

17.3. Fundamentals of a Compression Refrigeration System. All fluids behave in a manner similar to water in that they will evaporate if in the liquid state when heat is added or condense if in the vapor state when heat is taken away, at a temperature called the *saturation temperature*. The saturation temperature for any fluid depends upon the pressure in the containing tube or vessel. Any fluid for which the relationship between saturation temperature and vapor pressure is suitable may serve as the working medium in a compression refrigeration system.

Figure 17.1 is a flow diagram for a compression refrigeration system. All the principal parts are labeled, and the path of the refrigerant is indicated. The refrigerant passes through a circuit which is divided into a high-pressure portion and a low-pressure portion. The pressure is maintained at different levels in the two parts of the system by the expansion valve (high side float valve in Fig. 17.1) at one point and by the compressor at another. The function of the expansion valve is to allow the liquid refrigerant under high pressure to pass at a controlled rate into the low-pressure part of the system. Some of the liquid evaporates the instant that it passes the expansion valve, but the greater portion is vaporized in the evaporator at the low pressure which is maintained by the exhausting action of the compressor. In evaporating, each pound of liquid refrigerant absorbs its latent heat of vaporization, the greater part of which is conducted to it through the evaporator tubes from the air, water, or other material which is being cooled. The function of the compressor is to increase the pressure of the refrigerant vapor and discharge it into the condenser. Because of the high pressure in that chamber the vapor condenses at a comparatively high temperature. In passing through the condenser, the refrigerant gives up the heat which it absorbed in the evaporator plus the heat equivalent of the work done upon it by the compressor. This heat is transferred through tubes to the air or water which is used as

MECHANICAL REFRIGERATION

the condenser cooling medium. The diagrammatic layout of Fig. 17.1 shows a reciprocating compressor. However, the fundamental principle of the cycle would not be changed in any way if a rotary or a centrifugal compressor were used instead.

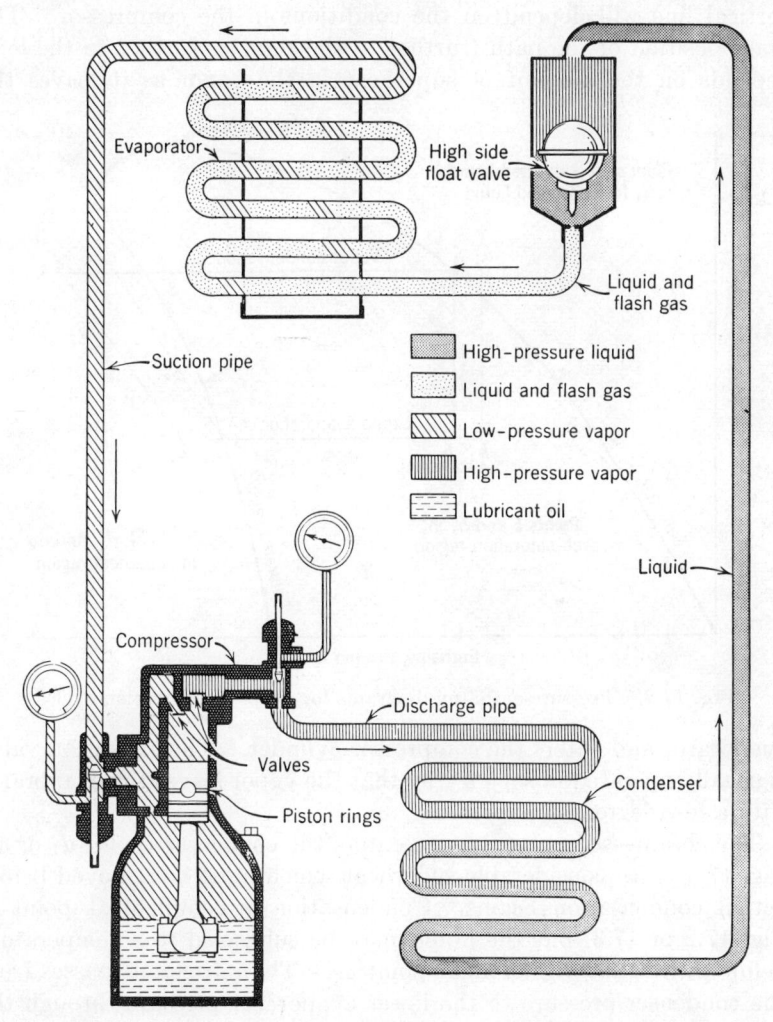

Fig. 17.1. Compression refrigeration system using high side float valve.

The different thermodynamic processes which are involved in the compression refrigeration cycle are shown on a pressure-enthalpy diagram in Fig. 17.2 and on a temperature-entropy diagram in Fig. 17.3. The processes of evaporation and condensation occur both at constant pressure, Fig. 17.2, and at constant temperature, Fig. 17.3. The compression process is shown on the (temperature-entropy) diagram as a vertical line (either cd_1 or c_1d_2), which assumes isentropic conditions in the cylinder. The deviation of the actual compression from a vertical line will depend on the conditions in the compressor. The exact position of the path (further to the right or further to the left) depends on the amount of superheat in the vapor as it leaves the

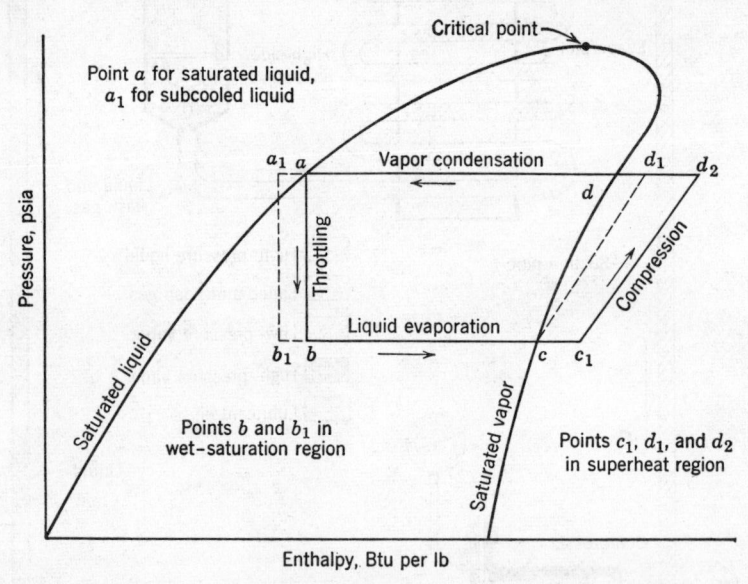

Fig. 17.2. Pressure-enthalpy diagrams for simple refrigeration cycles.

evaporator and enters the compressor cylinder. The expansion valve is usually operated in such a way that the vapor leaves the evaporator with a few degrees of superheat.

The compressed vapor *always* enters the condenser (point d_1 or d_2, Fig. 17.3) with considerable superheat which must be removed before actual condensation begins. Condensation is complete at point a, Fig. 17.2 or 17.3, but the liquid may be subcooled to a temperature below that of saturation as at point a_1. The refrigerant passes from the condenser pressure to the lower evaporator pressure through the expansion device; this process is termed *throttling* in which no work is done and the enthalpy remains constant. In this process some of the liquid becomes vapor (often referred to as *flash gas*) and the heat of vaporization is taken from the remaining liquid, causing a reduction in temperature but no change in enthalpy. Assuming that a subcooled liquid reached the expansion valve, the process is represented

MECHANICAL REFRIGERATION

by the line a_1b_1 and the ratio of the distances eb_1/ec (Fig. 17.3) is the fraction of the mixture entering the evaporator which is flash gas. The fraction of the mixture that is liquid still to be evaporated is given by b_1c/ec.

The compression refrigeration cycle is similar to the steam power cycle in that the working medium changes from liquid to vapor and back to liquid. However, in the steam power cycle evaporation takes

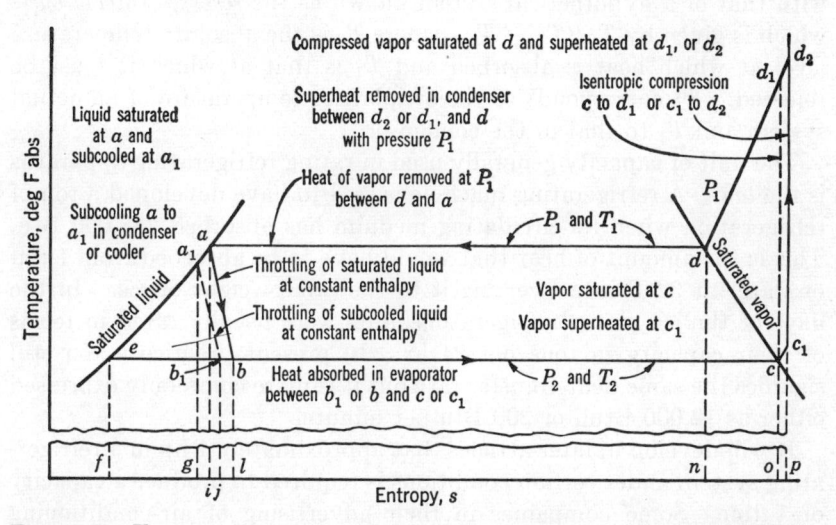

Fig. 17.3. Temperature-entropy diagram for a refrigerant with or without subcooling before entering the evaporator and with saturated or superheated vapor leaving the evaporator.

place at high temperature and pressure, condensation takes place at low temperature and pressure, and power is derived for use elsewhere; whereas in the compression refrigeration cycle evaporation takes place at low temperature and pressure, condensation occurs at high temperature and pressure, and power must be supplied from another source.

17.4. Coefficient of Performance, Ton. Since the purpose of a mechanical refrigeration system is to transfer heat from a low temperature to one that is higher through the expenditure of energy in the form of power, and since the amount of heat transferred is usually greater than the heat equivalent of the power supplied, the performance of such a system cannot logically be expressed as an efficiency, so the term coefficient of performance (COP) has been substituted.

The COP may be defined as the ratio of the heat absorbed by the refrigerant as it passes through the evaporator to the heat equivalent of the energy supplied to the compressor.

$$\text{COP} = \frac{\text{Heat absorbed in evaporator, Btuh}}{\text{Hp supplied to compressor} \times 2545} \quad (17.1)$$

The COP of an actual refrigerating system is sometimes compared with that of a hypothetical system known as the *reversed Carnot cycle* which is given by $T_2/(T_1 - T_2)$, where T_2 is the absolute temperature level at which heat is absorbed and T_1 is that at which it must be rejected. T_2 corresponds to the evaporator temperature of an actual system and T_1 to that in the condenser.

The unit of capacity generally used in rating refrigerating apparatus is the *ton*. A refrigerating machine is said to have developed a ton of refrigeration when its circulating medium has absorbed 288,000 Btu. This is the amount of heat that would have to be absorbed from 1 ton of water at 32 F in converting it to the same weight of ice. In the days of the ice plants refrigerating units were usually rated in terms of their capacity in tons per 24 hr. In present practice a ton still signifies the same heat-transfer rate but it is more commonly expressed either as 12,000 Btuh or 200 Btu per minute.

It will develop in later articles that approximately 1 hp in a refrigerating system under certain conditions is required to produce a capacity of 1 ton. Some companies in their advertising of air-conditioning units use the words *horsepower* and *ton* in such away that it is implied that 1 ton of refrigerating capacity always results when 1 hp of electric motor power is applied. This assumption is not correct because the actual relationship between motor hp and refrigerating capacity in tons depends on the electrical efficiency of the motor, the efficiency of the compressor, and the pressures that are maintained both in the evaporator and in the condenser. Small air-conditioning units should preferably be rated in terms of their cooling capacity in Btuh under specified conditions of operation.

17.5. Commercial Refrigerants for Compression Refrigeration Systems. An ideal refrigerant would be chemically stable, nontoxic, nonexplosive when mixed with air, low in cost, and have thermodynamic properties which result in minimum power requirements per unit of refrigeration effect produced. It is desirable that the temperature-pressure relationship be such that pressures lower than atmospheric will not be required in the evaporator. It is also desirable that the pressure required in the condenser will not be too high for lightweight compressors to handle.

MECHANICAL REFRIGERATION

In the early days of mechanical refrigeration few chemical compounds were available that could be used as refrigerants. Ammonia was used in most of the plants making artificial ice but carbon dioxide was used in marine, theater, and institutional air conditioning because it was nontoxic. Sulfur dioxide was used in the first household refrigerators because the condenser and evaporator pressures required could easily be handled by lightweight reciprocating compressors.

Physical chemists of the present era are able to synthesize almost any type of refrigerant needed for a specific use or for a specific type of equipment. Many new ones have recently appeared in the "Freon" group. Table 17.1, which includes data taken with permission from reference 1, gives several physical and thermal properties of the refrigerants that are used most commonly today.

The tendency of a refrigerant to leak is inversely proportional to the size of the molecules, which in turn is directly proportional to the square root of the *molecular weight*. The molecular weight is given for the refrigerants listed in column 3 of Table 17.1.

If the temperature required in the evaporator of a system is lower than the boiling temperature at 0 psig which is given in column 4, the evaporator pressure will be lower than that of the atmosphere and provision must be made for the removal of air which may leak into the system.

The *freezing temperature* given in column 5 is helpful in selecting a refrigerant for low-temperature operation.

The evaporator pressure at 5 F, column 8, and the condenser pressure at 86 F, column 9, show the comparative pressure differences against which the compressor must operate with different refrigerants when used under the same temperature range. The compression ratio may be easily obtained from these data by first changing the gage pressures given to the corresponding absolute pressures (assuming standard atmospheric pressure). For example, the compression ratio for 5 F and 86 F in the case of ammonia is $(154.5 + 14.7) \div (19.6 + 14.7) = 4.94$.

The *net refrigerating effect* of the liquid, column 10, is the enthalpy of the saturated vapor at 5 F minus the enthalpy of the saturated liquid at 86 F. The data given are based on the assumption that there is no subcooling of the liquid in the condenser and that there is no superheating of the vapor in the evaporator.

The refrigerant circulation per ton in pounds per minute, column 11, Table 17.1, was obtained by dividing 200 by the net refrigerating effect.

The specific volume of the vapor at 5 F, together with the refrigerant circulated per ton and the required capacity in tons, provides data in regard to the volume of gas to be handled by the suction piping.

The compressor displacement per ton in cubic feet per minute, column 13, is the product of the data in column 12 times that in column 11.

The standard cycle theoretical hp per ton of refrigeration produced column 14, is the best measure of the efficiency of the refrigerant.

TABLE 17.1. Comparative Refrigerant Characteristics
Performance based on 5 F evaporation and 86 F condensation

1	2	3	4	5	6	7	8	9
Refrigerant	Chemical Symbol	Molecular Weight	Boiling Temp. at 0 psig, deg F	Freezing Temp. at 0 psig, deg F	Critical Temp., deg F	Critical Pressure, psia	Evaporator Pressure, at 5 F, psig	Condensing Pressure at 86 F, psig
Ammonia	NH_3	17.03	−28.0	−107.9	271.4	1657.0	19.6	154.5
Freon-12	CCl_2F_2	120.9	−21.6	−252.0	232.7	582.0	11.8	93.2
Freon-22	$CHClF_2$	86.48	−41.4	−256.0	204.8	716.0	28.3	159.8
Freon-11	CCl_3F	137.38	74.7	−168.0	388.4	635.0	24.00§	3.6
Freon-21	$CHCl_2F$	102.93	48.0	−211.0	353.3	750.0	19.2§	16.5
Freon-13	$CClF_3$	104.40	−114.5	−296.0	83.9	561.3	177.1	‖
Freon-113	$C_2Cl_3F_3$	187.39	117.6	−31.0	417.4	495.0	27.9§	13.9§
Freon-114	$C_2Cl_2F_4$	170.93	38.4	−137.0	294.3	474.0	16.1§	22.0
Methyl chloride	CH_3Cl	50.48	−10.8	−144.0	289.4	968.7	6.5	80.0
Sulfur dioxide	SO_2	64.06	14.0	−103.9	314.8	1141.5	5.9§	51.8
Carrene 7	¶	99.29	−28.0	−254.0	221.1	631.0	16.4	113.4
Carbon dioxide	CO_2	44.00	−109.3	−69.9	87.8	1071.1	317.5	1031.0
Water at 40 F and 86 F	H_2O	18.02	212.0	32.0	706.1	3226.0	29.7§	28.6§

The COP given in column 15 is based on the theoretical hp given in column 14. For ammonia the COP was computed as $12,000 \div (0.989 \times 2545) = 4.76$.

Whether or not a particular refrigerant is miscible with oil is shown in column 16. When a refrigerant is used that is not miscible with oil, an oil separator must be provided in the discharge piping. Oil reaching a flooded evaporator in a case where the refrigerant is not miscible with it tends to remain there, causing the evaporator to become *oil-logged* and also causing a shortage of oil in the compressor. In a case where the oil is miscible with the refrigerant it will usually return to the compressor from the evaporator.

MECHANICAL REFRIGERATION

The specific heat at constant pressure is given in column 17 and that at constant volume is given in column 18 of Table 17.1. These data are necessary to certain computations involving compressor work and heat transfer.

Anhydrous ammonia, NH_3, is one of the oldest refrigerants and it

TABLE 17.1 (*Continued*)

10	11	12	13	14	15	16	17	18
Net Refrigerating Effect of Liquid, Btu/lb	Refrigerant Circulation per ton, lb/min	Specific Volume of Vapor at 5 F, cu ft/lb	Compressor Displacement per ton, cfm*	Theoretical Hp per ton, hp*	COP*	Miscible with Oil	Specific Heat at Constant Pressure, c_p†	Specific Heat at Constant Volume, c_v†
474.4	0.422	8.15	3.44	0.989	4.76	No	0.520	0.397
51.1	3.92	1.49	5.81	1.002	4.70	Yes	0.144	0.127
69.3	2.89	1.25	3.60	1.011	4.66	‡	0.157	0.133
67.5	2.96	12.27	36.32	0.927	5.09	Yes	0.137	0.121
89.4	2.24	9.13	20.43	0.943	5.05	Yes No	0.136	0.116
53.7	3.73	27.04	100.76	0.960	4.92	Yes	0.163	0.152
43.1	4.64	4.22	19.59	1.015	4.64	‡	0.163	0.150
150.2	1.33	4.47	5.95	0.962	4.90	Yes	0.198	0.159
141.4	1.41	6.42	9.09	0.968	4.87	No	0.154	0.123
61.1	3.27	1.52	4.97	1.022	4.61	Yes		
55.5	3.60	0.27	0.96	1.840	2.56	No	0.205	0.160
1025.3	0.195	2444	476.7	1.125	4.10	No		

* The data given apply only to the evaporator and condenser pressures for each refrigerant given in columns 8 and 9.
† From *Introduction to Heat Transfer*.[2]
‡ Miscible with lubricating oils at usual condensing pressures but not miscible at ordinary evaporator pressures.
§ Inches of mercury vacuum instead of pounds per square inch gage.
‖ Above critical—cannot operate at 86 F condensing temperature.
¶ Azeotropic (constant boiling temperature) mixture of Genetron 100 (CH_3-CHF_2) and Freon-12.

is still used extensively in applications where only trained personnel would be exposed to it in the case of a leak. It has excellent thermodynamic properties for ordinary refrigeration temperature ranges and it is relatively inexpensive. Ammonia is not used in air-conditioning systems because of its toxicity.

Freon-12 (dichlorodifluoromethane) is one of the most popular

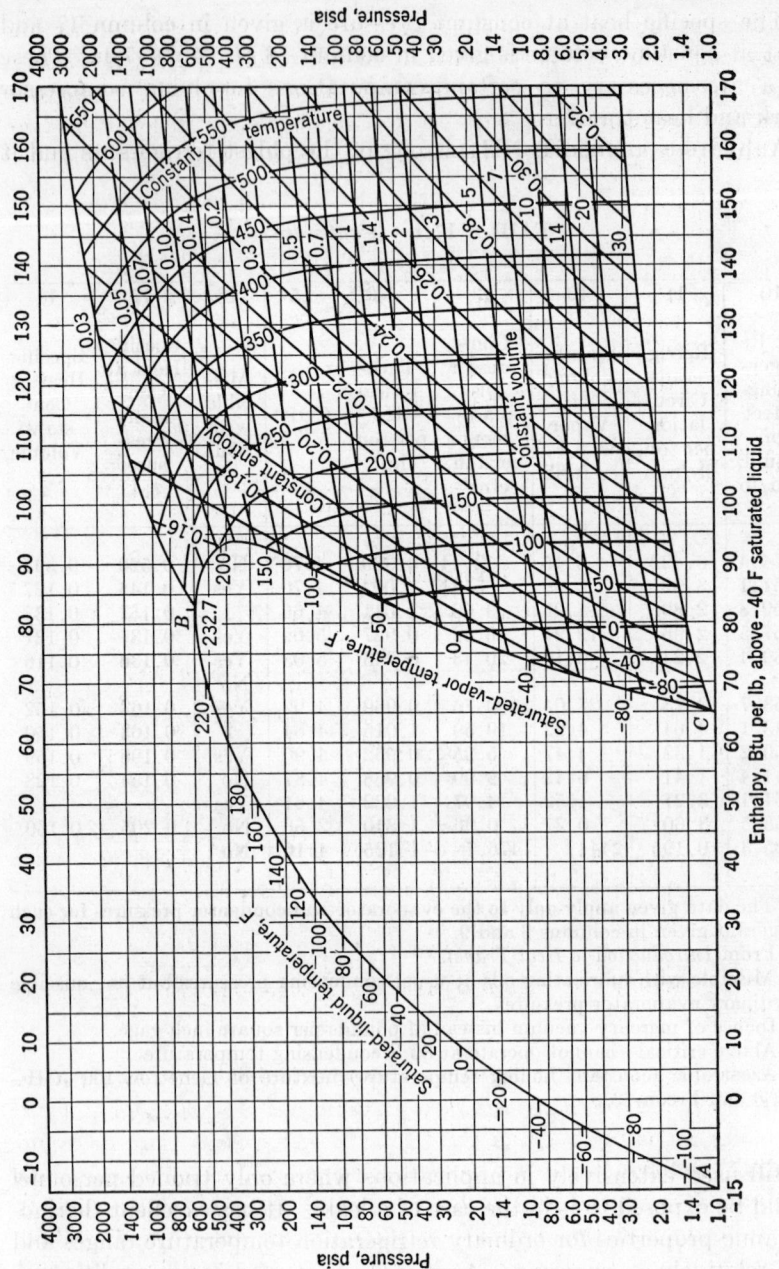

Fig. 17.4. Pressure-enthalpy diagram for dichlorodifluoromethane, Freon-12.

MECHANICAL REFRIGERATION

refrigerants for use in systems equipped with either reciprocating or rotary compressors. It is nontoxic, nonirritating, and nonexplosive. When it is used in connection with air conditioning, the evaporator operates at pressures above that of the atmosphere so that air leakage into the system is not a problem and the condenser pressure required

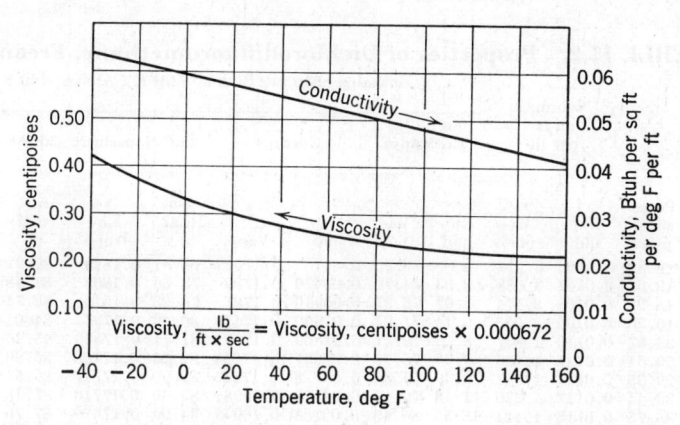

Fig. 17.5. Viscosity and thermal conductivity of liquid Freon-12.

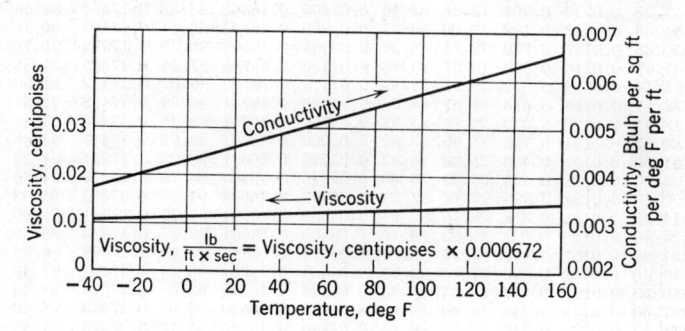

Fig. 17.6. Viscosity and thermal conductivity of Freon-12 vapor (pressure one atmosphere).

is not too high to be handled by a comparatively simple compressor. Figure 17.4 is a pressure-enthalpy diagram for this refrigerant from which the enthalpy in either the liquid or vapor state may be found for any condition likely to be encountered. Thermodynamic data in tabular form pertaining to this refrigerant are given in Table 17.2. The viscosity and the thermal conductivity are given for the liquid in Fig. 17.5 and for the vapor in Fig. 17.6. (A recently completed investigation of the properties of this refrigerant has resulted in the accumulation of new data which differ slightly from those given in

Table 17.2 and Figs. 17.4, 17.5, and 17.6. For precision work with this refrigerant it is recommended that the latest data be obtained from the manufacturer.) Space limitations do not permit the inclusion of similar data pertaining to the other refrigerants, but a brief discussion of their characteristics will be given in the following paragraphs.

TABLE 17.2. Properties of Dichlorodifluoromethane, Freon-12

Enthalpy, Btu per lb, and Entropy Above −40 F

Sat. Temp, deg F	Pressure, psia	Volume, cu ft per lb Liquid	Volume, cu ft per lb Vapor	Enthalpy Liquid	Enthalpy Vapor	Entropy Liquid	Entropy Vapor	25 F Superheat Enthalpy	25 F Superheat Entropy	50 F Superheat Enthalpy	50 F Superheat Entropy
−40	9.32	0.0106	3.911	0	73.50	0	0.1752	76.81	0.1829	80.21	0.1903
−30	12.02	0.0107	3.088	2.03	74.70	0.00470	0.1739	78.06	0.1808	81.48	0.1882
−20	15.28	0.0108	2.474	4.07	75.87	0.00940	0.1727	79.28	0.1815	82.74	0.1888
−10	19.20	0.0109	2.003	6.14	77.05	0.01400	0.1717	80.49	0.1797	84.01	0.1868
0	23.87	0.0110	1.637	8.25	78.21	0.01869	0.17091	81.71	0.17829	85.26	0.18547
5	26.51	0.0111	1.485	9.32	78.79	0.02097	0.17052	82.29	0.17786	85.89	0.18502
10	29.35	0.0112	1.351	10.39	79.36	0.02328	0.17015	82.90	0.17747	86.51	0.18460
15	32.44	0.0112	1.230	11.48	79.94	0.02556	0.16981	83.49	0.17710	87.13	0.18420
20	35.75	0.0113	1.121	12.55	80.49	0.02783	0.16949	84.09	0.17679	87.76	0.18382
25	39.33	0.0114	1.025	13.66	81.06	0.03008	0.16920	84.67	0.17643	88.37	0.18349
30	43.16	0.0115	0.939	14.76	81.61	0.03233	0.16887	85.25	0.17612	88.97	0.18315
35	47.28	0.0116	0.863	15.88	82.16	0.03458	0.16860	85.83	0.17582	89.56	0.18285
40	51.68	0.0116	0.792	17.00	82.71	0.03680	0.16833	86.41	0.17554	90.16	0.18256
45	56.38	0.0117	0.730	18.14	83.26	0.03904	0.16808	86.96	0.17528	90.76	0.18227
50	61.39	0.0118	0.673	19.27	83.78	0.04126	0.16785	87.54	0.17505	91.38	0.18203
55	66.74	0.0119	0.622	20.41	84.31	0.04348	0.16763	88.09	0.17482	91.93	0.18181
60	72.41	0.0119	0.575	21.57	84.82	0.04568	0.16741	88.64	0.17458	92.51	0.18155
65	78.44	0.0120	0.532	22.72	85.32	0.04789	0.16721	89.18	0.17436	93.11	0.18132
70	84.82	0.0121	0.493	23.90	85.82	0.05009	0.16701	89.72	0.17417	93.66	0.18114
75	91.60	0.0122	0.458	25.08	86.32	0.05229	0.16681	90.25	0.17397	94.23	0.18092
80	98.76	0.0123	0.425	26.28	86.80	0.05446	0.16662	90.78	0.17379	94.80	0.18075
85	106.40	0.0124	0.395	27.48	87.28	0.05665	0.16644	91.27	0.17361	95.33	0.18056
90	114.30	0.0125	0.368	28.70	87.74	0.05882	0.16624	91.77	0.17344	95.86	0.18040
95	122.80	0.0126	0.343	29.93	88.19	0.06100	0.16604	92.27	0.17323	96.39	0.18020
100	131.60	0.0127	0.319	31.16	88.62	0.06316	0.16584	92.75	0.17308	96.92	0.18004
105	140.90	0.0128	0.298	32.40	89.03	0.06534	0.16564	93.24	0.17291	97.46	0.17991
110	150.70	0.0129	0.277	33.65	89.43	0.06749	0.16542	93.66	0.17274	97.93	0.17976
115	161.00	0.0130	0.258	34.90	89.80	0.06965	0.16520	94.07	0.17253	98.40	0.17955
120	171.80	0.0132	0.240	36.16	90.15	0.07180	0.16495	94.47	0.17233	98.84	0.17939
125	183.10	0.0133	0.224	37.41	90.46	0.07395	0.16467	94.86	0.17210	99.26	0.17918
130	194.90	0.0134	0.208	38.69	90.76	0.07607	0.16438	93.25	0.17186	99.70	0.17897
135	207.40	0.0136	0.194	39.95	91.01	0.07816	0.16404	95.64	0.17161	100.13	0.17877
140	220.20	0.0138	0.180	41.24	91.24	0.08024	0.16363	96.03	0.17134	100.56	0.17856

Freon-22 (monochlorodifluoromethane) can be used in systems where the evaporator boiling temperature is as low as −41.4 F without the necessity of maintaining a vacuum in the evaporator.

Freon-11 (trichloromonofluoromethane), sometimes called Carrene 2, requires a vacuum in the evaporator for boiling temperatures below +75 F but the condenser pressure at 86 F condensing temperature is only 3.58 psig. It is a popular refrigerant for use in systems where centrifugal compressors are employed.

Freon-21 (dichloromonofluoromethane) has a saturation temperature of +48 F corresponding to standard atmospheric pressure and consequently does not require a vacuum in the evaporator that is as high as that required with Freon-11 for the same type of service. It is stated in reference 3 that a smaller centrifugal compressor can be used when the refrigerant is Freon-21 instead of Freon-11.

Freon-13 (monochlorotrifluoromethane) was developed especially for use in systems where extremely low boiling temperatures are necessary. A boiling temperature of -110 F can be produced without creating a vacuum in the evaporator. It is a nonexplosive replacement for ethane which has been the principal refrigerant for this type of service.

Freon-113 (trichlorotrifluoroethane) has the highest molecular weight of any of the refrigerants listed in Table 17.1 and it is suitable for use with centrifugal compressors.

Freon-114 (dichlorotetrafluoroethane) requires evaporator and condenser pressures which make it suitable for use in systems employing either centrifugal or rotary compressors. Its vapor has a higher dielectric strength than that of any other refrigerant except Freon-11. For this reason it is well suited for use in hermetically sealed units in which the motor windings are cooled by the low-temperature refrigerant vapor from the evaporator.

Methyl chloride is a colorless, nonirritating liquid which was introduced for refrigerating purposes around 1920. It can be handled satisfactorily by either reciprocating or rotary compressors. It is very flammable and for that reason is being replaced by refrigerants in the Freon group.

Sulfur dioxide is nonexplosive and works well in household refrigerating units equipped with either reciprocating or rotary compressors. It was an almost universal refrigerant for this type of service during the 1920's and 1930's but has gradually been replaced by Freon-12 since 1940.

Carrene 7[4] is an *azeotropic mixture* of Genetron-100 and Freon-12. (When an azeotropic mixture of two fluids is boiled, the composition of the vapor formed is the same as that of the liquid remaining, with the result that the boiling temperature does not change as the process continues to completion.) The pressure-temperature relationships are between those of Freon-12 and Freon-22. Changing the refrigerant in a system from Freon-12 to Carrene 7 increases both the refrigerating capacity and the hp requirements approximately 20 per cent.

Carbon dioxide was at one time used as a refrigerant for certain types of service principally because ammonia was the only other alternative. With it, extremely high pressures are required in the condenser and few if any new systems are being designed to use it.

Water has a higher latent heat of vaporization than any of the other refrigerants and it has many other properties which are desirable. However, it requires almost a perfect vacuum in the evaporator even for a boiling temperature of $+40$ F and it will not mix with oil. Because of these serious disadvantages it is not used in the ordinary compression system.

17.6. Types of Compressors for Vapor Compression Systems. The three types of compressors used in vapor compression systems are reciprocating, rotary, and centrifugal. In the reciprocating compressor, the pressure of the refrigerant vapor is increased by the co-ordinated action of pistons and valves. In both rotary and centrifugal compressors the refrigerant vapor is handled by rotating members but the principle of operation is different. The rotary compressor operates on a positive displacement principle similar to that of the reciprocating type; whereas the centrifugal compressor increases the pressure of the refrigerant vapor by first creating a very high velocity, then converting a part of the kinetic energy of the moving particles into the needed increase in static pressure.

Fig. 17.7. Self-contained motor-driven refrigeration unit with air-cooled compressor and condenser. (Courtesy Curtis Manufacturing Co.)

17.7. Reciprocating Compressors. This type of compressor is always used with refrigerants which require a condenser pressure greater than 100 psig and it is usually the preferred type for pressures exceeding 20 psig. It can be made in a range of sizes to handle capacities varying from a fraction of a ton, Fig. 17.7, to 100 tons or more. It is made in both single-cylinder and multicylinder styles. The arrangement of the cylinders in a multicylinder compressor may be in-line, or in V or W forms. In all arrangements in current practice the cylinders are single acting. During the suction stroke

MECHANICAL REFRIGERATION

of the piston 1-2-3 in Fig. 17.8, low-pressure refrigerant gas is drawn in through the suction valve which may be located either in the piston or in the cylinder head. If it were possible to build a compressor cylinder without clearance, the volume of gas taken in would be equal to the piston displacement, $V_3 - V_1$. However, the actual cylinder always includes some volume that is not swept through by the piston

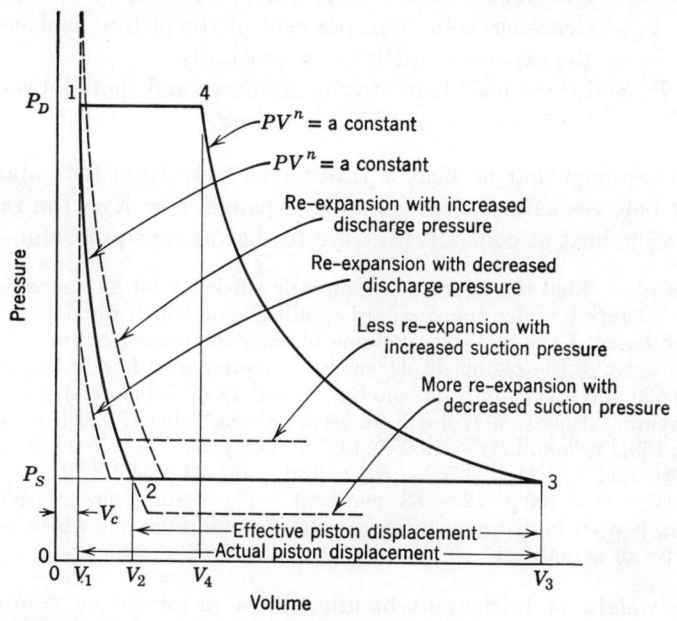

Fig. 17.8. Pressure-volume relationship in a cylinder of a reciprocating compressor.

and the high-pressure gas remaining in this space, represented by V_c, Fig. 17.8, will expand and the pressure will drop gradually as the piston starts the suction stroke. No suction gas can enter the cylinder until the piston has moved enough to drop the pressure to that in the suction line, the distance $V_2 - V_1$ (V_1 and V_c are equal). The volumetric efficiency is the volume pumped measured at suction conditions, $V_3 - V_2 \div$ the piston displacement, $V_3 - V_1$. It is obvious from Fig. 17.8 that for a fixed clearance volume and a fixed piston displacement the volume V_2 increases and the volumetric efficiency decreases as the discharge pressure is increased with fixed suction pressure or as the suction pressure is decreased with fixed discharge pressure. By expressing the clearance volume as a percentage of the piston displacement and assuming that the re-expansion

1-2 occurs along a curve represented by the expression $PV^n = $ a constant, the following equation for the volumetric efficiency may be derived

$$E_{vc} = 100 - V_c \left[\left(\frac{P_D}{P_S} \right)^{1/n} - 1 \right] \qquad (17.2)$$

where $E_{vc} = $ clearance volumetric efficiency, per cent.
$V_c = $ clearance volume in per cent of the piston displacement.
$n = $ the exponent in $PV^n = $ a constant.
P_D and P_S as used here are the discharge and suction pressures, pounds per square inch absolute.

If it is assumed that no heat is transferred to or from the vapor and that it behaves as a perfect gas as it expands, $n = K = $ the ratio of the specific heat at constant pressure to that at constant volume.

Example. Find the clearance volumetric efficiency for an ammonia compressor operating under the standard conditions on which the data of Table 17.1 are based if the clearance volume of each compressor cylinder is equal to 5 per cent of the piston displacement. Assume $n = K = 0.520/0.397 = 1.31$ (0.520 and 0.397 are from columns 17 and 18 of Table 17.1).

Solution. Equation 17.2 will be used. From Table 17.1, $P_D = 154.5 + 14.7 = 169.2$ psia and $P_S = 19.6 + 14.7 = 34.3$ psia. $V_c = 5.0$ per cent and $n = 1.31$ and $1/1.31 = 0.763$. $E_{vc} = 100 - 5[(169.2/34.3)^{0.763} - 1] = 100 - 5[3.41 - 1] = 100 - 12 = 88$ per cent. The term *clearance volumetric efficiency* is used to distinguish the calculated value from one which is determined by an actual test.

The weight of refrigerant handled by a reciprocating compressor may be computed from the following:

$$m_r = \frac{V_P E_{vc}}{V_{sp} \times 100} \qquad (17.3)$$

where $m_r = $ weight of refrigerant circulated, pounds per minute.
$V_P = $ total piston displacement volume, cubic feet per minute.
$E_{vc} = $ clearance volumetric efficiency (or actual volumetric efficiency if known from test data), per cent.
$V_{sp} = $ specific volume of the refrigerant vapor at suction conditions, cubic feet per pound.

Example. Compute the weight of ammonia that would be circulated by the compressor of the previous example under the same conditions if the piston displacement volume equals 40 cfm.

Solution. Equation 17.3 will be used. $V_P = 40$ cfm. V_{sp} from Table 17.1 $= 8.15$, E_{vc} from the previous example is 88 per cent. $M_r = (40/8.15) \times (88/100) = 4.32$ lb per min.

MECHANICAL REFRIGERATION

17.8. Hermetic Compressors.[5] The reciprocating compressor of Fig. 17.7 is classed as an open type in which the crankshaft passes through a sealed bearing and the driving motor is enclosed in a housing separate from that of the compressor. The small unit shown in Fig. 17.9 is arranged with both the compressor and the motor enclosed in a hermetically sealed enclosure. The principal advantage in this

Fig. 17.9. Small hermetic compressor and air-cooled condenser. (Courtesy Tecumseh Products Co.)

arrangement is that there are no bearings which must be sealed either against escape of refrigerant to the atmosphere or leakage of air into the refrigerant circuit. In the hermetic arrangement the motor windings are cooled by the refrigerant vapor on its way to the suction valves of the compressor. Practically all small motor-compressor units for home refrigerators or window air conditioners are of the hermetic type. In the case of larger hermetically sealed units, the cylinder heads are usually removable so that the valves and pistons can be serviced. This type of hermetic unit is called *semihermetic*.

17.9. Rotary Compressors.[6] Rotary compressors are common in household refrigerators and in other applications of refrigeration where the capacity needed is small. Two different arrangements of parts are known as roller type and vane type. In the roller type shown in Fig. 17.10 the center line of the drive shaft coincides with that of the cylinder. However, the rotor is mounted with an eccentricity slightly less than the difference between the radius of the cylinder and that of the rotor. The result of this arrangement is that one point on the rotor (the lowest point in Fig. 17.10) makes continuous contact with the cylinder wall (except for the slight clearance) as it is rotated by the motor shaft. The refrigerant vapor which enters the open space to the right of the spring-loaded divider is pushed out through the discharge valve at the left. *The displacement per revolution is the cross-sectional area of the cylinder minus that of the rotor times the length of the cylinder.*

In the vane type of rotary compressor the rotor is arranged coaxially with the motor drive shaft and revolves about its own center. However, in this case the shaft is not centered in the cylinder and sliding contact is maintained by vanes which move in and out of slots in the rotor.

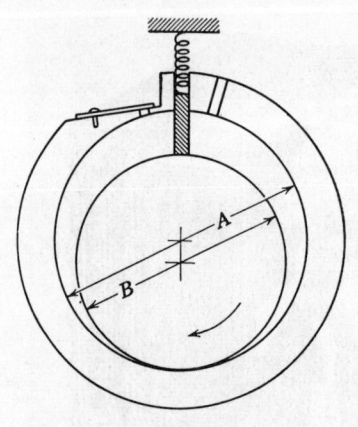

Fig. 17.10. Roller-type rotary compressor.

No suction valves are needed in either the roller type or vane type of rotary compressor, and there is very little pulsation in the suction line. Also, there is relatively little vibration. The principal disadvantage is the precision machining that is necessary in making the cylinders and revolving parts. Rotary compressors and their driving motors as now manufactured are usually made as hermetically sealed units.

17.10. Centrifugal Compressors. This type of compressor, the principle of which was mentioned in Art. 17.6, is well adapted to the handling of refrigerants such as Carrene 2, (Freon-11) and Freon-113 where the volume passing through the compressor is large and the required increase in pressure is small. One advantage of the centrifugal compressor is that its high rotative speed makes it suitable for direct connection to electric motors or steam turbines, thereby simplifying the drive and effecting a saving in floor space. Another advantage is that there are no piston rings or valves to wear out so that maintenance of this type of compressor is, in general, less costly. Some compressors of this type are equipped with vanes at the inlet

MECHANICAL REFRIGERATION

to each stage which provide capacity control in the range from 100 per cent to 10 per cent.

The efficiency of compression is usually lower with the centrifugal compressor than with the reciprocating type, but this disadvantage may be more than offset by the advantages which have been previously discussed.

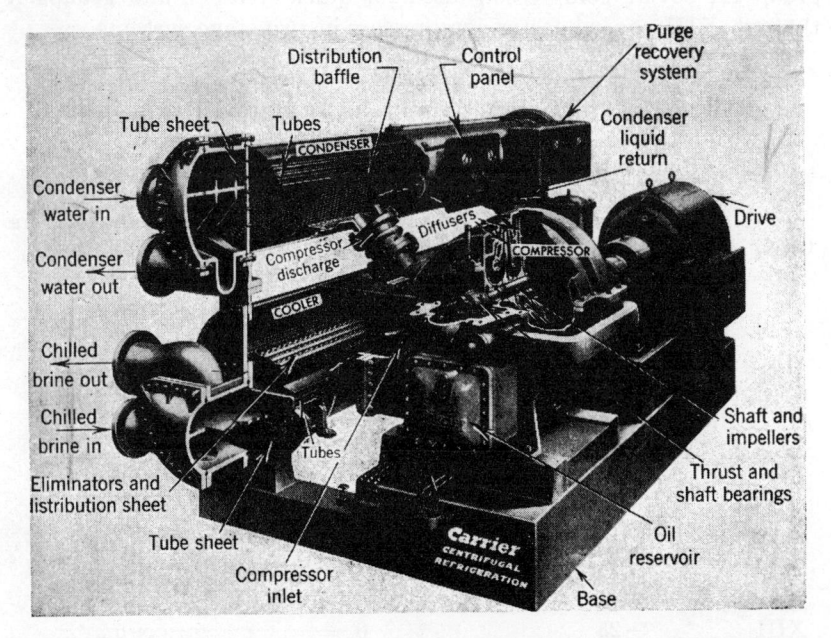

Fig. 17.11. Centrifugal-compressor refrigeration unit. (Courtesy Carrier Corp.)

Figure 17.11 shows the arrangement of a centrifugal compressor for use with Freon-11, (Carrene 2). The operation of the compressor under the standard conditions of Table 17.1 produces a vacuum of 24 in. of mercury in the evaporator (cooler) while maintaining a pressure of 3.58 psig in the condenser.

17.11. Conditions for Rating Compressors. Prior to 1949 the standard conditions for testing refrigerant compressors were based on the saturation pressure corresponding to 5 F at the inlet and to 86 F at the discharge. In December 1949 the American Society of Refrigerating Engineers set up 13 different conditions for testing compressors, each one identified by a group number as in Table 17.3.[7] The 13 different groups cover the wide range of uses now served by the refrigeration industry. The saturation temperature given for each group in the second column fixes the suction pressure that is to be maintained, and the saturation temperature in the fourth column fixes the discharge pressure. The third column stipulates the vapor temperature as it enters the compressor. A suitable means for controlling this temperature must be provided in the test arrangement. Groups I through VIII cover the general ranges of refrigeration, group IX covers compressors used for quick freezers, and groups X through XIII cover booster compressors in two-stage systems.

TABLE 17.3. Standard Conditions for Rating Refrigerant Compressors

Group No.	Temperature of Saturated Vapor at Pressure Entering Compressor, deg F	Controlled Temperature of Refrigerant Entering Compressor, deg F	Saturated Temperature of Refrigerant at Pressure Leaving Compressor, deg F
I	−10	65	100
II	5	65	100
III	20	65	105
IV	40	65	110
V	−10	15	95
VI	5	30	95
VII	20	40	100
VIII	40	55	100
IX	−40	−15	95
X	−40	−15	40
XI	−40	−15	0
XII	−25	0	40
XIII	−25	0	5

A manufacturer must choose the set of conditions under which any specific compressor model is to be tested and published ratings must indicate by group number the test conditions on which they are based. In many of the groups the temperature of the suction gas entering the compressor is considerably higher than the boiling temperature in the evaporator. Tonnage ratings based on these standards assume that the superheating of the vapor is useful refrigeration. In actual installations a part of this superheating may result from heat absorbed from the ambient air.

17.12. Horsepower Requirements of Refrigerant Compressors. The hp required by the compressor may be easily computed from thermodynamic data if the suction and discharge pressures are known and it is assumed that no heat is transferred from the vapor

MECHANICAL REFRIGERATION

as it is compressed (isentropic compression). If the temperature and pressure of the vapor at the compressor inlet are known, the enthalpy can be found from a table, such as Table 17.2 which is for Freon-12, or from a diagram such as Fig. 17.4 which is applicable to the same refrigerant. Assuming that the compression will be under isentropic conditions from the suction to the discharge pressure, the enthalpy of the discharge may also be read from a diagram such as Fig. 17.4 and the difference is the work done on each pound of refrigerant in the compressor, expressed in Btu. The following expression may be used:

$$\text{Theoretical hp} = \frac{m_r(h_d - h_s) \times 778}{33{,}000} \tag{17.4}$$

where m_r = weight of refrigerant compressed, pounds per minute.
h_d = enthalpy of the compressed vapor, assuming isentropic compression, Btu per pound.
h_s as used here is the enthalpy of the vapor at compressor inlet conditions, Btu per pound.

Example. Find the theoretical hp required to compress 10 lb per min of Freon-12 vapor under the conditions specified in group VII, Table 17.3.

Solution. Equation 17.4 will be used. The inlet pressure corresponding to 20 F, from Table 17.2, is 35.75 psia and h_s at that pressure and at 40 F (20 deg of superheat) by interpolation from the same table is 83.36 Btu per lb. The entropy at the same condition may also be found by interpolation in Table 17.2 as 0.175. The discharge pressure corresponding to a saturation temperature of 100 F from Table 17.2 is 131.6 psia. The enthalpy h_d after isentropic compression (entropy 0.175, pressure 131.6 psia) as read from Fig. 17.4 is 93.0 Btu per lb.

$$\text{Theoretical hp} = \frac{10(93.0 - 83.36) \times 778}{33{,}000} = 2.27 \text{ hp}$$

The following equation which has been derived from an analysis of the compression process, Fig. 17.8, may be used to compute the bhp required when the piston displacement, the suction conditions, the discharge pressure, and the mechanical efficiency are known:

$$\text{bhp} = \frac{n}{n-1} \frac{144 V_P P_S E_{vc}}{33{,}000 E_m} (R^{(n-1)/n} - 1) \tag{17.5}$$

Symbols are as in Equations 17.2 and 17.3. R = compression ratio P_D/P_S.

Experience has shown that the hp required by a refrigeration compressor may be greatest at some intermediate condition between that of start-up and that of equilibrium with continuous operation. Figure

17.12 from Thomsen[8] provides a means of determining the compression ratio which will result in the maximum bhp requirements.

Example. Using equation 17.5 and Fig. 17.12, determine the maximum bhp demand that could occur when a compressor having a clearance volume equal to 5 per cent of the piston displacement and a displacement equal to 100 cfm is to handle ammonia against a constant condenser pressure equal to 170 psia. Assume isentropic compression and a mechanical efficiency of 90 per cent.

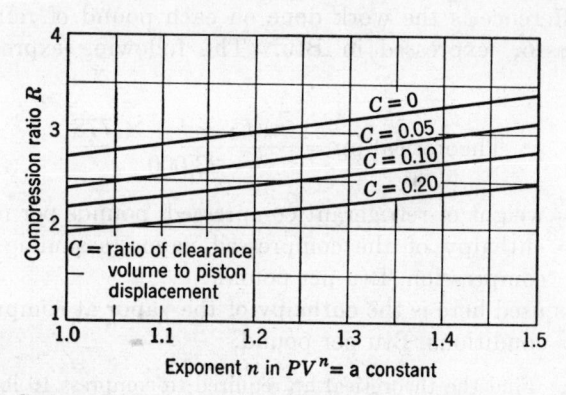

Fig. 17.12. Compression ratio as function of exponent n for maximum bhp of reciprocating compressors having various clearances but operating at constant discharge pressure.

Solution. For this assumed condition $n = k = c_p/c_v = 0.520/0.397 = 1.31$ (c_p and c_v from Table 17.1). From Fig. 17.12 it is determined that the maximum bhp will occur when the suction pressure is such that the compression ratio is 2.8. Under these conditions the suction pressure will be $170 \div 2.8 = 60.7$ psia. Equation 17.2 will be used to find the volumetric efficiency $E_v = 100 - 5[(170/60.7)^{1/1.31} - 1] = 100 - 5 \times 1.21 = 93.95$ per cent. From equation 17.5

$$\text{bhp} = \frac{1.31}{1.31 - 1} \frac{144 \times 100 \times 60.7 \times 93.95}{33,000 \times 90} (2.8^{(1.31-1)/1.31} - 1) =$$

$$4.23 \times 27.6 \times 0.275 = 32.1 \text{ bhp}$$

17.13. Condensers for Compression Refrigeration Systems. It is through the condenser that a compression refrigeration system disposes of the heat which is absorbed in the evaporator together with that which is added by the compressor in the form of work. Condensers may be divided into two groups, those cooled by air and those cooled with water. Air-cooled condensers are used in all household refrigerators and in many larger units. Water-cooled condensers usually provide a lower condensing pressure than the air-cooled type under design conditions and are generally preferred in systems larger than 5 tons except in cases where the problem of obtaining an adequate amount of cooling water is a major consideration.

17.14. Air-Cooled Condensers. Formerly, air-cooled condensers were limited to use with small systems under 2-ton capacity. However, with the present severe water restrictions in some areas the air-cooled condenser applied in multiple units has found application in

Fig. 17.13. Battery of air-cooled condensers. (Courtesy Bush Manufacturing Co.)

systems as large as 100 tons. A small air-cooled condenser is incorporated in the unit of Fig. 17.7. A larger type which incorporates a separate, motor-driven fan is shown in Fig. 17.13.

The compressor in a refrigeration system using an air-cooled condenser usually operates against a higher discharge pressure than in a system which uses a type that is water-cooled even when the water must be cooled by contact with air and recirculated. In an air-cooled condenser the dry-bulb temperature of the air controls the condensing temperature whereas the wet-bulb temperature is the prime factor in determining the available temperature of the cooling water. However, many owners of refrigeration systems prefer the air-cooled condenser in spite of somewhat greater operating costs because it requires no water and there are no scaling, corrosion, or freeze-up problems.

17.15. Water-Cooled Condensers. Water-cooled condensers are made in the following types: horizontal shell and tube, vertical shell and tube, shell and coil, double pipe, and the evaporative condenser. The condenser of Fig. 17.11 is a shell and tube type in which the cooling

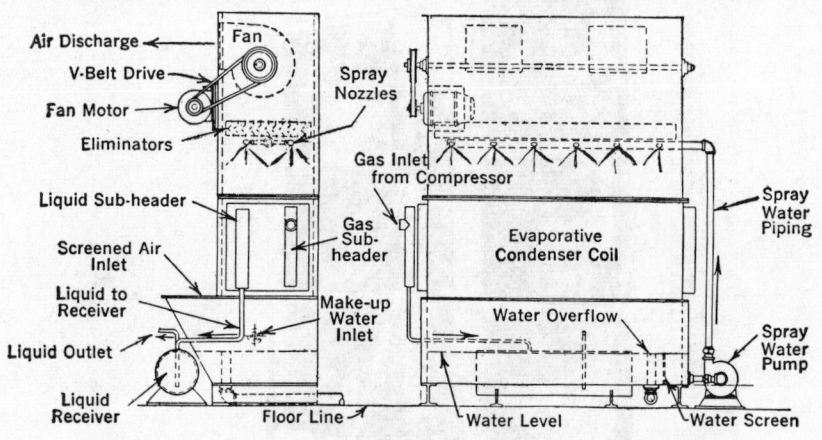

Fig. 17.14.   Evaporative condenser.   (Courtesy The Trane Co.)

water passes through the tubes and the refrigerant vapor is confined within the shell. The refrigerant condenses on the outside of the tubes and the liquid formed collects in the bottom of the shell. The condensing action is the same in the vertical shell and tube and in the shell and coil types. Double tube condensers have inner tubes through which water is circulated and outer tubes surrounding the inner tubes, the refrigerant being confined in the spaces between them.

Evaporative condensers, Fig. 17.14, are in effect a condenser and a cooling tower combined. They may be installed either indoors or outdoors. The action of an evaporative condenser is secured by spraying water upon the tubes of a refrigerant cooling coil located within the unit housing. The water trickles over the tubes in a direction opposite to the flow of air which is produced by a fan. Cooling of the water occurs as some of it evaporates into the air stream. The water eventually falls to the bottom of a reservoir from which it is taken by a pump and again discharged through the spray nozzles. The refrigerant vapor is distributed to the tubes by means of a header

MECHANICAL REFRIGERATION

and condensation occurs on the inner surfaces. The liquid refrigerant formed flows along the bottoms of the tubes into another header from which it is conducted to a receiver. When an interior installation is made a duct may be necessary to convey the cooling air from outdoors to the unit, and an exhaust duct is always required to carry away the heated and humidified air. Two heat-transfer processes (one from the refrigerant to the water and the other from the water to the air) occur simultaneously and an analysis of them is very difficult. Most condensers of this type are selected on the basis of manufacturers' test results.

17.16. Prediction of Condenser Capacity Needed. The amount of heat which must be transferred in the condenser is always greater than that absorbed by the refrigerant in the evaporator because of the heat added in the compression process. The actual amount per pound of refrigerant circulated will depend on the saturation temperature in the evaporator, that in the condenser, the number of degrees of superheat in the vapor leaving the evaporator, and the value of n in the compression process. However, the heat added by compression may be approximated by assuming isentropic compression. The following example will illustrate a method of estimating the heat-transfer capacity that is needed in a condenser.

Example. Find the amount of heat in Btuh which must be rejected in the condenser of a 20-ton system using Freon-12 if the following conditions are assumed: Saturation temperature in the evaporator -20 F, vapor leaves evaporator with 10 deg of superheat, isentropic compression, saturation temperature in the condenser 110 F, no subcooling of the liquid.

Solution. Referring to Fig. 17.3, the enthalpy per pound at c_1 (saturation temperature -20, 10 deg of superheat) by interpolation from Table 17.2 equals 77.23 Btu per lb and the entropy at the same point and from the same source equals 0.17603. The saturation pressure in the condenser (saturation temperature 110 F) will be 150.70 psia. The enthalpy at d_2, Fig. 17.3 (assuming isentropic compression), from Fig. 17.4 (entropy 0.17603, pressure 150.70) equals 95 Btu per lb. The enthalpy at point a, Fig. 17.3, assuming no subcooling of the liquid, from Table 17.2 equals 33.65 Btu per lb. The heat rejected per pound of refrigerant circulated is $95 - 33.65 = 61.35$ Btu per lb. The refrigeration effect per pound of refrigerant circulated (bc_1, Fig. 17.3) is $77.23 - 33.65 = 43.58$ Btu per lb. The ratio of heat rejected to refrigeration effect is $61.35/43.58 = 1.4$ and the heat-transfer rate required in the condenser is $20 \times 12,000 \times 1.4 = 336,000$ Btuh.

17.17. Selection of Condensers. Selection of condensers is usually made with the aid of manufacturers' rating data. Figure 17.15 gives data for a water-cooled condenser of the general form shown in Fig. 17.11 from which it is possible to determine the over-all coefficient of heat transfer for any given water flow rate. The tubes used in the condensers to which these data apply are an "all in one" finned copper design. The fins are formed by a special machine which rearranges the metal on the exterior of the tubes. The U values given in Fig. 17.15 are based on the effective surface of the tubes, which is the amount of plain exterior surface area that would transfer the same amount of heat as the finned tubes. The effective area per lineal foot

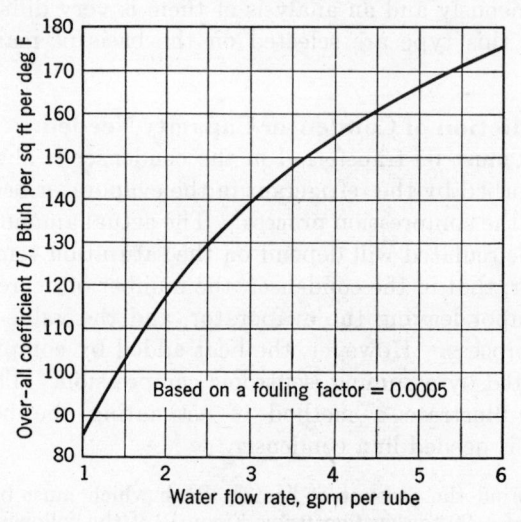

Fig. 17.15. Over-all coefficient U for one type of Freon-12 condenser. (Courtesy Acme Industries, Inc.)

is more than double that of plain tubing having the same external diameter. The effective areas provided by several different lengths in each of four different shell diameters are given in Table 17.4.

The following information is needed before rating data of the type provided in Fig. 17.15 and Table 17.4 can be used in selecting a refrigerant condenser: (1) the load in tons which the system will be expected to carry; (2) the saturation temperature which must be carried in the evaporator; (3) the saturation temperature which must be carried in the condenser; (4) the temperature of the cooling water into the condenser; (5) the water temperature out of the condenser or the available water volume in gallons per minute. The procedure will be illustrated by an example.

Example. From Table 17.4 select a Freon-12 condenser for the system of the example of Art. 17.16. The cooling water is available at a temperature of 80 F. The water will make four passes through the unit and will be supplied in the necessary amount to give a temperature rise of 15 deg.

MECHANICAL REFRIGERATION

Solution. From the example of Art. 17.16 the load is 20 tons, the evaporator temperature is -20 F, the condenser temperature is 110 F, and the heat to be transferred in the condenser has been determined as 336,000 Btuh. The weight of water required is $336,000/15 = 22,400$ lb per hr and the total flow in gallons per minute measured at the inlet temperature is $(22,400 \times 7.48)/(60 \times 62.19) = 44.9$. The shell diameter is tentatively selected as $8\frac{5}{8}$ in.

TABLE 17.4. Effective Surface of One Line of Freon Condensers*

Shell OD $8\frac{5}{8}$ In., 40 Tubes		Shell OD $10\frac{3}{4}$ In., 60 Tubes		Shell OD $12\frac{3}{4}$ In., 92 Tubes		Shell OD 14 In., 120 Tubes	
Model No.	Total Eff. Surface Area, Sq ft	Model No.	Total Eff. Surface Area, sq ft	Model No.	Total Eff. Surface Area, sq ft	Model No.	Total Eff. Surface Area, sq ft
STF 84	73.5	STF 104	110	STF 124	169	STF 144	220
85	93.4	105	140	125	215	145	280
86	113	106	170	126	260	146	339
87	133	107	200	127	306	147	399
88	153	108	229	128	352	148	458
89	173	109	259	129	397	149	518
810	193	1010	289	1210	443	1410	578
811	213	1011	319	1211	489	1411	637
812	232	1012	349	1212	535	1412	697
813	252	1013	378	1213	580	1413	757
814	272	1014	408	1214	626	1414	816
815	292	1015	438	1215	672	1415	876
816	312	1016	468	1216	718	1416	936
				1217	763	1417	995
				1218	809	1418	1055
				1219	854	1419	1115
				1220	900	1420	1174

* Acme Industries, Inc. (Table does not cover the manufacturer's complete line of condensers.)

which fixes the total number of tubes as 40. A four-pass condenser is used so there are 10 tubes in each pass. The flow rate in gallons per minute per tube therefore is 4.49. The heat-transfer rate from Fig. 17.15 is then found to be 160 Btuh per sq ft per deg F of ΔT.

$$\text{Log mean temperature difference} = \frac{(110 - 80) - (110 - 95)}{\log_e (110 - 80)/(110 - 95)} = 21.6 \text{ F}$$

(see Art. 16.4). The amount of effective heating surface required is $336,000/(160 \times 21.6) = 97.3$ sq ft. In Table 17.4 it may be noted that model STF 86 provides 113 sq ft so it will be selected since the next smaller size will not provide the necessary surface. The friction of the water circulating system must be estimated and a pump selected that will circulate at least 44.9 gpm against that resistance.

The catalogue from which the data of Fig. 17.15 and Table 17.4 were taken includes a table suggesting different fouling factors for unusual cooling water conditions and additional curves and tables which make the selection procedure somewhat simpler than that used in the preceding example. The catalogue includes rating data for

TABLE 17.5. Rating Data for the Air-Cooled Condensers Shown in Fig. 17.13*

Basic ratings are based on a saturation temperature in the evaporator of 40 F, a saturation temperature in the condenser or 120 F, and an air dry-bulb temperature of 90 F. Refrigerant, Freon-12

Model No.	Btuh at Evaporator	Tons	cfm	Fan	Motor Hp	rpm
PFC-220	26,640	2.22	2100	16 in.	$\frac{1}{6}$	1100
PFC-330	40,000	3.33	3150	22 in.	$\frac{1}{6}$	1100

Conversion Factors

Suction temp.	−30	−20	−10	0	+10	+20	+30	+40
Multiplier	1.32	1.24	1.18	1.13	1.09	1.05	1.02	1.00

Condensing Temperature

Entering Air, dbt	105	110	115	120	125	130
80	1.20	1.00	.85	.75	0.67	0.60
90	2.00	1.50	1.20	1.00	0.85	0.75
100		3.00	2.00	1.50	1.20	1.00

* Bush Manufacturing Co.

other condensers in addition to those listed in Table 17.4 and it also contains data on the drop in water pressure.

Table 17.5 includes data which may be used in the selection of the type of air-cooled condenser that is shown in Fig. 17.13. These condensers are made in two models rated as shown in the table. *The ratings are based on the conditions stated at the top of the table and give the capacity of an evaporator which can be served by the condenser.* This capacity is given both in Btuh and in tons. Conversion factors are given which should be used in selecting a condenser for operation at a condition other than that on which the tabular ratings are based. Greater condenser capacity is needed in the case of lower evaporator temperature and pressure since the condenser heat is proportionately greater because of increased compressor work. Greater condenser capacity is needed in the case of lower condensing temperature because

MECHANICAL REFRIGERATION 393

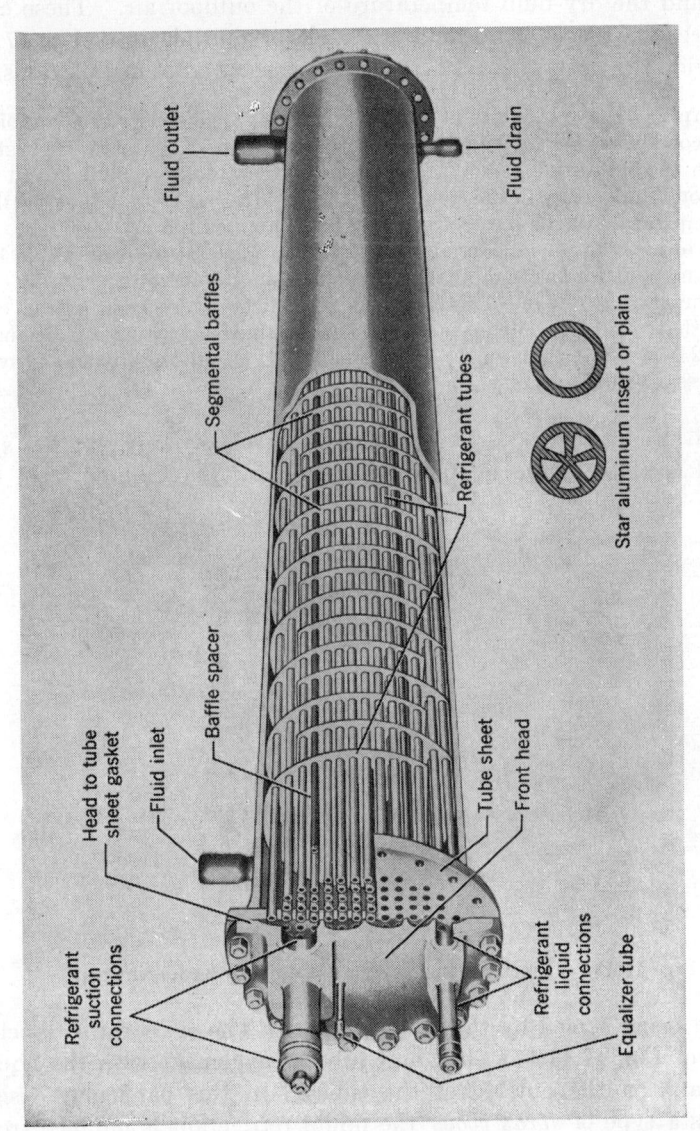

Fig. 17.16. Dry expansion type of evaporator.

of the reduced temperature differential between that of the refrigerant vapor and the dry-bulb temperature of the outdoor air. These condensers can be placed in one or more stacks of multiple units, Fig. 17.13, to provide the necessary capacity in the case of large installations.

Example. Using the data of Table 17.5, select a condenser or a combination of condensers that would serve a 4.25-ton Freon-12 refrigerating system in which the saturation temperature in the evaporator is to be 0 F and the saturation temperature in the condenser is to be 115 F. Assume that the dry-bulb temperature of the outdoor air may be as high as 100 F.

Solution. The conversion factor for suction temperature from Table 17.5 is 1.13 and that for the combination of condensing temperature and air dry-bulb temperature is 2.0. The condensing capacity under basic rating conditions that would be adequate under the assumed operating conditions is $1.13 \times 2.0 \times 4.25 = 9.6$ tons. Three model PFC-330 units would provide the necessary total capacity.

17.18. Evaporators. The cooling effect of any refrigerating system takes place in the evaporator where heat is extracted from the

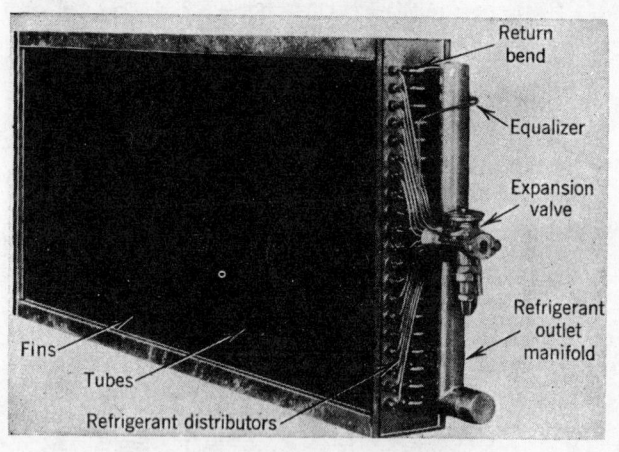

Fig. 17.17. Direct expansion coil. (Courtesy Kennard Corp.)

medium being cooled by the boiling liquid. The evaporator (labeled cooler) of Fig. 17.11 is a shell and tube arrangement with the liquid refrigerant on the outside of the tubes. In this particular design which is a type of *spray cooler* the liquid refrigerant is pumped from the sump in the bottom of the shell to the top of the distribution sheet which produces a spray effect on the tubes. In other designs of the same type spray nozzles replace the distribution sheet. Similar designs in which the shell is filled with liquid refrigerant to the height of the upper tubes are known as *flooded evaporators*. In the shell and

MECHANICAL REFRIGERATION

tube arrangement of Fig. 17.16, often called a *dry expansion cooler*, the refrigerant liquid evaporates inside of the tubes. A simple shell and coil arrangement is often used for cooling drinking water and in other applications where the required heat-transfer capacity is low. The *direct expansion coil* of Fig. 17.17 is a special type of dry expansion evaporator which is extensively used in summer air conditioners. In this arrangement liquid refrigerant (and flash gas) is distributed to the beginning of each tube circuit and slightly superheated vapor from all of the circuits is collected in the outlet manifold. The outlet manifold may be located on the same end of the coil as the refrigerant distributor, as shown in Fig. 17.17, or it may be on the opposite end.

17.19. Selection of Commercial Evaporators.[9] Figure 17.18 and Table 17.6 provide typical manufacturer's selection data and are applicable to the evaporator of Fig. 17.16. The following example will illustrate the method.

Example. Using the data of Fig. 17.18 and Table 17.6, select a water chiller for a refrigerating system using Freon-12 and carrying a load of 7.5 tons and find the water pressure drop. The water is to be cooled from 54 F to 46 F. Refrigerant temperature is to be 40 F.

Solution. *Step 1. Determine the capacity needed in tons.* The refrigerant load is given as 7.5 tons in this example.

Step 2. Determine the required water flow rate in gallons per minute. The flow rate in pounds per minute is $7.5 \times 200/(54 - 46) = 187.5$. The flow rate in gallons per minute is $187.5 \times 7.48/62.38 = 22.45$. (62.38 is the density of the water at the mean temperature, 50 F.) The flow rate and temperature drop may be specified, in which case the refrigeration capacity in tons, step 1, is calculated.

Step 3. Calculate the mean effective temperature difference (METD). This is customarily taken as the *log mean temperature difference* (equation 16.2).

$$\text{METD} = \frac{(54 - 40) - (46 - 40)}{\log_e \frac{54 - 40}{46 - 40}} = 9.43$$

Step 4. Enter Fig. 17.18 at 22.45 gpm on the lower vertical scale and move horizontally to the inclined line representing the type unit to be used. The possible choices are 8K, 8L, and 10K. Unit 8K will be selected. (The numbers indicate the shell diameter in inches and the letters refer to the spacing of the segmental baffles.) *From the point of intersection move vertically upward to the inclined line at the top of Fig. 17.18 representing the METD,* 9.43 as found in step 3, *and from this point of intersection move horizontally to the scale at the upper left and read the loading* as 1200 Btuh per sq ft. (The loading is the U value times the METD.)

Step 5. Determine the number of square feet of external tube surface required. Tube surface is the total load expressed in Btuh divided by the loading in Btuh per sq ft: $7.5 \times 12,000/1200 = 75$ sq ft.

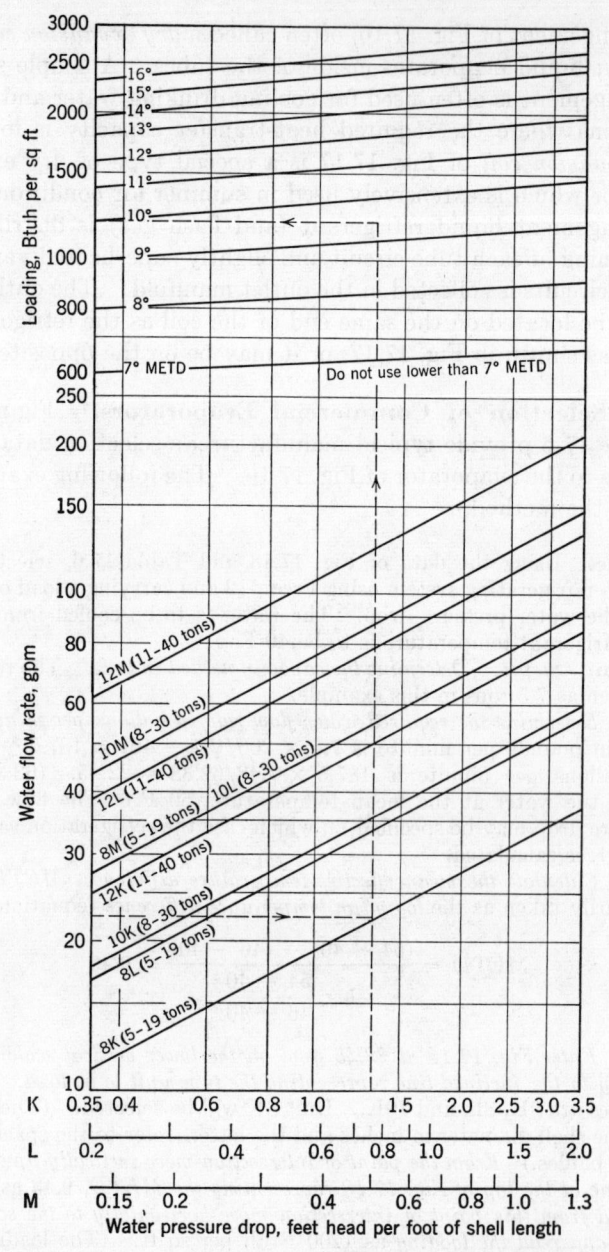

Fig. 17.18. Selection curves applying to dry expansion evaporators of Fig. 17.16 (refrigerant, Freon-12). (Courtesy Acme Industries, Inc.) (Dashed lines refer to the example of Art. 17.19.)

MECHANICAL REFRIGERATION

TABLE 17.6. Specifications for the Evaporator Shown in Fig. 17.16*

Model No.	Eff. Tube Area, sq ft	Capacity Range, tons	Std. No. of Circuits	No. of Tubes	Min. Lb. Freon-12 Charge	Volume of Shell Space, gal	Total Oper. Wt. with Water
DXH-805	41				15	8	460
806	50				18	9	520
807	59				21	11	590
808	67	5.4			24	13	650
809	76	to	1	44	27	15	710
810	85	19			30	16	770
811	94				33	18	835
812	102				36	19	900
813	111				39	21	970
814	119				42	23	1040
DXH-1005	64				23	13	660
1006	77				27	15	750
1007	91				32	18	850
1008	104				37	21	940
1009	118	8			42	24	1030
1010	131	to	1	68	47	26	1120
1011	145	30			52	29	1220
1012	158				56	32	1320
1013	171				61	35	1410
1014	184				65	37	1500
1015	197				70	40	1590
1016	211				75	42	1680
DXH-1206	105				37	22	1020
1207	123				44	26	1150
1208	141				50	30	1280
1209	159				56	34	1405
1210	177	11			63	38	1530
1211	195	to	2	92	69	42	1715
1212	213	40			76	46	1800
1213	231				83	50	1920
1214	249				89	53	2040
1215	267				96	57	2170
1216	285				102	61	2300

* Acme Industries, Inc. (This table covers only a portion of the manufacturer's line.)

Step 6. Read the required model number from Table 17.6. Model DXH-809 having 76 sq ft of effective tube area will be used. (The last two figures in the model number indicate the nominal tube length in feet.) The unit used will have a shell diameter of 8 in., the K type of baffle spacing, and a nominal tube length of 9 ft.

Step 7. Find the water pressure drop of the unit. The water pressure drop per foot of shell length is read on the K scale directly below the vertical line used in the graphical chart solution, 1.30 ft of water per ft of shell length or a total pressure drop of $9 \times 1.30 = 11.70$ ft of water.

If one desires to know the over-all coefficient of heat transfer for any condition of operation with this particular evaporator design, it may be obtained by dividing the loading by the METD, $1200 \div 9.43 = 127.2$ Btuh per sq ft per deg F in the case of the preceding example. The manufacturer's catalogue from which the data of Fig. 17.18 and Table 17.6 were taken includes nine sets of selection curves covering a range of capacities from 5 to 200 tons. Performance data for Freon-22 are also included as well as performance data for units in which special inserts are used in the tubes to increase the capacity.

17.20. Expansion Devices. An expansion device is needed in every compression refrigeration system to control the flow of liquid refrigerant into the evaporator. Full utilization of the heat-transfer surface requires the presence of liquid refrigerant in or on all parts of all of the tubes. It is also important that no liquid be carried out of the evaporator with the vapor that is returned to the compressor. The most commonly used types of expansion devices are the thermostatic expansion valve, the capillary tube, the float valve, and the automatic expansion valve.

Figure 17.19 is a sectional view of a *thermostatic expansion valve* which is designed to automatically regulate the flow of liquid refrigerant into the evaporator so that the vapor leaving is always superheated approximately 10 F. The liquid refrigerant enters at M and passes through strainer J to orifice Q. The needle valve P is attached to a yoke member which is rigidly attached to bellows I in the valve body. When the pressure in the body, which is the same as the pressure in the suction line to the compressor, increases, it actuates this bellows and pulls the valve against its seat. A decrease in this pressure tends to open the valve. The power element F contains a bellows U which is attached by means of capillary tube G to the feeler bulb K which may be attached to the compressor suction pipe at the evaporator outlet. The feeler bulb is charged with refrigerant so that the pressure exerted on bellows U is determined by its temperature. The bellows U works against bellows I through push rod S so that an increase in the temperature of the feeler bulb tends to cause the needle valve to open. Since the pressure in the valve body and in bellows I is determined by the evaporating temperature and the pressure in bellows U by the temperature of the vapor leaving the evaporator, the operation of the expansion valve is controlled by the difference between the two or the number of degrees of superheat. Tests have indicated that a

MECHANICAL REFRIGERATION

thermostatic expansion valve of this type will maintain practically a constant number of degrees of superheat in the vapor leaving the evaporator over a wide range of operating pressures.

The *capillary tube* serves almost all small refrigeration systems having capacities less than 1 ton. It is a small tube varying in length from 2 to 20 ft and having an internal diameter in the range 0.025 in. to 0.090 in. It is the ultimate in simplicity but must be carefully selected[10] if it is to perform satisfactorily.

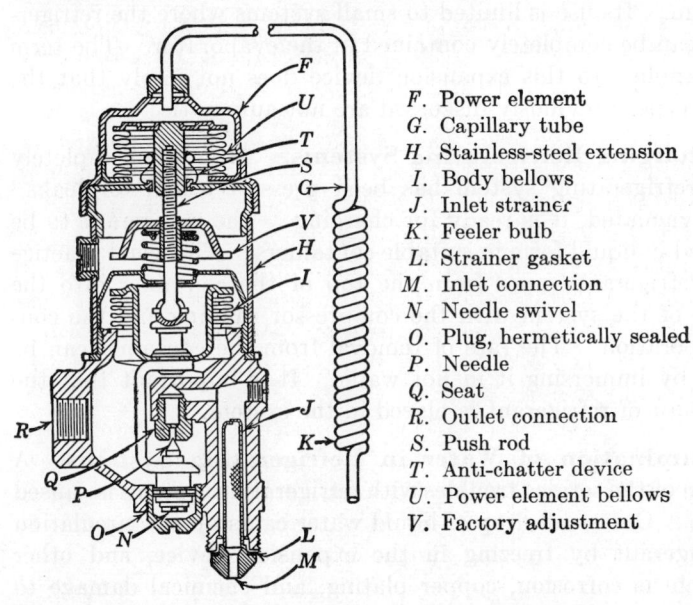

F. Power element
G. Capillary tube
H. Stainless-steel extension
I. Body bellows
J. Inlet strainer
K. Feeler bulb
L. Strainer gasket
M. Inlet connection
N. Needle swivel
O. Plug, hermetically sealed
P. Needle
Q. Seat
R. Outlet connection
S. Push rod
T. Anti-chatter device
U. Power element bellows
V. Factory adjustment

Fig. 17.19. Thermostatic expansion valve. (Courtesy Detroit Lubrication Co.)

Almost all capillary tubes are installed with a section in contact with the suction line. The liquid in the tube is cooled by the cold gas from the evaporator, thereby reducing the amount of flashing and increasing the refrigerating effect.

The capillary tube allows the pressure to equalize throughout the system when the compressor stops; therefore the load on the driving motor is low when it is again started. In this pressure-equalization process most of the refrigerant in the system may collect in the evaporator; consequently, the evaporator must be designed with adequate storage volume and the correct weight of refrigerant must be used in charging the system.

The flow of liquid refrigerant may be controlled by a float valve which maintains a constant level in the evaporator. If the float is actually located in the evaporator or in a chamber directly attached to it, the arrangement is called a *low side float*. Figure 17.1 shows a *high side float;* it also shows the portion of a compression refrigeration system that is normally filled with liquid refrigerant and the portion normally filled with vapor.

The expansion device, which is known as an *automatic expansion valve,* is operated by a pressure-sensing element located in the evaporator and its function is to maintain a constant pressure at that point in the system. Its use is limited to small systems where the refrigerant charge can be completely contained in the evaporator. The term automatic applied to this expansion device does not imply that the other three types previously discussed are not automatic.

17.21. Charging Refrigerating Systems. After the completely assembled refrigerating system has been pressure-tested for leaks,[1] dried and evacuated, it is ready for charging. The refrigerant to be used is stored in liquid form in suitable containers. The usual practice is to feed refrigerant vapor from the top of the container into the suction line of the system with the compressor running and the condenser in operation. The rate of removal from the container can be accelerated by immersing it in hot water. It is important that the correct amount of refrigerant is placed in the system.

17.22. Elimination of Water in Refrigerating Systems. A large portion of the service troubles with refrigerating systems is caused by moisture. One small drop of liquid water can stop the circulation of the refrigerant by freezing in the expansion device, and other troubles such as corrosion, copper plating, and chemical damage to insulation may result if moisture is present in the system. Water may occur in a refrigeration system due to one or more of the following causes: (1) incomplete drying of the equipment before charging; (2) low side leaks admitting air which contains moisture; (3) leakage in a water-cooled condenser; (4) water formation through oxidation of certain hydrocarbons in the oil; (5) water dissolved in either oil or liquid refrigerant when it is placed in the system; and (6) decomposition of the motor insulation in hermetically sealed units. Methods of drying[11–13] a system prior to charging include heating the assembly, circulating dehumidified warm air through it; creating a vacuum within it through a mechanical vacuum pump; and removing the water by the combined use of alcohol and carbon dioxide. Small quantities of water from any of the above causes excepting a leak in a water-cooled condenser can usually be removed satisfactorily by

MECHANICAL REFRIGERATION

incorporating a suitable drier at a strategic point in the path of the refrigerant. Combination filter-drier units are available in two styles. One type contains a replaceable cartridge. The other type is a sealed unit which must be replaced when it is no longer effective. Driers are usually placed in the liquid line between the condenser and the expansion device.

The drying agent is called a *dessicant*. It is a chemical compound which either adsorbs water or eliminates it from the refrigerant by reacting with it chemically. Commonly used adsorbents are granular aluminum oxide, called *activated alumina*, and silicon dioxide in the same form, which is called *silica gel*. Chemical driers include anhydrous calcium sulfate sold under the trade name *Drierite*, calcium oxide or quicklime, and calcium chloride. Adsorption-type dessicants remove moisture[14] more slowly than the chemical driers but they can be left in the system for a long period or permanently, whereas it is generally not recommended that any chemical drier be left in the system longer than 24 hr in one installation.

17.23. Removal of Noncondensable Gases. Air and other noncondensable gases may be present in a refrigerating system from one or more of the following sources: (1) incomplete removal of the gas used for pressure-testing the system prior to charging; (2) formation of noncondensable gas by chemical reactions within the system; (3) air or other noncondensable gases dissolved in the refrigerant when charged;[15] (4) leaks in the piping in the case of refrigerants which require a vacuum in any part of the system.

Noncondensable gases in a refrigerating system increase operating pressures and temperatures and decrease efficiency. The presence of a troublesome amount may be detected[1] by comparing the pressure and temperature in the condenser when the compressor is not in operation. If the pressure is higher than that corresponding to the temperature in a table giving the properties of the saturated refrigerant vapor, it may be assumed that the system contains noncondensable gases in objectionable amounts.

In the case of a small unit it may be practicable to eliminate noncondensable gases by wasting the charge, evacuating the system, and recharging it with fresh refrigerant. In the case of large systems the noncondensable gases can be removed[16] from the top of the condenser and from the top of the evaporator with an accompanying loss of only a small portion of the refrigerant charge. The process of removing noncondensable gases from a refrigerating system is called *purging*.

Special purging units are available which include a compressor and a condenser for recovery of the refrigerant that is unavoidably removed with the noncondensable gases.

17.24. Tests for Leakage of Refrigerant. Good maintenance procedures include frequent and regular checks for leaks at all possible leakage points in the system. Leakage of sulfur dioxide is easily detected by the white smoke which appears if a swab soaked with a 25 per cent solution of aqueous ammonia is brought near it. An ammonia leak may be similarly detected with a burning sulfur candle. Loss of any of the halogenated hydrocarbons (Freons) may be detected by the use of a *halide torch*. (The halide torch burns alcohol and the flame turns green in the presence of Freon vapors.)

17.25. Control of Compression Refrigeration Systems. Control of the refrigerating effect so as to equal the load is accomplished by varying the capacity of the compressor. In small systems such as household refrigerators and home air conditioners the compressor is stopped when the cooling thermostat has been satisfied and it is started again when further refrigeration is needed.

In large systems employing centrifugal compressors the capacity may be varied to match a changing load by: (1) varying the speed; (2) adjusting vanes at the inlet to the impellers; (3) throttling the suction gas; (4) varying the condenser pressure. Reducing the compressor capacity by lowering the speed or using inlet vanes does not adversely effect the efficiency. Control of capacity by either throttling the suction gas or raising the condenser pressure results in reduced system efficiency.

In large systems using reciprocating compressors, capacity control may be accomplished through: (1) regulators which throttle the suction gas so as to keep the evaporator pressure constant; (2) controlled bypass of gas from the compressor discharge to the suction pipe; (3) cylinder unloaders which hold the suction valves open, thereby stopping the compression process without stopping the compressor. Methods 1 and 2 seriously reduce efficiency. Multicylinder compressors may be arranged with a step control which unloads one, two, three, or more cylinders as needed to match the capacity with the load. Multicompressor installations may be provided similarly with a step-control arrangement which stops one or more of the compressors as the load is decreased.

Although the reduced condenser pressure which normally accompanies a reduced system load results in a saving in compressor power, it may interfere with the flow of liquid refrigerant through the expan-

MECHANICAL REFRIGERATION

sion device and cause unsatisfactory operation of the system. Suitable controls[17] are available for maintaining normal condenser pressure during periods of operation at reduced load.

17.26. Piping in Compression Refrigeration Systems. The combination of subcooling of the liquid and pressure drop between the condenser and the expansion device should be such that no flash gas will form in the liquid line. If the expansion device is located at an elevation higher than the liquid level in the condenser (or receiver) from which the liquid refrigerant flows, the hydrostatic head of liquid refrigerant must be allowed for in determining the minimum condenser pressure if flashing is to be avoided. The piping between the expansion device and the evaporator is usually very short and can be the same size as the liquid line in spite of the fact that it handles a much greater volume of fluid due to the flash gas formed when the pressure drops.

From the standpoint of cost for power, the friction loss in the suction pipe should be made low by designing for low velocity. However, in many systems, oil reaching the evaporator due to leakage past the piston rings must be returned to the compressor with the suction gas. In Freon-12 systems the velocity must not be less than 750 fpm in horizontal pipes and not less than 1500 fpm in vertical pipes where the flow is upward. Approximately the same velocities are also necessary in the discharge pipe between the compressor and the condenser. When the condenser is located above the compressor and a considerable distance from it, the piping should include a pocket or trap that will prevent oil or liquid refrigerant from draining back into the compressor during a period of shutdown.

The many different piping problems caused by special arrangements, such as compressors operating in parallel and evaporators operating in parallel, are beyond the scope of this book. Additional information on the subject may be obtained from references 1 and 18.

The friction loss in refrigerant piping may be computed by equation 6.3 in exactly the same way that it was found for water and steam in the examples of Art. 6.6. The viscosity of Freon-12 in both the liquid and vapor states is given in Figs. 17.5 and 17.6. The density may be obtained from Table 17.2. After the Reynolds number has been calculated, the friction factor may be read from Fig. 6.6. Several charts and tables designed to simplify the problem of sizing refrigerant piping will be found in papers mentioned in the bibliography.

17.27. Methods of Defrosting. In many compression refrigeration systems the evaporator is a direct expansion coil used to cool and dehumidify the air passing over it. If the conditions of operation require metal surface temperatures below 32 F, frost will collect on the tubes and fins and periodic defrosting will be necessary. The methods of defrosting include: (1) interrupted service, (2) water spray; (3) hot gas; (4) hot oil; (5) either electric resistance heaters adjacent to the coil or an arrangement for heating the coil by passing low-voltage current through the metal.

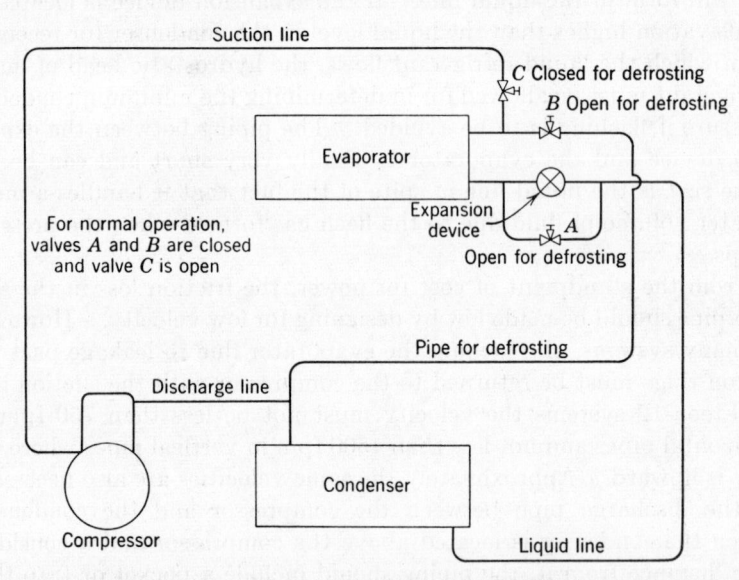

Fig. 17.20. Piping details for defrosting with hot gas.

Method 1 is the simplest but it cannot be used in applications such as cold-storage rooms where the air temperature in the room is at all times below 32 F. Method 2 can be used in all applications. However, when the coil to be defrosted is located in a low-temperature room, piping which supplies the spray nozzles must be arranged so that the water will drain out of it at the end of the defrosting operation. Method 3 involves passing hot discharge gas from the compressor through the direct expansion coil by a suitable special piping arrangement as in Fig. 17.20. The evaporator temporarily performs the function of the condenser during the process of defrosting. Method 4[19] involves pumping electrically heated lubricating oil through the evaporator while the compressor is stopped. The oil returns to the storage tank through the compressor suction pipe into which it is connected. The storage tank serves as a low side oil separator during normal operation of the system. Method 5[20] involves shutting down the compressor and activating the electric heating arrangement until the frost on the coils is melted.

17.28. Special Problems in Low-Temperature Refrigeration. Figure 17.21 from reference 1 shows the compression ratios required for suction temperatures below −40 F for three commonly used refrigerants when the condensing temperature is 100 F. Ordinary

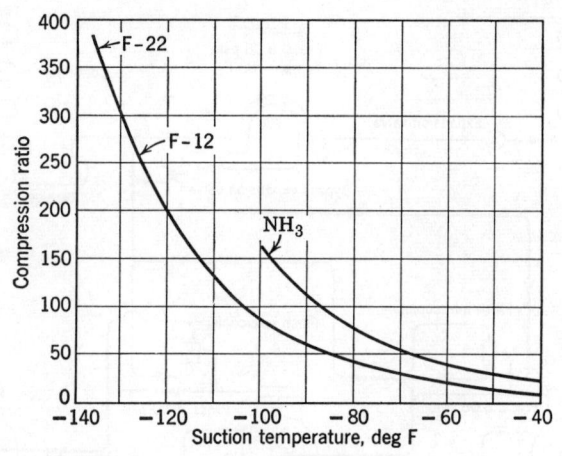

Fig. 17.21. Compression ratio versus suction temperature at 100 F condensing temperature for Freon-22, Freon-12, and NH_3.

single-stage compression systems cannot operate efficiently when a compression ratio greater than 10 is required. Multistage systems and cascade systems are now in common use for producing extremely low temperatures with mechanical refrigeration.

17.29. Multistage Compression Refrigeration. Figure 17.22 from reference 21 shows a simplified schematic diagram of a low-temperature refrigeration system using Freon-22 in which three compressors operate in series. A portion of the liquid refrigerant leaving the receiver is diverted through the bypass expansion valve and is used to subcool the remainder. The cold vapor resulting from the evaporation of this bypassed liquid is piped into the suction of the interstage compressor where it mixes with and helps to cool the hot gas from the first stage unit. Liquid refrigerant is passed through the *flash intercooler* and added to the cold vapor from the main line liquid subcooler in an amount necessary to maintain the desired temperature of the vapor entering the interstage compressor. Liquid refrigerant alone is used to control the temperature of the vapor entering the high stage compressor. In the over-all compression process, no compressor operates against a compression ratio greater than 4.96, whereas to maintain the same evaporator and condenser pressures in a single-stage system a compression ratio of 122 would be required. The compression ratio in each stage of two-stage compression equals the square root of the compression ratio required in a single-stage system operating with

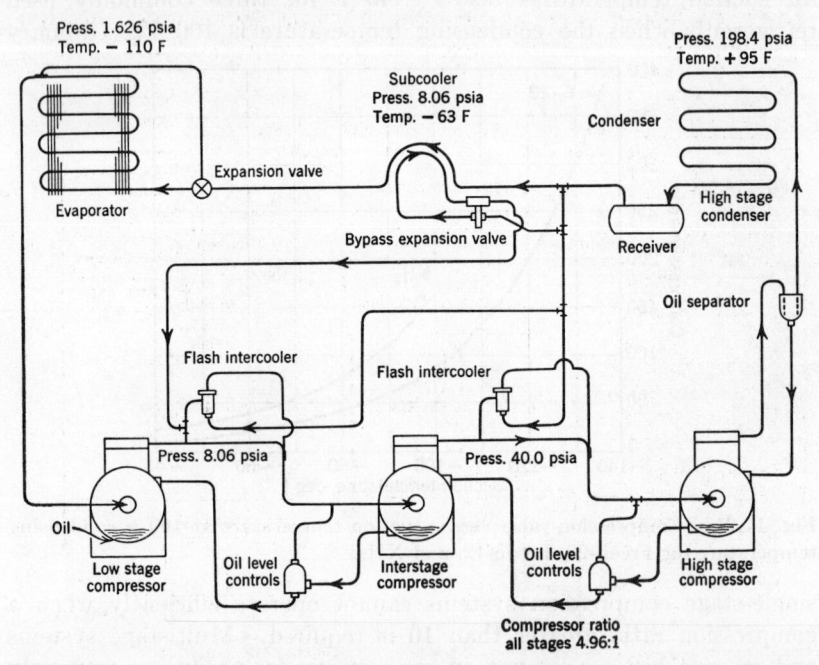

Fig. 17.22. Simplified piping diagram of a low-temperature refrigeration system employing three-stage compression. (Pressure-temperature data shown are for Freon-22.) (Courtesy Tenney Engineering, Inc.)

the same evaporator and condenser pressures. In the case of three-stage compression, the ratio in each stage is the cube root of the ratio that would be required if only one stage were used.

17.30. Cascade Systems. Figure 17.23, also from reference 21, shows a two-stage cascade arrangement which consists of two single-stage refrigeration systems in series. Heat is absorbed from the low stage condensing refrigerant by the evaporating high stage refrigerant. This heat exchanger is called a *cascade condenser*. It serves as an evaporator for the high stage and as a condenser for the low stage. The only useful refrigerating effect is produced in the evaporator of the low stage system. The principal advantage in the cascade arrangement is that it permits the use of two different refrigerants. A refrig-

MECHANICAL REFRIGERATION

erant such as Freon-13, which does not require a vacuum in the evaporator to produce a low suction temperature, is selected for the low stage system. Another refrigerant such as Freon-22, which does not require an excessive pressure in the condenser, is selected for use in the high stage system.

In the cascade system it is necessary that the operation of the high stage compressor be co-ordinated with the load on the low stage unit

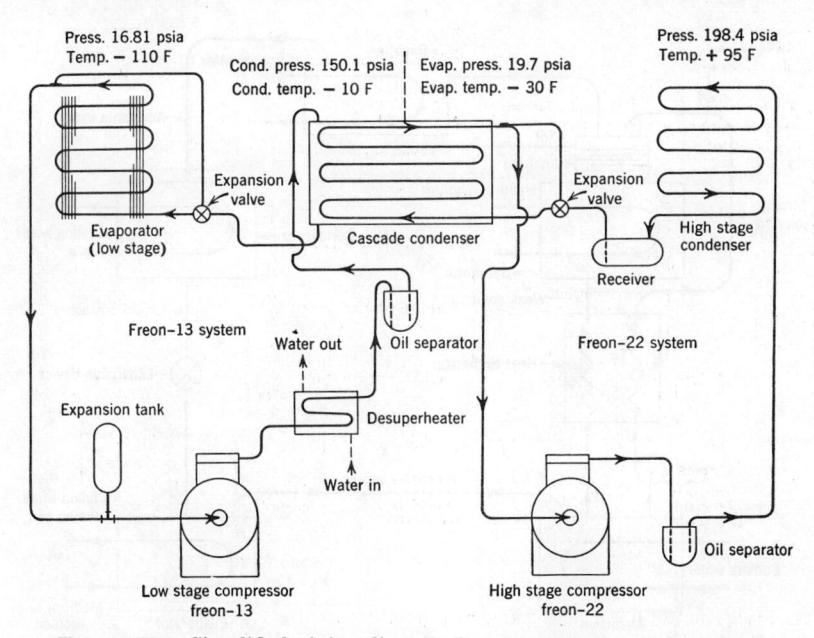

Fig. 17.23. Simplified piping diagram for a two-stage cascade system.

so as to maintain a constant refrigerant vapor pressure in the cascade condenser but there are no interconnecting pipes between the two stages and there are no special problems such as oil transfer between compressors.

17.31. The Ammonia Absorption System.[22] The absorption refrigeration system is a heat-operated unit which uses a refrigerant that is alternately absorbed by and liberated from the absorbent. Figure 17.24 from reference 23 shows a simplified flow diagram of an ammonia absorption system in which water serves as the absorbent. The principle of this system is identical with that of the single-stage compression system except in the manner in which the pressure of the refrigerant vapor is increased to the level required for condensation. The compressor of the compression system is replaced by the absorber, the pump, and the generator in the absorption system. Instead of compressing the low-pressure refrigerant vapor it is first absorbed by a weak solution of the refrigerant in water. The strong solution thus formed is then pumped into the generator where it is heated. In the heating process the refrigerant vapor is driven out of the solution and conducted to the condenser through the analyzer and rectifier under the necessary pressure.

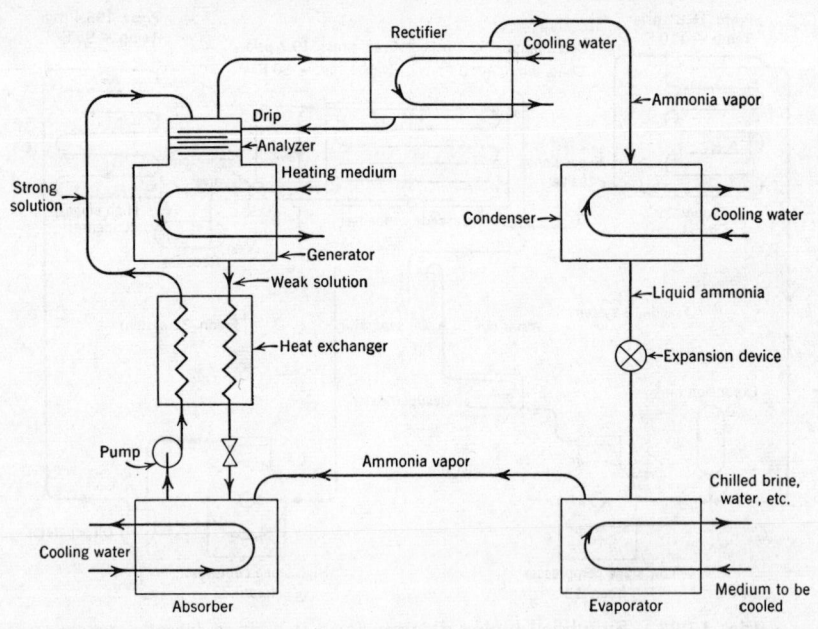

Fig. 17.24. Schematic diagram of absorption system showing addition of features to improve efficiency.

The cost of operation of this type of absorption system may be less than that for one of the same capacity employing a compressor, if exhaust steam or some other low-cost heating medium is available for use in the generator.

In driving the ammonia vapor out of solution in the generator it is impossible to avoid evaporating some of the water. It is the function of the analyzer and rectifier combination to remove the water vapor from the ammonia vapor. The analyzer is a direct-contact heat exchanger consisting of a series of trays mounted above the generator. The strong solution from the absorber flows downward over the trays to cool the outgoing vapors. This heat exchanger not only reduces the amount of heating medium needed in the generator but it also condenses some of the water vapor. The final reduction in the percentage of water vapor in the ammonia going to the condenser occurs in the rectifier, which is a water-cooled heat exchanger which condenses water vapor and returns it to the generator through the drip line shown in Fig. 17.24. The heat exchanger between the generator and the absorber cools the hot weak solution and heats the cool strong solution. This interchange of heat reduces the amount of heating medium required for the generator and also reduces the amount of cooling water required for the absorber.

17.32. Water Vapor Absorption System. In the absorption system of the previous article, water was used as the absorbent. However, in the system of Fig. 17.25, water serves as the refrigerant and a solution of lithium bromide salt in water serves as the absorbent. Referring to Fig. 17.25, water returned from the air-conditioning coil or other load (chilled water in) is passed through a coil and cooled by colder water from spray nozzles located above it. This coil is located in the evaporator chamber, and the spray water which is circulated by the evaporator pump is cooled by its own partial evaporation. This evaporative cooling process is able to reduce the temperature of the spray water to around 38 F because of the very low pressure (high vacuum) that is maintained in the evaporator chamber. The low pressure in the evaporator chamber is maintained by the spray of lithium bromide solution in the lower part of the shell which encases it. This lithium bromide solution has a strong affinity for water vapor because of its very low vapor pressure. It therefore absorbs the water vapor as fast as it is released in the evaporation process in the evaporator.

In absorbing water vapor, the solution is diluted and it is necessary to reconcentrate it by bypassing a portion of the discharge from the solution pump through the generator shell. The solution in the generator is heated by a coil supplied with steam, and the excess water is evaporated. The water vapor released in this heating process passes into the condenser chamber where it condenses on a coil that is cooled by water from a cooling tower, well, or other suitable source. This water also cools the solution spray in the bottom of the lower shell. After passing through the generator, the by-passed portion of the solution again enters the main circuit after passing through the heat exchanger. The condensed water vapor is returned to the evaporator through a pipe connecting the two chambers.

The capacity of the unit is controlled by regulating the flow of condenser water through a thermostat located in the leaving-chilled-water line. The solution boiling temperature is maintained at a constant value by maintaining a constant steam pressure. The amount of water that can be removed from the solution is determined by the amount of vapor which can be condensed, and that is determined by

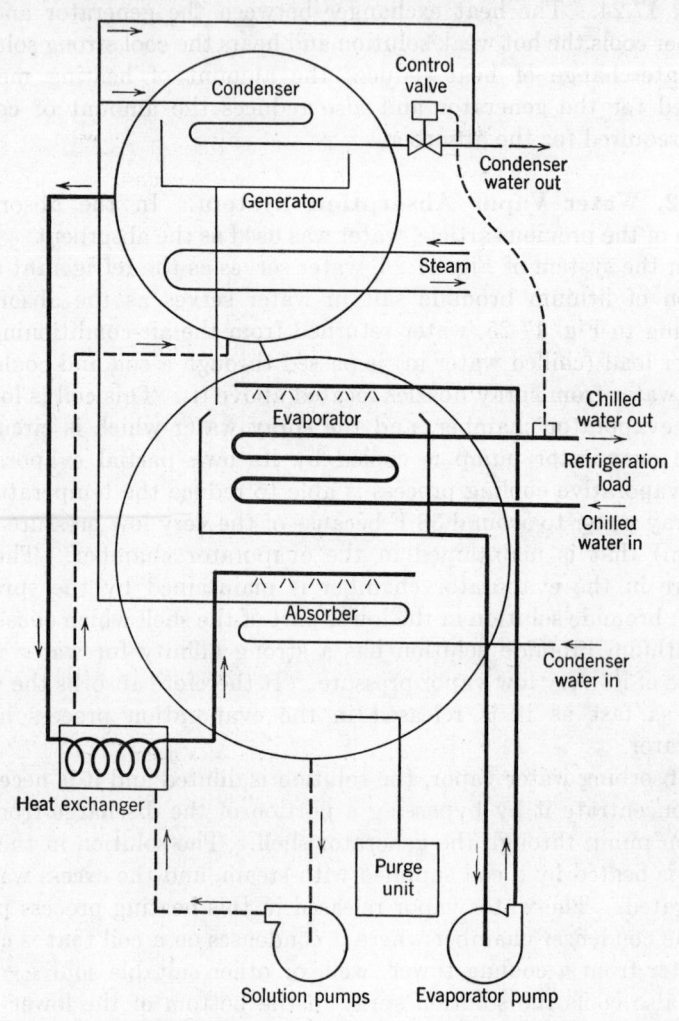

Fig. 17.25. Water vapor absorption system. (Courtesy Carrier Corporation.)

the amount of cooling water passing through the condenser coil. When the refrigeration load decreases, a drop in chilled water temperature results in throttling the control valve.

The apparatus shown in Fig. 17.25 may be assembled in a compact

MECHANICAL REFRIGERATION

unit which is free from serious noise or vibration. Its principal application is for chilling water for use in air conditioning. It is available in different sizes ranging in capacity from 100 to 700 tons. It is completely safe and can be located in any part of any building.

17.33. Steam-Jet Refrigeration Apparatus.[24] Commercial equipment for the production of refrigeration by means of steam jets, using water as the refrigerant, is shown in Fig. 17.26. Steam passing through the nozzles of the primary ejector acquires a high velocity and aspirates the vapor and gases from the evaporator chamber to

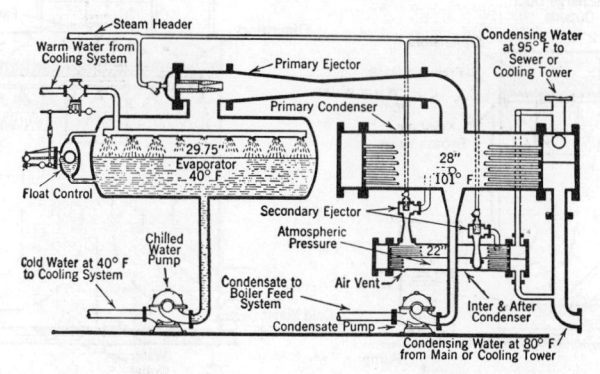

Fig. 17.26. Ross decalorator. (Courtesy American Blower Corp.)

produce a low pressure within it. The warmed water, returning from an air washer or a cooling coil, is broken into a spray in the evaporator chamber and a small part of it, usually less than 1 per cent, flashes into vapor. The remainder of the water is cooled, losing an amount of sensible heat equal to the latent heat taken up by the vapor that is formed. As a result, it is possible to cool water to 40 F if a vacuum of 29.75 in. of mercury, referred to a 30-in. barometer, can be maintained in the evaporating chamber.

The steam passing through the primary ejector and the vapor taken from the flash chamber or evaporator are condensed in the primary condenser, in which a vacuum is also maintained. In order to maintain a sufficiently low pressure in the primary condenser a secondary steam ejector is used to remove the air and noncondensable gases from it. This secondary ejector is served by the water-cooled intercondenser which reduces to water most of the steam which it handles. Any uncondensed vapor and the noncondensable gases removed from the intercondenser by the other secondary ejector are discharged to the atmosphere. Steam at any pressure above 12 to 15 psig may be used in the ejectors. The higher the initial steam pressure the less the steam requirements per unit of refrigeration.

17.34. The Condenser Cooling Water Problem. With the ever-increasing use of water in private homes, in commercial buildings, and by many different types of industries, water shortages have developed in practically all metropolitan areas and in many small communities. Many city governments have passed ordinances requiring that water

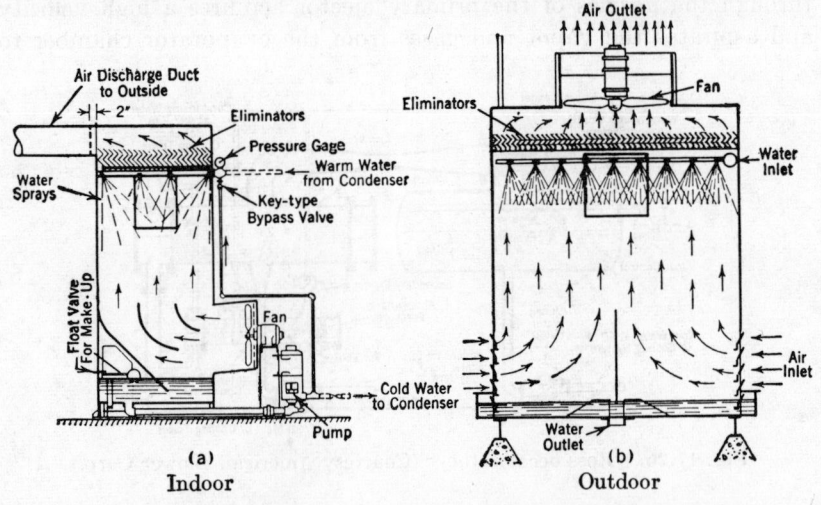

Fig. 17.27. Cooling towers. (Courtesy Binks Manufacturing Co.)

recovery apparatus be used in connection with the operation of any refrigerating system which has a capacity exceeding a stipulated figure, which varies greatly in the different regulatory measures. Even if no legal restrictions are imposed, the cost of the water may make it advisable to use conservation measures. The water recovery device most commonly used is the cooling tower which will be discussed in the next article. Evaporative condensers were discussed in Art. 17.15. Spray cooling ponds are sometimes used where the necessary ground area is available.

17.35. Cooling Towers. In cooling towers, the water transfers heat to air. In one arrangement, Fig. 17.27, the water is brought into contact with air by means of a suitable pattern of spray nozzles at the top of an enclosure. In addition to the eliminators shown in Fig. 17.27, the interior of a cooling tower usually contains a filling of suitable wood or metal latticework to provide wetted surfaces in addition to those of the water droplets. If the air flows through the

MECHANICAL REFRIGERATION

enclosure by gravity action the device is known as an *atmospheric cooling tower*. If a fan is used it is called a *forced convection cooling tower*. If the fan is located at the top, as in Fig. 17.27b, the term *induced draft tower* is applied; if the fan blows the air into the enclosure as in Fig. 17.27a, the arrangement is referred to as a *forced draft tower*.

Fig. 17.28. Forced draft cooling tower suitable for use with small air-conditioning systems. (Courtesy Binks Manufacturing Co.)

The cooling tower of Fig. 17.27a is arranged for location inside a building, but forced draft towers may also be located outside. The roof of the building often serves as the location for an outdoor tower. The cooling tower of Fig. 17.28 may be placed out-of-doors or it may be placed in a basement or utility room with a duct leading to the outside from the outlet collar near the top of the unit.

Small forced convection towers are available which use a water turbine to drive the fan, thereby eliminating the necessity for any electrical connections. The water pressure drop required to operate the turbine, together with the friction head of the system, is furnished by the circulating pump. In this arrangement the water is distributed uniformly over redwood slats with which the interior of the tower is packed, by slowly rotating, perforated tubes.

The performance of a cooling tower[23] is commonly expressed in terms of *range* and *approach*. The *range* is the reduction in water temperature and *approach* is the difference between the entering wet-bulb temperature of the air and the temperature of the leaving water. The heat-transfer process in a cooling tower consists of a transfer of heat from the water to unsaturated air. There are two driving forces, namely, the difference in dry-bulb temperature and the difference between the pressure of the water vapor mixed with the air and the vapor pressure of the water droplets or wetted surfaces. These two forces combined are called *enthalpy potential*. The heat-transfer rate that occurs in an actual tower is dependent on many factors, including water temperature, water pressure at the spray nozzles, nozzle design and spacing, arrangement of the interior of the tower, wet-bulb temperature of the air, velocity of the air, and height of the tower. Analysis[1] of the heat-transfer process is further complicated by the fact that the enthalpy potential is not the same in different sections of the tower.

Manufacturers' rating data applying to specific designs and sizes, usually obtained from actual tests, are generally given in the form of tables. Table 17.7 includes data for the unit of Fig. 17.28.

TABLE 17.7. Temperature Performance Table for the Cooling Tower of Fig. 17.28

	6.7-Deg Cooling Range				10-Deg Cooling Range			
Tons Refrigeration	Water over Tower, gpm	Wet Bulb Temperature, deg F	Temperature On	Temperature Off	Tons Refrigeration	Water over Tower, gpm	Wet Bulb Temperature, deg F	Temperature On	Temperature Off
2	9	70	82.9	76.2	3	9	70	88	78
2	9	72	84.8	78.1	3	9	72	90	80
2	9	75	87.5	80.8	3	9	75	92	82
2	9	76	88.4	81.7	3	9	76	93	83
2	9	78	90.2	83.5	3	9	78	95	85
2	9	80	92.0	85.3	3	9	80	97	87

In Table 17.7, the amount of water circulated is fixed and the temperature rise is 6.7 deg when the load on the refrigeration system is 2 tons, and 10 deg when the load is 3 tons. The water temperatures on and off the tower are given for the two different loads and for a

MECHANICAL REFRIGERATION

range of air wet-bulb temperatures. The heat rejection per ton on which ratings are based is 250 Btu per min per ton and it may be assumed that it is equal to the maximum likely to be encountered in air-conditioning service.

17.36. Use of an Intermediate Medium. In the majority of commercial and industrial refrigerating systems, heat is transferred in the evaporator from an intermediate medium which is then piped to another location where the actual desired cooling effect takes place. This practice reduces the weight of refrigerant in the system and lessens the possibility of leaks. Water is invariably used as the intermediate carrier in air-conditioning systems. In applications of refrigeration such as skating rinks, food freezers, low-temperature metal treatment, and low-temperature research investigations, where the service requires an intermediate medium at a temperature below 32 F, it is necessary to use some other fluid. Water solutions of alcohol, glycerine, ethylene glycol, proylene glycol, sodium chloride, and calcium chloride are generally satisfactory for this purpose.

17.37. Refrigeration System Used for Heating—Heat Pump. Any refrigeration system is a heat pump in that the refrigerant absorbs heat at a certain temperature and rejects it at a higher temperature. In the over-all process, the temperature level of heat is increased somewhat as the pressure of a liquid is increased by the action of a pump, hence the familiar term *heat pump*. Refrigeration systems are extensively used for pumping heat from refrigerated spaces and discharging it to any available cooling medium such as water or air. They can also be used for pumping heat from any available heat source, such as the outdoor air, the ground, or water, and discharging it to air or water at a higher temperature. The air or water that is heated in this manner may be used for warming buildings. It is current practice to refer to a refrigeration system as a heat pump only when it is used for both warming and cooling the space it serves.

The medium from which heat is extracted by a heat pump is called the *heat source* and the medium to which heat is rejected when the apparatus is used for cooling is called the *sink*. When a heat pump is used for both heating and cooling, as in all-year air conditioning, the same medium may serve alternately as a *source* and as a *sink*.

Because the heat transferred from the heat source to the heating medium may be a much greater amount than the heat equivalent of the power required by the compressor, the cost of operating such a system for the purpose of heating may not exceed the cost of operating systems using fuels, in spite of the fact that the electricity needed for the heat pump is a more expensive form of energy (Art. 5.15).

17.38. Possible Apparatus Arrangements in Heat Pumps. Figure 17.29 illustrates a commercially available "packaged heat pump" designed to abstract heat from a stream of outdoor air when

Fig. 17.29. Packaged air-to-air heat pump with panels removed. (Courtesy Westinghouse Electric Corp.)

heating is required and to dissipate heat to the same medium when there is need for cooling and dehumidification. The unit shown is provided with two fans and contains a switching valve and other necessary piping details to permit interchanging the functions of the coil in the outdoor-air stream and the coil in the path of the conditioned air. The coil in the outdoor-air stream (not shown) serves as the evaporator when heating is required and as the condenser when cooling is needed. In a similar manner, the coil in the air stream from and to the conditioned space (indoor coil) serves as the condenser during cold weather and as the evaporator during hot weather. *Air-to-air heat*

MECHANICAL REFRIGERATION

pumps must be provided with a means of defrosting the coil that is in the outdoor-air stream except when used in climates where the outdoor temperature is always well above 32 F. The need for defrosting is less frequent in extremely cold weather than is the case with outdoor temperatures around 32 F because of the low humidity ratio of the outdoor air under those conditions.

A suitable antifreeze solution may be used in place of outdoor air as a source of heat for winter heating and it may also serve as a satisfactory cooling medium for the condenser during hot weather. Figure 17.30 shows a possible layout of a system in which a suitable fluid is circulated through a coil buried in the ground. In this arrangement, the ground serves as the source of heat or as the sink for waste heat, depending on the need at the particular time. With the apparatus of Fig. 17.30 no change is required in the air-handling system when switching from winter heating to summer cooling or vice versa. The

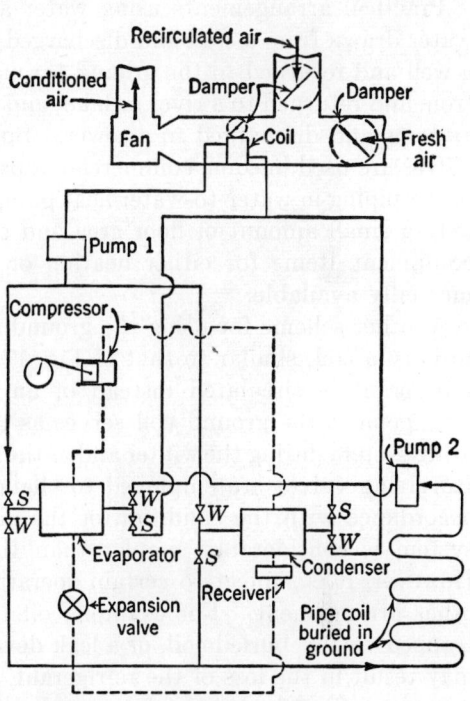

Fig. 17.30. Heat pump with ground coils through which an intermediate heat-transfer medium is circulated.

dampers shown are for varying the proportion of fresh air and their settings are independent of the type of processing to be done. A coil bypass is provided so that part of the air stream may be cooled sufficiently to produce dehumidification without delivering air to the conditioned space that is too cool for comfort. Likewise, the path of the refrigerant is not affected by the type of service performed by the system. Four valves labeled S and four others labeled W are provided for properly directing the flow of antifreeze solution to accomplish the desired purpose. The S valves are to be open during summer weather and closed during the wintertime. Similarly, the W valves are to be open in the winter and closed in the summer. During summer operation pump 2 circulates the solution from the buried coil through the condenser tubes and back to the coil. Pump 1 provides circulation in another closed circuit consisting of the evaporator and the coil in the air stream. In the over-all process, during summer weather heat is removed from air and transferred to the ground.

Practical arrangements using water as the source or sink include: water drawn from a well and discharged to a drain; water drawn from a well and returned to the ground through another well; water drawn from and returned to a river or lake; and water drawn from city supply systems and discharged to a sewer. Special three-way and four-way valves are used in some commercial apparatus to permit simplification of the piping in water-to-water heat pumps. Packaged units requiring only a small amount of floor area and containing all of the principal equipment items for either heating or cooling water are now commercially available.

Another scheme for using the ground as the source and as a sink is to bury a coil, similar to that of Fig. 17.30, through which a suitable refrigerant is circulated instead of an antifreeze solution. In this arrangement the ground coil serves as the evaporator of the refrigerating system during the winter and as the condenser during the summer. Reversing valves are employed to change the function of the coil in accordance with the condition of the air in the space served. This system is considerably simpler than the one shown in Fig. 17.30. However, it is subject to certain operating hazards not present in the other arrangement. For example, oil from the compressor may be trapped in the buried coil, or a leak developing in the external piping may result in the loss of the refrigerant.

17.39. Performance of Heat Pumps.[25] Although a heat pump is a refrigerating system, the COP (Art. 17.4) does not apply when it is operating on the heating cycle because it is the condenser heat that is useful instead of the refrigerating effect. The term *heat pump performance ratio* (HPPR) (sometimes called *performance factor* and *heating energy ratio*) is used to express the effectiveness of a heat pump.

$$\text{HPPR} = \frac{\text{Heat supplied at condenser, Btuh}}{\text{Compressor power consumption, Btuh}} \quad (17.6)$$

If heat loss from the pipes carrying the hot gas from the compressor to the condenser is neglected, the HPPR is always 1.0 greater than the COP of a given system under the same conditions of operation. The cost of heating a building with a heat pump is determined by its heat loss, the HPPR, and the cost of electricity. Figure 17.31, from reference 25, includes data based on a typical performance of a reciprocating compressor in a well-designed system in good condition and shows the effect of evaporating and condensing temperatures on the HPPR.

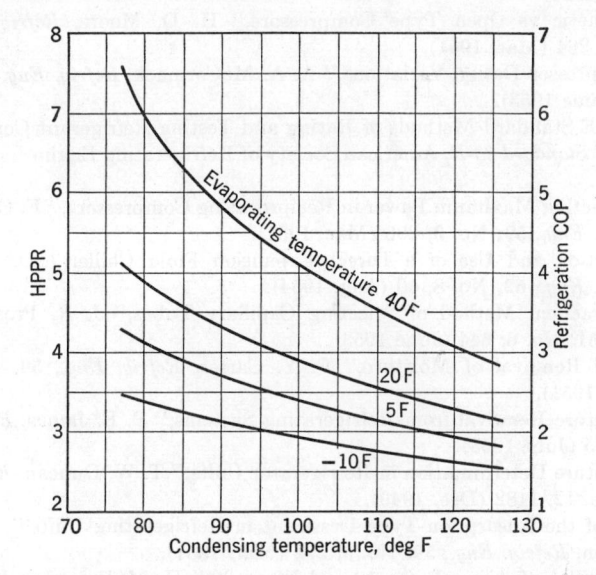

Fig. 17.31. Typical *HPPR*'s for various condensing and evaporating temperatures.

17.40. Applications for Heat Pumps. Heat pumps are generally regarded as being applicable to the heating and cooling of buildings. However, there are many other applications where substantial savings can be accomplished through their use. A separate heat pump may be used in a home for heating water for household use.[26, 27] A heat pump water heater located in the basement of a home provides hot water at low operating cost and it also dehumidifies the air in that part of the house.

Opportunity for substantial savings through the application of the heat pump principle exists in all commercial or industrial enterprises where both heating and cooling are needed at the same time.

REFERENCES

1. *Air Conditioning, Refrigeration Data Book, Design Volume*, The American Society of Refrigerating Engineers, New York, 9th ed. (1956).
2. *Introduction to Heat Transfer*, Brown and Marco, McGraw-Hill Book Co., New York, 2nd ed., p. 245 (1951).

3. "How to Select Refrigerants for Centrifugal Air Conditioning Systems," S. L. Soo, *Refrig. Eng.*, **63**, No. 11, 43 (Nov. 1955).
4. "Carrene 7, New Refrigerant," C. M. Ashley, *Refrig. Eng.*, **58**, No. 6, 553 (June 1950).
5. "Hermetic vs Open Type Compressors," H. D. Moore, *Refrig. Eng.*, **59**, No. 3, 264 (Mar. 1951).
6. "Compressor Design Variations," A. A. McCormack, *Refrig. Eng.*, **61**, No. 6, 622 (June 1953).
7. "ASRE Standard Methods of Rating and Testing Refrigerant Compressors," *ASRE Standard 23-R*, American Society of Refrigerating Engineers, New York (1949).
8. "Predicting Maximum Power in Reciprocating Compressors," E. G. Thomsen, *Refrig. Eng.*, **59**, No. 3, 296 (Mar. 1951).
9. "Selection and Use of a Direct Expansion Fluid Chiller," D. G. Merrill, *Refrig. Eng.*, **62**, No. 8, 60 (Aug. 1954).
10. "A Practical Method of Selecting Capillary Tubes," J. R. Prosek, *Refrig. Eng.*, **61**, No. 6, 644 (June 1953).
11. "Field Removal of Moisture," F. Y. Carter, *Refrig. Eng.*, **59**, No. 6, 547 (June 1951).
12. "Moisture Removal from Refrigerating Systems," P. E. James, *Refrig. Eng.*, **61**, 625 (June 1953).
13. "Moisture Determination in Refrigerator Units," T. W. Duncan, *Refrig. Eng.*, **57**, No. 12, 1182 (Dec. 1949).
14. Role of the Adsorption-Type Dessicant in Refrigerating Units," W. A. Pennington, *Refrig. Eng.*, **59**, No. 3, 272 (Mar. 1951).
15. "Solubility of Air in Freon-12 and Freon-22," H. M. Parmelee, *Refrig. Eng.*, **69**, No. 6, 573 (June 1951).
16. "Preventive Maintenance is a Must," F. E. Peltier, *Refrig. Eng.*, **62**, No. 10, 58 (Oct. 1954).
17. "Control of Discharge Pressures Prevents Refrigeration Troubles," H. A. Blair, *Heating and Ventilating*, **48**, No. 11, 61 (Nov. 1951).
18. "Fundamentals of Refrigerant Piping for Freon-12 and Freon-22 Systems," C. W. Leegard and W. E. Dodson, *Refrig. Eng.*, **60**, No. 5, 473 (May 1952).
19. "An Automatic Defrosting System for Low Temperature Evaporators," I. C. White, Jr., *Refrig. Eng.*, **57**, No. 11, 1088 (Nov. 1949).
20. "Automatic Electric Defrosters for Freezers and Chill Rooms," *Modern Refrig.*, **57**, No. 671, 48 (Feb. 1954).
21. "Freon-13 Operates in Cascade System at -110 F," F. Hermann, *Power*, **98**, No. 8, 120 (Aug. 1954).
22. "An Absorption Refrigeration System May Be Your Answer," C. L. Rescorla, *Refrig. Eng.*, **61**, No. 3, 276 (Mar. 1953).
23. "Notes on Refrigeration," W. F. Stoecker, University of Illinois (in mimeographed form).
24. "Refrigeration by Steam Jet," D. W. R. Morgan, *Power*, **78**, No. 9, 506 (Sept. 1934).
25. "Design Factors for Industrial Heat Pump Installations," R. J. Martin, *Mech. Eng.*, **74**, 280 (Apr. 1952).
26. "Performance of a Heat Pump Water Heater," E. B. Penrod, *Bull. No. 23, Eng. Exp. Sta.*, University of Kentucky (1952).
27. "Heat Pump Water Heater," A. W. Ruff, *Refrig. Eng.*, **59**, No. 2, 153 (Feb. 1951).

BIBLIOGRAPHY

Refrigerants

1. "The New Thermodynamic Properties of Freon-12," R. C. McHarness, B. J. Eiseman, Jr., and J. J. Martin, *Refrig. Eng.*, **63**, No. 9, 32 (Sept. 1955).
2. "Water Solubility of Freon Refrigerants," H. M. Parmelee, *Refrig. Eng.*, **61**, No. 12, 1341 (Dec. 1953).
3. "Use of Freon-13 in Cooling High Pressure Air," J. J. Fritz, J. G. Aston, and L. F. Shultz, *Refrig. Eng.*, **62**, No. 5, 48 (May 1954).
4. "Pressure-Volume-Temperature Properties of Freon Compounds," B. J. Eiseman, Jr., *Refrig. Eng.*, **60**, No. 5, 496 (May 1952).
5. "How to Handle Ammonia Safely," L. Brandt, *Power*, **94**, No. 7, 85 (July 1950).
6. "Solubility of Air in Freon-12 and Freon-22," H. M. Parmelee, *Refrig. Eng.*, **69**, No. 6, 573 (June 1951).
7. "The Operating Engineer Looks at the Common Refrigerants," E. L. Forbes, *Refrig. Eng.*, **60**, No. 11, 1182 (Nov. 1952).

Compressors

1. "Theory on Performance of Reciprocating Compressors," E. G. Thomsen, *Refrig. Eng.*, **60**, No. 1, 61 (Jan. 1952).
2. "Influence of Speed on Compressor Volumetric Efficiency," G. Lorentzen, *Refrig. Eng.*, **60**, No. 3, 272 (Mar. 1952).
3. "Testing and Comparison of Small Refrigeration Compressors," H. M. Meacock, *Modern Refrig.*, **56**, No. 658, 6 (Jan. 1953).
4. "Is It Better To Use a Reciprocal or Centrifugal Compressor?" E. T. Neubauer, *Air Conditioning, Heating, Ventilating*, **54**, No. 1, 70 (Jan. 1957).

Refrigerant Condensers

1. "A Simple Method of Rating Water Cooled Condensers," H. Soumerai, *Refrig. Eng.*, **63**, No. 10, 52 (Oct. 1955).
2. "Adaptability of Air Cooled Condensers for Large Tonnage Systems," O. J. Nussbaum, *Refrig. Eng.*, **62**, No. 3, 43 (Mar. 1954).
3. "The Effect of Water Distribution on Evaporator Condenser Efficiency," H. Vetter, *Refrig. Eng.*, **59**, No. 4, 366 (Apr. 1951).
4. "Evaporator Condenser Performance Factors," D. D. Wile, *Refrig. Eng.*, **58**, No. 1, 55 (Jan. 1950).

Refrigerant Evaporators

1. "Heat Transfer Coefficients in Horizontal Tube Evaporators," W. L Bryan and G. W. Quaint, *Refrig. Eng.*, **59**, No. 1, 67 (Jan. 1951).
2. "Brine Film Resistance to Heat Flow of Calcium Chloride, Sodium Chloride and Ethelene Glycol," C. T. Shields, *Refrig. Eng.*, **59**, No. 9, 880 (Sept. 1951).
3. "A Study of Boiling Heat Transfer," W. M. Rohsenow and J. A. Clark, *ASME Trans.*, **73**, No. 5, 609 (July 1951).
4. "Photographic Study of Surface-Boiling Heat Transfer to Water with Forced Convection," F. C. Gunther, *ASME Trans.*, **73**, No. 2, 115 (Feb. 1951).
5. "The Heat Transfer of Evaporating Freon," C. M. Ashley, *Refrig. Eng.*, **43**, No. 2, 89 (Feb. 1942).

Expansion Devices

1. "Theory and Use of the Capillary Tube Expansion Device," N. E. Hopkins, *Refrig. Eng.*, **56**, No. 6, 519 (Dec. 1948).
2. "Rating the Restrictor Tube," N. E. Hopkins, *Refrig. Eng.*, **58**, No. 11, 1087 (Nov. 1950).
3. "Capillary Tubes, A Guide to their Application on Freon-12 Hermetic Condensing Units," L. A. Staebler, *Refrig. Eng.*, **58**, No. 1, 48 (Jan. 1950).
4. "Method of Testing Capillary Tubes—Proposed ASRE Standard No. 28-53," *Refrig. Eng.*, **61**, No. 8, 876 (Aug. 1953).
5. "How is Capillary Tubing Made?" L. E. Gibbs, *Refrig. Eng.*, **58**, No. 3, 252 (Mar. 1950).
6. "Capillary Tube Refrigerant Flows in Theory and in Practice," H. G. Goldstein, *Modern Refrig.*, **55**, No. 647, 43 (Feb. 1952).
7. "Tips on Selecting Thermal Expansion Values for Refrigeration," M. L. Hoglund, *Mill and Factory*, **51**, No. 6, 120 (Dec. 1952).
8. "Capillary Tube Improves Heat Pump Design," G. L. Biehn, *Heating and Ventilating*, **51**, No. 2, 75 (Feb. 1954).

Moisture Removal from Refrigerating Systems

1. "Dehydration of Refrigeration Components," O. B. Branson, *Modern Refrig.*, **54**, No. 634, 9 (Jan. 1951).
2. "Dessicant-Refrigerant Moisture Equilibria," A. J. Gully, H. A. Tooke, and L. H. Bartlett, *Refrig. Eng.*, **62**, No. 4, 62 (Apr. 1954).
3. "Moisture Determination in Refrigerant-Oil Solutions by the Karl Fisher Method," F. T. Reed, *Refrig. Eng.*, **62**, No. 7, 65 (July 1954).
4. "Determination of Water Content in Freon-12 Circulating in a Refrigeration System," A. W. Diniak, E. E. Hughes, and M. Fujic, *Refrig. Eng.*, **62**, No. 2, 56 (Feb. 1954).
5. "Moisture Migration in Hermetic Refrigeration Systems," W. R. Briskin, *Refrig. Eng.*, **63**, No. 7, 42 (July 1955).

Refrigerant Piping

1. "Eleven Piping Tips for Smart-Running Small Refrigeration Plants," *Power*, **96**, No. 10, 112 (Oct. 1952).
2. "Determining Pressure Drop in Freon Systems," W. L. Holladay, *Refrig. Eng.*, **62**, No. 9, 55 (Sept. 1954).
3. "Refrigerant Piping for Air Conditioning Systems," S. C. Dietrick, *Heating and Ventilating*, **50**, No. 3, 96 (Mar. 1953).
4. "Suction Line Selector Chart and ΔP Computer," H. E. Ferrill, *Refrig. Eng.*, **53**, No. 5, 436 (Nov. 1947).
5. "Determination of Refrigerant Pipe Size," H. M. Hendrickson, *Refrig. Eng.*, **52**, No. 4, 317 (Oct. 1946).
6. "Freon-12 Refrigerant Pipe Sizing," M. B. Goddard, *Refrig. Eng.*, **58**, No. 8, 753 (Aug. 1950).
7. "Successful Refrigerant Piping for Air Conditioning Systems," F. K. Ladewig, *Air Conditioning, Heating, Ventilating*, **53**, No. 11, 85 (Nov. 1956).

Low-Temperature Refrigeration

1. "Low Temperature Refrigeration Systems," T. Lopiccolo, *Refrig. Eng.*, **59**, No. 3, 239 (Mar. 1951).
2. "Refrigeration Design Considerations Below -20 F," E. R. Michel, *Refrig. Eng.*, **59**, No. 6, 552 (June 1951).
3. "Simplified Analysis of Two Stage Low Temperature Refrigeration Systems," H. P. Soumerai, *Refrig. Eng.*, **61**, No. 7, 746 (July 1953).
4. "Use of Freon-13 in Cooling High Pressure Air," J. J. Fritz, J. G. Aston, and L. F. Schultz, *Refrig. Eng.*, **62**, No. 5, 48 (May 1954).

Cooling Towers

1. "Efficient Hookup of Your Cooling Tower or Evaporative Condenser," M. Metz, *Power*, **98**, No. 7, 126 (July 1954).
2. "Prediction of Cooling Tower Performance," W. W. Smith, *ASHVE Trans.*, **60**, 27 (1954).
3. "Performance of a Forced-Draft Cooling Tower," B. H. Spurlock, Jr., *ASHVE Trans.*, **59**, 61 (1953).
4. "Selecting Cooling Towers for Air Conditioning Service," H. E. Degler, *Heating, Piping, Air Conditioning*, **22**, No. 3, 89 (Mar. 1950).
5. "Silencing Cooling Towers," Martin Hirschorn, *Heating, Piping, Air Conditioning*, **24**, No. 8, 95 (Aug. 1952).
6. "Corrosion Characteristics of Aluminum Cooling Towers," *Heating and Ventilating*, **51**, No. 1, 88 (Jan. 1954).

Heat Pumps

1. *Heat Pumps*, P. Sporn, E. R. Ambrose, and T. Baumeister, John Wiley & Sons, New York (1947).
2. "Conditions that Promote Heat Pump Evaporator Frosting," D. Goldenberg, *Heating and Ventilating*, **49**, No. 1, 61 (Jan. 1952).
3. "Design and Economics of Solar Energy Heat Pump Systems," R. C. Jordan and J. L. Threlkeld, *ASHVE Trans.*, **60**, 212 (1954).
4. "Moisture Movement in Soils Due to Temperature Difference," W. A. Hadley and R. Eisenstadt, *ASHVE Trans.*, **59**, 395 (1953).
5. "Thermal Conductivity of Soils for Design of Heat Pump Installations," G. S. Smith and T. Yamauchi, *ASHVE Trans.*, **56**, 355 (1950).
6. "Earth Heat Pump Research, Part IV," E. B. Penrod and P. L. Pfennigwerth, *Bull. No. 35, Eng. Exp. Sta.*, University of Kentucky (1955).
7. "Heat Pump Results in Equitable Building," J. D. Kroeker, R. C. Chewning, and C. E. Graham, *ASHVE Trans.*, **55**, 375 (1949).
8. "Ground Temperatures as Affected by Weather Conditions," A. B. Algren, *ASHVE Trans.*, **55**, 363 (1949).
9. "Economic and Technical Aspects of the Heat Pump," W. E. Johnson, *ASHVE Trans.*, **54**, 201 (1948).
10. "Theory of the Ground Pipe Heat Source for the Heat Pump," L. R. Ingersoll and H. J. Plass, *ASHVE Trans.*, **54**, 339 (1948).
11. "Field Test of Water to Air Heat Pump," S. L. Pappas and F. R. O'Brien, *Heating and Ventilating*, **51**, No. 2, 79 (Feb. 1954).

12. "Heating and Cooling Residence with 3-HP Air Source Heat Pump," W. F. Stoecker and R. R. Herrick, *Refrig. Eng.*, **60**, No. 11, 1172 (Nov. 1952).
13. "Design and Performance of Residential Earth Heat Pump," M. Baker, *ASHVE Trans.*, **59**, 371 (1953).
14. "Thermodynamic Study of Vapor Compression Heat Pump Cycles," *Refrig. Eng.*, **61**, No. 11, 1202 (Nov. 1953).
15. "Two Year Performance of Heat Pump System Furnishing Year-Round Air Conditioning in Modern Office Building," *ASHVE Trans.*, **57**, 483 (1951).
16. "Theory of Earth Heat Exchangers for the Heat Pump," L. R. Ingersoll, F. T. Adler, H. J. Plass, and A. C. Ingersoll, *ASHVE Trans.*, **57**, 167 (1951).
17. "Heat Pump Application to a Newspaper Plant," J. D. Kroeker, J. H. Bonebrake, and J. A. Melvin, *ASHVE Trans.*, **57**, 467 (1951).
18. "Climatology as an Aid in Heat Pump Design," G. S. Smith, *ASHVE Trans.*, **57**, 499 (1951).

PROBLEMS

1. Using the data of Table 17.1 and assuming standard atmospheric pressure, compute the compression ratio required with (a) Freon-12; (b) Freon-113; (c) sulfur dioxide; (d) carbon dioxide.

2. Using the data of Table 17.1, calculate the volume in cubic feet per minute of ammonia vapor flowing in the suction line of a system handling a load of 100 tons.

3. Using Fig. 17.4, determine the enthalpy, entropy, and specific volume of superheated dichlorodifluoromethane at a pressure of 20 psia and a temperature of 250 F.

4. Calculate the net refrigerating effect per pound of Freon-12 circulated, based on the following assumptions: evaporator pressure, 23.87 psia; condenser pressure, 131.6 psia; no subcooling of the liquid; vapor leaving the evaporator superheated 10 F; neglect heat gain or loss from or to the ambient air around the piping.

5. Calculate the net refrigerating effect in Problem 4 if the liquid is subcooled 20 F. What is the mean specific heat of the liquid in the temperature range through which it passes in the subcooling process?

6. Referring to Problem 4, assuming that the refrigerant leaves the compressor with 50 deg of superheat and using the data of Table 17.2, calculate the compressor work in Btu per pound of refrigerant circulated. What is the COP? Neglect pressure loss in the piping between the compressor and the condenser.

7. Calculate the clearance volumetric efficiency which may be expected when a reciprocating compressor having a clearance volume equal to 6 per cent of the piston displacement is to increase the pressure of Freon-22 vapor from 28.3 psig to 159.8 psig. Also calculate the weight of refrigerant handled in pounds per minute if the total displacement volume is 150 cfm. Assume standard atmospheric pressure.

8. Calculate the weight of Freon-12 in pounds per minute that would be handled by a compressor operating under the conditions of group V, Table 17.3, if the actual volumetric efficiency is 85 per cent and the piston displacement is 175 cfm.

MECHANICAL REFRIGERATION

9. Compute the theoretical hp required to compress 50 lb per min of Freon-12 vapor under the conditions of group VI, Table 17.3.

10. A sulfur dioxide compressor has two 5-in. diameter pistons and the stroke of each one is 5 in. The clearance volume is 6 per cent of the piston displacement. Assuming isentropic compression, a speed of 500 rpm, and a mechanical efficiency of 92 per cent, estimate the maximum bhp that can be required if the condensing pressure is 80 psia.

11. Assuming isentropic compression, calculate the amount of heat in Btuh that must be disposed of in the condenser and liquid subcooler of a 10-ton Freon-12 system, if the saturation temperature in the evaporator is -10 F and that in the condenser is 100 F. Assume that the refrigerant enters the expansion device as a liquid at 80 F and that the vapor leaves the evaporator at 15 F.

12. Calculate the condenser heat rejection rate in Btuh for the low-pressure part of a cascade system assuming the following: refrigerant, Freon-12; load, 10 tons; saturation temperature in the evaporator, -40 F; actual temperature of vapor leaving evaporator, -15 F; saturation temperature in the condenser, 40 F; no subcooling of the liquid refrigerant; isentropic compression.

13. Select a condenser from Table 17.4 that would be suitable for serving an air-conditioning system which requires a refrigerating capacity of 50 tons. The saturation temperature in the evaporator will be 35 F and that in the condenser 110 F. Cooling water will be supplied at a temperature of 90 F and in sufficient volume to give a temperature rise of 15 F. Assume that the water connections will provide a four-pass circuit. Assume that the refrigerant vapor leaving the evaporator is superheated 12.5 deg and that the compression is isentropic. Assume no subcooling of the liquid. The refrigerant is Freon-12.

14. Select a condenser from Table 17.4, given the following data: load, 30 tons; saturation temperature in the evaporator, 0 F; actual temperature of vapor leaving the evaporator, 15 F; pressure in the condenser, 140.9 psia; refrigerant liquid leaving the condenser at the saturation temperature; 4.95 Btu per lb removed from the liquid and added to the suction gas in a heat exchanger; assume isentropic compression; cooling water supplied at 85 F; total volume of cooling water 75 gpm; water connections arranged to give four passes. The refrigerant is Freon-12.

15. Using the data of Table 17.5 for air-cooled condensers, specify the model number and the number of units that would be needed to serve a Freon-12 refrigerating system, given the following: evaporator load, 4.85 tons; saturation temperature in the evaporator, $+10$F; saturation temperature in the condenser, 110 F; dry-bulb temperature of the outdoor air, 95 F.

16. Using the data of Fig. 17.18 and Table 17.6, find the required length of a 12-in. diameter chiller that would be suitable for cooling 36 gpm of water from 55 F to 45 F (measured at the higher temperature). Assume that the refrigerant is Freon-12, and that the pressure in the evaporator is maintained at 51.68 psia. Assume that the K type of baffle spacing is used. Find the pressure differential in feet that will be required to overcome the frictional resistance in the water circuit. Find the over-all coefficient of heat transfer U on which the selection is based.

17. Referring to Table 17.6 and Fig. 17.18, a model DXH-1010 water chiller having the M type of baffle spacing is to deliver chilled water at 40 F and serve a load amounting to 14.2 tons. The water to be chilled is to be circulated at a rate such that the METD will be 10 F. Find the required circulation rate in gallons per minute and the pressure loss in the water circuit expressed in feet of water.

18. What capacity in tons may be expected from a DXH-812 chiller with type-L baffle spacing when operated under the following conditions: pressure in the evaporator, 47.28 psia; inlet water temperature, 51 F; outlet water temperature, 44 F; volume of water circulated, 29 gpm.

19. The condenser and receiver of a certain Freon-12 refrigerating system have a maximum liquid storage volume of 5 cu ft and the temperature of the liquid refrigerant in the space may reach 110 F. What weight of Freon-12 in pounds should be used in charging the system?

20. Using the method of Art. 6.6 and the data of Fig. 17.5 and Table 17.2, find the friction loss in pounds per square inch per 100 ft of equivalent length when 50 lb of liquid Freon-12 per min flows through nominal $\frac{5}{8}$-in. type-L copper tubing. Assume the liquid refrigerant is at a temperature of 100 F.

21. If the expansion device of a Freon-12 refrigeration system is located 20 ft above the liquid level in the condenser and the friction loss in the liquid line amounts to 10 psi, what condenser pressure in pounds per square inch absolute is necessary to prevent flashing in this line if the maximum temperature of the liquid reaching the expansion device is 95 F?

22. Using the method of Art. 6.6 and the data of Fig. 17.6 and Table 17.2, find the friction loss in pounds per square inch per 100 ft in the suction pipe of a Freon-12 system in which 80 lb of saturated vapor at 40 F is handled per min. Assume that 2-in. (nominal) type-L copper tubing is used.

23. Determine the proper nominal size of type-L copper tubing to be used as a suction line in a Freon-12 refrigerating system designed to carry a load of 10 tons. The pressure in the line is 29.35 psia. Assume that the vapor is saturated and that the flow is upward in a vertical pipe. Assume that the liquid refrigerant reaches the expansion device at a temperature of 85 F.

24. A certain low-temperature Freon-12 system using an air-cooled condenser must operate with saturation temperatures of -40 F in the evaporator and $+120$ F in the condenser. Find the compression ratio required in a single-stage system. What compression ratio will be required if two-stage compression is used? If two-stage compression is used, what interstage pressure will result in the same compression ratio in each stage?

25. Determine the range and approach from the following data for the performance of a cooling tower: temperature of the water going to the tower, 100 F; temperature of the water coming from the tower, 90 F; wet-bulb temperature of the air entering the tower, 80 F.

26. (a) From the data of Table 17.7 for the cooling tower of Fig. 17.28, determine the range and approach that may be expected when it serves a refrigerating load of 3 tons and the wet-bulb temperature of the air is 76 F. (b) What range and approach may be expected if the system refrigeration load is 2 tons and the wet-bulb temperature of the air is 78 F?

18

Estimation of Cooling Loads

18.1. The Components of a Cooling Load. The two main divisions of a load imposed upon an air-conditioning plant operating during hot weather are: (1) the sensible heat to be removed, and (2) the water vapor to be abstracted.

Sensible heat to be removed in the process of summer air conditioning is derived from any or all of the following sources.

1. Heat flowing into the building by conduction through exterior walls, floors, and ceilings due to temperature difference.

2. Heat conducted through exterior walls, roofs, windows, and doors due to the absorption of solar radiation by exterior surfaces.

3. Heat conducted through interior partitions from rooms in the same building which are not conditioned.

4. Heat given off by lights, motors, machinery, cooking operations, industrial processes, etc.

5. Heat liberated by the occupants.

6. Heat carried in by outside air which leaks in or is brought in for ventilation.

7. Heat gain through the walls of ducts carrying conditioned air through unconditioned spaces in the building.

8. Heat gain from fan work.

427

The water vapor which must be removed by the dehumidifying apparatus may come from any or all the following sources.

1. Moisture in the outside air entering by infiltration or introduced for the purpose of ventilation.
2. Moisture from occupants.
3. Moisture from any process such as cooking foods which takes place within the conditioned space.
4. Moisture passing directly into the conditioned space through permeable walls or partitions from the outside or from adjoining regions where the water vapor pressure is higher.

In the process of removing water vapor by condensation the latent heat of that which is removed must be absorbed by the cooling apparatus. The amount of this heat absorption in Btuh is often referred to as the *latent-heat load* on the apparatus. The sum of the sensible heat and latent heat is called the total heat. The required capacity of any cooling and dehumidifying unit is determined by the *total-heat load* which it must handle.

The moisture gain expressed in pounds per hour may be converted to latent-heat load in Btuh or vice versa by applying a conversion factor. The factor used in reference 1 is 1076. This factor is the approximate number of Btu which must be abstracted from 1 lb of water vapor as it enters the cooling unit to effect its removal from the air stream. It includes some sensible heat in addition to the latent-heat as the water vapor is initially superheated.

18.2. Outdoor Design Temperatures. The outdoor temperatures which should be assumed for the purpose of estimating cooling loads should not be the highest ever recorded, as such temperatures occur infrequently and only for periods of a few hours' duration. In all large cities certain design dry-bulb and wet-bulb temperatures have become well established and may be used with confidence by engineers designing air-conditioning machines for use in that vicinity. Table 18.1 gives this information for several of the large cities of the country. Reference 1, from which the data of Table 18.1 were extracted, gives similar information for many cities which are not listed in Table 18.1.

It may be noted from Table 18.1 that a design dry-bulb temperature of 95 F is used in many cities and that this figure is close to the temperature that is used in nearly all of the cities for which data are given. Design wet-bulb temperatures range from 64 F to 80 F, but the majority are within 3 deg of 76 F. In estimating cooling loads for smaller cities for which no data are available, the design dry- and wet-bulb temperatures given for the nearest large city may be used

ESTIMATION OF COOLING LOADS

TABLE 18.1. Design Dry- and Wet-Bulb Temperatures and Wind Velocities in Local Use at Principal Cities

State	City	Design Dry-Bulb, deg F	Design Wet-Bulb, deg F	Summer Wind Velocity, mph	Highest Temperature Ever Recorded*
Ala.	Birmingham	95	78	5.4	107
Ariz.	Phoenix	105	76	6.0	118
Ark.	Little Rock	95	78	6.2	110
Calif.	Los Angeles	90	70	5.8	109
	San Francisco	85	65	10.7	101
Colo.	Denver	95	64	6.9	105
D.C.	Washington	95	78	5.9	106
Fla.	Tampa	95	78	7.4	98
Ga.	Atlanta	95	76	7.9	102
Idaho	Boise	95	65	5.8	112
Ill.	Chicago	95	75	9.5	105
Ind.	Indianapolis	95	76	8.9	106
Iowa	Des Moines	95	78	8.6	111
Kan.	Wichita	100	75	11.8	114
Ky.	Louisville	95	78	7.2	107
La.	New Orleans	95	80	6.9	102
Mass.	Boston	92	75	12.5	104
Mich.	Detroit	95	75	9.5	104
Minn.	Minneapolis	95	75	10.2	108
Mo.	St. Louis	95	78	9.5	110
Mont.	Helena	95	67	8.1	103
Neb.	Lincoln	95	78	9.7	115
N.M.	Albuquerque	95	70	7.8	99
N.Y.	Buffalo	93	73	12.1	95
	New York	95	75	12.5	102
N.C.	Asheville	93	75	5.6	99
N.D.	Bismarck	95	73	9.5	114
Ohio	Cleveland	95	75	11.1	100
Okla.	Oklahoma City	101	77	9.8	113
Oreg.	Portland	90	68	6.5	107
Pa.	Philadelphia	95	78	9.7	106
S.C.	Charleston	95	78	9.8	104
Tenn.	Chattanooga	95	76	5.6	103
Tex.	Galveston	95	80	9.7	101
	San Antonio	100	78	7.8	107
Utah	Salt Lake City	95	65	9.8	105
Va.	Richmond	95	78	6.4	107
Wash.	Seattle	85	65	7.7	100
Wis.	Milwaukee	95	75	9.8	105
Wyo.	Cheyenne	95	65	9.2	100

* Temperatures abstracted from "Heating Ventilating Air-Conditioning Guide 1957." Used by permission.

except in cases where there is a considerable difference in elevation or these are unusual differences caused by factors such as intervening mountains or adjacent bodies of water. In general, wet-bulb temperatures are relatively low at high elevations.

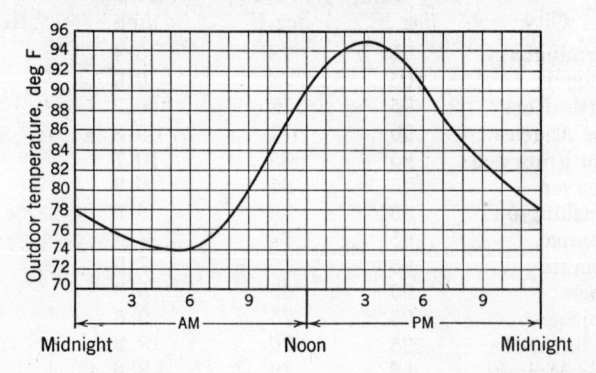

Fig. 18.1. Typical daily outdoor-air temperature variation during hot weather.

In all areas the outdoor dry-bulb temperature is lower at night than during the sunlit hours. Figure 18.1 shows a typical 24-hr plot of the outdoor dry-bulb temperature when the maximum temperature is 95 F and the range is 21 F. It may be noted that the maximum temperature characteristically occurs at 3:00 P.M. and that the minimum is likely

TABLE 18.2. Daily Range of Dry-Bulb Temperature in the United States During July*

Area	Temperature Range, deg F	Area	Temperature Range, deg F
East sea shore	12 to 18	East of Mississippi river	19 to 24
Gulf sea shore	12 to 18	Mississippi river to Rocky Mts.	24 to 33
Great Lakes shore	18 to 21	Rocky Mountain area	33 to 42
West sea shore	15 to 20	West coastal states	20 to 36

* From "Heating Ventilating Air-Conditioning Guide 1956." Used by permission.

to occur at 5:00 A.M. The daily range may be an important consideration when estimating cooling loads. This is particularly true when the peak load occurs in the early morning. Table 18.2 from reference 1 gives the typical daily range for the principal areas of the United States.

18.3. Inside-Air Temperatures. There is no general agreement among authorities in regard to what constitutes ideal conditions for

ESTIMATION OF COOLING LOADS

summer comfort. In the majority of installations which are operated solely for the purpose of producing human comfort the dry-bulb temperature is controlled at 76 F. The relative humidity is usually maintained at 50 per cent if the apparatus is arranged to control it. Because the cost of cooling equipment is proportional to its capacity and because design conditions prevail for relatively short periods, an inside dry-bulb temperature that is somewhat higher than 76 F is usually assumed in computing the cooling load. Table 18.3 from reference 1 offers typical design conditions applicable to the greater portion of the United States.

TABLE 18.3.* Design Room Conditions usually Specified for Summer Average Peak Load in Comfort Air Conditioning†

Type of installation	Dry-Bulb Temp., deg F	Wet-Bulb Temp., deg F	Relative Humidity, per cent	Grains per Lb†	Effective Temp., deg F§
Ample capacity	78	65	50	72.7	72.2
Practical application	80	67	51	78.5	74.0
Occupancy 15 to 40 min	82	68	49	80	75.3

* From "Heating Ventilating Air-Conditioning Guide 1957." Used by permission.

† Values in table are for peak load conditions. It is general practice to operate a system at 76 F and 50 per cent relative humidity at other than peak load conditions.

‡ Psychrometric data for standard atmospheric pressure.

§ Fig. 3.1.

For brief occupancy the dry-bulb temperature should be higher and the relative humidity lower than in the case of offices or homes. Dry-bulb temperatures as high as 84 F have been found to be satisfactory when accompanied by a relative humidity around 35 per cent. In the field of industrial air conditioning, indoor design conditions are established primarily with regard to their effects on goods or processes.

The cooling load calculations determine the conditions which can be guaranteed. If the thermostat is set to maintain a dry-bulb temperature that is lower than that used in the calculations, the setting can be maintained during the greater portion of the cooling season but not at peak load.

18.4. Solar Radiation. Solar radiation striking the outside surfaces of a building may contribute appreciably to the peak load on summer air-conditioning apparatus and must therefore be considered. The amount of heat that flows toward the interior of a building owing to solar radiation depends on the altitude angle of the sun, the clearness of the sky, the position of the surface with respect to the direction of the sun's rays, the absorptivity of the surface, the ratio of the overall coefficient of heat transfer of the wall to the coefficient of heat transfer of the outside-air film, and the temperature of the ground and surrounding objects with which the heated surface may interchange radiant heat.[2] The altitude angle of the sun in turn depends upon the latitude of the locality, season of year, and hour of day.

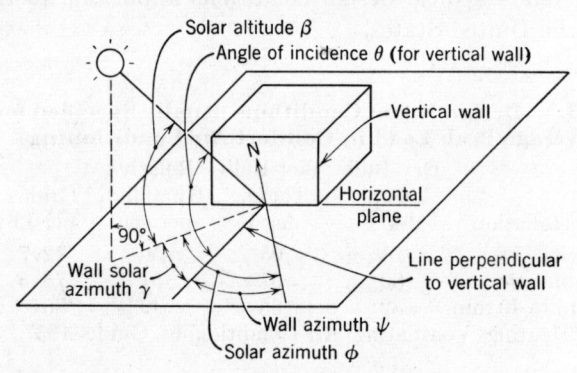

Fig. 18.2. Definitions of solar angles. (From "Heating Ventilating Air-Conditioning Guide 1957." Used by permission.)

Heat from solar radiation is received by building surfaces in two forms, namely, *direct radiation* and *sky radiation*. Direct radiation is the impingement of the sun's rays upon the surface. Sky radiation (or diffuse radiation) is received from moisture and dust particles in the atmosphere which absorb part of the energy of the sun's rays, thereby becoming heated to a temperature above that of the air. Sky radiation is received by surfaces which do not face the sun.

The various angles which affect the proportion of the direct radiation that is absorbed by a square foot of any particular surface of any specified building at any stated time are depicted in Fig. 18.2. The *solar altitude angle* β is the angle between the sun's rays and a horizontal line directly under them. The *angle of incidence* θ for a vertical wall is the angle between the sun's rays and a horizontal line normal to the plane of the wall. At exactly 12:00 noon the angle of incidence for a wall facing south is the same as the solar altitude angle. At any other time of the day the wall-solar azimuth angle γ affects the angle of incidence. The angle γ is the difference between the horizontal projection of the sun's rays and a line normal to the vertical wall of interest. For a wall facing south the angle γ is the same as ϕ, which

ESTIMATION OF COOLING LOADS

is the angle made by a horizontal projection of the sun's rays with a line pointing south. Table 18.4 gives values of γ on August 1 for 12 wall orientations, 3 different latitudes, and all of the sunlit hours. Table 18.4 also includes data on the solar altitude angle β.

In the case of a wall which has an orientation other than any of those listed in Table 18.4, it is necessary to *calculate* the wall-solar azimuth angle γ. In that situation the wall azimuth angle ψ (Fig. 18.2) must be considered. This is the angle between a line pointing south and a line normal to the wall of interest. (The angle ψ for either an east wall or a west wall would be 90 deg.) The wall-solar angle γ may be computed from the data of Table 18.4 in any case where it is not given directly. Three simple examples will illustrate the procedure.

Example 1. Calculate the angle γ at 10:00 A.M. on August 1 for a wall facing 10 deg east of south. The wall is located at 50 deg north latitude.

Solution. The angle ψ is 10 deg east of south. The angle ϕ (column "S" in Table 18.4) is 48 deg east of south. Therefore $\gamma = 48 - 10 = 38$ deg.

Example 2. Find the angle γ for the wall of Example 1 at 2:00 P.M.

Solution. The angle ψ as in Example 1 is 10 deg east of south. The angle ϕ from column "S," Table 18.4, is 48 deg west of south. The angle γ is then $48 + 10 = 58$ deg.

Example 3. Calculate the angle γ at 9:00 A.M. August 1 for a wall facing 20 deg south of east. The location is in 30 deg north latitude.

Solution. In this case the south wall is in the shade and the angle ϕ cannot be determined from column "S" in Table 18.4. However, Table 18.4 gives the angle γ at 9:00 A.M. for a wall facing east as 6 deg south of east. The angle γ for a wall facing 20 deg south of east at this time is $20 - 6 = 14$ deg.

In Example 3 it was determined that the angle γ at 9:00 A.M. is south of east by noting the variation of the figures given in the column. The same procedure and the same reasoning may be used for obtaining the angle γ in the afternoon for a wall oriented at any angle north or south of a westerly direction (see Example 8).

Following are definitions of the terms from reference 1 that will be used later in the discussion of solar radiation:

I_{Dn} = direct solar radiation normal to the sun's rays, Btuh per square foot.

I_d = sky radiation, Btuh per square foot.

I_D = direct solar radiation on any actual surface, Btuh per square foot. $I_D = K I_{Dn}$. In the case of vertical surfaces, K is the cosine of the effective angle of incidence θ (see Fig. 18.2). For horizontal surfaces K is the sine of the solar altitude angle β.

I_T = total incident solar radiation, Btuh per square foot. $I_T = I_D + I_d$.

TABLE 18.4. Values of the Wall Solar Azimuth γ for Variously Oriented Walls and Solar Altitude*

Computed for 18-deg declination, north (August 1)

Latitude	Sun Time A.M.→		Solar Altitude β, deg	Azimuth Angle γ, deg					
				N	NE	E	SE	S	SW
30 Deg North	6 A.M.	6 P.M.	9.0	74	29	16	61		
	7	5	21.5	81	36	9	54		
	8	4	34.5	88	43	2	47	shade	
	9	3	47.5	shade	51	6	39	84	
	10	2	60.0		62	17	28	73	
	11	1	72.0		83	38	7	52	shad
	12		78.0		shade	90	45	0	45
40 Deg North	5 A.M.	7 P.M.	0.5	66	21	24	69		
	6	6	11.5	76	31	14	59		
	7	5	23.0	85	40	5	50	shade	
	8	4	34.5	shade	50	5	40	85	
	9	3	45.5		61	16	29	74	
	10	2	56.0		76	31	14	59	shade
	11	1	64.5		shade	55	10	35	80
	12		68.0			90	45	0	45
50 Deg North	5 A.M.	7 P.M.	4.5	67	22	23	68		
	6	6	13.5	78	33	12	57		
	7	5	23.5	90	45	0	45	90	
	8	4	33.0	shade	57	12	33	78	
	9	3	42.0		70	25	20	65	
	10	2	50.0		87	42	3	48	shade
	11	1	56.0		shade	64	19	26	71
	12		58.0			90	45	0	45

↑ P.M. → N NW W SW S SE

*From "Heating Ventilating Air-Conditioning Guide 1957." Used by permission.

The value of I_{Dn} outside of the earth's atmosphere is thought to be approximately 420 Btuh per sq ft. The intensity of the sun's rays is considerably reduced in creating the diffuse or sky radiation as they approach the earth's surface. The amount of this reduction depends on the amount of moisture present and the degree of atmospheric pollution. Table 18.5 from reference 1 gives standardized values of I_{Dn} and I_d for various solar altitude angles and for both a clear atmosphere and one that is typical of industrial areas. Table 18.6 gives

ESTIMATION OF COOLING LOADS

TABLE 18.5. Values of I_{Dn}, Direct Solar Radiation Received at Normal Incidence at the Earth's Surface, and of I_d, Diffuse or Sky Solar Radiation, Received by Variously Oriented Surfaces*

Solar Altitude β, deg	For Clear Atmospheres, Btuh						For Industrial Atmospheres, Btuh					
	Direct† Normal Radiation	Diffuse or Sky Radiation‡ I_d					Direct§ Normal Radiation	Diffuse or Sky Radiation‡ I_d			
A.M.→ ↓	I_{Dn}	N	E	S	W	Horiz.	I_{Dn}	N	E	S	W	Horiz.
5	67	6	11	4	4	7	34	4	11	5	3	9
10	123	11	20	8	7	14	58	8	22	9	7	18
15	166	14	27	11	10	19	80	11	28	13	9	24
20	197	15	32	13	12	23	103	13	36	17	12	31
25	218	16	35	15	13	26	121	16	43	21	16	38
30	235	17	36	17	15	28	136	18	47	24	18	44
35	248	17	36	19	16	30	148	19	50	27	21	48
40	258	18	36	21	17	31	158	20	50	30	23	52
45	266	19	35	23	18	32	165	21	49	31	25	55
50	273	19	33	25	19	33	172	22	47	34	27	58
60	283	21	28	27	21	34	181	22	41	37	30	63
70	289	22	23	29	23	35	188	22	34	41	34	69
80	292	—	—	—	—	—	195	—	—	—	—	—
90	294	—	—	—	—	—	200	—	—	—	—	—
↑ P.M.→		N	W	S	E	Horiz.		N	W	S	E	Horiz.

* From "Heating Ventilating Air-Conditioning Guide 1957." Used by permission.

† Moon's proposed standard for sea level, 20 mm precipitable water vapor, 300 dust particles per cu cm, 2.8 mm Hg partial pressure of ozone.

‡ For 40 deg north latitude on about Aug. 1. Based on observations by ASHAE Laboratory at Cleveland on cloudless days during which the observed normal incidence values closely approximated the normal incidence values tabulated.

§ Derived from recommended design sol-air temperatures for New York City for a horizontal surface with absorptivity of 1.0.

values of K for variously oriented vertical surfaces and for horizontal surfaces.

The values of the solar altitude angle β and of the wall-solar azimuth angle γ given in Table 18.4 for 30, 40, and 50 deg north latitude were computed for August 1 when the *declination angle* of the sun is 18 deg north. The declination angle of the sun is the angle between the

TABLE 18.6.* Values of *K* for Variously Oriented Walls and a Horizontal Surface†

Computed for 18-Deg declination, north (August 1)

Latitude	Sun Time A.M.→ ↓		N	NE	E	SE	S	SW	Horiz.
30 Deg North	6 A.M.	6 P.M.	0.267	0.862	0.952	0.484			0.156
	7	5	0.144	0.752	0.919	0.548			0.367
	8	4	0.030	0.604	0.824	0.561			0.566
	9	3		0.427	0.672	0.524	0.068		0.737
	10	2		0.234	0.476	0.438	0.144		0.866
	11	1		0.039	0.246	0.310	0.192		0.951
	12				0.000	0.147	0.208	0.147	0.978
40 Deg North	5 A.M.	7 P.M.	0.406	0.934	0.914	0.358			0.009
	6	6	0.237	0.840	0.951	0.505			0.199
	7	5	0.079	0.705	0.919	0.594			0.391
	8	4		0.533	0.824	0.631	0.069		0.566
	9	3		0.337	0.673	0.614	0.196		0.713
	10	2		0.129	0.475	0.542	0.292		0.829
	11	1			0.246	0.424	0.354	0.076	0.903
	12				0.000	0.265	0.375	0.265	0.927
50 deg North	5 A.M.	7 P.M.	0.385	0.922	0.920	0.378			0.078
	6	6	0.199	0.813	0.951	0.532			0.233
	7	5	0.010	0.656	0.918	0.643			0.399
	8	4		0.465	0.824	0.700	0.166		0.545
	9	3		0.252	0.673	0.699	0.316		0.669
	10	2		0.030	0.475	0.642	0.433		0.766
	11	1			0.247	0.532	0.505	0.183	0.829
	12				0.000	0.375	0.530	0.375	0.848

↑
P.M.→ N   NW   W   SW   S   SE   Horiz.

* Data from "Heating Ventilating Air-Conditioning Guide 1957." Used by permission.
† For use in the expression $I_D = K I_{Dn}$.

direction of the sun's rays with respect to the earth and a plane perpendicular to the earth at the equator.

The use of the data of Tables 18.4, 18.5, and 18.6 in finding the radiation received by variously oriented surfaces on August 1 will be illustrated by several additional examples.

Example 4. Find the solar radiation received at 11:00 A.M. on August 1 by 1 sq ft of a flat roof located at 40 deg north latitude. The atmosphere at the location is clear.

ESTIMATION OF COOLING LOADS

Solution. The solar altitude angle from Table 18.4 at 11:00 A.M. is 64.5 deg. The intensity of the sun's rays on a surface normal to the sun in a clear atmosphere from Table 18.5 by interpolation is 285.7 Btuh per sq ft. The value of K from Table 18.6 is 0.903 (this is the sine of the altitude angle in the case of a horizontal surface) and the intensity of the direct beamed radiation on the surface I_D is $KI_{Dn} = 0.903 \times 285.7 = 257.5$ Btuh per sq ft. The intensity of the sky radiation on a horizontal surface, from Table 18.5 (using $\beta = 64.5$), is 34.45 Btuh per sq ft, and the magnitude of the total radiation I_T is $257.5 + 34.45 = 291.95$ Btuh per sq ft.

Example 5. Find the solar radiation received at 3:00 P.M. August 1 on a south wall located at 30 deg north latitude. The location is in an industrial atmosphere.

Solution. From Table 18.4 the solar altitude angle β at 30 deg north latitude at 3:00 P.M. is 47.5. From Table 18.5, I_{Dn} for an industrial atmosphere, by interpolation, is 168.5 Btuh per sq ft and I_d is 32.5 Btuh per sq ft. The value of K from Table 18.6 is 0.068 and I_D is $168.5 \times 0.068 = 11.46$ Btuh per sq ft. The total radiation received I_T is $11.46 + 32.5 = 43.96$ Btuh per sq ft.

In the case of any vertical surface having an orientation different from any of those listed in Table 18.6, K must be calculated. $K = \cos\beta, \cos\gamma$ (Fig. 18.2). The wall-solar azimuth angle γ can be *determined in* any case from the data of Table 18.4 as has been previously explained in this article.

Example 6. Find the value of K for a wall located at 40 deg north latitude and facing 10 deg east of south. The computation is to be made for 9:00 A.M. August 1.

Solution. The solar azimuth angle ϕ from column "S," Table 18.4, is 74 deg east (the same as the wall-solar azimuth for a wall facing south). The wall azimuth angle is 10 deg east. The wall-solar angle γ is $74 - 10 = 64$ deg. The solar altitude angle β from Table 18.4 is 45.5 deg. $K = \cos\beta \cos\gamma = \cos 45.5 \times \cos 64 = 0.7009 \times 0.4384 = 0.307$.

Example 7. Find the value of K for a wall facing 10 deg west of south with all other conditions as in the preceding example.

Solution. In this case the angle γ is $74 + 10 = 84$ deg and $K = \cos 45.5 \times \cos 84 = 0.7009 \times 0.1045 = 0.0732$.

Example 8. Find the value of K for a wall facing 20 deg south of west at 5:00 P.M.; the location is at 30 deg north latitude.

Solution. The south wall is in the shade at this time, so γ will be found by the method of Example 3. At 5:00 P.M., Table 18.4 gives γ for a west wall in latitude 30 deg north as 9 deg *north* of west. Since the wall faces 20 deg south of west the applicable angle is $9 + 20 = 29$ deg. The solar altitude angle β from Table 18.4 is 21.5 deg. $K = \cos 21.5 \times \cos 29 = 0.930 \times 0.875 = 0.813$.

Exterior surfaces do not absorb all of the solar energy which reaches them because some of it is reflected. The portion absorbed depends on the absorptivity of the material which constitutes the surface. Materials which have high emissivity values (Table 4.2) also have a high absorptivity and vice versa. Part of the solar energy which is absorbed passes to the outside air, the ground, surrounding objects, and the sky because of the resulting rise in surface temperature. The remainder of the absorbed energy passes to the interior of the building. A complete analysis of the heat gain through exterior walls and roofs due to the combined effects of air temperature differential and solar radiation is very difficult. Most designers prefer to use the equivalent temperature differential method of Art. 18.8.

18.5. Estimation of Sensible-Heat Gain Through Glass Areas. In many cases the heat gain through the glass areas in windows, doors, and panels constitutes a major portion of the load on cooling apparatus. When a sheet of glass is exposed to the sun, a portion of the direct radiation I_D is transmitted directly to the interior of the building of which it is a part. Of the portion which is not transmitted, part is absorbed and the remainder is reflected. In the absorption process, the temperature of the glass increases until it is in a position to lose heat at the same rate in an interchange of energy with surfaces and gases both inside and outside of the building. A simultaneous interchange of radiant energy takes place between the exterior surface of the glass and the heated particles in the atmosphere and between the interior surface and the various objects in the room. In addition to the radiant effects, the net heat gain into the interior of a building through a sheet of glass is affected by convection air currents on both sides. If the temperature of the glass is higher than that of the outdoor air, there will be no gain by conduction in spite of the fact that the outdoor air is warmer than that inside. In that case, a portion of the radiation absorbed by the glass will be removed by inside-air currents, another portion by outside-air currents, and the remainder by the net loss in interchanges of radiant energy between the glass and all of the surfaces it can "see."

The complete heat balance can be most simply expressed as follows:

Net heat gain = transmitted solar radiation ± heat flow by
 convective and radiative exchanges at the indoor surface (18.1)

Neglecting the heat which may be stored in the glass or which may be given up by the glass under changing conditions, the second term in equation 18.1 may be expressed as follows:

Heat flow by convective and radiative exchanges at the indoor surface
 = absorbed solar radiation ± radiative and convective exchanges at
 the outdoor surface (18.2)

ESTIMATION OF COOLING LOADS

TABLE 18.7. Instantaneous Rates of Heat Gain due to Transmitted Direct and Diffuse or Sky Solar Radiation by a Single Sheet of Unshaded Common Window Glass*

For clear atmospheres and 18-deg declination, north (August 1)
Note: For total instantaneous heat gain, add these values to the Table 18.8 values.

Latitude	Sun Time A.M.→ ↓		N	NE	E	SE	S	SW	W	NW	Horiz.
			\multicolumn{9}{c}{Instantaneous Heat Gain, Btuh per sq ft}
30 Deg North	6 A.M.	6 P.M.	25	98	108	52	5	5	5	5	17
	7	5	23	155	190	110	10	10	10	10	71
	8	4	16	148	205	136	14	13	13	13	137
	9	3	16	106	180	136	21	15	15	15	195
	10	2	17	54	128	116	34	17	16	16	241
	11	1	18	20	59	78	45	19	18	18	267
	12		18	19	19	35	49	35	19	19	276
40 Deg North	5 A.M.	7 P.M.	3	7	6	2	0	0	0	0	1
	6	6	26	116	131	67	7	6	6	6	25
	7	5	16	149	195	124	11	10	10	10	77
	8	4	14	129	205	156	18	12	12	12	137
	9	3	15	79	180	162	42	14	14	14	188
	10	2	16	31	127	148	69	16	16	16	229
	11	1	17	18	58	113	90	23	17	17	252
	12		17	17	19	64	98	64	19	17	259
50 Deg North	5 A.M.	7 P.M.	20	54	54	20	3	3	3	3	6
	6	6	25	128	149	81	8	7	7	7	34
	7	5	12	139	197	136	12	10	10	10	80
	8	4	13	107	202	171	32	12	12	12	129
	9	3	14	54	176	183	72	14	14	14	173
	10	2	15	18	124	174	110	16	15	15	206
	11	1	16	16	57	143	136	42	16	16	227
	12		16	16	18	96	144	96	18	16	234
	↑ P.M.→		N	NW	W	SW	S	SE	E	NE	Horiz.

* From "Heating Ventilating Air-Conditioning Guide 1957." Used by permission.

Values of the transmitted solar radiation (representing the first term on the right side of equation 18.1) for a single sheet of unshaded common glass are given in Table 18.7 for three latitudes for August 1. Heat flow rates into the interior of a building due to the effects described in the second term of equation 18.1 are given in Table 18.8. These values may be positive or negative as indicated in the table. The values given in Table 18.8 are based on an inside-air dry-bulb temperature of 80 F and on an outside-air dry-bulb temperature cycle with a 95 F maximum. The total heat gain is the sum of the applicable data from Table 18.7 and those from Table 18.8 (see item 1 of example of

TABLE 18.8. Instantaneous Rates of Heat Gain by Convection and Radiation from a Single Sheet of Unshaded Common Window Glass*
For clear atmospheres and 18-deg declination, North (August 1)
For 80 F indoor temperature
Note: For total instantaneous heat gain, add these values to the Table 18.7 values.

Sun Time	Dry-Bulb, deg F	North Latitude deg	\multicolumn{9}{c}{Instantaneous Heat Gain, Btuh per sq ft}								
			N	NE	E	SE	S	SW	W	NW	Hor.
5 A.M.	74		−6	−6	−6	−6	−6	−6	−6	−6	−6
6	74		−5	−4	−4	−5	−5	−6	−6	−6	−5
7	75		−5	−2	−2	−3	−5	−5	−5	−5	−3
8	77		−3	0	1	0	−2	−3	−3	−3	0
9	80		0	2	4	3	1	0	0	0	3
10	83		3	4	6	6	5	3	3	3	8
11	87		8	8	10	11	10	9	8	8	13
12	90	30, 40, 50	12	12	12	13	14	13	12	12	16
1 P.M.	93		15	15	15	16	17	17	17	15	20
2	94		16	16	16	16	18	19	19	17	21
3	95		17	17	17	17	19	21	21	19	21
4	94		16	16	16	16	17	20	20	19	19
5	93		15	15	15	15	15	18	19	18	17
6	91		13	13	13	13	13	14	15	15	13
7	87		8	8	8	8	8	8	8	8	8
8	85		6	6	6	6	6	6	6	6	6
9	83		3	3	3	3	3	3	3	3	3

* From "Heating Ventilating Air-Conditioning Guide 1957." Used by permission.

Art. 18.17). If the estimated cooling load is to be based on different conditions and it is essential that the calculations be made as accurately as possible, reference 1 recommends adding 1 Btuh per sq ft for every degree that the assumed indoor dry-bulb temperature is below 80 F and for every degree that the assumed maximum outdoor dry-bulb temperature is above 95 F. Tables giving factors for extending the data of Tables 18.7 and 18.8 to windows containing special glasses or more than one sheet of common glass are included in reference 1.

ESTIMATION OF COOLING LOADS

Information is also given in reference 1 from which the total heat gain through specially figured glass and glass blocks can be estimated.

18.6. Effect of Shading of Glass Areas on Heat Gains. Heat gains through glass of any sort can be greatly reduced by shading. The fraction of heat transmitted by unshaded glass that would be transmitted by that which is shaded in different ways is given for a variety of devices in the ASHAE Guide. *Data for three types are: canvas awnings, sides open, 0.25; inside roller shade, fully drawn, white or cream color, 0.71; inside Venetian blinds, white or cream, slats at 45 deg, 0.56; and outside Venetian blinds, white or cream, slats at 45 deg, 0.15. The shading factor is applied to the transmitted (Table 18.7) portion of the total heat gain.*

Windows and doors which are set in from the plane of the weather side of a vertical wall will have a portion of their areas shaded. The portion of a window facing the sun that is actually sunlit is given by the following equation from reference 1:

$$G_f = 1 - \frac{r_1 \tan \beta}{\cos \gamma} - r_2 \tan \gamma + \frac{r_1 r_2 \tan \beta \tan \gamma}{\cos \gamma} \qquad (18.3)$$

where G_f = the portion of the window which is sunlit.
  $r_1 = s/l$; $r_2 = s/w$.
  s = the setback of the window from the weather face of the wall, feet.
  l = the length (or height) of the window, feet.
  w = the width of the window, feet.

The altitude angle β and the wall-solar azimuth angle γ are explained in Art. 18.4.

The application of equation 18.3 will be illustrated in two examples.

Example 1. Find the portion of a south window 3 ft wide and 6 ft high which will be sunlit at 2.00 P.M. on August 1 if the location is near 40 deg north latitude. The window is set back 6 in. from the weather face of the wall.

Solution. Equation 18.3 will be used. $s = 0.5$ ft, $l = 6$ ft, and $w = 3$ ft. $r_1 = 0.5/6.0 = 0.0833$. $r_2 = 0.5/3.0 = 0.166$. β from Table 18.4 = 56 deg, $\tan \beta = 1.483$. γ from Table 18.4 = 59 deg. $\tan \gamma = 1.664$, $\cos \gamma = 0.515$.

$$G_f = 1 - \frac{0.0833 \times 1.483}{0.515} - 0.166 \times 1.664$$
$$+ \frac{0.0833 \times 0.166 \times 1.483 \times 1.664}{0.515}$$

$$= 1 - 0.240 - 0.276 + 0.066 = 0.550$$

Just a little more than one-half of the window area is actually sunlit.

Example 2. Find the heat gain through the window of Example 1 if it is assumed that it consists of one sheet of common glass. Assume an inside-air temperature of 80 F and a maximum outdoor dry-bulb temperature of 95 F.

Solution. The shading of a window due to recessing in the wall reduces the transmission of direct sunlight in proportion to the amount shaded but it does not appreciably affect that portion of the heat gain which is given by Table 18.8. The total heat gain in this case is [0.55 × 69 (Table 18.7) + 18 (Table 18.8)]3 × 6 = 1009 Btuh. (A small error results from the assumption that the diffuse sky radiation included in the data of Table 18.7 is affected by shading to the same extent as the direct radiation from the sun.)

18.7. Sol-Air Temperature. A convenient way of combining all of the factors which affect the heat entry into the weather side of a sunlit wall or roof is by the *sol-air temperature*.[3] This is the dry-bulb temperature of outdoor air which in the absence of all interchanges of radiant heat would give the same rate of heat entry into a building surface as would occur with the actual combination radiant and convective exchanges which take place between this surface and the air, sun, and sky to which it is exposed. Summer design sol-air temperatures which were used in the computation of the data of Tables 18.9 and 18.10 may be found in reference 1.

18.8 Heat Gain Through Outside Roofs and Walls. Absorption of solar radiation by the weather side of a building wall or roof does not result in an immediate gain in the interior. In the case of a heavy masonry wall, the maximum rate of heat flow from it to the inside air may occur at a time that is 12 or more hours later than that at which solar radiation was received at the maximum rate on the weather face. Because of this lag, the outer surface may be shaded or the sun may have set before equilibrium flow conditions through the wall become established. In this case, the inside surface temperature never reaches that which would result from equilibrium conditions for maximum heat entry on the outside. A convenient method of considering solar radiation and also allowing for the effect of lag is to use an *equivalent temperature differential*. Equivalent temperature differentials for roofs are given in Table 18.9 and those applying to walls are included in Table 18.10. These data from reference 1 are explained in references 4 and 5.

The equivalent temperature differentials of Tables 18.9 and 18.10 are to be used (in place of the actual difference between outside and inside temperatures) with the over-all coefficient U, and the area of the roof or wall, in computing the maximum heat gain. It is suggested in reference 1 that the outside surface conductance be taken as 4.0 instead of 6.0 in calculating U. It is also suggested that the conduc-

ESTIMATION OF COOLING LOADS

TABLE 18.9. Total Equivalent Temperature Differentials for Calculating Heat Gain Through Sunlit and Shaded Roofs*

	\multicolumn{8}{c}{Sun Time}
	\multicolumn{3}{c}{A.M.}	\multicolumn{5}{c}{P.M.}
Description of Roof Construction†	8	10	12	2	4	6	8	10	12

Light Construction Roofs—Exposed to Sun									
1″ wood‡ or									
1″ wood‡ + 1″ or 2″ insulation	12	38	54	62	50	26	10	4	0
Medium Construction Roofs—Exposed to Sun									
2″ concrete or									
2″ concrete + 1″ or 2″ insulation or	6	30	48	58	50	32	14	6	2
2″ wood‡									
2″ gypsum or									
2″ gypsum + 1″ insulation									
1″ wood‡ or									
2″ wood‡ or } + 4″ rock wool	0	20	40	52	54	42	20	10	6
2″ concrete or } in furred ceiling									
2″ gypsum }									
4″ concrete or									
4″ concrete with 2″ insulation	0	20	38	50	52	40	22	12	6
Heavy Construction Roofs—Exposed to Sun									
6″ concrete	4	6	24	38	46	44	32	18	12
6″ concrete + 2″ insulation	6	6	20	34	42	44	34	20	14
Roofs Covered with Water—Exposed to Sun									
Light construction roof with 1″ water	0	4	16	22	18	14	10	2	0
Heavy construction roof with 1″ water	−2	−2	−4	10	14	16	14	10	6
Any roof with 6″ water	−2	0	0	6	10	10	8	4	0
Roofs with Roof Sprays—Exposed to Sun									
Light construction	0	4	12	18	16	14	10	2	0
Heavy construction	−2	−2	2	8	12	14	12	10	6
Roofs in Shade									
Light construction	−4	0	6	12	14	12	8	2	0
Medium construction	−4	−2	2	8	12	12	10	6	2
Heavy construction	−2	−2	0	4	8	10	10	8	4

* From "Heating Ventilating Air-Conditioning Guide 1957." Used by permission. (A portion of the footnotes have been omitted.)
† Includes ⅜ in. felt roofing with or without slag. May also be used for shingle roof.
‡ Nominal thickness of the wood.

Notes for table:
1. *Source.* Calculated by Mackey and Wright method (see reference 4). Estimated for July in 40 deg north latitude. For typical design day where the maximum outdoor temperature is 95 F and minimum temperature at night is approximately 75 F (daily range of temperature, 20 F) mean 24-hr temperature 84 F for a room temperature of 80 F. All roofs have been assumed a dark color which absorbs 90 per cent of solar radiation, and reflects only 10 per cent.
2. *Application.* These values may be used for all normal air-conditioning estimates; usually without correction, in latitude 0 deg to 50 deg north or south when the load is calculated for the hottest weather.
3. *Peaked Roofs.* If the roof is peaked and the heat gain is primarily due to solar radiation, use for the area of the roof, the area projected on a horizontal plane.
4. *Attics.* If the ceiling is insulated and if a fan is used in the attic for positive ventilation, the total temperature differential for a roof exposed to the sun may be decreased 25 per cent.
5. See footnote 3 of Table 18.10 if correction for temperature differential or range is necessary.

TABLE 18.10. Total Equivalent Temperature Differentials for Calculating Heat Gain Through Sunlit and Shaded Walls*

	A.M.			P.M.						
	8	10	12	2	4	6	8	10	12	

North Latitude Wall Facing	Exterior Color of Wall: D = Dark, L = Light																		South Latitude Wall Facing
	D	L	D	L	D	L	D	L	D	L	D	L	D	L	D	L	D	L	
Frame																		
NE	22	10	24	12	14	10	12	10	14	14	14	14	10	10	6	4	2	2	SE
E	30	14	36	18	32	16	12	12	14	14	14	10	10	6	6	2	2	E
SE	13	6	26	16	28	18	24	16	16	14	14	10	10	6	4	2	2	NE
S	−4	−4	4	0	22	12	30	20	26	20	16	14	10	10	6	6	2	2	N
SW	−4	−4	0	−2	6	4	26	22	40	28	42	28	24	20	6	4	2	2	NW
W	−4	−4	0	0	6	6	20	12	40	28	48	34	22	22	8	8	2	2	W
NW	−4	−4	0	−2	6	4	12	10	24	20	40	26	34	24	6	4	2	2	SW
N (shade)	−4	−4	−2	−2	4	4	10	10	14	14	12	12	8	8	4	4	0	0	S (shade)
4-In. Brick or Stone Veneer + Frame																		
NE	−2	−4	24	12	20	10	10	6	12	10	14	14	12	12	10	10	6	4	SE
E	2	0	30	14	31	17	14	14	12	14	14	12	12	10	8	6	6	E
SE	2	−2	20	10	28	16	26	16	18	14	14	14	12	12	10	8	6	6	NE
S	−4	−4	−2	−2	12	6	24	16	26	18	20	16	12	12	8	8	4	4	N
SW	0	−2	0	−2	2	2	12	8	32	22	36	26	34	24	10	8	6	6	NW
W	0	−2	0	0	4	2	10	8	26	18	40	28	42	28	16	14	6	6	W
NW	−4	−4	−2	−2	2	2	8	6	12	12	30	22	34	24	12	10	6	6	SW
N (shade)	−4	−4	−2	−2	0	0	6	6	10	10	12	12	12	12	8	8	4	1	S (shade)
8-In. Hollow Tile or 8-In. Cinder Block																		
NE	0	0	0	0	20	10	16	10	10	6	12	10	14	12	12	10	8	8	SE
E	4	2	12	4	24	12	26	14	20	12	12	10	14	12	14	10	10	8	E
SE	2	0	2	0	16	8	20	12	20	14	14	12	14	12	12	10	8	6	NE
S	0	0	0	0	2	0	12	6	24	14	26	16	20	14	12	10	8	6	N
SW	2	0	2	0	2	0	6	4	12	10	26	18	30	20	26	18	8	6	NW
W	4	2	4	2	4	2	6	4	10	8	18	14	30	22	32	22	18	14	W
NW	0	0	0	0	2	0	4	2	8	6	12	10	22	18	30	22	10	8	SW
N (shade)	−2	−2	−2	−2	−2	−2	0	0	6	6	10	10	10	10	10	10	6	6	S (shade)
8-In. Brick or 12-In. Hollow Tile or 12-In. Cinder Block																		
NE	2	2	2	2	2	10	2	16	8	14	8	10	6	8	10	10	10	8	SE
E	8	6	8	6	14	8	18	10	18	10	14	8	14	10	14	10	12	10	E
SE	8	4	6	4	6	4	14	10	18	12	16	12	12	10	12	10	12	10	NE
S	4	2	4	2	4	2	4	2	10	6	16	10	16	12	12	10	10	8	N
SW	8	4	6	4	6	4	8	4	10	6	12	8	20	12	24	16	20	14	NW
W	8	4	6	4	6	4	8	6	10	6	14	8	20	16	24	16	24	16	W
NW	2	2	2	2	2	2	2	4	2	6	4	8	6	10	8	16	14	18	14	SW
N (shade)	0	0	0	0	0	0	0	2	2	6	6	8	8	8	8	6	6	S (shade)
12-In. Brick																		
NE	8	6	8	6	8	4	8	4	10	4	12	6	12	6	10	6	10	6	SE
E	12	8	12	8	12	8	10	6	12	8	14	10	14	10	14	8	14	8	E
SE	10	6	10	6	10	6	10	6	10	6	12	8	14	10	14	10	12	8	NE
S	8	6	8	6	6	4	6	4	6	4	8	4	10	6	12	8	12	8	N
SW	10	6	10	6	10	6	10	6	6	10	6	10	8	10	8	14	10	NW
W	12	8	12	8	12	8	10	6	10	6	10	6	10	6	12	8	16	10	W
NW	8	6	8	6	8	4	8	4	8	4	8	4	8	6	10	6	10	6	SW
N (shade)	4	4	2	2	2	2	2	2	2	2	2	2	2	4	4	6	6	S (shade)
8-In. Concrete or Stone or 6-In. or 8-In. Concrete Block																		
NE	4	2	4	0	16	8	14	8	10	6	12	8	12	10	10	8	8	6	SE
E	6	4	14	8	24	12	18	10	14	10	14	10	12	10	10	8	E
SE	6	2	6	4	16	10	18	12	18	12	14	12	10	12	10	10	8	N
S	2	1	2	1	4	1	12	6	16	12	18	12	14	12	10	8	8	6	NE
SW	6	2	4	2	6	2	8	4	14	10	22	16	24	16	22	16	10	8	NW
W	6	4	6	4	6	4	8	6	12	8	20	14	28	18	26	18	14	10	W
NW	4	2	4	0	4	2	4	4	6	6	12	10	20	14	22	16	8	6	SW
N (shade)	0	0	0	0	0	0	2	2	4	4	6	6	8	8	6	6	4	4	S (shade)

ESTIMATION OF COOLING LOADS

TABLE 18.10 (Continued)
Sun Time

	A.M.					P.M.				
	8	10	12	2	4	6	8	10	12	

North Latitude Wall Facing	\multicolumn{9}{c}{Exterior Color of Wall: D = Dark, L = Light}	South Latitude Wall Facing
	D	L	D	L	D	L	D	L	D	L	D	L	D	L	D	L	D	L	

12-In. Concrete or Stone

NE	6	4	6	2	6	2	14	8	14	8	10	8	10	8	12	10	10	8	SE
E	10	6	8	6	10	6	18	10	18	12	16	10	12	10	14	10	14	10	E
SE	8	4	8	4	6	4	14	8	16	10	16	10	14	10	12	10	12	10	NE
S	6	4	4	2	4	2	4	2	10	6	14	10	16	12	14	10	10	8	N
SW	8	4	8	4	6	4	6	4	8	6	10	8	18	14	20	14	18	12	NW
W	10	6	8	6	8	6	10	6	10	6	12	8	16	10	24	14	22	14	W
NW	6	4	6	2	6	2	6	4	6	4	8	6	10	8	18	12	20	14	SW
N (shade)	0	0	0	0	0	0	0	0	2	2	4	4	6	6	8	8	6	6	S (shade)

* From "Heating Ventilating Air-Conditioning Guide 1957." Used by permission. (A part of footnote 3 has been omitted.)

Notes for Table:
1. *Source.* Same as Table 18.9. A north wall has been assumed to be a wall in the shade; this is practically true. Dark colors on exterior surface of walls have been assumed to absorb 90 per cent of solar radiation and reflect 10 per cent; white colors absorb 50 per cent and reflect 50 per cent. This includes some allowance for dust and dirt since clean, fresh white paint normally absorbs only 40 per cent of solar radiation.
2. *Application.* These values may be used for all normal air-conditioning estimates, usually without corrections, when the load is calculated for the hottest weather. Correction for latitude (note 3) is necessary only where extreme accuracy is required. There may be jobs where the indoor room temperature is considerably above or below 80 F, or where the outdoor design temperature is considerably above 95 F, in which case it may be desirable to make correction to the temperature differentials shown. The solar intensity on all walls other than east and west varies considerably with time of year.
3. *Corrections. Outdoor minus room temperature.* If the outdoor maximum design temperature minus room temperature is different from the base of 15 deg, correct as follows: When the difference is greater (or less) than 15 deg, add the excess to (or subtract the deficiency from) the above differentials.
Outdoor daily range temperature. If the daily range of temperature is less than 20 deg, add 1 deg to every 2 deg lower daily range; if the daily range is greater than 20 deg, subtract 1 deg for every 2 deg higher daily range. For example, the daily range in Miami, Fla., is 12 deg, or 8 deg less than 20 deg; therefore, the correction is +4 deg.
Color of Exterior surface of wall. Use temperature differentials for light walls only where the permanence of the light wall is established by experience. For cream colors use the values for light walls. For medium colors interpolate halfway between the dark and light values. Medium colors are medium blue, medium green, bright red, light brown, unpainted wood, natural color concrete, etc. Dark blue, red, brown, green, etc., are considered dark colors.
For latitudes other than 40 *deg north; and in other months.* These table values will be approximately correct for the east or west wall in any latitude (0 deg to 50 deg north or south) during the hottest weather. In the lower latitudes when the maximum solar altitude is approximately 80 to 90 deg (the maximum occurs at noon) the temperature differential for either a south or north wall will be approximately the same as a north, or shade, wall.
4. *For insulated walls* use same temperature differentials as used for uninsulated walls.

tance of the inside film be taken as 1.2 in calculating the U for a flat roof. The gain through exterior walls is usually a small portion of the total maximum heat gain of a room or building and the error in calculating the total load, caused by using the over-all coefficient applying to winter conditions, where it is available, is usually negligible. The equivalent temperature differentials given in Tables 18.9 and 18.10 are based on an outside-inside design temperature differential of 15 F (usually 95–80) and a daily temperature range of 20 F.

Example. Calculate the heat gain at 2:00 P.M. on August 1 through 1 sq ft of a horizontal roof consisting of 4 in. of stone concrete covered with $\frac{3}{8}$ in. of felt roofing with slag. The location is 40 deg north latitude; the outdoor design dry-bulb temperature is 95 F; the design room temperature is 80 F; and the swing of the outdoor temperature is 20 deg.

Solution. The design conditions are those for which the equivalent temperatures differentials of Table 18.9 have been computed and no correction for either temperature differential or temperature range is necessary. From Table 18.9 the equivalent temperature differential for a 4-in. concrete roof at 2:00 P.M. is 50 F. The following data will be used in calculating the over-all coefficient of heat transfer for summer conditions (symbols and method are as in Chap. 4): $f_i = 1.2$, $f_o = 4.0$, K for concrete-12.0 (Table 4.1), C for built-up roofing $\frac{3}{8}$ in. thick, 3.0 (Table 4.1).

$$U = \frac{1}{\frac{1}{4.0} + \frac{1}{3.0} + \frac{4}{12.0} + \frac{1}{1.2}} = 0.573$$

The gain per square foot is $0.573 \times 50 = 28.7$ Btuh.

18.9. Conduction Heat Gains Through Interior Partitions, Floors, or Ceilings. It seldom happens that the entire interior volume of a building is air-conditioned, so it is necessary to consider the heat that may be gained through either vertical or horizontal partitions separating the conditioned spaces from other portions of the building. Heat gain from this source may be computed by application of the following equation:

$$H = UA(t_n - t_c) \tag{18.4}$$

U may be calculated or it may be obtained from a suitable table. In every case it will be necessary for the designer to estimate the temperature in the unconditioned space on the opposite side of each interior partition. *The temperature in a well ventilated room above ground level may be assumed to be the same as that of the outdoor air, the heat gain will be periodic in nature, and the partition may be treated as an outside wall facing north.*

The air temperature in a poorly vented attic may exceed 130 F.[6] The temperature[7] in an attic for which ventilation openings are provided in the roof overhangs in addition to louvers in the gable ends is not likely to exceed 115 F. Mechanical ventilation of an attic space will further reduce the air temperature. Footnotes 3 and 4 of Table 18.9 suggest a method of estimating the heat gain through the roof

ESTIMATION OF COOLING LOADS

and ceiling combination in the case of an attic that is mechanically ventilated.

Example. Find the heat gain at 4:00 P.M. through each square foot of an insulated ceiling for which the calculated U is 0.07. The attic space above, formed by a peaked roof, is mechanically ventilated. The location is at 40 deg north latitude.

Solution. From Table 18.9 the equivalent temperature differential for the roof alone at 4:00 P.M. is 50 F. The value that will be used for the ceiling below the roof (see footnote 4, Table 18.9) is $0.75 \times 50 = 37.5$ F. The calculated heat gain per square foot of ceiling area is $0.07 \times 37.5 = 2.63$ Btuh.

TABLE 18.11. Allowances to Be Made in the Estimation of Cooling Loads for the Heat and Moisture Emission of Appliances

Appliance	Sensible Heat Emitted, Btuh	Moisture Emitted, grains per hr	Increase in Latent Heat Load, Btuh
Electric lights*	Total wattage × 3.416	0	0
Electric baking ovens, toasters, food-warming receptacles, etc.	Total wattage × 2.733, assuming 80% of total	Total wattage × 4.45	Total wattage × 0.683, assuming 20% of total
Instrument sterilizer	650	7800	1200
Small Bunsen burner using natural gas	1680	2730	420
Natural gas burned in unvented burner, per cu ft of gas burned per hr	900	650	100
Manufactured gas burned in unvented burner, per cu ft gas burned	540	390	60
Continuous-flame cigar lighter	900	650	100
Coffee urn, per gal of capacity	1025	6670	1025
Heat liberated by food in a restaurant, per person	30	195	30
Steam table, per sq ft of top surface	300	5200	800
Insulated steam pipes, per sq ft surface	110	0	0

* Multiply by 1.2 in the case of fluorescent lights to allow for power consumed in the ballast.

18.10. Heat and Moisture Emission of Appliances. Appliances frequently used in air-conditioned spaces which may liberate heat or water vapor may be electrical, gas-fired, or steam-heated. Table 18.11 gives a tabulation of several of the commonly used appliances together with approximate values for the sensible heat and moisture emitted when they are used. Additional data on heat and moisture emission from appliances may be found in reference 1.

For a concentrated source of heat or moisture or both, such as a steam table in a restaurant, it may be advisable to install a hood over the heated area to collect the warm air or water vapor rising from it,

TABLE 18.12. **Sensible Heat Gain from Electric Motors**
Heat Gain, Btu per Hr per Hp of Rating

Name-Plate Rating, Hp	Connected Load in Air-Conditioned Space, Motor Outside	Motor in Air-Conditioned Space, Load Outside	Motor and Connected Load Both in Air-Conditioned Space
$\frac{1}{20}$ to $\frac{1}{8}$	2546	2354	4900
$\frac{1}{8}$ to $\frac{1}{2}$	2546	1374	3920
$\frac{1}{2}$ to 3	2546	724	3220
3 to 20	2546	454	3000
20 to 150	2546	284	2830

thus preventing it from becoming an added load on the cooling and dehumidifying equipment.

When electric motors are used to operate equipment within the conditioned space, at least a part of the heat equivalent of the power consumed will be added to the cooling load. The general equation for calculating this load is

$$H_{em} = \left(\frac{\text{Horsepower rating}}{\text{Motor efficiency}}\right) \times (\text{load factor}) \times 2545 \quad (18.5)$$

The *load factor* is the fraction of rated load actually used under normal continuous operation. Motor efficiencies vary from about 0.50 for motors smaller than $\frac{1}{8}$ hp to 0.88 for those having a rating greater than 10 hp. Table 18.12 gives recommended allowances for heat gain from electric motors in different positions with respect to the conditioned space.

18.11. Heat and Moisture Liberated by People Occupying a Conditioned Space. When a conditioned space is occupied by a large number of people, the heat and moisture emitted from their bodies may constitute the major portion of a summer cooling load. Table 18.13 gives the information needed for the computation of this portion of the total load. Table 18.13 does not give the moisture gain in terms of weight but it may be obtained in pounds per hour by dividing the latent-heat gain in Btuh by the factor 1076 (see Art. 18.1).

ESTIMATION OF COOLING LOADS

TABLE 18.13.* Rates of Heat Gain from Occupants of Conditioned Spaces†

Degree of Activity	Typical Application	Total Heat Adults, Male, Btuh	Total Heat Adjusted,‡ Btuh	Sensible Heat, Btuh	Latent Heat, Btuh
Seated at rest	Theater—Matinee	390	330	180	150
	Theater—Evening	390	350	195	155
Seated, very light work	Offices, hotels, apartments	450	400	195	205
Moderately active office work	Offices, hotels, apartments	475	450	200	250
Standing, light work; or walking slowly	Department store, retail store, dime store	550	450	200	250
Walking; seated Standing; walking slowly	Drugstore Bank	550	500	200	300
Sedentary work	Restaurant§	490	550	220	330
Light bench work	Factory	800	750	220	530
Moderate dancing	Dance hall	900	850	245	605
Walking 3 mph; Moderately heavy work	Factory	1000	1000	300	700
Bowling‖	Bowling alley				
Heavy work	Factory	1500	1450	465	985

* From "Heating Ventilating Air-Conditioning Guide 1957." Used by permission.

Note: Tabulated values are based on 80 F dry-bulb temperature. For 78 F room dry-bulb, the total heat remains the same, but the sensible-heat values should be increased by approximately 10 per cent, and the latent-heat values decreased accordingly.

‡ *Adjusted total-heat gain* is based on normal percentage of men, women, and children for the application listed, with the postulate that the gain from an adult female is 85 per cent of that for an adult male, and that the gain from a child is 75 per cent of that for an adult male.

§ Adjusted total-heat value for *sedentary work, restaurant,* includes 60 Btuh for food per individual (30 Btu sensible and 30 Btu latent).

‖ For *bowling* figure one person per ally actually bowling, and all others as sitting (400 Btuh) or standing (550 Btuh).

18.12. Heat and Moisture Brought into Conditioned Spaces by Outdoor Air.
Outdoor air may be brought in deliberately for the purpose of providing ventilation or it may come in by infiltration through cracks in the building structure. Ventilation air, deliberately introduced into an air-conditioned space, is usually brought in through the cooling unit, in which case the heat and moisture gains due to it are *not* parts of the *direct room load* and do not affect the required

state of the conditioned air. Nevertheless, they must be considered in determining the refrigerating capacity that will be required to serve the air-conditioning unit. The following expressions may be used to determine the gains in total heat and moisture due to the introduction of outdoor air:

$$H_t = M_a(h_o - h_i) \tag{18.6}$$
$$M_w = M_a(w_o - w_i) \tag{18.7}$$

In a case where the calculations are made for a time of day other than 3:00 P.M. the design wet-bulb temperature and the total heat per pound of dry air corresponding to it will not apply. In that event, determine the state of the outdoor air at the time for which the calculations are made from a psychrometric chart as illustrated in the next example.

Example. An air conditioner serves a certain room which requires fresh air at the rate of 100 cfm, measured under standard conditions. The design outdoor dry-bulb and wet-bulb temperatures commonly used in the area are 95 F and 75 F respectively. Because of a large window facing east, the peak load will occur at 8:00 A.M. The typical variation in outdoor temperature is as shown in Fig. 18.1. Find the increase in the refrigeration load in Btuh and the additional moisture which must be removed because of the ventilation air. The room dry-bulb temperature and relative humidity are to be 80 F and 50 per cent respectively.

Solution. The weight of dry air involved will be taken as $100 \times 60 \times 0.075 = 450$ lb per hr. The dry-bulb temperature of the outdoor air at 8:00 A.M. from Fig. 18.1 is 77 F and the enthalpy at that time, from Fig. 18.3, is 34.0 Btu per lb of dry air. The solution for the enthalpy of the outdoor air at 8:00 A.M. from the design conditions given for 3:00 P.M. is based on the assumption that the humidity ratio remains constant throughout the day. This is essentially true during design weather except when a storm occurs. The enthalpy of the room air, Fig. 18.3, is 31.2 and the increase in the refrigeration load is $450(34.0 - 31.2) = 1260$ Btuh. The additional moisture to be removed is $450(98.5 - 76.5) = 9900$ grains per hr or 1.415 lb per hr. The humidity ratios of the outdoor and room air were obtained from the psychrometric chart, Fig. 2.2.

Infiltration may be estimated by the method given in Chap. 4. In general, the weight of outdoor air entering a building by infiltration is less in summer than during cold weather. Infiltration of outdoor air can usually be neglected if large amounts of ventilation air are used because most mechanical ventilation systems are designed to slightly pressurize the building.

18.13. Moisture Gain Through Permeable Walls, Ceilings, and Floors. During summer weather the vapor pressure of the outdoor air is greater than that in air-conditioned spaces and there will be

ESTIMATION OF COOLING LOADS

some passage of water vapor into these spaces through any wall, ceiling, or floor which is not provided with an effective vapor barrier (see Art. 4.26). Any water vapor gaining entry by this means will add to the latent-heat load on the conditioner. In general, the moisture gain by transmission through the walls, floor, and ceiling of a conditioned space is a negligible portion of the total moisture gain.

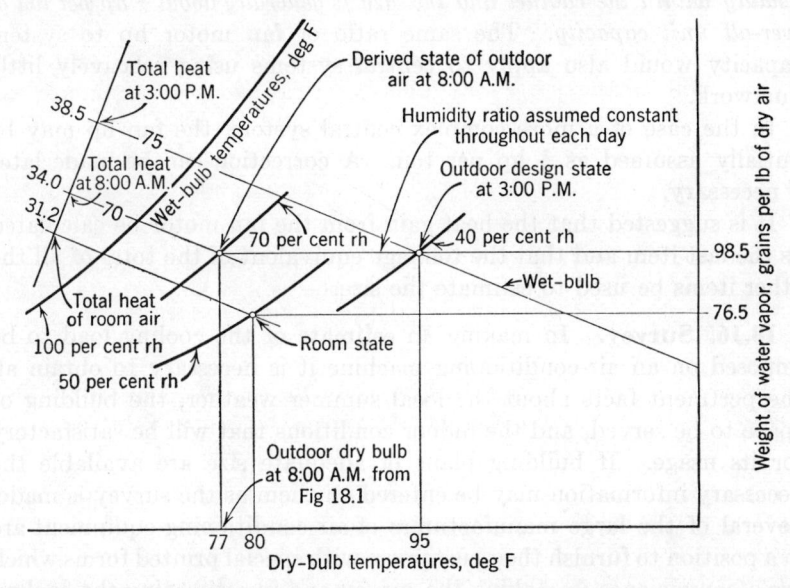

Fig. 18.3. Skeleton psychrometric chart illustrating a method of determining the enthalpy of the outdoor air at a time of day other than that for which the design dry-bulb and wet-bulb temperatures are stated.

18.14. Heat Gain Through Duct Walls. In general, heat gains through duct walls may be neglected, but where long ducts must pass through unconditioned spaces or where ducts pass through an enclosure such as a boiler room or an unventilated attic they should be estimated and included in the total cooling load (see Art. 14.24).

18.15. Heat Gain from Fan Work. The energy delivered by the blades of a fan in an air-conditioning system is eventually converted into sensible heat by friction. This heat becomes part of the sensible heat which must be removed by the cooling unit. The heat gain from fan work is equal to the shaft hp times 2545 expressed in Btuh.

If the motor driving the fan is outside of the conditioned space there will be no additional heat gain from it. If this motor is located within the conditioned area or in the air stream passing through the conditioner, it should be considered as a motor and connected load within the conditioned space (Table 18.12). *In this situation the heat gain from Table 18.12 or from equation 18.5 includes the fan work.*

The exact magnitude of the fan work and the exact size of the fan motor cannot be determined until the design of the entire system has been completed. *In packaged conditioners (Fig. 20.1) the fan motor is usually within the cabinet and the size is generally about $\frac{1}{8}$ hp per ton of over-all unit capacity.* The same ratio of fan motor hp to system capacity would also apply to central systems using relatively little ductwork.

In the case of a more complex central system, the fan hp may be initially assumed as $\frac{1}{8}$ hp per ton. A correction can be made later if necessary.

It is suggested that the heat gain from the fan motor be calculated as the last item and that the tonnage equivalent of the total of all the other items be used to estimate the size.

18.16. Survey. In making an estimate of the cooling load to be imposed on an air-conditioning machine it is necessary to obtain all the pertinent facts about the local summer weather, the building or space to be served, and the indoor conditions that will be satisfactory for its usage. If building plans of adequate size are available the necessary information may be entered on them as the survey is made. Several of the large manufacturers of air-conditioning equipment are in a position to furnish their customers with special printed forms which are a convenience in making the survey and in estimating the cooling load.

18.17. Example of Cooling-Load Calculations. There is no standard procedure which may be used in estimating the cooling load of a proposed summer air-conditioning plant because each individual installation is likely to present problems which are peculiar to it alone. The purpose of the following illustrative calculations is to indicate a method of approach which in a general way is applicable to all problems of this type.

Example. Figure 18.4 shows a plan of a restaurant which is to be air-conditioned. The following information was obtained from a thorough preliminary survey. Design conditions: outside 95 F dbt, 76 F wbt, range 20 deg; inside 80 F dbt, 50 per cent relative humidity. Building data: location near 40 deg north latitude; ceiling height, 12 ft; outer walls 8 in. thick, 4 in. of face brick and 4 in. of common brick, with $\frac{1}{2}$ in. of gypsum plaster on the inside face (U value, either computed as in Chap. 4 or obtained from a table in reference 1, is 0.46); common wall with adjacent store, 8 in. of common brick with $\frac{1}{2}$ in. of gypsum plaster on both faces ($U = 0.318$); partition between

ESTIMATION OF COOLING LOADS

the serving area and the kitchen constructed of three-cell gypsum partition tile ($U = 0.33$); windows of single-thickness common glass, 8 ft high and equipped with canvas awnings, sides open; and horizontal roof of 6-in. concrete (sand and gravel aggregate), provided with 2 in. of glass fiber insulation on top and covered with built-up felt roofing $\frac{3}{8}$ in. thick ($U = 0.10$). The building includes a basement. Equipment located in the area to be conditioned; three coffee urns, 3-gal capacity each; one 5000-watt toaster; and total lighting 6000 watts, fluorescent type. Miscellaneous information: the restaurant is open from 6:00 A.M. until 10:00 P.M. The rush hours are from 7:30 to 8:30

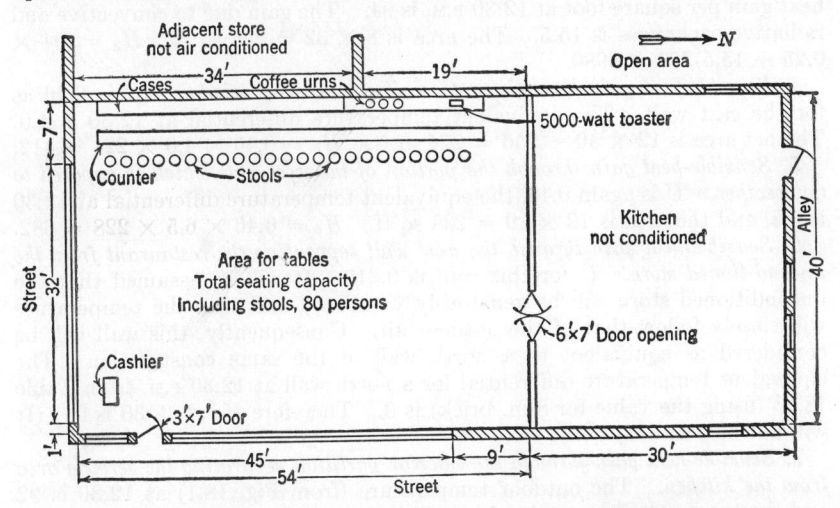

Fig. 18.4. Plan of restaurant to be air-conditioned.

A.M., 12:00 noon to 1:00 P.M., and 5:30 to 6:30 P.M. The entire seating capacity (80 persons) may be used during the rush hours. One cashier and eight waitresses work in the serving area during those periods. The temperature in the kitchen was found to be approximately 10 deg above that of the outdoor air during rush hours. Both the circulating fan and its motor are located in the conditioned area.

Solution. Because of the large heat and moisture gain from the occupants during the rush hours, the peak load on the cooling apparatus is certain to occur at one of those periods. The awnings on the east windows will minimize the solar effect on them during the morning rush hour, so it appears likely that the peak load will occur either around 12:30 P.M. or 6:00 P.M. The total-heat and moisture gain from all sources will be calculated for each of these two periods for August 1. The calculations for the mid-point of the lunch-time rush hour are made as follows (all of the heat flow rates are in Btuh per sq ft and the final answer in the case of each item is in Btuh):

1. *Sensible-heat gain through east window and door.* (The door will be considered as part of the window.) The heat gain due to transmitted solar radiation through unshaded single common glass, facing east, at 12:30, from Table 18.7 (40 deg north latitude), is 18. The shading factor due to the canvas awnings, from Art. 18.6, is 0.25. The heat flow into the room due to convec-

tive and radiative exchanges, from Table 18.8, is 13.5. The area is $8 \times 45 = 360$ sq ft. $H_1 = (18 \times 0.25 + 13.5)360 = 6480$. This is also the total heat gain from this source as there is no moisture gain.

2. *Sensible-heat gain through east wall exclusive of window and door.* The U valve has been previously determined as 0.46. The equivalent temperature differential by interpolation from Table 18.10 (assume color is dark) is 15 deg. The net area is $(12 \times 54) - 360 = 288$ sq ft. $H_2 = 0.46 \times 15 \times 288 = 1985$ Btuh.

3. *Sensible-heat gain through south window.* The transmitted portion of the heat gain per square foot at 12:30 P.M. is 94. The gain due to convective and radiative exchanges is 15.5. The area is $8 \times 32 = 256$ sq ft. $H_3 = (94 \times 0.25 + 15.5)256 = 9980$.

4. *Sensible-heat gain through south wall exclusive of window.* U is 0.46 as for the east wall. The equivalent temperature differential at 12:30 is 4.0. The net area is $12 \times 40 - 256 = 224$ sq ft. $H_4 = 0.46 \times 4.0 \times 224 = 412$.

5. *Sensible-heat gain through the portion of the west wall which is exposed to the weather.* U is again 0.46, the equivalent temperature differential at 12:30 is 6.5, and the area is $12 \times 19 = 228$ sq ft. $H_5 = 0.46 \times 6.5 \times 228 = 682$.

6. *Sensible-heat gain through the west wall separating the restaurant from the unconditioned store.* U for this wall is 0.318. It will be assumed that the unconditioned store will be generously ventilated and that the temperature will closely follow that of the outdoor air. Consequently, this wall will be considered as equivalent to a north wall of the same construction. The equivalent temperature differential for a north wall at 12:30 P.M. from Table 18.10 (using the value for 8-in. brick) is 0. Therefore H_6 at 12:30 is 0. (It will not be 0 at 6:00 P.M.).

7. *Sensible-heat gain through the interior partition separating the serving area from the kitchen.* The outdoor temperature (from Fig. 18.1) at 12:30 is 92 and the temperature in the kitchen will be assumed as $92 + 10 = 102$. The U valve is 0.33. The door will be considered as part of the partition and the area is $12 \times 40 = 480$ sq ft. $H_7 = 0.33 \times 480 \times (102 - 80) = 3490$.

8. *Sensible-heat gain through roof.* The U value for the roof is 0.10. The roof area is $40 \times 54 = 2160$ sq ft. The equivalent temperature differential from Table 18.9 at 12:30 P.M. is 23.5. $H_8 = 0.10 \times 2160 \times 23.5 = 5130$.

9. *Sensible-heat gain from lighting.* The sensible-heat equivalent of a kilowatt-hour is given in Table 18.11 as 3.416 Btu. As mentioned in the footnote, a factor of 1.2 must be used in the case of fluorescent lighting. $H_9 = 6000 \times 1.2 \times 3.416 = 24,650$.

10. *Total-heat and moisture gain from toaster.* Food-warming receptacles such as toasters evaporate some of the moisture contained in the material placed in them. Consequently, some of the electrical energy enters the room in the form of water vapor. Assuming that the toaster will be used continuously during the rush hours and using the data of Table 18.11, the total-heat gain H_{10} is $5000(2.733 + 0.683) = 5000 \times 3.416 = 17,080$. The moisture emitted into the room is $5000 \times 4.45 = 22,250$ grains per hr or $22,250 \div 7000 = 3.18$ lb per hr.

11. *Total-heat and moisture gains from three coffee urns.* Each urn has a capacity of 3 gal and from Table 18.11 the total-heat gain per gallon of capacity is $1025 + 1025 = 2050$ Btuh. $H_{11} = 3 \times 3 \times 2050 = 18,450$. The moisture gain is $3 \times 3 \times 6670 = 60,000$ grains per hr or 8.57 lb per hr.

ESTIMATION OF COOLING LOADS

12. *Total-heat and moisture gain from customers and cashier.* From Table 18.13 the total-heat gain from restaurant customers, including the gain from the warm food served, is 550 Btuh for each, consisting of 220 Btu of sensible and 330 Btu of latent heat. The cashier will be included with the customers. $H_{12} = 81 \times 550 = 44{,}650$. The moisture gain is $81 \times 330/1076 = 24.85$ lb per hr.

13. *Total-heat and moisture gain from workers.* The waitresses will be walking rapidly and carrying loaded trays, so they will be classified as doing moderately heavy work (see Table 18.13). $H_{13} = 8 \times 1000 = 8000$. Table 18.13 gives the latent-heat gain for this type of activity as 700 Btuh. The moisture gain is $8 \times 700/1076 = 5.20$ lb per hr.

14. *Total-heat and moisture gain due to the introduction of outside air for the purpose of ventilation.* If the ventilation air is brought in through the conditioner, this item is not a *direct gain* (see Art. 18.12). It is always a part of the total load on the cooling unit and must be estimated. The fresh-air intake of the system will be designed to admit 12.5 cfm per person measured under standard conditions (Art. 15.3). A check shows that this will be more than one air change per hour. The weight of ventilation air is $12.5 \times 89 \times 60 \times 0.075 = 5010$ lb per hr. (For convenience this will be taken as the weight of dry air.) The enthalpy of the outdoor air for 12:30 P.M. when the dry-bulb temperature is 92 (Fig. 18.1) obtained by the method illustrated in Fig. 18.3 (95 F dtb and 76 F wbt at 3:00 P.M.) is 38.65 Btu per lb of dry air. The enthalpy of the room air at 80 F and 50 per cent relative humidity is 31.2 and the total-heat gain H_{14} is $5010 (38.65 - 31.2) = 37{,}300$. The humidity ratio of the outdoor air (assumed constant throughout the day) is 105 grains per lb of dry air and that of the room air is 76.5. The moisture gain is $5010(105 - 76.5)/7000 = 20.4$ lb per hr.

15. *Sensible-heat gain from fan motor.* It will be assumed that the general arrangement of the apparatus will be as indicated in Fig. 20.1, in which case the heat equivalent of the power supplied to the fan motor will appear as a direct sensible-heat gain. A preliminary summation of all of the preceding items indicates a total of approximately 178,289 Btuh which would require 14.8 tons of refrigerating capacity. The size of the fan motor will be taken as $14.8/8 = 1.85$ (Art. 18.15). $H_{15} = 1.85 \times 3220 = 5950$ (see Table 18.12).

Because of the outdoor air introduced for the purpose of ventilation, a slight pressure will be maintained in the conditioned area, and infiltration of outside air may be neglected. Moisture gain through the walls would be a negligible factor in this restaurant. Since all the area to be served by the machine is in one room, there will be no long ducts, and heat gain through the walls of the distributing ducts can be neglected.

By duplicating the foregoing procedure, calculations may be made for 6:00 P.M. The greater of the two totals will be the heat gain used in the design of the conditioning unit. All the pertinent data from the two sets of calculations are given in Table 18.14.

By a coincidence the estimated total-heat gain for this restaurant is practically the same at 6:00 P.M. as at 12:30 P.M. Some of the heat gain items are larger at the later hour, but the increase in these items is almost exactly balanced by the decrease in certain others. The maximum heat gain of any room or building which is to be air-conditioned can be estimated by application of the same general procedure that was used in the foregoing example.

However, the detailed items must be suited to the particular case at hand. Items which are extremely important in one space may be negligible in another used for a different purpose, and vice versa. In making the calculations for certain other types of buildings, it may be necessary to estimate the total heat gain for more than two periods during the day in order to definitely determine the maximum.

TABLE 18.14. Tabulation of Heat Gains for Restaurant Shown in Fig. 18.4

	At 12:30 P.M.		At 6:00 P.M.	
Item No. Referring to Previous Calculations	Total Heat, Btuh	Moisture, lb per hr	Total Heat, Btuh	Moisture, lb per hr
1	6,480	0	5,220	0
2	1,985	0	1,855	0
3	9,980	0	3,780	0
4	412	0	1,648	0
5	682	0	1,470	0
6	0	0	778	0
7	3,490	0	3,250	0
8	5,130	0	9,500	0
9	24,650	0	24,650	0
10	17,080	3.18	17,080	3.18
11	18,450	8.57	18,450	8.57
12	44,650	24.85	44,650	24.85
13	8,000	5.20	8,000	5.20
14	37,300	20.40	36,050	20.40
15	5,950	0	5,950	0
Totals	184,239	62.20	182,331	62.20

18.18. Estimating Heat and Moisture Gain of a Residence. The heat gain of a residence is relatively small in relation to the mass of the interior walls, floors, ceilings, and furnishings which warm and cool with changes in the inside temperature. Studies[8] have shown that the temperature in some houses which are not air-conditioned (doors and windows closed) varies only 5 to 7 F during a 24-hr period of hot weather while the variation in outdoor temperature is as much as 30 deg. Due to this *flywheel effect* which tends to retard change in interior temperature, an air conditioner which has a capacity considerably less than that equal to the peak load, will permit a rise in temperature of only a few degrees above the thermostat setting under the most severe conditions. Two prominent trade associations have published tabulated data and simplified procedures[9, 10] for estimating the heat and moisture gains of residences. Both methods take into consideration this flywheel effect and result in the selection of refrigerating equipment that has a capacity less than the actual peak load. When this method of sizing a cooling unit is used, it should be realized that the indoor-air temperature will rise above the thermostat setting when the total-heat gain is at a maximum rate. From limited experience with this method of sizing cooling equipment it appears at the time of this writing that the results are generally acceptable to the average home owner.

REFERENCES

1. "Heating Ventilating Air-Conditioning Guide," ASHAE, New York (1956).
2. "Radiative Environment of Buildings and Its Effect on Heat Transmission," A. C. Pallot, *J. Inst. Heating and Ventilating Engrs. (London)*, **22**, No. 224, 1 (Apr. 1954).
3. "Summer Weather Data and Sol-Air Temperature—Study of Data for New York City," C. O. Mackey and E. B. Watson, *Heating, Piping, Air Conditioning*, **16**, No. 11, 651 (Nov. 1944).
4. "Periodic Heat Flow—Homogeneous Walls or Roofs," C. O. Mackey and L. T. Wright, Jr., *ASHVE Trans.*, **50**, 293 (1944).
5. "Solar Heat Gain Through Walls and Roofs for Cooling Load Calculations," J. P. Stewart, *ASHVE Trans.* **54**, 361 (1948).
6. "Cooling a Small Residence Using a Perimeter-Loop Duct System," D. R. Bahnfleth, C. F. Chen, and H. T. Gilkey, *ASHVE Trans.*, **60**, 271 (1954).
7. "Cooling a Small Residence with a Two-Horsepower Mechanical Condensing Unit," H. T. Gilkey, D. R. Bahnfleth, and R. W. Roose, *ASHVE Trans.*, **59**, 283 (1953).
8. "Short Cut for Residential Cooling Load Calculations," W. A. Grant, *Refrig. Eng.*, **61**, No. 5, 519 (May 1953).
9. "Design and Installation of Summer Air Conditioning," *Manual No. 11*, National Warm Air Heating and Air Conditioning Association, Cleveland (1955).
10. "Cooling Load Calculation Guide," *Guide C-30*, Institute of Boiler and Radiator Manufacturers, New York (1956).

BIBLIOGRAPHY

1. "Analogue Computer Analysis of Residential Cooling Loads," T. N. Willcox, C. T. Oergel, S. G Reque, C. M. toeLaer, and W. R. Briskin, *ASHVE Trans.*, **60**, 505 (1954).
2. "Circuit Analysis Applied to Load Estimating," H. B. Nottage and G. V. Parmalee, *ASHVE Trans.*, **60**, 59 (1954).
3. "Selection of Outside Design Temperature for Heat Load Estimation," M. L.

Ghai and R. Sundaram, *Heating, Piping, Air Conditioning*, **26,** No. 10, 137 (Oct. 1954).
4. "Radiant Energy Emission of Atmosphere and Ground—A Design Factor in Heat Gain and Heat Loss," G. V. Parmalee and W. W. Aubele, *Heating, Piping, Air Conditioning*, **23,** No. 11, 120 (Nov. 1951).
5. "Handling Sun Loads on Buildings," W. B. Foxhall, *Heating and Ventilating*, **48,** No. 4, 76 (Apr. 1951).
6. "Solar Energy Transmittance of Figured Rolled Glass," G. V. Parmalee and W. W. Aubele, *Heating, Piping, Air Conditioning*, **23,** No. 2, 124 (Feb. 1951).
7. "Heat Gain Through Glass Skylight Fenestrations," D. J. Vild and G. V. Parmalee, *Heating, Piping, Air Conditioning*, **28,** No. 1, 201 (Jan. 1956).
8. "Reducing Heat Gain Through the Roof," A. W. Carroll, *Air Conditioning, Heating, Ventilating*, **53,** No. 7, 61 (July, 1956).
9. "A New Concept for Cooling Degree Days," E. C. Thom, *Air Conditioning, Heating, Ventilating*, **54,** No. 6, 73 (June 1957)

PROBLEMS

1. Find the solar radiation received on 1 sq ft of a flat roof located at 30 deg north latitude at 3:00 P.M. on August 1. The roof is located in an industrial atmosphere.

2. Find the solar radiation in Btuh per sq ft received at 3:00 P.M., August 1, on a west wall located at 50 deg north latitude. The location is in a small town with no heavy industry.

3. Find the solar radiation received in Btuh per sq ft at 10:00 A.M., August 1, on a wall facing east. The location is in Pittsburgh, Pa. (40 deg north latitude).

4. Find the solar radiation in Btuh per sq ft received on a wall facing 10 deg east of south at 2:00 P.M. on August 1. The location is at 30 deg north latitude and the building is in an industrial type of atmosphere.

5. Same as Problem 4 except that the wall is facing 20 deg north of west.

6. Estimate the total-heat gain at 9:00 A.M. on August 1 through 1 sq ft of unshaded common window glass facing east and located near 50 deg north latitude. The design outside and inside dry-bulb temperatures are 95 F and 80 F respectively.

7. Same as Problem 6 except that the window is shaded by a canvas awning with the sides open.

8. Find the heat gain at 3:00 P.M. on August 1 through a window of common glass in a west wall of a building located near 50 deg north latitude. The window is $3\frac{1}{2}$ ft wide, 7 ft high, and is recessed 8 in. from the plane of the exterior wall surface. There is no other shading. The design temperatures are as in Problem 6.

9. Calculate the heat gain in Btuh per sq ft at 4:00 P.M. on August 1 through a flat concrete roof 6 in. thick. The concrete was made with a lightweight aggregate and weighed approximately 80 lb per cu ft. The design outdoor dry-bulb temperature is 95 deg (daily range 20 deg) and the design room temperature is 80 F. The concrete deck is covered with built-up roofing $\frac{3}{8}$ in. thick.

ESTIMATION OF COOLING LOADS

10. Same as Problem 9 except that the outdoor design dry-bulb temperature is 100 F and the inside design dry-bulb temperature is 75 F.

11. Same as Problem 9 except that the daily range in outside dry-bulb temperature is 30 F.

12. Calculate the heat gain in Btuh per square foot at 4:00 P.M. on August 1 through 1 sq ft of a 12-in. brick wall furred on the inside with metal lath and plaster. The wall is located in a north latitude and faces south. The outside design dry-bulb is 95 F and the inside design dry-bulb is 80 F. The U value for the wall is 0.25 under winter conditions.

13. Same as Problem 12 except that the wall is frame consisting of wood siding and insulation board $2\frac{5}{3\frac{1}{2}}$ in. thick on the outside, and metal lath and plaster on the inside. There is no insulation between the studs. The exterior is painted white. The U value for winter conditions is 0.20.

14. Estimate the heat gain in Btuh per sq ft at 6:00 P.M. on August 1 into a conditioned space maintained at 80 F from an unconditioned area through an 8-in. concrete block partition. The unconditioned space is freely ventilated and the outside design dry-bulb temperature is 95 F. The daily range in outdoor dry-bulb temperature is 20 F. The concrete blocks were made with cinder aggregate and the partition is plastered on both sides with $\frac{1}{2}$ in. of gypsum plaster (sand aggregate) and is painted a light color on both sides.

15. Find the heat gain at 2:00 P.M. on August 1 through the ceiling of an air-conditioned home 24 ft wide and 32 ft long. The house has a peaked roof of light construction wood with no insulation. The summer coefficient of heat transfer for the ceiling construction is 0.06 Btuh per sq ft per deg F. The outside design dry-bulb temperature is 95 F and the daily range characteristic of the area is 20 F. The attic is mechanically ventilated.

16. Calculate the increase in the refrigeration load that will result if the volume of ventilation air supplied to a certain air-conditioned factory building is increased by 1000 cfm. The design conditions based on 3:00 P.M. are 95 F dbt and 78 F wbt. Due to a certain process carried out only in the morning, the peak load on the system occurs at 9:00 A.M. The daily variation in outdoor dry-bulb temperature is similar to that shown in Fig. 18.1. The inside conditions are to be maintained at 85 F dbt and 30 per cent relative humidity. Base the weight of ventilation air on the actual outdoor conditions at the time of peak load on a design day. Assume standard barometric pressure.

17. Furnish all of the detailed calculations for the heat gains at 6:00 P.M. in the example of Art. 18.17.

18. Estimate the refrigeration capacity in tons that will be needed to serve the restaurant of the example of Art. 18.17 at 8:00 A.M. Assume the same occupancy and the same use of appliances as at 12:30 P.M. Estimate the weight of moisture in pounds per hour that will be removed from the air by the dehumidifying action of the conditioner at 8:00 A.M.

19

Apparatus for Producing Comfort in Summer

19.1. Use of Fans for Producton of Comfort in Summer. Circulation of night air, induced by a suitable fan placed in the attic, can bring a considerable measure of relief from the discomfort of hot summer evenings. Figure 19.1 shows a feasible arrangement and suggests a possible schedule for a two-story home in which the living room is located on the first story and the bedrooms are on the second story. The fan draws comparatively cool night air through the open windows of the rooms that are being used, and the combination of lowered air temperature and increased air movement usually produces a tolerable effective temperature in the path of the flowing air. Reference 1 suggests that the fan selected should be able to produce 1 air change every 2 to 3 min in the northern part of the United States and 1 air change per min in the south.

Attic fans are now available which are designed to operate in a horizontal position (vertical discharge) which makes it possible to mount them in a specially provided opening in the ceiling structure. Attic fans of this type are usually provided with an automatic shutter.

Adequate area of the openings in the attic structure through which the air handled by the fan can escape to the outside is essential to the satisfactory performance of the fan. The free area velocity through

APPARATUS FOR COOLING AIR

these openings must never exceed 1000 fpm and should preferably not exceed 600 fpm. The required area may be provided by louvered openings in the gable ends as indicated or by open windows in dormers or by openings in overhanging eaves called *soffits*. The free area in every case is less than that of the opening if there is any protective device to keep out rain, birds, or insects. The free area in percentage of the area of the opening may be assumed to be 50 per cent for wooden louvers only and 70 per cent for metal louvers only. If 16-mesh

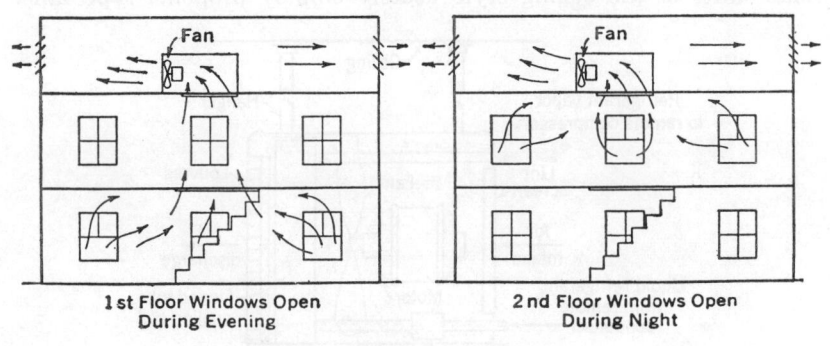

1st Floor Windows Open
During Evening

2nd Floor Windows Open
During Night

Fig. 19.1. Night-air cooling system.

screen is used to exclude insects the free area should be taken as 50 per cent of the gross area regardless of the type of louver used outside of it.

If the use of an attic fan is not feasible, circulation of night air through a home or an apartment may be produced by a fan of suitable design placed in a window and directed so as to exhaust air from the room in which it is located. The room selected for the fan location should preferably be the one that is used the least during the evening and night, and all windows in this room must remain closed except the one occupied by the fan.

Although circulation of night air is a very practicable method of reducing hot-weather discomfort, completely satisfactory summer air conditioning requires the employment of equipment which is much more intricate.

19.2. Remote Room-Cooling Units. The simplest type of summer air conditioner is the remote room-cooling unit, the principal parts of which are a motor-driven fan and a finned-tube coil. The cooling medium supplied to the coil may be either liquid refrigerant, chilled water, or simply cold water from either a well or a city water system. Satisfactory results can be achieved with water from a well if it is available at temperatures which do not exceed 60 F at any time. In addition to the principal parts which have been mentioned, remote room coolers usually include some sort of air filter, a set of louvers to direct the flow of the cooled air, and a suitable pan with drain connection for handling the water which is removed from the air stream.

Remote units are made in two distinctly different types. One style is arranged for suspension from the ceiling as shown in Fig. 19.2; the other is designed for placement on the floor as shown in Fig. 19.3. Small units of the ceiling style usually employ propeller-type fans.

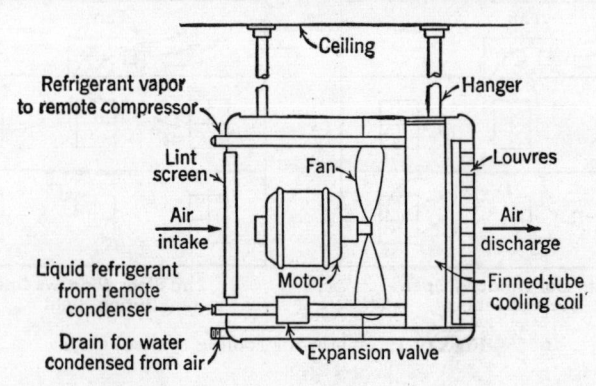

Fig. 19.2. Suspended propeller-fan-type remote unit air conditioner. (From "Heating Ventilating Air-Conditioning Guide 1957." Used by permission.)

However, many horizontal suspended units incorporate centrifugal fans as in the case of the all-year conditioner of Fig. 20.1. Floor-type units invariably employ centrifugal fans and discharge the cooled air upward as illustrated in Fig. 19.3. Units of the suspended type are well adapted to the cooling of stores, restaurants, and other commercial establishments.

Both of the units shown in Figs. 19.2 and 19.3 are provided with direct expansion coils which must be supplied with liquid refrigerant from a remote condenser. The compressor and condenser may be located in the basement of the building as shown in Fig. 19.3a, in a closet having air inlet and outlet openings in an exterior wall as in Fig. 19.3b, in a well-ventilated attic, or outdoors if the casing is designed for this location. (The condenser of Fig. 19.3a is water-cooled.)

The direct expansion coil of any remote unit may be replaced by a coil which is designed to circulate either chilled water or chilled brine, provided that the co-ordinating refrigerating unit is adapted to the cooling of these circulating media.

APPARATUS FOR COOLING AIR

19.3. Self-Contained Room-Cooling Units. A self-contained room-cooling unit incorporates a small mechanical refrigerating plant and does not require a supply of either liquid refrigerant or chilled water. However, some means must be provided for carrying away the heat that is given up by the refrigerant as it passes through the condenser. This may be accomplished by a stream of outdoor air, handled by a serparate fan in an air-cooled condenser, or by means of cooling water. The air-cooled condenser is the type in most common use in self-contained room-cooling units. The unit of Fig. 19.4 is

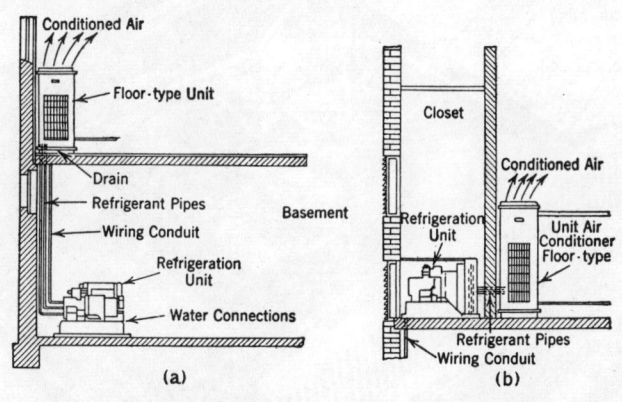

Fig. 19.3. Floor-type remote unit air conditioner. (Courtesy Westinghouse Electric Corp.)

equipped with an air-cooled condenser and is especially arranged for placement in the lower part of a window opening.

Figure 19.5 shows a typical self-contained room-cooling unit which employs a water-cooled condenser. This arrangement can be located at any point in the room which it serves. A water pipe from a suitable source of cooling water is required, and a conduit must be provided to carry the warmed water either to a drain or to a cooling device.

19.4. Forced-Circulation Air Coolers. A forced-circulation air cooler may have its parts arranged essentially the same as is shown in Fig. 19.5 except that collars for the attachment of ducts are provided in place of the supply-air grilles shown. This type of system is intended to serve more than one room either in homes or in commercial buildings. The fan used must be a design which is capable of creating the static pressure needed for overcoming the resistance of the ducts. An air-cooled condenser, placed outside of the conditioned area, may be used in place of the water-cooled type that is shown in Fig. 19.5. Figure 19.6a shows a downflow type of forced-circulation air cooler that is suitable for use in a basementless house which has a perimeter heating system. The schematic diagram in Fig. 19.6b shows this type of cooling unit as a part of a year-round air-conditioning system for a home. The cooling unit is located in a utility room, and the air-cooled

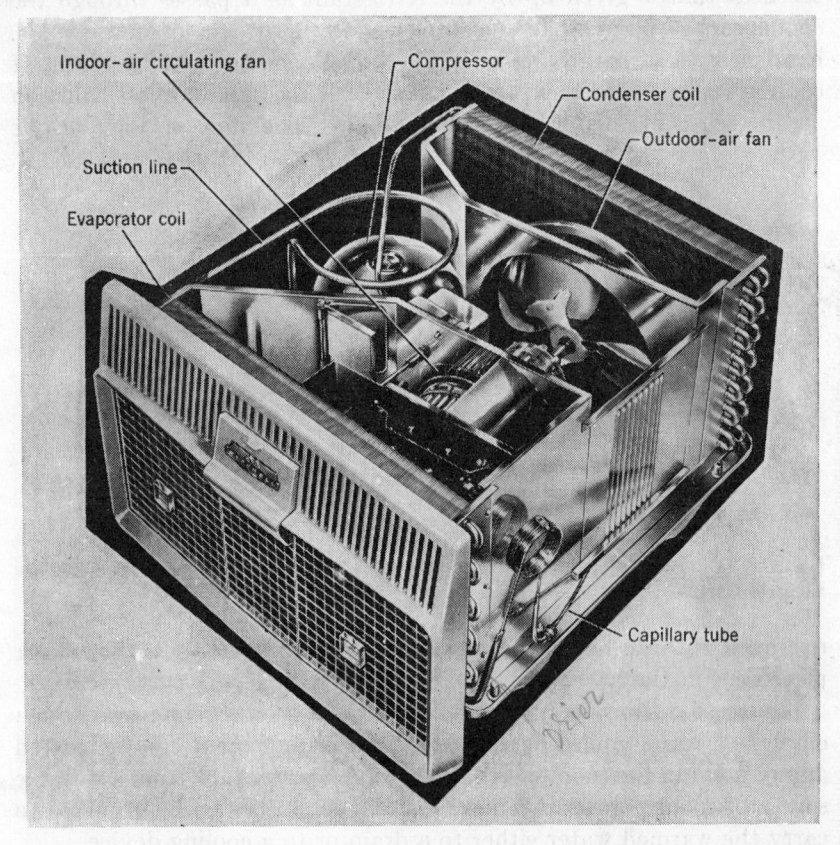

Fig. 19.4. Window type of self-contained room air conditioner. (Courtesy Mitchell Manufacturing Co.)

condenser which supplies liquid refrigerant to the air-cooling coil is located on a small concrete slab outside the house.

19.5. Summer Central Systems. A central system for summer air conditioning may employ any one of many different arrangements of apparatus. When the building usage is such as to require 100 per cent fresh air, as might be the case with a theater, arrangement 1 in Fig. 19.7 is the one most commonly used. When more than one air-conditioning system is to be installed in a building, installation and

APPARATUS FOR COOLING AIR

maintenance costs may be reduced by substituting a chilled water coil for the direct expansion coil in each system, in which case a central water-chilling unit such as the one shown in Fig. 17.11 supplies chilled water to several cooling coils. The compressors, condensers, and expansion coils would then be eliminated from the individual units.

Arrangement 2 in Fig. 19.7 illustrates the use of a chilled water spray for the cooling and dehumidification of the air supply to all the zones. This arrangement employs a separate fan for each zone and provides for the recirculation of different proportions of air in the different zones.

Arrangement 3 in Fig. 19.7 shows a major digression from the principles of the other two in that dehumidification is by either the process known as adsorption or that known as absorption. Either process must be followed by after-cooling. The details of the apparatus required to produce dehumidification by absorption or adsorption will be discussed in the following article. Because the air does not have to be cooled to a low temperature to effect the necessary dehumidification, water for the cooling coils may have a higher initial temperature, thus making it practicable to use city water in most cases where it is available. Even where it is necessary to chill the water, a lesser amount is required for a system in which the air is dehumidified in this manner.

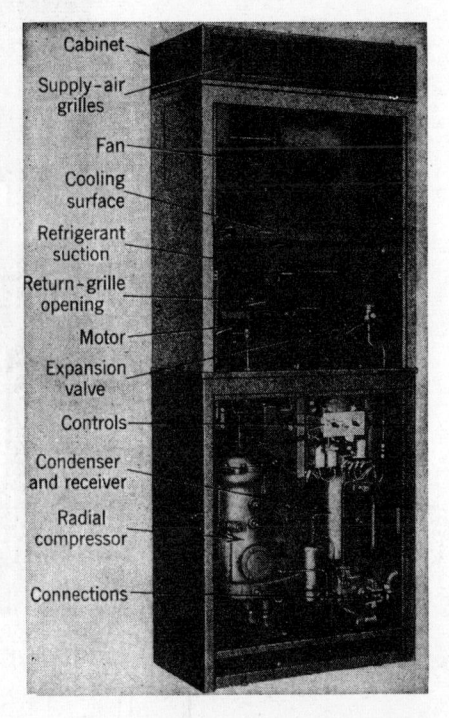

Fig. 19.5. Open view of a self-contained room-cooling unit employing a water-cooled condenser. (Courtesy Chrysler Corp.)

19.6. Apparatus for Dehumidification by Absorption or Adsorption. Special salt solutions capable of absorbing water vapor from air may be used as agents for dehumidification, but since they may also be used for humidification the necessary apparatus for

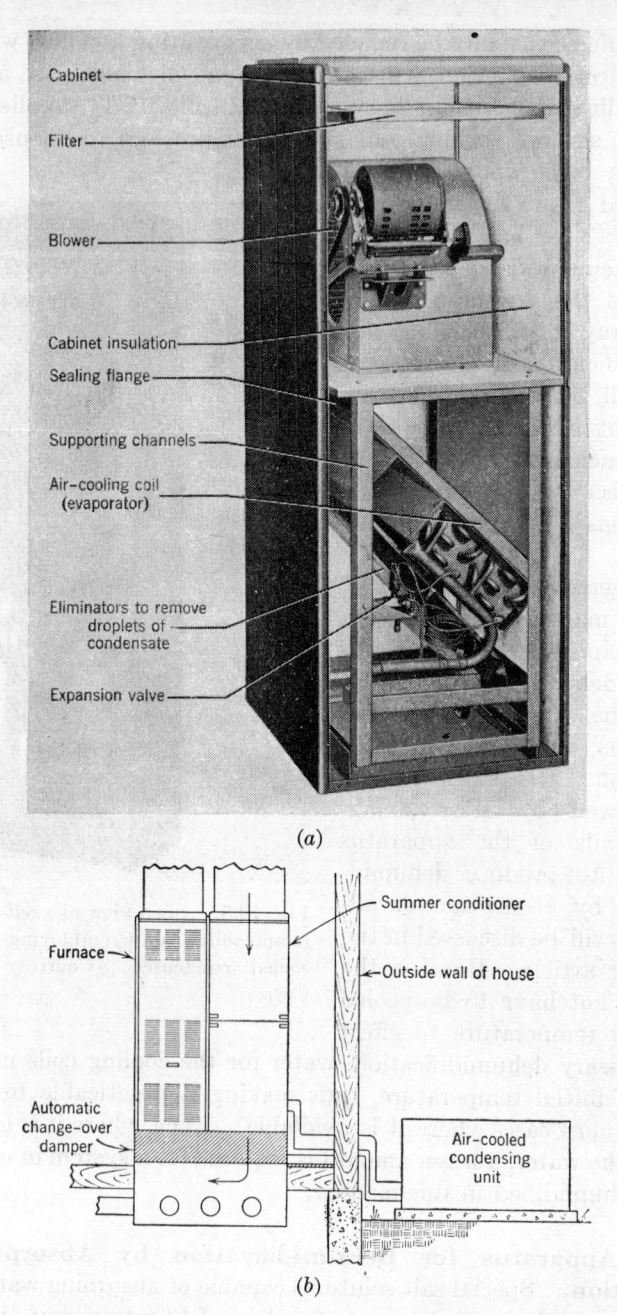

Fig. 19.6. (a) Forced-circulation air cooler adapted to use in a basementless home which has a perimeter heating system. (b) Typical method of installation. (Courtesy Lennox Industries, Inc.)

APPARATUS FOR COOLING AIR 467

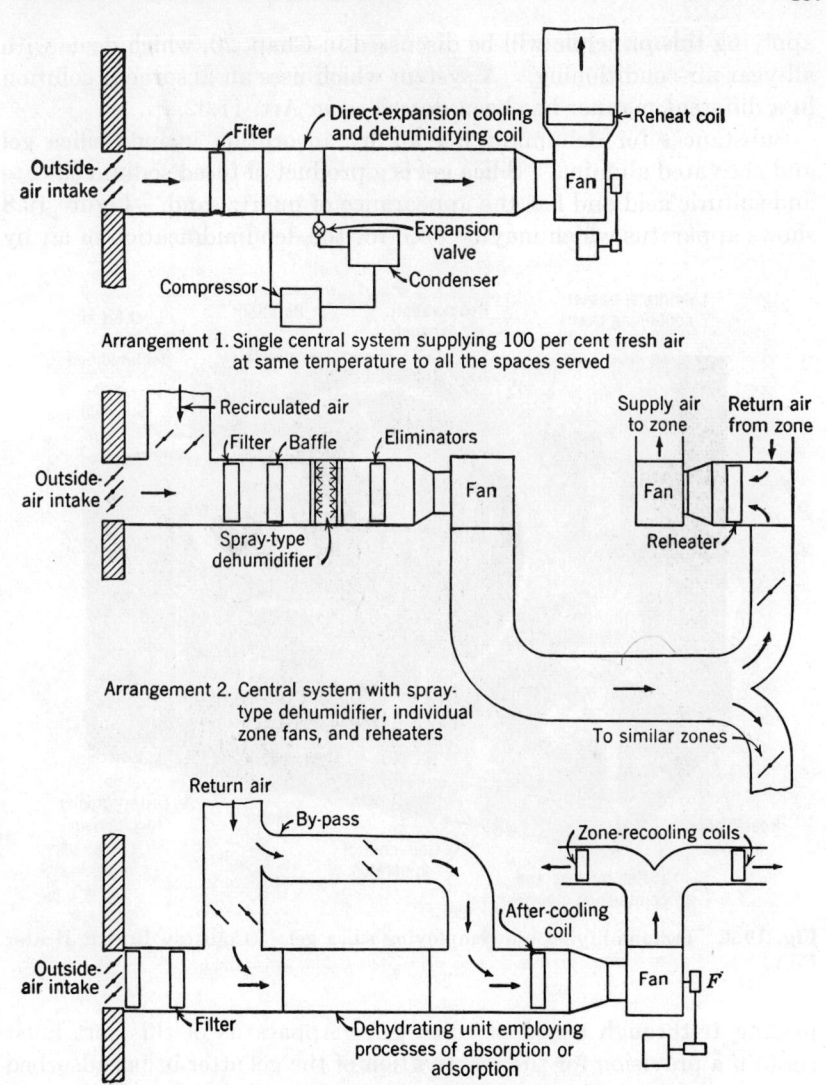

Arrangement 1. Single central system supplying 100 per cent fresh air at same temperature to all the spaces served

Arrangement 2. Central system with spray-type dehumidifier, individual zone fans, and reheaters

Arrangement 3. Central system in which air is humidified by absorption or adsorption and then cooled. Final cooling is controlled by zone thermostats

Fig. 19.7. Arrangements of apparatus in central summer air-conditioning systems.

applying this principle will be discussed in Chap. 20, which deals with all-year air-conditioning. A system which uses an absorbent solution in a different manner has been described in Art. 17.32.

Substances for dehumidifying air by adsorption include silica gel and activated alumina. Silica gel is a product of fused sodium silicate and sulfuric acid and has the appearance of quartz sand. Figure 19.8 shows apparatus which may be used for the dehumidification of air by

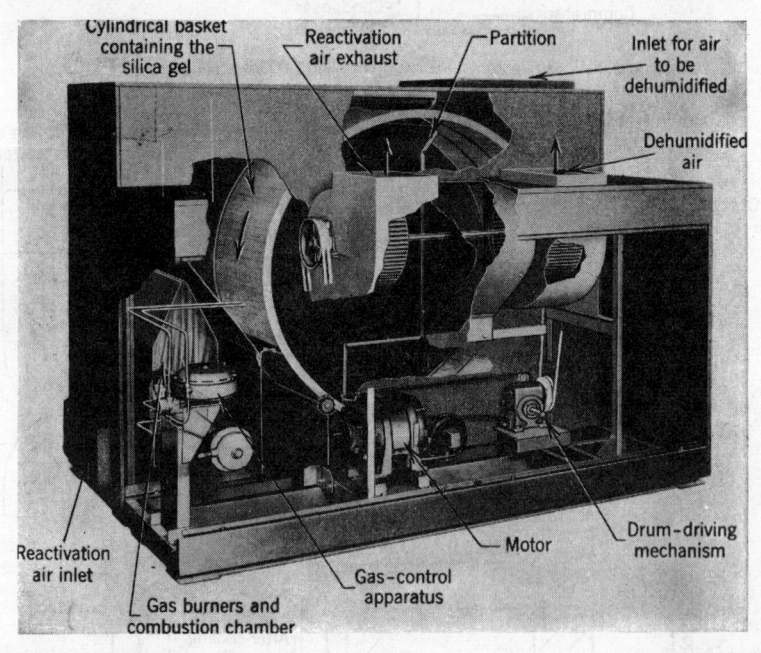

Fig. 19.8. Dehumidifying unit employing silica gel. (Courtesy Bryant Heater Co.)

passing it through a bed of silica gel. Apparatus of this sort must contain a provision for the reactivation of the gel after it has adsorbed moisture equal to approximately 25 per cent of its own weight. In the arrangement shown in Fig. 19.8 the gel in the form of small particles is contained in a cylindrical basket the interior of which is divided into two compartments. Partitions outside the drum together with the one inside make it possible to direct one stream of air through the gel in one half the basket while a separate stream is directed through that in the other half. The right half of the apparatus shown dries the air that is being conditioned while a stream of air from a heater at a temperature of approximately 350 F is passed through the gel in the left half of the basket to cause its reactivation.

APPARATUS FOR COOLING AIR

Since silica gel reduces the moisture content of a stream of air passing through it to a lower value than is desirable in most rooms, a summer air-conditioning machine equipped with a dehumidifying unit employing this principle should provide a means of bypassing a portion of the air around it as indicated in arrangement 3, Fig. 19.7.

The adsorption system of dehumidification has its most advantageous application in certain fields of industrial air conditioning where extremely low relative humidities are desired in the conditioned space. Since the moisture is removed in a process that is entirely separate from the removal of sensible heat, almost any desired relative humidity can be maintained without difficulty or excessive cost.

19.7. Types of Cooling Coils. Cooling coils in summer air-conditioning machines may transfer heat directly from air to an evaporating refrigerant, in which case they are called direct expansion coils, or they may transfer it to chilled water which in turn is cooled by mechanical refrigeration. Direct expansion coils are discussed in Art. 17.18. Figure 19.9 shows an example of a finned-tube coil suitable for inclusion in a summer air-conditioning machine and designed for the use of chilled water as the cooling medium.

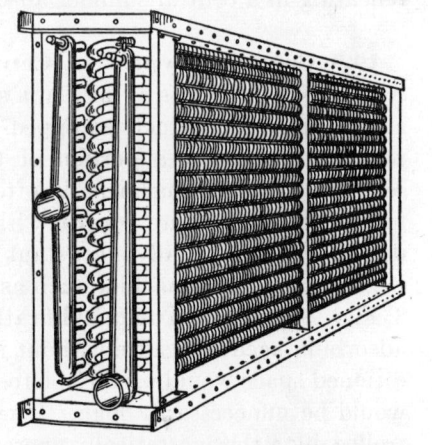

Fig. 19.9. Aerofin continuous-tube cooling coil. (Courtesy Aerofin Corp.)

Because one heat transfer is eliminated when direct expansion cooling coils are used, it is possible for the refrigerant to evaporate at a higher pressure while effecting the same final results. However, when several systems are to be installed in one building, there is an important advantage in using chilled water coils in order that one mechanical refrigerating system may serve the entire building by supplying liquid refrigerant to a central water cooler. Liquid refrigerant from the condenser of a central refrigerating plant may be piped to several different direct expansion coils in as many different air-conditioning machines, but the danger of leaks developing in long pipes transporting liquid refrigerant causes many designers to specify chilled water coils.

19.8. Coils for Reheating. Where dehumidification is accomplished by cooling, the air temperature must be reduced to a value that is considerably below that required in the space served. Since the air leaving the washer or coil is usually at least 20 deg cooler than that desired in the conditioned rooms, the air must be reheated after it has been dehumidified unless the sensible-heat gain in the space served is sufficient for this purpose. Machines which are equipped with reheating coils are capable of maintaining accurate control of both temperature and relative humidity in the conditioned space. However, reheating is expensive and generally it is not employed where accurate control of the relative humidity is not essential. Heat-transfer coils suitable for use in a heating system may be used for reheating in a central summer air-conditioning machine.

19.9 Use of a Bypass in Central Summer Air-Conditioning Systems. A bypass as used in a summer air-conditioning machine is an arrangement of ducts equipped with dampers and provided for the purpose of passing a portion of the air around the dehumidifying section. This arrangement is often highly desirable as it affords a means of temperature control without resorting to reheat and while delivering substantially a constant total-air quantity. A bypass is an essential part of a machine such as the one illustrated in arrangement 3, Fig. 19.7, in which dehumidification is brought about by means of an adsorbing agent, because without it the relative humidity in the conditioned space would usually be too low and the cost of cooling the air would be unnecessarily high. The final air temperature may be controlled by a thermostatically operated damper in the bypass, in which case the process is known as automatic bypass. This method of temperature control must, however, be used with caution as the moisture content of the air supplied to the conditioned space will increase with the proportion that is bypassed. Some reheating may be necessary in order to achieve a satisfactory control of the final relative humidity. Control of the damper in the bypass duct by means of an instrument that is sensitive to the relative humidity in the conditioned space is particularly well adapted to a machine utilizing the adsorption process for dehumidification of the air.

19.10. Zoning in Summer Air Conditioning. When a summer air-conditioning machine serves more than one room, the sensible- and latent-heat gains in the different spaces served are likely to vary in such a manner that different temperatures of the air supplied are required. Zoning consists of an arrangement of equipment such that the temperature of each space served by the system may be individually controlled. Accurate control of the relative humidity in the different zones is impossible, but usually the variation is not enough to affect the comfort of the occupants. Rooms intended for similar usage at all

APPARATUS FOR COOLING AIR

times which are situated so as to receive similar or negligible sun effect may be served as one zone.

Five different methods may be used for achieving separate zone control with a central unit.

1. Volume control.
2. Separate reheating when necessary, following dehumidification by cooling.
3. Separate recooling following dehumidification by adsorption or absorption.
4. Multiple fans with individual bypass.
5. Double duct system.

Volume control is achieved by means of dampers in each of the branch ducts so that the quantity of conditioned air delivered to each room can be varied. The dampers may be air-operated through remote control by the building engineer or by room thermostats. This system of zoning is the least expensive and most frequently utilized but its use may result in insufficient ventilation or unsatisfactory distribution.

Zone control by separate reheating is accomplished by omitting the reheat coil from the central apparatus and placing a smaller reheat coil in each of the zone supply ducts. The heat output of each coil is then controlled by a thermostat located in the zone which it serves.

Zone control by separate recooling after dehumidification by adsorption or absorption also affords a satisfactory method of compensating for varying sensible-heat gains in the different parts of the building. When the air is dehumidified in this manner the resulting temperature is always higher than that desired in the air supplied so that a thermostatically regulated cooling coil in each branch duct can be effective in the control of conditions in the space served by that duct.

Zone control may be achieved by means of separate fans each one of which receives cool and dehumidified air from a central conditioning unit as illustrated in arrangement 2, Fig. 19.7. Each fan is arranged to recirculate a controlled portion of the air it handles from the zone it serves. In this way the amount of air that is bypassed around the conditioner can be regulated according to the requirements of each individual zone.

The double duct system is a relatively new development which is producing excellent results in many installations. It can be used in winter heating as well as in summer cooling and will be discussed in Chap. 20.

19.11. Evaporative Coolers. Evaporative cooling as a method of summer air conditioning is impractical except in dry climates. However, it does provide a very low cost method of cooling where it is applicable. Air at 100 F dbt and 60 F wbt can be cooled to approximately 62 F dbt by passing it through a spray of recirculated water. Air coming from the washer would be nearly saturated, but under certain circumstances satisfactory conditions may be maintained by introducing this low-temperature moist air in proper amounts into a

Fig. 19.10. Evaporative cooler. (Courtesy Lennox Industries, Inc.)

room in which the sensible-heat gain is high and the latent-heat gain is low.

Evaporative cooling has been previously discussed in Art. 2.16, and it is exactly the same process as adiabatic saturation.

Attractively packaged summer air conditioners operating on the evaporative cooling principle are available from several manufacturers. A typical unit suitable for use in a residence in a dry area is shown in Fig. 19.10. Partial adiabatic saturation of the air (or evaporative cooling) is brought about in the unit of Fig. 19.10 by passing it through a pad of aspen excelsior which is continuously wetted by water returning to the sump at the bottom from a perforated tray at the top. A pump keeps the tray supplied with water. It is stated in reference 2 that manufacturers usually make the pads for a unit of this type of a thickness to give a humidifying efficiency (or saturating efficiency) of around 80 per cent. When aspen excelsior pads are used, it is common

APPARATUS FOR COOLING AIR

practice to use a face area velocity of 230–300 fpm and a pad thickness of 2 in.

19.12. Procedure in Designing a Summer Air-Conditioning Machine. Following is an outline of the customary procedure in designing a central system for summer air conditioning.

1. Select inside design conditions.
2. Determine outside design conditions from weather-bureau records.
3. Determine ventilation requirements of spaces to be served.
4. Estimate the maximum sensible-heat gain and maximum moisture gain in the spaces which are to be served.
5. Study the layout and uses of the spaces served, and decide upon the most practicable method of zoning, the number of zones required, and the rooms to be served by each zone.
6. Select the equipment and design the duct work.

A numerical example will be given in the next chapter for an all-year air-conditioning system which will include all the steps required in the design of apparatus for summer conditioning only.

19.13. Most Economical Use of an Ideal Air Washer for Cooling and Dehumidification, Apparatus Dew Point. When dehumidification of summer air is brought about by passing it through a spray of chilled water, the air may leave the washer in practically a saturated state and at a dry-bulb temperature that is usually lower than that desired in the conditioned space. If the air leaving the washer is saturated, its dry-bulb temperature is also its dew-point temperature. In any case the dew-point temperature of the air leaving the dehumidifying chamber fixes its moisture content. When the air is delivered from the conditioner its temperature will gradually increase to the equilibrium dry-bulb temperature in the room. As the temperature of the processed air increases it absorbs heat from the air already in the room, thus tending to reduce the room temperature. This tendency of the cool air to reduce the room temperature just balances the tendency for the sensible-heat gains to increase it, and the desired temperature is maintained, provided that the proper amount of air is delivered.

If there were no moisture liberation within the conditioned room and all the air entered through the conditioner, the state point would move horizontally to the right on a psychrometric chart (line AB in Fig. 19.11) as it approaches the constant room temperature represented by point B. Under these conditions the correct dew-point temperature for the air leaving the ideal washer would be that corresponding to the specified room condition, represented by point A. However, if there is moisture liberation within the conditioned space, it will tend to increase the moisture content of the air in the room and raise its dew-point temperature. The only way this tendency can be counteracted is to introduce conditioned air having a lower moisture content and a lower dew-point temperature than that corresponding to the specified room condition so that the absorption of moisture from the air in the room by the air from the conditioner will just balance the moisture gain. When the dew-point temperature of the conditioned

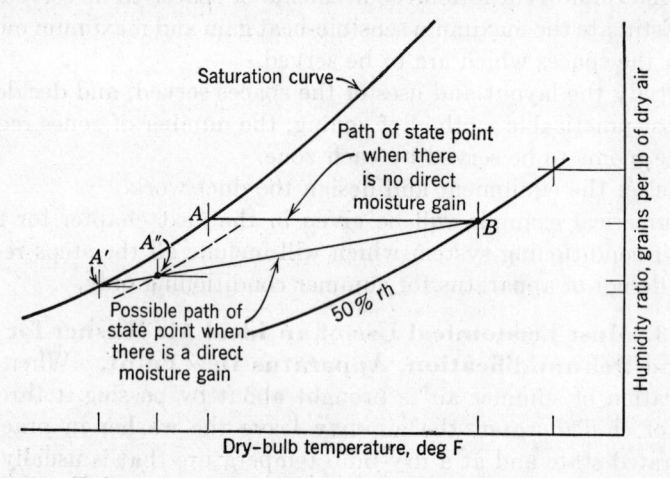

Fig. 19.11. Skeleton psychrometric chart showing path of state point or load-ratio line for two conditions of operation.

air is lower than that of the room air, the state point will no longer move horizontally to the right but will move along a condition line (often called the *load-ratio line*) which slopes upward from the horizontal (line $A'B$ in Fig. 19.11). The slope of this line is determined by the ratio of the sensible-heat gain in the space served to the moisture gain in that space. It is also proportional to the ratio of the total-heat gain to the moisture gain. The required slope of this line for any particular problem in order that the cooled and dehumidified air may absorb total heat and moisture in the ratio as the gains of these items may be easily obtained if the *direct* total heat and moisture gains are known. After the slope of the room load-ratio line has been established, the proper temperature of the air leaving an ideal washer may be determined by drawing it on a psychrometric chart from the specified state of the room air to the saturation curve. The weight of dry air which must be processed may then be computed. The procedure will be illustrated by an example.

APPARATUS FOR COOLING AIR

The symbols which will be used for the various heat and moisture quantities in the example and in later discussions are as follows:

h_r = total heat (or enthalpy) of the room air, Btu per pound of dry air.
w_{gr} = humidity ratio of the room air, grains per pound of dry air.
h_x = total heat at state point X (point X is used solely for the purpose of drawing the room load-ratio line).
w_{gx} = humidity ratio at state X.
h_e = total heat of the air leaving the cooling and dehumidifying apparatus (this is also the total heat of the air as it enters the space served).
w_{ge} = humidity ratio of the air leaving the cooling and dehumidifying apparatus.

Example. The air in a room is to be maintained at 78 F dbt and 50 per cent relative humidity. The estimated sensible-heat gain directly into the space under design conditions is 51,970 Btuh, and the maximum moisture liberation within the space from all sources is estimated to be 150,000 grains per hr. Find (a) the correct temperature of the air leaving an ideal washer, and (b) the weight of dry air which must be delivered into the space served.

Solution. The moisture gain expressed as a latent-heat gain is (150,000 × 1076)/7000 = 23,030 Btuh, and the total-heat gain is 51,970 + 23,030 = 75,000 Btuh. The ratio of the total-heat gain to the moisture gain, often called the *enthalpy-humidity difference ratio*, is 75,000/150,000 = 0.5, which means that cooled and dehumidified air that will satisfy the conditions of this problem must be in a position to absorb total heat and moisture in the ratio of 0.5 Btu per grain.

The next part of the solution is determining the proper slope of the load-ratio line, and it is carried out on the psychrometric chart (see Fig. 19.12).

1. Locate the room state (dry-bulb temperate = 78, relative humidity = 50, h_r = 30.0, w_{gr} = 72.0).

2. Find a point X on the correct room load-ratio line. (Any point on the correct line will have the proper ratio of heat-absorbing capacity to moisture-absorbing capacity.)

(a) Arbitrarily assume the humidity ratio of point X to be determined (60 will be used in this example). If this point is to be on the correct line the following relationship must be satisfied: $(30.0 - h_x)/(72 - 60) = 0.5$, or $h_x = 24.0$. Point X ($h_x = 24.0$, $w_{gx} = 60$) can then be plotted and the room load-ratio line drawn through it and the room state. If an ideal washer is used, the air leaving it and entering the room will be saturated and the proper state will be at the intersection of the room load-ratio line and the saturation curve on the psychrometric chart. This point of intersection (Fig. 19.12) is at 50 F dbt, 50 F wbt, h_e = 20.3, and w_{ge} = 53. [The answer to part (a) of the example is 50 F.]

(b) The weight of conditioned air required may be found by dividing the total-direct heat gain by the heat-absorbing capacity per pound of dry air as it leaves the cooling and dehumidifying apparatus, 30.0 − 20.3 = 9.7 Btu per lb in the case of this example (assuming an ideal washer). The weight of dry air to be processed or delivered into the space served is 75,000/9.7 = 7730 lb per hr. As a check on the solution, the weight of dehumidified air that is required to absorb the direct moisture gain may also be found as 150,000/(72 − 53) = 7890 lb per hr. The weight will be taken as the mean of the two determinations, or 7810 lb per hr.

(If the room load-ratio line is perfectly drawn, the weight of dry air needed to absorb the total-heat gain will be exactly the same as the weight required to absorb the moisture gain. However, in the solution of an actual problem

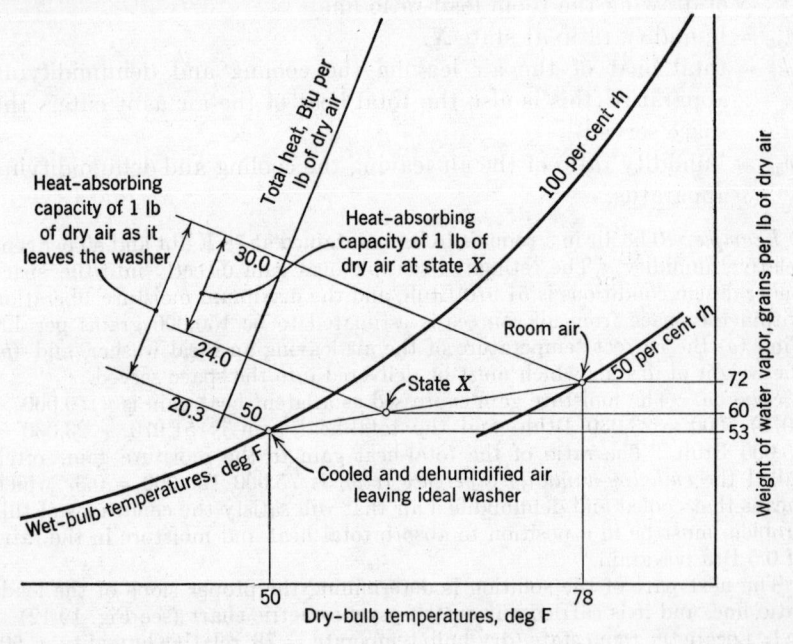

Fig. 19.12. Skeleton psychrometric chart showing data used in the example of Art. 19.13.

it is difficult to plot the points and draw the room load-ratio line with a precision sufficient to give a better check than was obtained in the foregoing example.)

The temperature of the air leaving a perfect washer is called the *apparatus dew point*. The term is often used in connection with cooling problems involving an actual washer or a cooling and dehumidifying coil. In any case it is the temperature at the point of interesection of the room load-ratio line and the saturation curve.

19.14. Design of an Actual Washer for Cooling and Dehumidification. Actual washers do not deliver air in a saturated state, though it is closely approached in some designs under favorable conditions of operation. Hendrickson[3] suggests that the performance of air washers when used with chilled water can best be correlated by

APPARATUS FOR COOLING AIR

means of a *performance factor* (PF), which is defined for the purpose of this discussion as the ratio of the heat actually removed from the entering air to the heat that would be removed in an ideal washer. In equation form:

$$\text{PF} = (h_1 - h_2)/(h_1 - h_s) \tag{19.1}$$

where h_1 and h_2 are the actual entering and leaving enthalpies and h_s

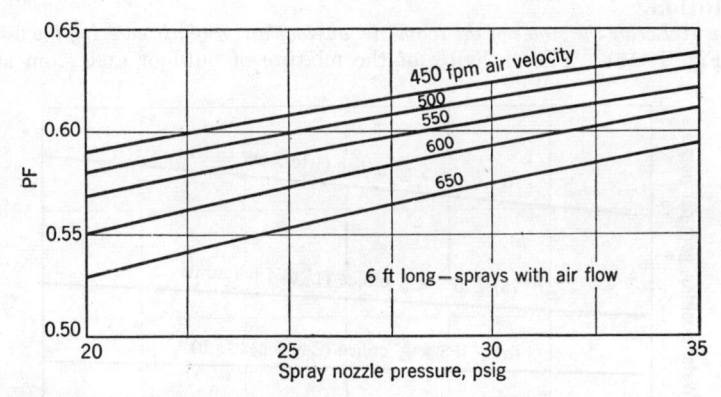

Fig. 19.13. Typical PF data for a washer with one bank of spray nozzles.

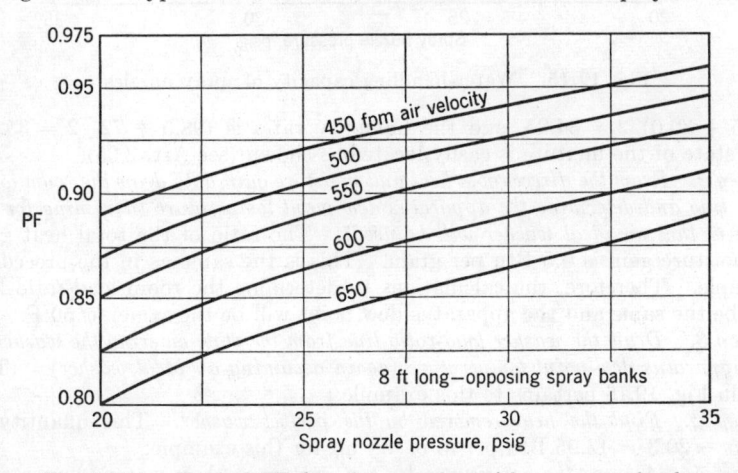

Fig. 19.14. Typical PF data for a washer with two spray banks.

is the enthalpy of the saturated air which would leave a perfect washer; all in Btu per pound of dry air. Figures 19.13 and 19.14 from reference 3 give typical PF data for two types of commercial air washers when used for the cooling and dehumidyfing of air. Figure 19.15 gives performance data for nozzles used in air washers. A step-by-step procedure for the design of an air washer for cooling and dehumidyfing is given in the following example.

Example. Design a two-bank (opposed) washer for a building for which the direct total heat and moisture gains have been estimated at 750,000 Btuh and 1,500,000 grains per hr respectively. The spray water will be supplied at 40 F. The washer is to handle 50 per cent fresh air. Outdoor design conditions are 95 F dbt and 75 F wbt. The room conditions are to be 78 F dbt with 50 per cent relative humidity.

Solution.

Step 1. Locate the states of the room air, outdoor air, and air entering the washer (see Fig. 19.16). The enthalpy of the mixture of outdoor and room air is

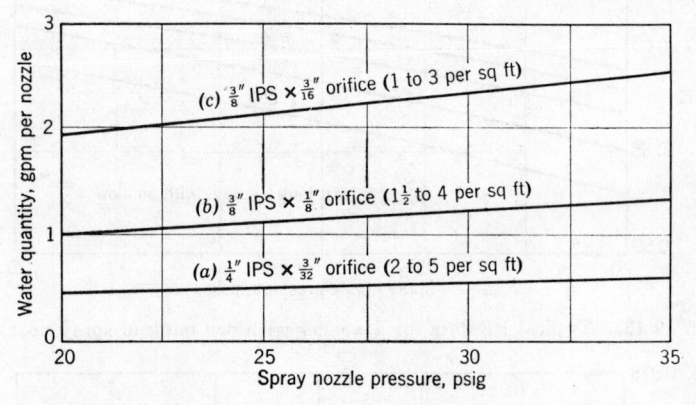

Fig. 19.15. Water-handling capacity of spray nozzles.

$(38.5 + 30.0)/2 = 34.25$, and the humidity ratio is $(98.5 + 72)/2 = 85.25$. The state of the mixture is easily located as shown (see Art. 2.25).

Step 2. From the direct room heat and moisture gain data draw the room load-ratio line and determine the apparatus dew-point temperature (assuming for the moment that an ideal washer will be used). The ratio of the total heat gain to moisture gain is 0.5 Btu per grain. This is the same as in the preceding example. Therefore, the calculations to determine the room load-ratio line will be the same and the apparatus dew point will be the same, or 50 F.

Step 3. Draw the washer load-ratio line from the state entering the washer to the apparatus dew-point temperature (again assuming an ideal washer). (The line in Fig. 19.16 pertains to this example.)

Step 4. Find the heat removed by the perfect washer. This quantity is $34.25 - 20.3 = 13.95$ Btu per lb of dry air for this example.

Step 5. Assume the entering water temperature. (Any water temperature above 36 F that is lower than the apparatus dew-point temperature may be used.) A temperature of 40 F will be used in this example.

Step 6. Determine the water-to-air ratio for the theoretically perfect washer by equating the heat given up by the air to the heat absorbed by the water. $W(50 - 40) = 13.95$. $W = 1.395$ lb of water per lb of dry air for the ideal washer.

Step 7. From performance curves (or catalogue data) for the type of washer to be used, determine the PF of the actual washer after assuming the face velocity

APPARATUS FOR COOLING AIR 479

and the spray nozzle pressure. A velocity of 500 fpm and a spray nozzle pressure of 25 psig will be assumed for this washer, and it is to have two opposing banks of spray nozzles. From Fig. 19.14 the PF is 0.905.

Step 8. Find the heat per pound of dry air that will actually be removed and the heat content of the air leaving the washer. The heat actually removed will be the PF × the heat removed by the perfect washer = 0.905 × 13.95 = 12.62. The heat content of the air leaving the washer will be 34.25 − 12.62 = 21.63 Btu per lb of dry air.

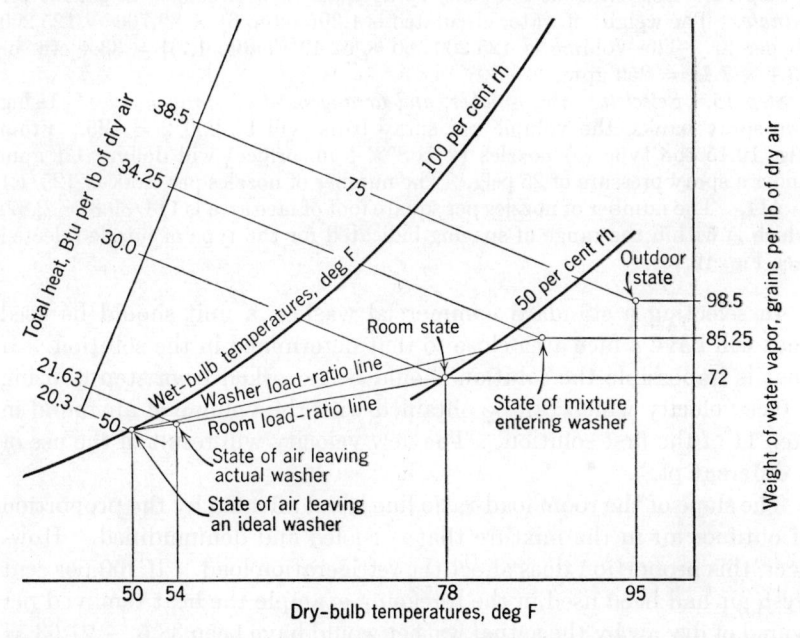

Fig. 19.16. Skeleton psychrometric chart showing data used in the example of Art. 19.14.

Step 9. Locate a point on the room load-ratio line having the heat content found in step 8. The intersection of the line representing a constant total heat of 21.63 with the room load-ratio line for this example is shown in Fig. 19.16. If the chart used does not include lines of constant total heat, assume that wet-bulb lines are also lines of constant total heat (see Art. 2.14). In this procedure the state of the air leaving the actual washer is transferred from the washer load-ratio line to the room load-ratio line as indicated in Fig. 19.16. The error made in so doing is usually slight because the two lines are always close together at the point where the transfer is made.

Step 10. Find the weight of dry air needed by dividing the direct total heat gain of the space served by the heat-absorbing capacity per pound of dry air as it leaves the actual washer. The direct-total-heat gain for this example is 750,000 Btuh and the weight of conditioned air needed is 750,000/(30.0 − 21.63) = 89,700 lb of dry air per hr.

Step 11. Find the volume of standard air entering the washer in cubic feet per minute and the washer face area. $89{,}700/(60 \times 0.075) = 89{,}700/4.5 = 19{,}900$ cfm. The face area is $19{,}900 \div 500 = 39.8$ sq ft.

Step 12. Calculate the total refrigerating load in tons. Refrigerating load = $[89{,}700 \times 12.62 \text{ (step 8)}]/12{,}000 = 94.5$ tons. (See Art. 17.4.)

Step 13. Determine the temperature of the water leaving the actual washer using the water-to-air ratio found for the ideal washer. $1.395(t - 40) = 12.62$, $t = 49.0$ F.

Step 14. Determine the volume of spray water to be circulated in gallons per minute. The weight of water circulated is 1.395 (step 6) $\times 89{,}700 = 125{,}200$ lb per hr. The volume is $125{,}200/[60 \times 62.42$ (Table 1.7)$] = 33.4$ cfm or $33.4 \times 7.48 = 250$ gpm.

Step 15. Select the size, number, and arrangement of spray nozzles. Using two spray banks, the volume per spray bank will be $250/2 = 125$. From Fig. 19.15 the type (b) nozzles ($\frac{3}{8}$ IPS $\times \frac{1}{8}$ in. orifice) will deliver 1.1 gpm under a spray pressure of 25 psig. The number of nozzles per bank is $125/1.1 = 114$. The number of nozzles per square foot of face area is $114/39.8 = 2.87$, which is within the range of spacing indicated for the type of nozzle selected (see Fig. 19.15).

In selecting a standard commercial washer, a unit should be used that will have a face area close to that determined in the solution. If that is impossible the solution should be reworked from step 6, using a face velocity which can be obtained with the volume of air found in step 11 of the first solution. The new velocity will result in the use of a different pf.

The slope of the room load-ratio line is not affected by the proportion of outdoor air in the mixture that is cooled and dehumidified. However, this proportion does affect the refrigeration load. If 100 per cent fresh air had been used in the foregoing example the heat removed per pound of dry air by the actual washer would have been $38.5 - 21.63 = 16.87$ Btu per lb of dry air instead of 12.62.

19.15. Selection of Cooling and Dehumidifying Coils. When cooling coils reduce the humidity ratio of the air passing through them by condensing some of the water vapor on the chilled metal surfaces, the performance can be predicted accurately only from actual tests. Most coil manufacturers have prepared such data for all cooling and dehumidifying coils which they manufacture. Each manufacturer's line of coils usually includes units with a differing number of rows of tubes as well as differing dimensions of their face areas. Table 19.1 for direct expansion cooling coils has been abridged with permission from one of several tables given in reference 4. The complete table includes data on refrigerating capacity required per square foot of coil area. The other tables not included here cover face area velocities to 600 fpm and refrigerant temperatures ranging from 35 to 50 F. Refer-

APPARATUS FOR COOLING AIR

ence 4 also includes a method of selecting water-cooling coils together with the data needed to employ it.

A procedure which can be used in selecting a direct expansion cooling coil is illustrated in the following example.

Example. Specify the number of rows of tubes, the face area needed in square feet, and the refrigerating capacity required in the case of a Trane direct expansion coil which is to serve a room into which the sensible-heat gain is 100,000 Btuh and into which the moisture gain is 180,000 grains per hr. The

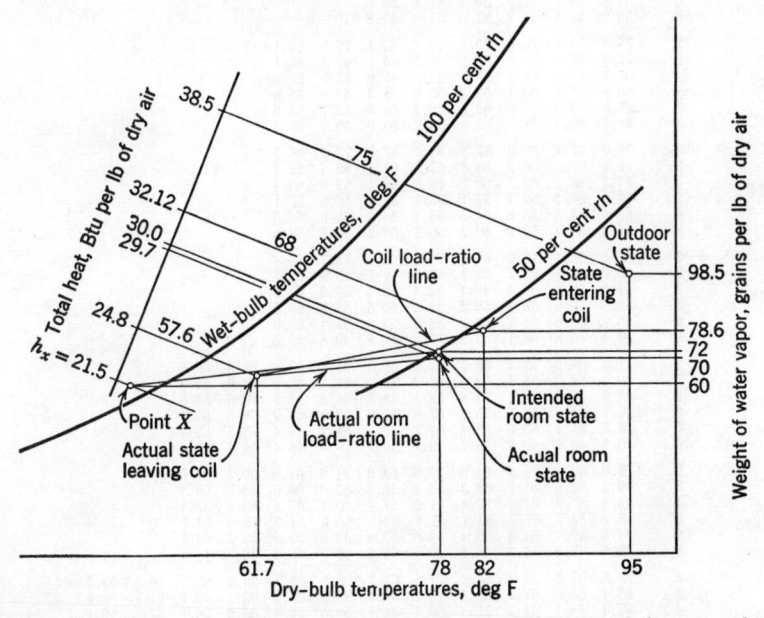

Fig. 19.17. Skeleton psychrometric chart showing data used in example of Art. 19.15.

room is to be maintained at 78 F with 50 per cent relative humidity. The conditioner is to take 25 per cent of the air handled from the outside. Outside design conditions are 95 F dbt and 75 F wbt. Infiltration is to be neglected.

Solution.

Step 1. Calculate the enthalpy-humidity difference ratio and draw the room load-ratio line on the psychrometric chart (see example of Art. 19.13 for more details). The direct latent-heat gain in this example is $180,000 \times 1076/7000 = 27,700$ Btuh. The direct total-heat gain is $100,000 + 27,700 = 127,700$ Btuh, and the enthalpy-humidity difference ratio is $127,700/180,000 = 0.709$ Btu per grain. Assume the point X used for defining the room load-ratio line will have a humidity ratio of 60 grains per lb of dry air. Then (Fig. 19.17) $(30.0 - h_x)/(72 - 60) = 0.709$, $h_x = 21.5$. Plot $h_x = 21.5$, $w_{gx} = 60$, and draw the room load-ratio line. (It will be noted that point X plots to the left of the saturation curve but this does not matter.)

TABLE 19.1. Actual Performance Data for Trane Type-DE Direct Expansion Coils*

Data presented are based on a velocity of 400 fpm and a refrigerant saturation temperature of 40 F with 5 deg of superheat at the outlet

Initial Wet-Bulb	Rows of Tubes	Initial Dry-Bulb 100 Final dbt	Final wbt	Initial Dry-Bulb 96 Final dbt	Final wbt	Initial Dry-Bulb 92 Final dbt	Final wbt	Initial Dry-Bulb 88 Final dbt	Final wbt	Initial Dry-Bulb 84 Final dbt	Final wbt	Initial Dry-Bulb 80 Final dbt	Final wbt	Initial Dry-Bulb 76 Final dbt	Final wbt
80	2	79.3	70.9	77.4	70.9	75.5	70.9								
	3	72.2	66.8	70.8	66.8	69.5	66.8								
	4	66.5	63.1	65.6	63.1	64.7	63.1								
	5	61.8	59.7	61.1	59.7	60.4	59.7								
	6	57.8	56.7	57.3	56.7	57.1	56.7								
	8	52.0	51.8	51.8	51.8	51.8	51.8								
78	2	78.7	69.1	76.6	69.1	74.7	69.1	72.7	69.1						
	3	71.4	65.1	70.0	65.1	68.7	65.1	67.4	65.1						
	4	65.7	61.6	64.7	61.6	63.7	61.6	62.9	61.6						
	5	60.8	58.3	60.1	58.3	59.6	58.3	58.9	58.3						
	6	57.0	55.5	56.6	55.5	56.2	55.5	55.7	55.5						
	8	51.3	50.9	51.1	50.9	50.9	50.9	50.9	50.9						
76	2	78.0	67.3	76.0	67.3	74.1	67.3	72.2	67.3						
	3	71.4	63.5	69.2	63.5	67.9	63.5	66.5	63.5						
	4	64.7	60.1	63.7	60.1	63.0	60.1	62.0	60.1						
	5	59.9	57.1	59.4	57.1	58.7	57.1	58.0	57.1						
	6	56.4	54.5	55.8	54.5	55.4	54.5	55.0	54.5						
	8	51.0	50.2	50.8	50.2	50.5	50.2	50.3	50.2						
75	2	77.5	66.4	75.6	66.4	73.7	66.4	71.7	66.4						
	3	70.0	62.7	68.7	62.1	67.3	62.7	65.9	62.7						
	4	64.3	59.4	63.4	59.4	62.4	59.4	61.5	59.4						
	5	59.6	56.5	58.9	56.5	58.3	56.5	57.7	56.5						
	6	55.7	53.9	55.2	53.9	54.9	53.9	54.4	53.9						
	8	50.4	49.8	50.1	49.8	50.0	49.8	49.8	49.8						
74	2	77.3	65.6	75.3	65.6	73.4	65.6	71.4	65.6	69.5	65.6				
	3	69.9	62.0	68.5	62.0	67.2	62.0	65.7	62.0	64.4	62.0				
	4	68.9	58.8	63.1	58.8	62.1	58.8	61.6	58.8	60.0	58.8				
	5	59.1	55.9	58.5	55.9	57.8	55.9	57.2	55.9	56.6	55.9				
	6	55.5	53.4	55.0	53.4	54.7	53.4	54.2	53.4	53.7	53.4				
	8	50.2	49.4	49.9	49.4	49.8	49.4	49.6	49.4	49.4	49.4				
72	2	77.0	64.0	74.9	64.0	72.9	64.0	71.0	64.0	69.1	64.0				
	3	69.3	60.5	67.7	60.5	66.4	60.5	65.0	60.5	63.7	60.5				
	4	63.2	57.4	62.2	57.4	61.2	57.4	60.1	57.4	59.4	57.4				
	5	58.4	54.7	57.7	54.7	57.1	54.7	56.5	54.7	55.8	54.7				
	6	54.8	52.4	54.3	52.4	54.0	52.4	53.6	52.4	53.1	52.4				
	8	49.6	48.6	49.3	48.6	49.1	48.6	48.9	48.6	48.7	48.6				
70	2	76.5	62.2	74.4	62.3	72.5	62.3	70.3	62.3	68.5	62.3	66.5	62.3		
	3	68.4	55.9	67.2	59.1	65.7	59.1	64.4	59.1	63.1	59.1	61.7	59.1		
	4	62.4	56.1	61.2	56.1	60.1	56.1	59.5	56.1	58.5	56.1	57.6	56.1		
	5	57.6	53.5	56.8	53.5	56.1	53.5	55.5	53.5	54.9	53.5	54.3	53.5		
	6	54.0	51.3	53.7	51.4	53.2	51.4	52.7	51.4	52.3	51.4	51.9	51.4		
	8	49.0	47.9	48.8	47.9	48.6	47.9	48.3	47.9	48.2	47.9	48.0	47.9		
68	2	76.0	60.0	73.8	60.5	72.0	60.7	69.9	60.7	67.9	60.7	66.0	60.7		
	3	68.0	56.9	66.3	57.4	65.0	57.6	63.6	57.6	62.3	57.6	61.0	57.6		
	4	61.9	54.3	60.6	54.7	59.8	54.9	58.7	54.9	57.8	54.9	57.0	54.9		
	5	57.0	52.0	56.3	52.4	55.5	52.4	55.0	52.5	54.4	52.5	53.8	52.5		
	6	53.5	50.1	53.0	50.4	52.5	50.5	52.1	50.5	51.6	50.5	51.3	50.5		
	8	48.5	47.0	48.4	47.2	48.2	47.3	48.0	47.3	47.7	47.3	47.6	47.3		
66	2	76.0	57.6	73.6	58.2	71.2	58.8	69.2	59.1	67.2	59.1	65.3	59.1	63.5	59.1
	3	67.9	54.4	66.0	55.2	64.1	55.9	62.8	56.1	61.3	56.1	60.0	56.1	58.7	56.1
	4	61.6	51.8	60.1	52.8	59.0	53.4	57.7	53.5	57.1	53.6	56.2	53.6	55.1	53.6
	5	56.7	49.7	55.6	50.8	54.8	51.2	54.8	51.4	53.7	51.4	53.1	51.4	52.4	51.4
	6	52.9	47.9	52.0	48.9	51.8	49.4	51.4	49.5	51.1	49.6	50.6	49.6	50.1	49.6
	8	47.8	45.4	47.7	46.2	47.6	46.5	47.4	46.6	47.2	46.6	47.0	46.6	46.8	46.6
65	2	76.0	56.4	73.6	57.0	71.2	57.6	69.1	58.2	67.1	58.3	65.1	58.3	63.2	58.3
	3	67.9	53.1	66.0	54.0	64.1	54.9	62.7	55.3	61.3	55.5	60.0	55.5	58.7	55.5
	4	61.6	50.4	60.1	51.5	58.8	52.5	57.4	52.9	56.6	53.0	55.7	53.0	54.9	53.0
	5	56.7	48.3	55.6	49.6	54.8	50.5	53.9	50.8	53.2	50.9	52.6	50.9	52.0	50.9
	6	52.9	46.5	51.9	47.9	51.5	48.7	51.0	49.0	50.7	49.1	50.1	49.1	49.8	49.1
	8	47.8	44.0	47.3	45.4	47.3	46.1	47.2	46.3	47.0	46.3	46.8	46.3	46.6	46.3

* Abridged with permission from Table 25 of Trane Co. *Bull. No. DS-865*.[4] This bulletin contains other tables covering velocities to 600 fpm and a range of refrigeration saturation temperatures from 35 F to 50 F.

APPARATUS FOR COOLING AIR

Step 2. Find the state of the air entering the coil. The enthalpy of the mixture entering the coil is (3 × 30.0 + 38.5)/4 = 32.12 Btu per lb. The humidity ratio is (3 × 72 + 98.5)/4 = 78.6 grains per lb. The plotted point is shown on Fig. 19.17: 82 F dbt and 68 F wbt.

Step 3. Using the dry-bulb and wet-bulb temperatures of the air entering the coil and the data of Table 19.1 (and that of additional tables if available), find the number of rows of tubes that will give a condition that will plot on, or close to, the room load-ratio line. If it is impossible to select a coil arrangement that will give a state that will plot close to the room load-ratio line, select one that will give a point below the line rather than above it because a room relative humidity less than that specified is less objectionable than one that is higher. In this example, Table 19.1 is used. The entering dry-bulb temperature is 82 F and the entering wet-bulb temperature is 68 F. It is found by interpolation that a coil with three rows of tubes will give 61.7 F dbt and 57.6 F wbt. This state plots slightly below the room load-ratio line, indicating that a coil with three rows of tubes will be satisfactory.

Step 4. Find the heat that will be absorbed by each pound of dry air that is cooled and dehumidified. In this example, the state of the air delivered by the coil is slightly below the room load-ratio line so the enthalpy of the room air will be 29.7 Btu per lb of dry air instead of 30.0 and the humidity ratio will be 70 grains per lb of dry air instead of 72. The enthalpy or total heat of the air leaving the coil, 57.6 F wbt, is 24.8 Btu per lb of dry air. The heat-absorbing capacity per pound of dry air is 29.7 − 24.8 = 4.9 Btu.

Step 5. Find the weight of dry air to be processed in pounds per hour and its volume measured under standard conditions in cubic feet per minute. In this example the weight of dry air to be processed is 127,700/4.9 = 26,050 lb per hr and the volume measured under standard conditions is 26,050/4.5 = 5800 cfm. (See step 11 of the previous example.)

Step 6. Find the required coil face area in square feet using the velocity on which the performance data were based. In this example the data of Table 19.1 are based on a face velocity of 400 fpm. The required face area is 5800/400 = 14.5 sq ft. The standard coil having a face area closest to this amount would be used.

Step 7. Find the refrigerating capacity in tons. In this example the refrigerating capacity required is (32.12 − 24.8)26,050/12,000 = 15.9 tons.

19.16. Conditions of Operation Which Require Reheating.

In a case where the moisture gain of a space is unusually high the room load-ratio line may be so steep that it does not intersect the saturation curve. In this case it would be impossible for an ideal washer to produce air in a position to absorb total heat and moisture in the right proportions. In a situation of this kind, which might occur in a restaurant or in a theater, it would be necessary to select a feasible temperature for the cooled and dehumidified air, calculate the weight of dry air required on the basis of its moisture-absorbing capacity, then provide for the necessary amount of reheat. The procedure will be illustrated in the following example.

Example. A certain space receives 50,000 Btuh of sensible heat under design conditions and gains 333,000 grains of moisture in the same period. It is desired to maintain 80 F dbt and 50 per cent relative humidity. The washer is to handle 100 per cent fresh air which under design conditions will be at 91 F dbt and 76 F wbt. Select the temperature to which the air should be cooled by an ideal washer and determine: the weight of dry air that will be needed, in pounds per hour; the refrigerating capacity required, in tons; the

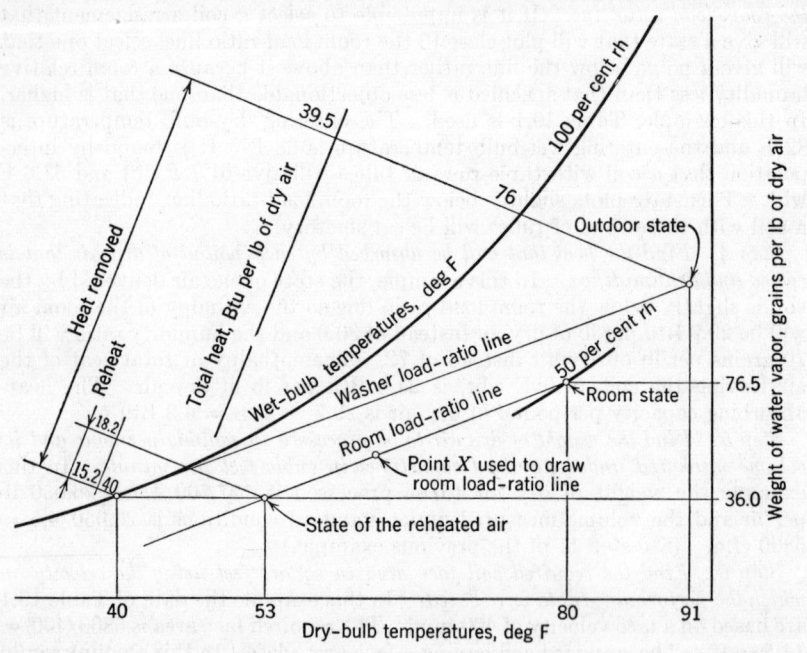

Fig. 19.18. Skeleton psychrometric chart showing data used in the example of Art. 19.16.

temperature to which the air must be reheated; and the amount of reheat, in Btuh.

Solution. The direct latent-heat load is $333,000 \times 1076/7000 = 51,300$ Btuh. The total heat load is then 101,300 Btuh and the enthalpy-humidity difference ratio is $101,300/333,000 = 0.3045$ Btu per grain. Let w_{gx} (see example in Art. 19.13) = 50, then $(31.2 - h_x)/(76.5 - 50) = 0.3045$, $h_x = 23.13$ (see Fig. 19.18). (h_x and w_{gx} are not shown on Fig. 19.18 but the location of point X may be verified by use of an actual psychrometric chart.) It may be noted that the room load-ratio line does not intersect the saturation curve. It will be assumed that it is impractical to cool the air to a temperature lower than 40 F. Air from an ideal washer (saturated) at 40 F contains 36 grains of water vapor per lb of dry air and each pound is in a position to absorb $76.5 - 36 = 40.5$ grains before exceeding the humidity ratio corresponding to 80 F and 50 per cent relative humidity. The weight of dry air from an ideal washer that is needed to absorb the moisture gain of the space is $333,000/(76.5 - 36) = 8220$ lb per hr. The refrigerating capacity required with 100 per cent fresh air is $8220(39.5 - 15.2)/12,000 = 16.64$ tons. The

APPARATUS FOR COOLING AIR

reheating process will be at constant humidity ratio and the air must be reheated until its state point lies on the room load-ratio line. The required dry-bulb temperature after reheating, from the psychrometric chart, Fig. 19.18, is 53 F. The reheat required is 8220(18.2 − 15.2) = 24,660 Btuh.

The principle involved when reheating is necessary would be the same regardless of the type of cooling unit used. In the case of an actual washer it would be necessary to determine the state of the air entering the washer, draw the washer load-ratio line for a perfect washer, decide on the type of washer, the face velocity, and the spray nozzle pressure, determine the performance factor, and locate the state of the air leaving the actual washer before proceeding to the conclusion of the problem as in the foregoing example.

In the case of a coil it may be possible to find a combination of refrigerant temperature, face velocity, and number of rows of tubes such that a condition will be given that will plot on a room load-ratio line which does not intersect the saturation curve. If that is impossible the best possible combination should be selected and the state it produces plotted on a psychrometric chart to determine the amount of reheat that will be required.

19.17. Possible Variations from Specified Room Conditions.

In the actual operation of a summer air conditioner both the sensible-heat gain and the moisture gain of the space served are bound to change with varying weather, varying time of day, and varying occupancy. The room load-ratio line will continue to be the same as under design conditions as long as the heat and moisture gains are in the same proportion regardless of their magnitudes. Whenever the room load-ratio line is the same as under design conditions but the load is lower, the room temperature can be controlled by supplying a lesser weight (or volume) of air, treated as under design conditions, and the relative humidity in the space will remain at that which was specified. However, if the ratio of total-heat gain to moisture gain is different than under design conditions, the slope of the room load-ratio line will be different, and it will be impossible to maintain the relative humidity at the desired level without changing the state of the conditioned air.

Most self-contained or packaged cooling and dehumidifying units are controlled by stopping the compressor when the load is not sufficient to require continuous operation. If the fan is also stopped the conditions in the room served are likely to be unsatisfactory due to lack of air circulation. If the fan continues to operate after the compressor stops, all of the condensate adhering to the fins and tubes of the cooling and dehumidifying coil re-evaporates and comes back into the room, thus raising the relative humidity of the air there. One solution of this problem is the use of a conditioner which includes two separate refrigeration systems in one casing. One unit operates continuously as long as there is any need for cooling. When the temperature in the space rises slightly because the load is greater than the capacity of the operating machine, the other unit is either automatically or manually started. The circulating fan operates continuously and there is some re-evaporation of condensate when the high stage unit stops. However, this situation will occur a relatively small number of times during a cooling season.

19.18. Ice Storage Air Conditioning. In the case of churches or other buildings where use is periodical, and for relatively short periods, a considerable reduction in the initial investment in summer air conditioning may be effected by use of the *ice storage principle*.[5,6] In this arrangement the refrigerating capacity is far below the peak load requirements of the system. The peak load, which occurs only when the space served is occupied by a crowd, is handled by building up an accumulation of ice ahead of time by the operation of the refrigerating equipment. This accumulation of ice is melted during a period of peak load operation. In this way a peak load of short duration may be handled by refrigerating equipment with a capacity as little as one-tenth of that which would be required without the use of the ice storage principle.

REFERENCES

1. "Heating, Ventilating, Air-Conditioning Guide," ASHAE, New York (1957).
2. "Evaporative Cooling—A Symposium," W. T. Smith, Chairman, *Heating, Piping, Air Conditioning*, **27**, No. 8, 141 (Aug. 1955).
3. "How Air Washers Perform When Cooling," H. M. Hendrickson, *Heating, Piping, Air Conditioning*, **26**, No. 3, 119 (Mar. 1954), continued in No. 9, 116 (Sept. 1954).
4. "Cooling Coils," *Bull. No. DS-365*, The Trane Company, La Crosse, Wis. (Mar. 1954).
5. "How to Air Condition Churches," J. F. Schmidt, *Air Conditioning, Heating, Ventilating*, **53**, No. 3, 83 (Mar. 1956).
6. "Ice Storage Air Conditioning," J. A. Wilkerson, *Heating, Piping, Air Conditioning*, **25**, No. 6, 90 (June 1953).

BIBLIOGRAPHY

1. "An Analysis Method for Predicting Behavior of Solid Adsorbants in Solid Sorption Dehumidifiers," W. L. Ross and E. R. McLaughlin, *Heating, Piping, Air Conditioning*, **27**, No. 5, 169 (May 1955).

APPARATUS FOR COOLING AIR

2. "Air Drying by Solid Granular Adsorbents," B. L. Rothmell and P. J. Batemen, J. *Inst. Heating Ventilating Engrs. (London)*, **19**, No. 198, 4711 (Feb. 1952).
3. "Air Cooling Coil Performance," D. D. Wile, *Refrig. Eng.*, **61**, No. 7, 727 (July 1953).
4. "How to Calculate Air Washer Performance," H. M. Hendrickson, *Heating, Piping, Air Conditioning*, **26**, No. 9, 116 (Sept. 1954).
5. "How to Cool Churches Economically," J. A. Wilkerson and A. P. Boehmer, *Refrig. Eng.*, **62**, No. 2, 36 (Feb. 1954).
6. "Cost Comparison of Air Conditioning Refrigerant Condensing Systems," C. E. Groseclose, *Refrig. Eng.*, **62**, No. 6, 54 (June 1954).
7. "Evaporative Cooling for Comfort," R. S. Farr, *Refrig. Eng.*, **61**, No. 5, 527 (May 1953).
8. "Operation and Maintenance of Air Conditioning Equipment," F. E. Ince, *Heating and Ventilating*, **49**, No. 12, 80 (Dec. 1952).
9. "Design Corrections for Altitude," N. L. Vinson, *Air Conditioning, Heating, Ventilating*, **52**, No. 2, 91 (Feb. 1955).
10. "Packaged Air Conditioner Performance at Partial Load," G. F. Keane, *Refrig. Eng.*, **60**, No. 2, 135 (Feb. 1952).
11. "Design of Air Conditioning Systems to Prevent High Humidity at Partial Load," A. J. McFarlan, *Refrig. Eng.*, **60**, No. 3, 247 (Mar. 1952).
12. "Design of Small Air Conditioning Systems to Maintain Comfort During Mild Humid Weather," D. D. Wile, *Refrig. Eng.*, **59**, No. 5, 447 (May 1951).
13. "The Rating and Testing of Air-Conditioners," *Standard 16-53*, American Society of Refrigerating Engineers, New York.
14. "Architectural and Engineering Factors in Air Conditioning Existing Big Buildings," C. L. Ringquist, *Air Conditioning, Heating, Ventilating*, **52**, No. 10, 87 (Oct. 1955).
15. "Cooling a Small Residence Using a Perimeter-Loop Duct System," D. R. Bahnfleth, C. F. Chen, and H. T. Gilkey, *ASHVE Trans.*, **60**, 271 (1954).
16. "The Operating Cost of Residential Cooling Equipment," S. F. Gilman, L. A. Hall, and E. P. Palmatier, *ASHVE Trans.*, **60**, 525 (1954).
17. "Cooling Studies in a Research Home," W. S. Harris and P. J. Waibler, *ASHVE Trans.*, **60**, 487 (1954).
18. "Air Conditioning the Automobile," M. Kalfus, *Heating and Ventilating*, **50**, No. 4, 97 (Apr. 1953).
19. "Automobile Air Conditioning and Its Problems," P. J. Kent, *Heating and Ventilating*, **51**, No. 3, 99 (Mar. 1954).
20. "Development of Automobile Air Conditioning Systems for Underhood Installation," J. R. Holmes, *General Motors Eng. J.*, **2**, No. 3, 2 (May–June 1955).
21. "Sizing Coils for Near-Freezing Air," M. A. Ramsey, *Air Conditioning, Heating, Ventilating*, **54**, No. 8, 75 (Aug. 1957).

PROBLEMS

1. An attic fan is to be installed in a home having an internal volume of 9000 cu ft and located in New Orleans. Specify its air-handling capacity in cubic feet per minute. Assuming that the ventilation openings in the attic are to be covered with 16-mesh wire screens and protected by metal louvers, specify the total area that should be provided, in square feet.

2. An air washer used as an evaporative cooler has one bank of spray nozzles which direct the water against the moving air. Find the probable dry-bulb temperature and relative humidity of the delivered air at a time when the outdoor air entering the unit has a dry-bulb temperature of 103 F and a relative humidity of 20 per cent.

3. The same as Problem 2 except that an evaporative cooling unit similar to Fig. 19.10 will be used. Assume that the pad thickness and air velocity are representative of good manufacturing practice.

4. Assuming that an ideal washer will be used to supply cooled and dehumidified air to a room for which the estimated direct sensible-heat gain is 60,000 Btuh and the estimated direct moisture gain is 100,000 grains per hr: (a) Find the proper temperature of the saturated air leaving the washer if the room conditions are to be 75 F dbt and 60 per cent relative humidity. (b) Find the weight of dry air in this condition that would be required to absorb both the total heat and moisture gain of the room.

5. A washer which has two banks of spray nozzles (opposed arrangement) is to be used to cool and dehumidify air which is to be supplied to a lecture room having 200 seats. The room is in the interior of a building, and it can be assumed that for practicable purposes the only sources of direct heat and moisture gains are the occupants and the lights. The room dry-bulb temperature is to be 76 F and the relative humidity is to be 56 per cent. All of the air handled by the washer is to be taken from the outside. Outside design conditions are 95 F dbt and 75 F wbt. Assume that the room may be filled at 3:00 P.M. (assume occupancy same as matinee theater). The room is provided with 10,000 watts of fluorescent lighting. Assume that the air will pass through the washer at a velocity of 500 fpm and that the spray water will enter the nozzles at a temperature of 42 F and at a pressure of 30 psig. (a) Find the apparatus dew-point temperature. (b) Determine the probable actual temperature and relative humidity of the air leaving the actual washer. (c) Compute the weight of dry air which must be handled in pounds per hour. (d) Calculate the refrigerating capacity needed in tons. (e) Determine the required cross-sectional area of the washer in square feet. (f) Find the total number of type (c) (Fig. 19.15) nozzles that will be required.

6. A washer having a single bank of nozzles which spray with the air flow is to serve a space which has a direct sensible-heat gain amounting to 200,000 Btuh and a direct moisture gain equal to 25 lb per hr. The outside design conditions are 101 F dbt and 30 per cent relative humidity, and the room conditions are to be 80 F dbt and 50 per cent relative humidity. The air entering the washer is to consist of equal parts by weight of outdoor air and recirculated air. Design the washer for a face area velocity of 500 fpm, a supply water temperature of 50 F, and a nozzle pressure of 25 psig. Assume that type (c) (Fig. 19.15) nozzles will be used. (a) Determine the desired face area of the washer in square feet. (b) Specify the capacity of the refrigerating apparatus in tons. (c) Calculate the required total number of spray nozzles and the number per square foot of washer face area.

7. A Trane direct expansion coil supplied with evaporating liquid refrigerant at 40 F (except for 5 deg of superheat at the exit) is to cool and dehumidify 100 per cent outdoor air at 92 F dbt and 76 F wbt and serve a room where the direct sensible-heat gain and direct moisture gain under design conditions are expected to be 48,420 Btuh and 14.29 lb per hr respectively. The room

conditions desired are 80 F dbt and 50 per cent relative humidity. Assume that the face velocity will be 400 fpm. (*a*) Specify the number of rows of tubes that should be used. (*b*) Calculate the weight of air that should be handled. (*c*) Find the coil face area in square feet that will be needed. (*d*) Compute the refrigerating capacity needed, in tons. (*e*) What apparatus dew-point temperature is required for this load?

8. The same as Problem 7 except that the direct sensible-heat gain has been estimated at 31,620 Btuh instead of 48,420 Btuh.

9. All data and requirements the same as in the example of Art. 19.16, except that the room conditions are to be 76 F dbt with 52.5 per cent relative humidity instead of those given and only 50 per cent of the air handled by the washer is to be fresh air, the remainder to be recirculated from the space served. Use the same spray-water temperature (40 F).

10. The air in an industrial process room is to be maintained at 85 F dbt, with 40 per cent relative humidity. The direct total-heat gain under design conditions is expected to be 80,000 Btuh accompanied by a direct moisture gain amounting to 200,000 grains per hr. Outside air at 96 F dbt and 75 F wbt is to be cooled and dehumidified by an eight-row Trane direct expansion coil operating under the conditions for which the data of Table 19.1 are applicable. (*a*) Find the dry-bulb temperature of the air leaving the coil and the temperature to which the air must be reheated. (*b*) Calculate the required coil face area in square feet. (*c*) Compute the refrigerating capacity required in tons and the reheater capacity needed in Btuh.

20

All-Year Air-Conditioning Methods and Equipment

20.1. Requirements of a Machine for All-Year Air Conditioning. All-year air-conditioning machines must be capable of maintaining a specified temperature and relative humidity within the spaces they serve regardless of the outdoor weather. In addition to the apparatus necessary for properly conditioning the air and distributing it through the spaces served, most all-year air-conditioning machines include filters for dust removal. Many also include special equipment for the removal of odors. The term *comfort air conditioning* applies to spaces in which human comfort and economy of operation are the only considerations. *Industrial air conditioning* applies to all spaces used for the carrying on of processes which require special air conditions for best results.

20.2. Applications for All-Year Air-Conditioning Machines. All-year air conditioning is desirable for practically every building intended for human occupancy. Control of the temperature and relative humidity in all seasons is a necessary part of many research investigations and the condition of the air in the spaces where the processes are carried out affect the quality of many manufactured articles. Table 20.1[1] shows optimum temperatures and relative humidities for several industrial processes.

ALL-YEAR AIR CONDITIONING

TABLE 20.1. Optimum Dry-Bulb Temperatures and Relative Humidities for Certain Industrial Processes*

Industry	Process	Optimum Dry-Bulb Temperature, deg F	Optimum Relative Humidity, per cent
Baking	Bread dough mixing	75–80	40–50
	Dried ingredients storage	70	55–65
Candy manufacture	Hand-dipping room	60–65	50–55
Ceramics	Molding room	80	60–70
Cereal	Packaging	75–80	45–50
Electrical products	Coil and transformer winding	72	15
Fruit	Banana ripening	68	90–95
Fur	Fur storage	40–50	55–65
Leather	Leather storage	50–60	40–60
Libraries	Book storage	70–80	40–50
Mushroom	Mushroom storage	32–35	80–85
Optical	Grinding lenses	80	80
Paint application	Air-drying lacquers	70–90	60
Pharmaceutical	Powder storage prior to mfg.	70–80	30–35
Printing	Newspaper printing	75–80	50–55
Rubber goods	Dipping surgical articles	75–90	25–30
Textile	Cotton spinning	80–85	70
	Linen weaving	80	80
	Rayon spinning	80–90	50–60
	Wool spinning	80–85	50–60
Tobacco	Cigar and cigarette making	70–75	55–65

* Extracted from Table 1, Chapter 45, of the ASHAE Guide.[1] Used by permission.

20.3. Possible Combinations of Processes in All-Year Air Conditioning. An all-year air-conditioning machine may be designed to deliver properly conditioned air during winter weather by employing the following combinations of processes.

1. (a) Preheating to a dry-bulb temperature approximating the dew-point temperature of the desired room condition, (b) humidifying by means of a spray of heated water maintained at that temperature, and (c) reheating to whatever temperature is necessary in order to maintain the desired room temperature (see Fig. 2.8).

2. (a) Preheating to a wet-bulb temperature approximating the dew-point temperature of the desired room condition, (b) humidifying by a spray of water that is neither heated nor cooled, and (c) reheating as in combination 1 (see Figs. 2.5 and 2.6).

3. (a) Preheating as in combination 1, (b) humidifying by controlled evaporation from an open pan of boiling water or by the introduction

of controlled jet steam from properly designed nozzles, and (c) reheating as in combinations 1 and 2 (see Fig. 2.10).

4. (a) Preheating to whatever temperature is required to maintain the desired room temperature, and (b) humidifing either by a pan of boiling water or with steam jets. With this combination of processes it is *not* necessary to provide a reheat coil ($AB''D$ in Fig. 2.10).

5. (a) Preheating when necessary to a temperature well above freezing, (b) humidifying by bringing the air in contact with a suitable solution of lithium chloride or calcium chloride maintained at the proper temperature and density, and (c) heating or cooling when and as required. Provision must be made for the regeneration of the absorbent solution by the addition of water to replace that which is given up to the air. (See Art. 20.9.)

In mild climates where the temperature of the air entering the machine is never lower than 32 F, preheating of the air may be eliminated from combinations 1, 3 and 5.

The final temperature of the air delivered during winter usually approximates that desired in the space in the case of split systems where disseminators such as radiators under separate thermostatic control are available. The only purpose of the winter air conditioner in this case is to deliver *tempered ventilation air*. The air is often delivered at a temperature which is a few degrees lower than that desired in the space so that a slight cooling effect is available for rooms in which the heat gain may be more than the heat loss. In heating and ventilating systems which do not include separate disseminators, the reheat coil (or the preheat coil in combination 4) must be controlled by a room thermostat and the final temperature of the air will depend on the net heat loss from the space served. When the air conditioner handles the heating load some kind of zone control of the reheating process is usually necessary for satisfactory results except in homes and certain commercial buildings such as a small store.

During summer weather the following combinations of processes may be used.

1. Cooling to achieve dehumidification, followed by reheating, if necessary, to maintain the desired dry-bulb temperature and relative humidity in the space served.

2. Dehumidifying by passing the air through a bed of adsorbent material, followed by cooling to whatever temperature is required to maintain the desired comfort in the room.

3. Dehumidifying by bringing the air in contact with a suitable hygroscopic solution, followed by cooling if required.

ALL-YEAR AIR CONDITIONING

The thermodynamics of all of the above summer and winter processes have been discussed in Chap. 2.

In designing an all-year air-conditioning machine, it is necessary to select a combination of apparatus that is capable of performing one sequence of processes during winter weather and another sequence of processes in summer. Certain parts of the apparatus that are essential during summer operation will be inoperative when the weather is cold, and vice versa. Specific arrangements will be discussed in later articles.

20.4. Unit All-Year Air Conditioners. A unit all-year air-conditioning machine is one in which all of the necessary apparatus for

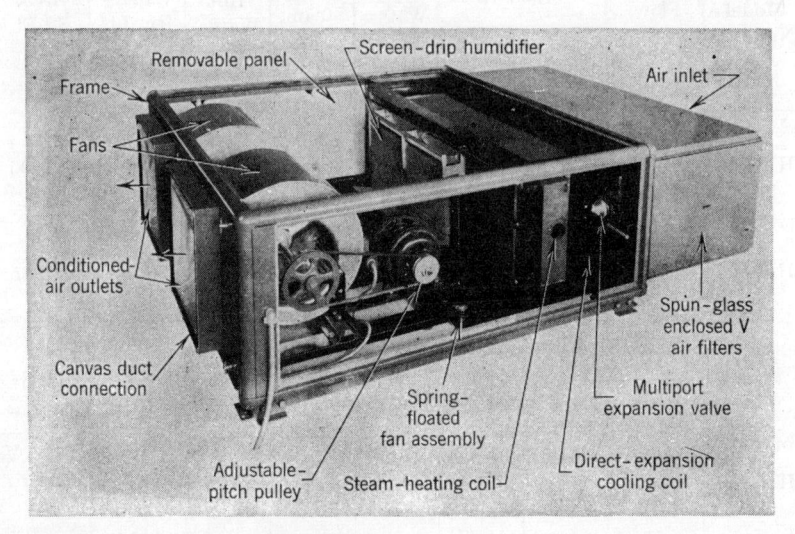

Fig. 20.1. Unit assembly air conditioner with top and side panels removed. (Courtesy General Electric Co.)

heating, humidifying, cooling, and dehumidifying are included in a single cabinet along with one or more fans and a filter. A *remote room-cooling unit*, Art. 19.2, may be converted to a unit all-year air conditioner by the addition of a heating coil and a humidifier. Remote units which are equipped with a water-type coil may be arranged to cool air in summer and heat it in winter with the same coil. Figure 20.1 shows a remote type of all-year air conditioner which is made in capacities up to 33 tons, and Table 20.2 includes data for units of the type shown. Units of this type may be obtained with a chilled water cooling coil in place of the one shown. The steam coil shown for heating may be replaced by one adapted to the circulation of heated water. *Self-contained air-conditioning units* incorporate a small refrigerating plant and may be classified according to : (1) the method of rejecting the condenser heat, i.e., air-cooled, water-cooled or cooled

TABLE 20.2. Rating and Other Data for General Electric Company Remote-Type All-Year Air Conditioners

Air Conditioner Model Number	Range of Air Flow, Cfm	Range of Cooling Capacity — Freon 12 Tons	Range of Cooling Capacity — Freon 12 Btu per Hr	Range of Cooling Capacity — Water, Btu per Hr	Range of Heating Capacity — Steam, Btu per Hr	Range of Heating Capacity — Hot Water, Btu per Hr	Humidifying Capacity, Lb per Hr	Face Area, Sq Ft
HD-200	1200 to 2400	2.82 to 8.45	34,000 to 102,000	29,000 to 99,000	62,000 to 156,000	75,000 to 182,000	15	4.06
HD-300	2300 to 4700	5.38 to 16.5	65,000 to 199,000	57,000 to 192,000	121,000 to 302,000	145,000 to 353,000	30	7.85
	3500 to 7100	8.1 to 24.8	98,000 to 300,000	86,000 to 288,000	182,000 to 455,000	218,000 to 530,000	48	11.8
HD-500	4600 to 9400	10.8 to 33.0	130,000 to 398,000	114,000 to 384,000	242,000 to 604,000	290,000 to 706,000	48 to 96	15.7

Rating conditions: Cooling { Entering air—84 F dbt, 68 F wbt; Freon 12—40 F, 12 F superheat; Water—44 F }

Heating and Humidifying { Entering Air—70 F dbt, 30 per cent relative humidity; Steam—2 psig; Hot water—180 F entering }

by evaporation; (2) the way ventilation air is supplied by the unit, i.e., either 100 per cent recirculation with no deliberate ventilation, ventilation by drawing air from outside and discharging it into the room, or a combination of outside air and recirculated air; (3) the type of air delivery, i.e., free delivery with no ducts from the unit and forced

ALL-YEAR AIR CONDITIONING

circulation through distributing ducts; or (4) the placement of the unit in the room, i.e., floor type or suspended type.

The principal advantages of unit conditioners over central systems are reduction in initial cost through standardization and mass production and reduction in installation cost. The engineering involved in making an installation of unit conditioners consists of (1) estimating the cooling and heating loads, (2) choosing the proper types of units, (3) selecting the proper sizes, and (4) arranging them in the proper locations.

20.5. Central All-Year Air-Conditioning Plants. When several rooms in the same building are intended for uses which require air having approximately the same temperature and relative humidity, they can usually be air-conditioned more economically from a central system than from a number of self-contained units. One of the important advantages in a central system is that the peak summer load does not usually come at the same time in all the spaces served so that the capacity of the cooling and dehumidifying apparatus can be considerably lower than the combined capacity that would be needed if an individual self-contained unit were used in each space.

A great many different arrangements of apparatus may be successfully used to perform the functions of a central all-year air-conditioning machine. Figure 20.2 shows a simple arrangement for a central system which provides a positive control of the dry-bulb temperature and relative humidity of the air in the space served. The arrangement of inlet dampers permits taking any desired proportion of the air from outside. The apparatus shown in Fig. 20.2 is designed to employ the processes of preheating, humidifying with heated spray water, and reheating during cold weather. Summer air is cooled and dehumidified by chilled spray water in the same chamber that is used for humidification in winter, and the steam flow to the reheater coil may be controlled by a room thermostat to provide just the right amount of reheating during both seasons. The recirculated air enters the machine from the equipment room which serves as a plenum chamber. Suitable arrangements must be made for conducting the air to be recirculated from the spaces served to this room.

The apparatus of Fig. 20.2 would be satisfactory for serving one large room or a group of rooms having the same heat-gain and heat-loss characteristics. In general, it would not be satisfactory in a multiroom installation because it does not provide for zone control of the final air temperature.

Figure 20.3 shows schematic diagrams of three different arrange-

496   AIR CONDITIONING AND REFRIGERATION

ments of all-year air-conditioning apparatus which may be used when the building served must be divided into zones.

Arrangement 1 provides a positive control of the dry-bulb temperature in the different zones by using a separate reheat coil in each branch of the system instead of one in the central machine. The combination of a preheat coil and a tempering coil shown in arrangement 1 of Fig. 20.3 is necessary in some installations where ventilation requirements

Fig. 20.2. All-year central air-conditioning system. (Courtesy Carrier Corp.)

may demand that a large proportion of outdoor air be used even in extreme cold weather. The first steam-heating coil in a stream of air at a very low temperature must be "fully on" or "completely off" to prevent freezing of the condensate in the bottom. In order to achieve a satisfactory control of the air temperature entering the humidifying section of the conditioner under those conditions, it is then necessary to provide a preheat coil of low capacity which may be operated with its steam valve wide open, followed by a tempering coil the flow of steam to which may be thermostatically controlled. The open-pan type of humidifier indicated in this arrangement and shown in Fig. 2.9 is inexpensive in first cost and will provide satisfactory humidification if properly controlled. The flow of steam to the coils in the humidifying pan is usually controlled by a humidistat located in the return-air duct. Dehumidification during summer operation is accomplished

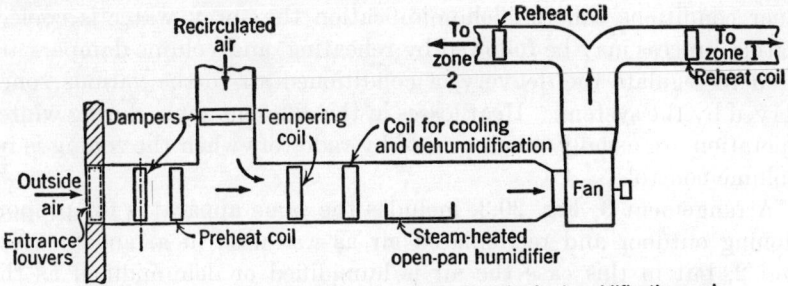

ALL-YEAR AIR CONDITIONING

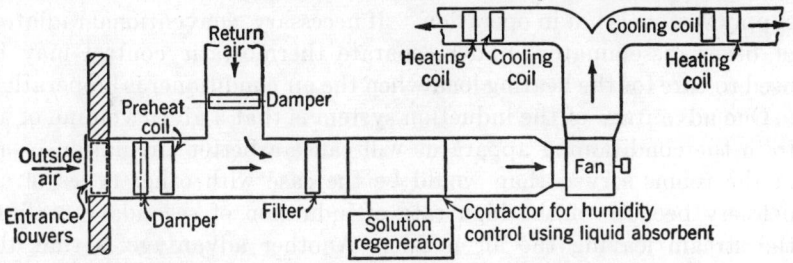

Fig. 20.3. Various arrangements of apparatus in all-year central systems.

by means of a finned-tube coil which may be supplied with either chilled water or a liquid refrigerant.

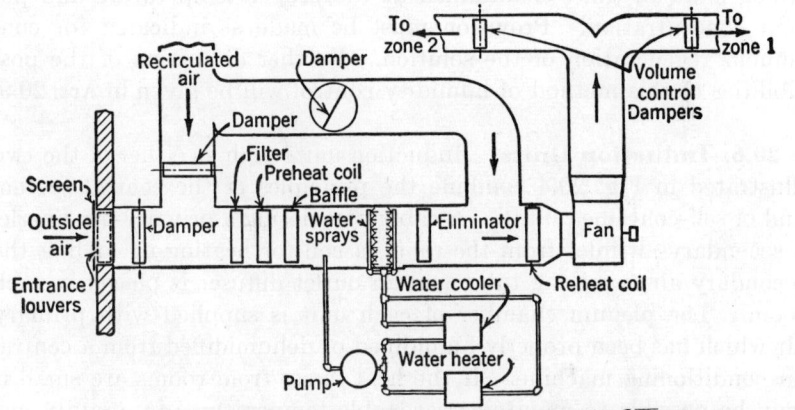

Arrangement 2, Fig. 20.3, utilizes some form of air washer for both humidification and dehumidification as in Fig. 20.2. When humidification of the air is required the spray water is heated, and when the outdoor conditions call for dehumidification the spray water is cooled. Either process may be followed by reheating, and volume dampers are used to regulate the delivery of conditioned air to the various zones served by the system. Heat losses in the different zones during winter operation are usually taken care of by radiators when the zoning is by volume control.

Arrangement 3, Fig. 20.3, includes the same apparatus for proportioning outdoor and recirculated air as was used in arrangements 1 and 2, but in this case the air is humidified or dehumidified as the occasion demands by bringing it in contact with an absorbent solution which is at all times maintained at the proper temperature and per cent concentration. Provision must be made as indicated for continuous regeneration of the solution. Further discussion of the possibilities of this method of humidity control will be given in Art. 20.9.

20.6. Induction Units. Induction units such as either of the two illustrated in Fig. 20.4 combine the principles of the central system and of self-contained units. A cabinet enclosing a primary-air nozzle, a secondary-air inlet from the room, a coil for heating or cooling the secondary air, a mixing tube, and an outlet diffuser is placed in each room. The plenum chamber of each unit is supplied with primary air which has been properly humidified or dehumidified from a central air-conditioning machine. If the heat losses from rooms are small it may be possible to maintain reasonable temperatures by gravity circulation of the secondary air over the heating coil when the central supply system is not in operation. If necessary, conventional radiators or other disseminators under separate thermostatic control may be used to care for the heating load when the air conditioner is inoperative.

One advantage of the induction system is that a given volume of air from the conditioning apparatus will cause a better circulation of air in the rooms served than would be the case with other types of air delivery because of the high rate of induction of secondary air into the stream leaving the air nozzle. Another advantage is that the admixture of secondary air with the conditioned air, in summer operation, raises it's temperature before it is delivered to the room, thus reducing the chance that cold drafts will be noticed by the occupants. Also, a minimum amount of cold air is transported through the ducts of the central system,, thereby reducing power costs below the amount that would be required if the full delivery of the diffusers were handled by the central fan.

Control of induction units during summer operation is usually through either manual or automatic control of the primary air. It

ALL-YEAR AIR CONDITIONING

may be advisable to bypass some air around the cooling and dehumidifying apparatus of the central system during periods of low cooling load so that the volume of primary air in the induction units will not be reduced below the point where satisfactory circulation is produced in the individual rooms.

20.7. High-Velocity Systems. As construction costs increase, architects and building owners are more aware of the space occupied

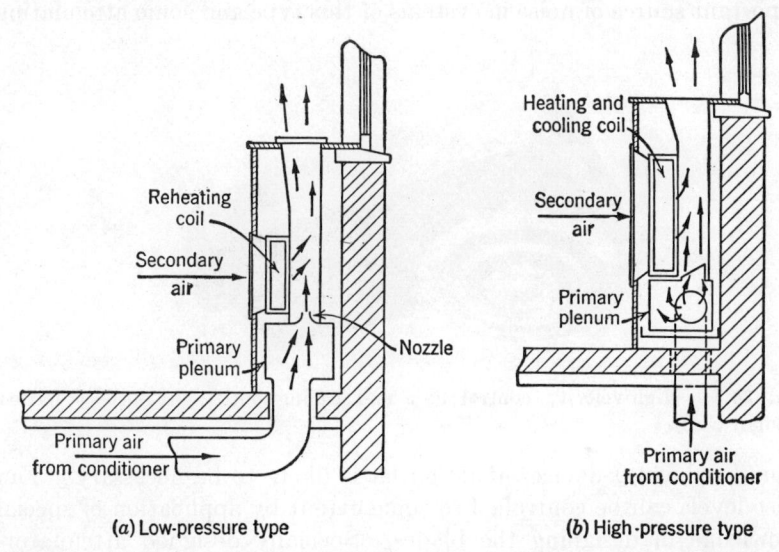

Fig. 20.4. Induction-type room air conditioners. (From "Heating Ventilating Air-Conditioning Guide 1948." Used by permission.)

by the ducts which distribute the air from central air-conditioning systems. Space for these ducts is usually provided by false ceilings. Decreasing the size by designing the system for higher velocities decreases the required depth of these spaces and may decrease the cost of the building per square foot of usable floor space.

Design procedures are the same regardless of the intended velocity. Phillips[2] suggests sizing trunk ducts of high-velocity systems by the equal pressure-loss method (see Art. 14.22) using a 1-in. drop in pressure per 100 ft of equivalent length in systems of moderate size. In general, the over-all system resistance should not exceed 2.5 in. of water. However, where under special conditions the savings in construction costs effected by the small ducts warrant the increased operating expense for power, over-all system resistance as high as 5 in. of water may be used. Phillips suggests that one-half of the theoretical static-pressure regain be subtracted from the total system resistance in determining the static pressure that will be needed at the fan outlet (see Art. 14.22). Wilson[3] suggests that fan static pressure can be conserved by selecting an air diffuser for the branch in the most remote location which requires a lower static pressure than those outlets which are closer to the fan.

The noise levels produced in all parts of an air-handling system will be higher with greater velocities. The fan is likely to be the most important source of noise in systems of this type and some attenuating

Fig. 20.5. High-velocity control unit and ceiling outlet. (Courtesy Barber Colman Co.)

(sound-reducing) device[3] at its outlet is likely to be necessary. Fan noise levels can be controlled to some extent by application of special knowledge in designing the blades. Specially designed attenuator-diffuser combinations must be used at all points in high-velocity systems where the air enters the spaces that are being served if objectionable noise is to be avoided. Such a device is shown in Fig. 20.5. The velocity of the air entering the room from such a unit is of the same order of magnitude as that from diffusers used with low-velocity systems.

20.8. The Double Duct System. In many commercial buildings certain interior areas may require cooling when the outdoor temperature is as low as 40 F while all rooms with windows in exterior walls require heating at the same time. West rooms may require cooling in the afternoon due to heat gain through windows at the same time that east rooms require heating. South rooms may require cooling through much of the working hours of a sunlit winter day. An excellent solution to the problem of supplying heating to some rooms and cooling to others at the same time, by means of a single central system, is through the use of a double duct arrangement as illustrated in Fig.

ALL-YEAR AIR CONDITIONING

20.6 from reference 4. A typical mixing unit for use in a double duct system is shown in Fig. 20.7. These units are usually located under windows in the rooms served and are connected to both the hot-air and cold-air ducts of the central system. The proportions of cooled and heated air are varied by a pair of dampers operated by a mechanism which is controlled by a room thermostat. The fan silencer which is shown in Fig. 20.6 is necessary only in a case where the double duct system is also a high-velocity system.

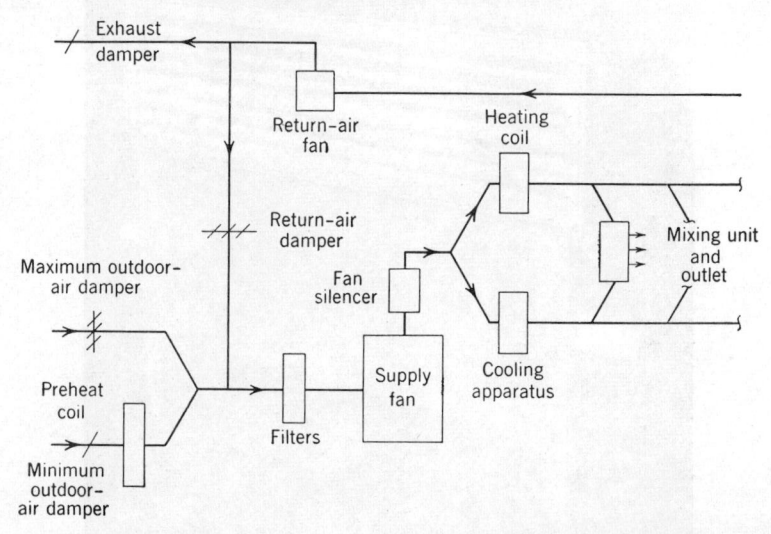

Fig. 20.6. Schematic layout of a typical double duct air-conditioning system.

One undesirable characteristic of the arrangement shown in Fig. 20.6 is that the static pressure tends to vary in the two supply ducts with changing load conditions. If both supply ducts are designed to handle the same volume of air, equal static pressure at both inlets to all of the mixing units will occur only when their thermostats are demanding equal volumes of hot and cold air. If all of the mixing units are using cold air only, the friction loss in the cold-air supply duct will be high and the static pressure toward its end will be low. At the same time there will be no friction loss in the hot-air supply duct and the static pressure near its end will be equal to that at the fan outlet (or at the silencer outlet in Fig. 20.6). Fluctuation of the static pressure in the hot and cold supply ducts can be reduced by the proper application and operation of dampers.

A fan selected to deliver the required total volume of air through only one of the two supply ducts will deliver a greater total system volume when the flow is divided between the two because of the lowered system resistance. This condition results in a greater total volume from each of the mixing units which increases the noise level and changes the air-distribution pattern in each room. This situation can be avoided by using an automatic fan volume control. Fan capacity can be controlled either by means of inlet vanes or by drive mechanisms which vary its speed.

Fig. 20.7. Medium pressure mixing plenum periphery unit. (Courtesy Tuttle and Bailey, Inc., Division of Allied Thermal Corp.)

The mixing units which are used with high-velocity double duct systems include an interior arrangement which attenuates the noise produced by the air passing at high velocity through the openings in the two dampers. Mixing units of this type are sometimes referred to as *attenuator-diffusers*. Mixing units for medium-and high-velocity systems (main duct velocities over 2000 fpm) are selected on the basis of volume to be handled and noise level which can be tolerated. Data for the type of unit shown in Fig. 20.7 are given in Table 20.3, which is abridged from Table XXVII of reference 5. The complete table covers units up to 58 in. long.

20.9. The Use of Absorbents in Central All-Year Air-Conditioning Systems. Solutions of certain salts such as lithium chloride are very strongly hygroscopic and can therefore be used to dehumidify

ALL-YEAR AIR CONDITIONING

an air stream. An absorbent salt solution of appreciable concentration will have a water-vapor pressure that is lower than the vapor pressure of pure water at the same temperature. Consequently, a spray of such a solution or a surface wetted with it will dehumidify air having a given water-vapor pressure while at a higher temperature than would be the case if plain water were used. As long as the vapor pressure of the solution is less than that of the air, water vapor from the air will enter the solution, causing the air to be dehumidified. Addition of water to a given amount of a solution will lower the concentration of

TABLE 20.3. Selection Data for the Mixing Diffuser Shown in Fig. 20.7

Cu Ft per Min	UNIT LENGTH—26 In.					UNIT LENGTH—34 In.					UNIT LENGTH—42 In.				
	Static Pressure, In. H_2O		Decibel Level			Static Pressure, In. H_2O		Decibel Level			Static Pressure, In. H_2O		Decibel Level		
	min.	max.	at min. s.p.	at 2″ s.p.	at 3″ s.p.	min.	max.	at min. s.p.	at 2″ s.p.	at 3″ s.p.	min.	max.	at min. s.p.	at 2″ s.p.	at 3″ s.p.
50	0.14	3.5	36	37	38										
70	0.20	3.5	37	38	39	0.15	3.5	36	37	38					
90	0.28	3.5	37	39	40	0.22	3.5	36	38	39	0.18	3.5	36	37	38
110	0.40	3.5	38	40	41	0.30	3.5	37	39	40	0.24	3.5	36	38	39
130	0.57	3.5	39	42	43	0.40	3.5	38	40	41	0.33	3.5	37	39	40
150	0.80	3.5	40	43	44	0.52	3.5	39	41	42	0.41	3.5	38	40	41
170	1.00	3.5	40	44	45	0.68	3.5	39	42	43	0.51	3.5	38	41	42
190						0.88	3.5	40	43	44	0.63	3.5	39	42	43
210						1.13	3.5	41	44	45	0.77	3.5	40	42	44
230											0.92	3.5	41	43	44
250											1.10	3.5	42	45	45

salt and increase the vapor pressure so that regeneration of the solution by boiling off the accumulated water is necessary. The latent heat that is released from the water vapor condensed raises the temperature of the solution so that in addition to regeneration the solution used in the contact chamber must be circulated through a cooler to maintain the proper temperature.

Figure 20.8 shows an air-conditioning unit of this type. The flow diagram is for the processing of summer air. In the arrangement illustrated the pump draws brine from the sump at the bottom of the regenerating chamber and forces approximately 85 per cent of it through a cooler on the way to the spray nozzles above the contactor cells. The remaining 15 per cent passes through a heater, then through the regenerating spray where the heated solution gives up water vapor to an auxiliary air stream which carries it out of the building. By heating the brine that is delivered to the contact chamber instead of cooling it, air may be humidified by the same unit. When used for humidification it is not necessary to use the regenerating portion of the apparatus, but water must be continually added to the solution to compensate for that which is given up to the air in the humidification process. At all times the brine flows from the contactor cells to the regenerator sump by gravity action.

Lithium chloride has many properties which make it a satisfactory material for use in humidity-control apparatus of this type. It is less

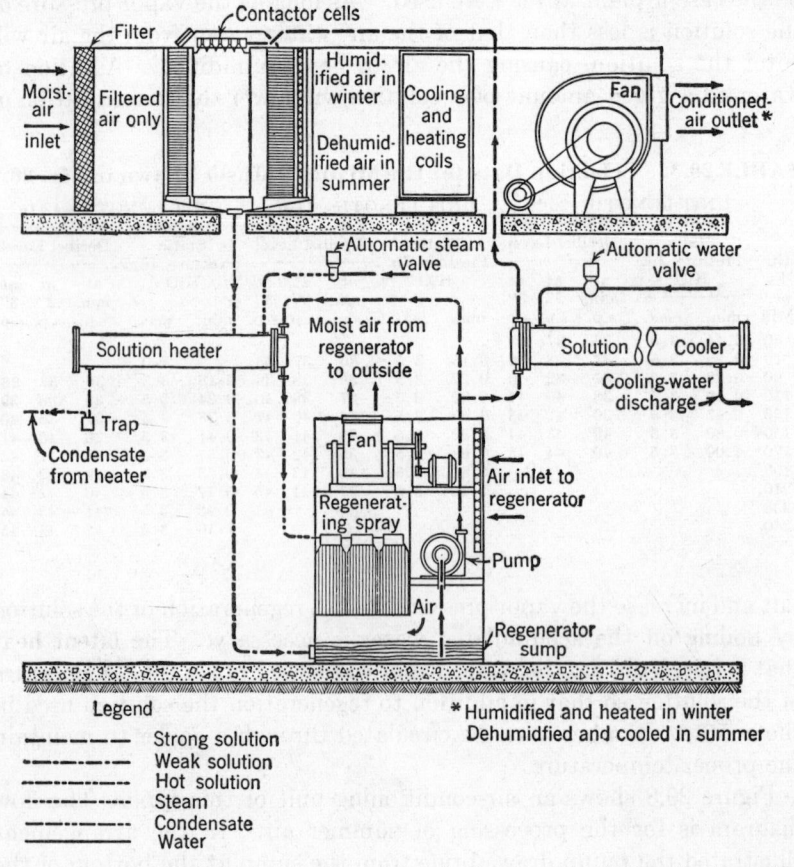

Fig. 20.8. Air-conditioning apparatus employing a salt solution for humidity control. (Courtesy Surface Combustion Corp.)

corrosive than clear water, is perfectly stable, does not react with the CO_2 in the air, does not evaporate at the temperature required to regenerate the solution, does not crystallize in any concentrations that are needed, and is nontoxic. Figure 20.9 gives the vapor pressure for different concentrations of this salt in water when the solution is held at different temperatures ranging from 20 F to 320 F. The chart also gives the relative humidity of air in equilibrium with this solution

ALL-YEAR AIR CONDITIONING

for any combination of temperature and concentration covered. By referring to Fig. 20.9 it is possible to specify a combination of solution concentration and solution temperature for any air condition which may be desired. For example, air which has come to equilibrium with a 40 per cent solution held at 80 F will have a relative humidity

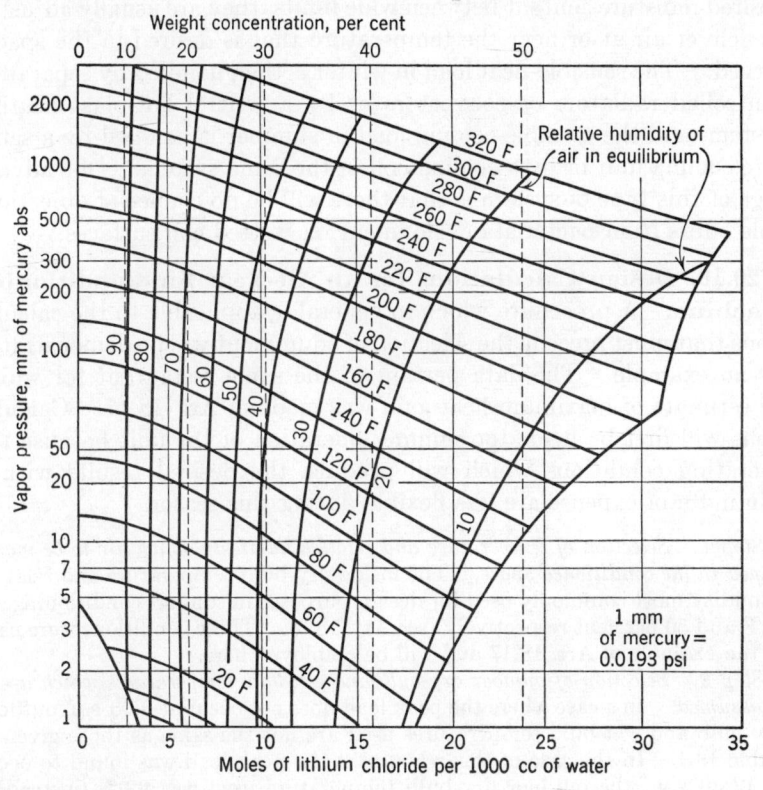

Fig. 20.9. Water-lithium chloride equilibria. (From *Air Conditioning and Refrigerating Data Book.*[6] Used by permission.)

of approximately 17 per cent. Likewise, air in equilibrium with a 30 per cent solution held at 100 F will have a relative humidity of 40 per cent. The vapor pressure of a solution of any concentration held at any given temperature within the range of Fig. 20.9 may be read in millimeters of mercury from the scale at the left. Comparison of the vapor pressure of the solution with that of water when at the same temperature may be made by converting it to equivalent pressure in pounds per square inch and referring to Table 1.7. For example, a 40 per cent concentration of lithium chloride at 102 F has a vapor pressure of 10 mm of mercury or 0.193 psi. From Table 1.7 it is found that pure water at approximately 52.2 F has the same vapor pressure. This means that a 40 per cent solution of lithium chloride at 102 F is as effective a dehumidifier for air as is pure water at 52.2 F.

Packaged units employing salt solutions for humidity control are available. These units can be adjusted to deliver air having any desired moisture content between wide limits, they are usually adjusted to deliver air at or near the temperature that is desired in the spaces served. The sensible-heat load in winter is then handled by separately controlled radiators or convectors or by a central hot-blast heating system, and the sensible-cooling load in summer is handled by a separate cooling unit or by a cooling coil in the same system. One advantage of this type of system is that there will be no chance of objectionable odors from bacterial development on wetted coil surfaces.

20.10. Design Calculations for an All-Year Air-Conditioning Machine.

A procedure which in general is applicable to the calculations that must precede the selection of equipment will be demonstrated by an example. The data pertain to the same restaurant for which an estimate of maximum heat gain was made in Art. 18.17. Calculations will first be made for summer operation of the unit because the operating conditions which will produce the desired results with a minimum of expense are less flexible during this season.

Step 1. Selection of temperature and relative humidity which are to be maintained in the conditioned space. The inside dry-bulb temperature and relative humidity most commonly used for design purposes in comfort conditioning are 80 F and 50 per cent respectively (see Art. 18.3). These conditions were used in the example of Art. 18.17 and will be employed here.

Step 2. Selection of outdoor dry-bulb and wet-bulb temperatures which are to be assumed. In a case where the peak load does not occur at 3:00 P.M. outdoor dry-bulb and wet-bulb temperatures used are not the same as those given in Table 18.1. In the case of this restaurant the peak load was found to occur at 12:30 P.M., the outdoor dry-bulb temperature used was 92 F (instead of 95 F), and the enthalpy of the outdoor air was derived to be 38.62 Btu per lb of dry air by the method which was explained in Art. 18.12.

Step 3. Estimation of maximum heat and moisture gains. The maximum heat gain for this restaurant has been estimated in the example of Art. 18.17 as 184,239 Btuh, and the maximum moisture gain as 62.20 lb per hr. However, the gains from the ventilation air amounting to 37,300 Btuh and 20.40 lb per hr (Table 18.14) are included in the above figures. Assuming that the ventilation air will be brought in through the cooling and dehumidifying unit, the *direct* total-heat gain is 184,239 − 37,300 = 146,939 Btuh and the *direct* moisture gain is 62.20 − 20.40 = 41.80 lb per hr or 41.80 × 7000 = 292,600 grains per hr.

Step 4. Select the type of unit which is to be used. It will be assumed that a suspended unit similar to that shown in Fig. 20.1 will be used. It will be presumed for the purpose of these computations that the data of Table 19.1

ALL-YEAR AIR CONDITIONING

apply to the direct expansion coils used in these machines and that the manufacturer will install a coil having the number of rows of tubes best suited to this installation.

Step 5. Draw the room load-ratio line and determine the apparatus dew-point temperature for design conditions. The room conditions are 80 F dbt, 50 per cent relative humidity, $h_r = 31.2$, and $w_{gr} = 76.5$. Using the direct heat and moisture gain data from step 3 and the method of the example of Art.

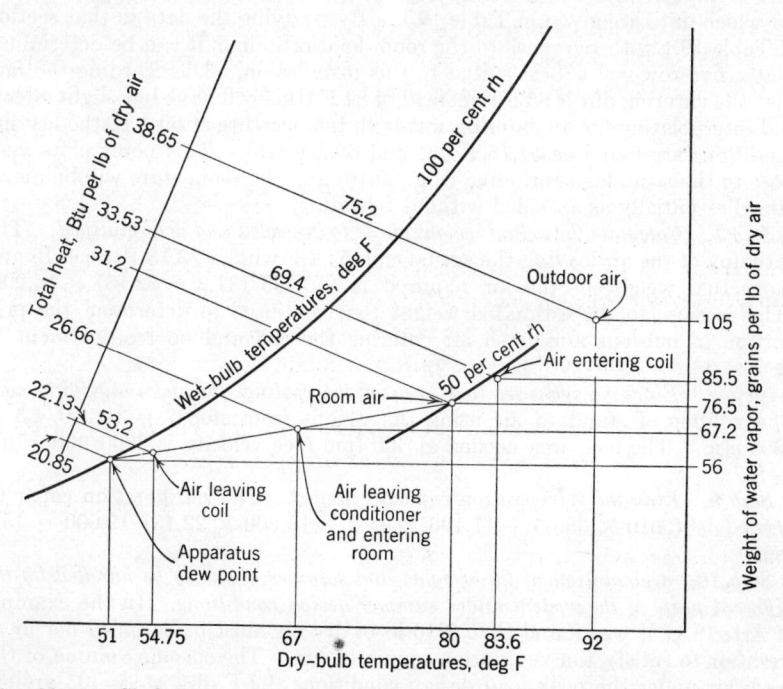

Fig. 20.10. Skeleton psychrometric chart showing data used in the example of Art. 20.10.

19.13, the enthalpy-humidity difference ratio is $146{,}939/292{,}600 = 0.502$ and the apparatus dew-point temperature is 51 F (see Fig. 20.10). The weight of saturated air at this temperature that would be required to absorb the direct total-heat gain is $146{,}939/(31.2 - 20.85) = 14{,}190$ lb per hr and the weight required to absorb the direct moisture gain is $292{,}600/(76.5 - 56) = 14{,}270$ lb per hr. This is a good check and it may be assumed that 51 F is the correct apparatus dew-point temperature and that the room load-ratio line that was drawn has the correct slope.

Step 6. Determine the state of the air entering the coil and select the number of rows of tubes that will give the best results. The weight of ventilation air needed for this installation (item 14 in the example of Art. 18.17) is 5010 lb per hr. Since it is known that the weight of conditioned air from the actual coil will be somewhat more than that obtained in step 5, it will be tentatively estimated at 16,000 lb per hr. On that basis $\tfrac{5}{16}$ of the air will be outdoor air and $\tfrac{11}{16}$ will be recirculated. The enthalpy and the humidity ratio of the outdoor air from the aforementioned example are 38.65 Btu per lb and 105 grains per lb respectively. Therefore, the enthalpy of the mixture entering the coil will be $(38.65 \times 5 + 31.2 \times 11)/16 = 33.53$ Btu per lb and the humidity ratio of the mixture will be $(105 \times 5 + 76.5 \times 11)/16 = 85.5$ grains per lb. Plotting these data on the psychrometric chart, Fig. 20.10, the dry-bulb and wet-bulb temperatures of the air entering the coil are found to be 83.6 F and 69.4 F respectively. These temperatures are close to 84 F dbt and 70 F wbt for which data are given in Table 19.1. By studying the data in that section of Table 19.1 with reference to the room load-ratio line, it can be determined that a five-row coil is best suited to this installation. Disregarding the fact that the entering dbt is 83.6 F instead of 84 F (this will have but slight effect) and interpolating for an entering wet-bulb temperature of 69.4 F, the leaving conditions are found as 54.75 F dbt and 53.2 F wbt. This point plots very close to the room load-ratio line (Fig. 20.10) and the room state will be maintained essentially as specified without reheating.

Step 7. Calculate the actual weight of air to be cooled and dehumidified. The enthalpy of the air leaving the actual coil, 53.2 F wbt, is 22.13 Btu per lb and the actual weight of the air required is $146,939/(31.2 - 22.13) = 16,200$. (This is close to the estimated weight that was used to determine the proportion of outdoor and fresh air entering the coil and no readjustment is necessary.)

Step 8. Find the required face area of the cooling and dehumidifying coil. The volume of standard air using the weight from step 7 is $16,200/4.5 = 3600$ cfm. The face area needed at 400 fpm face velocity is $3600/400 = 9.0$ sq ft.

Step 9. Find the refrigeration capacity needed. The refrigeration capacity needed is $(5010 \times 38.65 + 11,190 \times 31.2 - 16,200 \times 22.13)/12,000 = 15.4$ tons.

Step 10. Calculation of the weights and volumes of air to be handled by the different parts of the system under summer design conditions. In the example of Art. 18.17 it was found that 5010 lb of dry air must be supplied per hr as fresh air to satisfy the ventilation requirements. The specific volume of the fresh air under the peak load design conditions (92 F dbt, $w_g = 105$ grains) may be read from a psychrometric chart as 14.24 cu ft per lb. The volume of fresh air handled if exact design conditions are met will be $5010 \times 14.24/60 = 1190$ cfm. The weight of the recirculated air should be $16,200 - 5010 = 11,190$ lb per hr and the volume under exact design conditions $11,190 \times 13.84/60 = 2580$ cfm. The actual volume entering the coil will be $1190 + 2580 = 3770$ cfm. (The volume of 3600 cfm found in step 8 is based on standard conditions, density 0.075 lb per cu ft. This is the volume customarily used in determining the specified face area velocity but it is not the actual volume entering the coil.) The actual volume of air leaving the coil is $16,200 \times 13.14/60 = 3550$ cfm.

Cooled and dehumidified air at 54.75 F must be mixed with room air before contacting any occupant of the space served. This may be accomplished through the use of special diffusion devices or by mixing the cooled and dehumidified air with bypassed recirculation air, before it is delivered to the discharge grilles. Bringing in a weight of bypassed air equal to that which is cooled and dehumidified would provide air in a condition suitable for introduction into the space through ordinary grilles. The enthalpy of the delivered

ALL-YEAR AIR CONDITIONING

air would be $(31.2 + 22.13)/2 = 26.66$ Btu per lb and its humidity ratio would be $(76.5 + 58)/2 = 67.2$ grains per lb. (58 is the humidity ratio of the air leaving the coil; this figure is not shown in Fig. 20.10.) By plotting these data it is found that the dry-bulb temperature of the mixture will be 67 F (see Fig. 20.10). The volume of the bypassed air is $16,200 \times 13.84/60 = 3740$ cfm. The total weight of air delivered from the grilles would be 32,400 lb per hr and the volume would be $32,400 \times 13.47/60 = 7270$ cfm. The volume of air flowing in each of the various parts of the apparatus together with the

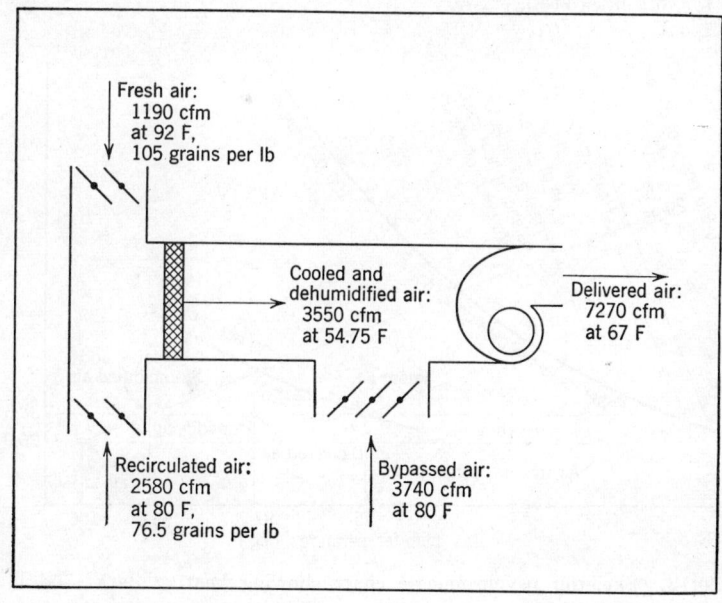

Fig. 20.11. Flow diagram for the example of Art. 20.10.

dry-bulb temperature and humidity ratio at several points of interest is shown in Fig. 20.11.

Step 11. Selection of indoor temperature and relative humidity for winter operation. A dry-bulb temperature of 75 F accompanied by a relative humidity of 30 per cent would be satisfactory for this installation. A higher relative humidity than 30 per cent under winter design conditions would be impractical because the windows are not provided with double glass. The enthalpy of the room air will be 24.2 Btu per lb of dry air and the humidity ratio will be 38.5 grains per lb (see Fig. 20.12).

Step 12. Selection of winter outdoor design temperature. It will be assumed that the winter design temperature in common use in the locality is 0 F. For convenience it will be assumed that the outdoor air is saturated. (The humidity ratio is extremely small at 0 F in any case.) The enthalpy and the humidity ratio of the outdoor air from Table 2.1 are, respectively, 0.8317 Btu per lb of dry air and 5.5 grains per lb of dry air. The volume per pound of dry air from the same source is 11.59 cu ft.

510 AIR CONDITIONING AND REFRIGERATION

Step 13. Estimation of heat losses. Following the procedure that has been outlined in Chap. 4, the heat losses due to conduction through exposed wall, window, and roof areas may be estimated at 101,000 Btuh. The heat required to raise the temperature of the cold air which enters the restaurant by opening and closing of the door can be neglected because heat gains from people, lights, etc., were not considered in estimating the heat losses. Heat gain from these and other interior sources will be more than adequate for the warming of this air when the serving area is occupied. The heat required to warm the air introduced through the machine for the purpose of ventilation will be provided for in a later step.

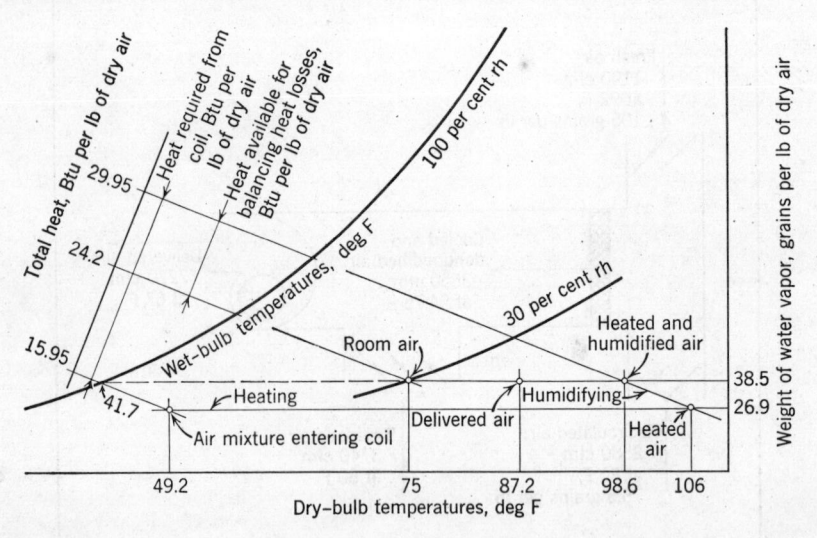

Fig. 20.12. Skeleton psychrometric chart showing heating data used in the example of Art. 20.10.

Step 14. Calculation of the weights and volumes of air handled under winter design conditions. (The following analysis is based on the assumption that the fan speed and damper settings needed to handle the volume of air required in all parts of the apparatus for summer design conditions will not be changed when the system is used for heating.) The weight of fresh air under winter design conditions is $1190 \times 60/11.59 = 6170$ lb per hr. The volume of the room air in cubic feet per pound of dry air at 75 F and 30 per cent relative humidity, from a psychrometric chart, is 13.59 cu ft, and the weight of recirculated air under winter conditions is $2580 \times 60/13.59 = 11,390$ lb per hr. The total weight of air processed is $6170 + 11,390 = 17,560$ lb per hr, and the weight of the bypassed air is $3740 \times 60/13.59 = 16,520$ lb per hr.

Step 15. Estimation of the humidifying capacity needed. The humidity ratio of the indoor air at the condition that has been specified for winter operation is 38.5 grains per lb of dry air and that of the outdoor air under design conditions is 5.5 grains. It can be assumed that the only outdoor air entering the space served is that brought in through the conditioner, and the humidify-

ALL-YEAR AIR CONDITIONING

ing capacity needed when there are no occupants and no appliances in operation which release moisture is 6170(38.5 − 5.5)/7000 = 29.1 lb per hr. The moisture gain at a rush hour from sources other than the ventilation air (example of Art. 18.17) amounts to more than the above weight but the humidifier will still be needed at times when there is little release of moisture within the space from other sources. If the humidifier is controlled by a humidistat (Art. 21.10) water will be supplied to the screen (Fig. 20.1) only when it is needed.

Step 16. Estimation of the heating capacity required in the heating coil. Assuming that there will be times when there are no interior sources of heat, the heating coil must have the capacity to heat the mixture of outdoor air and recirculated air an amount such that it will compensate for the heat losses from the room while maintaining the temperature at 75 F. The total weight of air processed must liberate 101,000 Btuh in cooling from the delivered temperature to that of the room air, 75 F. The required enthalpy of the air from the heating coil therefore is 24.2 + 101,000/17,560 = 29.95 Btu per lb of dry air. In the arrangement, Fig. 20.1, which is to be used, the humidification process is one approximating partial adiabatic saturation which means that the air must be heated at the entering humidity ratio until the enthalpy is equal to that of the final condition with the necessary water added (see Fig. 20.12). The enthalpy of the mixture entering the coil is (6170 × 0.8317 + 11,390 × 24.2)/17,560 = 15.95 Btu per lb, and the humidity ratio is (6170 × 5.5 + 11,390 × 38.5)/17,560 = 26.9 grains per lb. From the psychrometric chart, Fig. 20.12, it may be noted that the air under winter design conditions must be heated to 106 F. The cooling effect of the liquid water added by the screen drip humidifier reduces the temperature to 98.6 F. The required heating capacity of the coil is 17,560 (29.95 − 15.95) = 246,000 Btuh. The mixing of bypassed air with that from the humidifier before delivery to the room will result in a delivered state that is approximately midway between the room air and that of the heated and humidified air (Fig. 20.12).

The portion of the system that is shown in Fig. 20.1 could be installed on a properly supported false ceiling near the rear of the serving area and arranged to discharge the conditioned air toward the front. A suitable duct for conducting the mixture of recirculated air and fresh air from the point of mixing to the air intake of the unit could be located above a drop ceiling at the rear of the serving area. The logical location for the fresh-air intake of such a unit would be in the west wall adjacent to the partition which separates the serving area from the kitchen. The recirculated air could be removed from the conditioned space through a grille set with its lower edge at floor level in the northwest corner of the serving area. This grille and the vertical duct necessary for conducting the recirculated air to the mixing chamber could easily be masked, for decorative effect. A unit of this type can be arranged for bypassing air around the cooling and dehumidifying section by replacing a removable panel with a suitable inlet grille. The refrigeration unit consisting of the compressor and the condenser would have to be separately located and could be placed in the basement. If it is necessary to recover the condenser cooling water an evaporative condenser can be used; its discharge air duct should be located as far as practicable from the fresh-air intake for the air-conditioning unit.

An all-year conditioner of the type shown in Fig. 20.2 could be used in place of the arrangement for which the foregoing computations have been made. In that case the procedure of Art. 19.14 applicable to an actual washer would be used to determine the state of the cooled and dehumidified air instead of the calculations of step 6. All other computations in connection with summer operation of the unit would be the same. If a washer were included to cool and dehumidify the air in summer it would also be used for humidification in winter. Since the temperature of the mixture of outdoor air and recirculated air under winter design conditions is 49.2 F (Fig. 20.12), it would not be necessary to provide a preheat coil. It may also be noted from Fig. 20.12 that the wet-bulb temperature of this mixture is equal to the dew-point temperature of the room air, which means that if complete adiabatic saturation could be achieved it would not be necessary to heat the spray water. Although the humidity ratio of the leaving air may fall slightly short of the desired level under winter design conditions because of incomplete adiabatic saturation, it would not be necessary to provide for heating the spray water in this installation. At all times when the need for humidification is appreciably less than under design conditions, continuous operation of the washer even with no heat supplied to the water would result in the evaporation of more moisture than needed and it would be necessary to operate it intermittently. If humidification of the air were provided by a washer it would precede the heating process instead of following it as in the arrangement of Fig. 20.1.

20.11. Selection of Packaged Units. The calculations of the previous article were carried out on the basis of being able to specify the design of all the parts of an arrangement similar to the one shown in Fig. 20.1. Many units of that type are factory-assembled and one can only select the model best suited to the needs of a particular room or building. In selecting a unit from manufacturers' data such as Table 20.2, the only calculated data which can be used are the refrigerating capacity needed in summer and the heating and humidifying capacities needed in winter. The refrigerating capacity required is equal to the total-heat gain of the space to be served including that due to the ventilation air (184,239 Btuh in the case of the restaurant used as an

ALL-YEAR AIR CONDITIONING

example in the previous article). The heating capacity used in the equipment selection must include the heat required to warm the ventilation air that is to be brought in through the unit (246,000 Btuh in the case of the aforementioned example). According to the data of Table 20.2 Model HD-300 would be able to supply both the cooling capacity that would be required in summer and the heating capacity that would be needed in winter. The screen drip humidifier incorporated in the unit would be capable of evaporating the maximum amount of water needed for humidification.

It would be a coincidence if a packaged unit contained a cooling coil design that would produce air in a state that would plot on the room load-ratio line that applies to a specific room under design conditions. However, manufacturers of packaged units usually develop types that are well adapted in their arrangement of parts to certain applications such as offices, showrooms, homes, and stores. After a preliminary study certain conditions are assumed as being representative of the type of space a certain unit is likely to serve and it is then designed to handle those conditions in an economical and satisfactory manner.

20.12. All-Year Residence Air Conditioners. The majority of homes which are equipped for all-year air conditioning at the time of this writing use either a warm-air system or a steam or hot-water heating plant during cold weather and a window type (Fig. 19.4) of self-contained cooling and dehumidifying unit during hot weather. However, many combination fuel-fired warm-air systems and packaged central conditioners, Fig. 19.6a, are now being installed in new homes.

Certain manufacturers of radiators and convectors have developed special forced convectors, Fig. 8.7, which can be supplied with hot water from a boiler for heating during the winter and with chilled water from a water cooler for cooling and dehumidifying during the summer. The same circulating pump and the same piping system can be used throughout the year. However, the circulating pipes must be insulated and provided with a vaporproof covering if they are to handle chilled water whereas insulation is usually not required if the system is to be used for heating only. The possibilities of semicentral cooling and dehumidifying units in conjunction with hot-water or steam heating systems are now being studied at the University of Illinois.

The heat pump which has been discussed in some detail in Chap. 17 is a promising possibility for all-year air conditioning in homes to be built in the future.

REFERENCES

1. "Heating Ventilating Air-Conditioning Guide," ASHAE, New York (1957).
2. "High Velocity System Design," L. R. Phillips, *Air Conditioning, Heating, Ventilating*, **52**, No. 1, 92 (Jan. 1955).
3. "Handbook on High Velocity Air Distribution Design," C. M. Wilson, *Heating, Piping, Air Conditioning*, **26**, No. 11, 95 (Nov. 1954).
4. "How to Control High Velocity Double Duct Systems," N. J. Janisse, *Heating, Piping, Air Conditioning*, **27**, No. 11, 122 (Nov. 1955) and No. 12, 100 (Dec. 1955).
5. Medium and High Pressure Air Distribution Systems, Richard D. Tutt, *Bull. No. 110*, Tuttle and Bailey, Inc., New Britain, Conn. (1954).
6. *Air Conditioning and Refrigerating Data Book*, Design Volume, The American Society of Refrigerating Engineers, New York (1955–56).

BIBLIOGRAPHY

1. "Simplified Method of Maintaining Constant Relative Humidity," A. D. Benjamin, *Air Conditioning, Heating, Ventilating*, **52**, No. 5, 83 (May 1955).
2. "19 Zone, Double Duct System Conditions Underground Offices," J. E. Leary, *Heating, Piping, Air Conditioning*, **28**, No. 3, 130 (Mar. 1956).
3. "Elements of Dual-Duct Design and Performance," N. S. Shataloff, *Heating, Piping, Air Conditioning*, **27**, No. 9, 143 (Sept. 1955).
4. "Humidity—Its Measurement and Control," J. C. Hawkins, *Refrig. Eng.*, **62**, No. 3, 68 (Mar. 1954).
5. "Underwindow Fan-Coil-Filter Units for Year-Round Air Conditioning," G. B. Priester, *Air Conditioning, Heating, Ventilating*, **52**, No. 12, 98 (Dec. 1955).
6. "Remote Unit Central System Heats, Cools Luxury Apartments," N. Feder, *Heating, Piping, Air Conditioning*, **28**, No. 1, 158 (Jan. 1956).
7. "Air Conditioning of Multi-Room Buildings," R. W. Waterfill, *Heating, Piping, Air Conditioning*, **26**, No. 11, 141 (Nov. 1954).
8. "New Application Techniques Improve Residential Air Conditioning," E. P. Palmatier, *Heating and Ventilating*, **51**, No. 1, 67 (Jan. 1954).
9. "Room Air Distribution Research for Year Round Air Conditioning," H. E. Straub and S. F. Gilman, *Heating, Piping, Air Conditioning*, **25**, No. 11, 145 (Nov. 1953).
10. "Year Round Air Conditioners," A. E. Diehl, *Elec. Mfg.*, **54**, No. 1, 130 (July 1954).
11. "Application Techniques for All Year Residential Air Conditioning Systems," H. C. Pierce, *Sheet Metal Worker*, **43**, No. 12, 58 (Sept. 1952).
12. "Humidity Control for Small Air Conditioning Systems," W. T. Smith, *Refrig. Eng.*, **59**, No. 7, 653 (July 1951).
13. "Potential of Window Air Conditioner as Heat Pump," L. Taub, *Refrig. Eng.*, **60**, No. 7, 722 (July 1952).
14. "Cut Heating, Cooling Costs with Revolving Doors," A. M. Simpson, *Heating, Piping, Air Conditioning*, **26**, No. 11, 109 (Nov. 1954).

ALL-YEAR AIR CONDITIONING

PROBLEMS

1. A winter air-conditioning unit which is part of a split system consists of a preheat coil, washer, and reheat coil. There is no provision for heating the spray water. The unit is to deliver air to the space served at 68 F dbt and 40 per cent relative humidity. (a) Assuming that the air will leave the washer with a relative humidity of 90 per cent, find the temperature to which the air must be preheated when the dry-bulb and wet-bulb temperatures of the air entering the unit are 30 F and 25 F respectively. (b) Find the heat in Btu to be added to each pound of dry air by the preheat coil.

2. A winter air conditioner which is part of a split system consists of a preheat coil only and a pan-type humidifier. The unit is to be designed to handle 5000 lb of dry air per hr, 100 per cent of which is to be fresh air. The outside design conditions are to be 0 F with 100 per cent relative humidity and the unit is to deliver air at 76 F dbt with 30 per cent relative humidity. Assume that the pan will deliver saturated steam at 212 F. (a) Using a psychrometric chart and Tables 2.1 and 1.7, find the temperature to which the air must be heated prior to humidification. (b) Find the heat-transfer capacity that will be needed in the preheat coil in Btuh. (c) Find the weight of make-up water that will be required by the pan humidifier in pounds per hour.

3. The same as Problem 2 except that the air must be heated so as to take care of a heat loss from the space served amounting to 40,000 Btuh. Find the temperature to which the air must be heated prior to humidification.

4. A building is to be heated and cooled by a double duct system and a certain room will require 110 cfm from the diffuser. The usage of the room requires that the noise level from the diffuser be kept below 37 db. From the data of Table 20.3 select the length of the unit which should be used. What is the minimum static pressure in the branch lines feeding the unit?

5. A winter air conditioner using a solution of lithium chloride is to deliver air at 75 F dbt with a relative humidity of 30 per cent. Assuming perfect contact between the air and the solution, what weight concentration of the solution will be required?

6. A winter air conditioner using a 40 per cent (by weight) solution of lithium chloride under design conditions must deliver air at 110 F dbt having a humidity ratio of 40 grains per lb. Assuming perfect contact between the air and the solution, determine the required solution temperature and the heat in Btu per lb of dry air that will be required from the heating coil. Assuming no infiltration of outside air into the space served and no moisture release from appliances within the space, find the relative humidity therein if the equilibrium temperature is 75 F dbt. Assume the barometric pressure equals 29.92 in. of mercury at 32 F.

7. A summer air conditioner using a 45 per cent (by weight) concentration of lithium chloride in water as the dehumidifying agent is to deliver air at 60 F dbt having a humidity ratio of 60 grains per lb of dry air. (a) Assuming perfect contact between the solution and the air, determine the proper temperature of the solution in the contactor cells. (b) Find the heat in Btu per lb of dry air which must be removed by the cooling coil. (c) Find the relative humidity of the air as it leaves the contact cell and as it leaves the cooling coil. (d) Find the relative humidity in the space served if the equilibrium tem-

perature is 78 F and the enthalpy-humidity difference ratio is 0.50 Btu per grain. (Use Fig. 2.2 or a full-scale G.E. psychrometric chart for part of the solution.)

8. Rework the example of Art. 20.10, through step 9, using the following data: inside summer design conditions 78 F dbt, 50 per cent relative humidity; outside summer design conditions 95 F dbt, 77 F wbt, direct total-heat gain 100,000 Btuh; direct moisture gain 30 lb per hr; weight of fresh air required 4000 lb of dry air per hr. Make calculations for 12:30 P.M. as in the example.

9. Assume that the weight of fresh air used in the example of Art. 20.10 will be 50 per cent of the total weight needed. Use the same inside and outside summer air conditions and the same direct total-heat and moisture gains. Assume that a washer 8 ft long with two opposing spray banks will be used for cooling and dehumidifying the air and that the spray-water temperature will be 40 F during the summer. Assume a face velocity of 500 fpm, that type (b) nozzles (Fig. 19.15) will be used, and that the nozzle pressure will be 30 psig. (a) Find the total weight of dry air to be processed under summer design conditions. (b) Determine the refrigeration capacity required in tons. (c) Compute the required face area of the washer and the number of nozzles that will be needed. Assume the same inside and outside winter air conditions as in the example and same heat loss due to conduction. Assume that the volumes of outdoor and recirculated air will be the same in winter as in summer. Assume that the same washer will be used to humidify the air and that both a preheat and a reheat coil will be provided as in Fig. 20.2. There will be no provision for heating the spray water. (d) Find the temperature to which the mixture of fresh air and recirculated air must be heated prior to humidification under winter design conditions when the building is unoccupied and there are no moisture-releasing appliances in operation. (e) Find the heat-transfer capacity which must be provided in the preheat coil in Btuh. (f) Estimate the maximum rate of evaporation in the washer in pounds per hour. (g) Determine the heating capacity in Btuh that will be needed in the reheat coil.

21

Automatic Controls for Air-Conditioning Systems

21.1. Essential Parts of a Control System. Automatic control of any central air-conditioning system requires instruments which perform three different functions, namely: (1) thermostats, humidistats, and pressure controllers which measure changes in the temperature, relative humidity, or pressure in the space where they are located (these devices are commonly called controllers); (2) actuators which operate valves or dampers to regulate the flow of fuels, heating mediums, or cooling mediums in response to the demand of the controllers; (3) limit or safety controls which control the actuator in place of the controller when a dangerous condition exists in the system.

21.2. Electric and Pneumatic Systems. Control systems may be classified with respect to the source of power through which the controller-actuator combination brings about the proper positioning of valves and dampers. Although the expansion or contraction of either a liquid or a gas may be used as the source of power in certain simple controls (Fig. 21.3), the valves and dampers of central air-conditioning systems are usually operated either by electricity or compressed air. The design of controllers, actuators, and auxiliary apparatus is greatly influenced by the source of power which is to be used. Either source of power may be used with any type of air-conditioning system but in general pneumatic controls predominate in large systems serving schools, hotels, institutions, and other large commercial and industrial buildings whereas electric controls are used almost exclusively in homes and in small commercial applications of summer, winter, and all-year air conditioning.

Electric control systems usually employ low-voltage current (20 to 25 volts) provided by a transformer in at least some of the circuits involved. Where electric motors requiring line voltage must be started and stopped by control action and the thermostat is designed for low voltage, a relay (Art. 21.19) must be used.

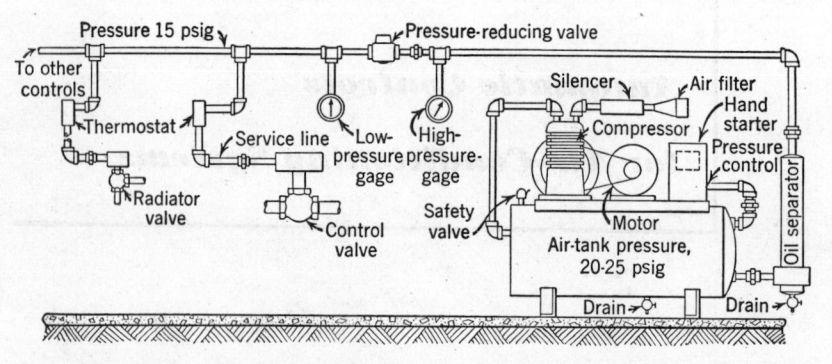

Fig. 21.1. Pneumatically operated temperature-control system.

The basic equipment generally used in pneumatic control systems[1] is shown diagramatically in Fig. 21.1. The source of the air supply is the air tank which in turn is supplied by the compressor. The air tank should be located in a cool place to encourage condensation of moisture at that point in the system rather than elsewhere. The pressure-reducing valve assures that a constant pressure will be maintained in the pipe lines to the controllers. The thermostats or other controllers incorporate a means of regulating the pressure in the lines connecting them with the actuators. The air pressure in an actuator, such as a radiator valve, works against a spring and the device is positioned according to the need for corrective action as sensed by the thermostat.

One desirable characteristic of a pneumatic system is that different springs may be incorporated in two or more actuators which are to be co-ordinated to achieve the desired programming. For example, a winter air conditioner may have a fresh-air intake with a preheat coil located in it. The steam valve which supplies the coil must open wide before the damper starts to open during severely cold weather to avoid the possibility of condensate freezing in some parts of the coil. The valve may be opened fully before starting to open the damper by

AUTOMATIC CONTROLS

installing a 3 to 8-psi spring in the valve actuator and an 8 to 13-psi spring in the damper motor operating both with compressed air from a single controller. As the thermostat (or other controller) increases the air pressure in the common circuit, the valve begins to open when the pressure rises to 3 psig, and becomes fully open when it reaches 8 psig. The damper starts to open when the increasing air pressure reaches 8 psig and it is fully opened when it reaches 13 psig. The pneumatic type of control system offers almost limitless possibilities for any desired co-ordination of relays, dampers, and valves.

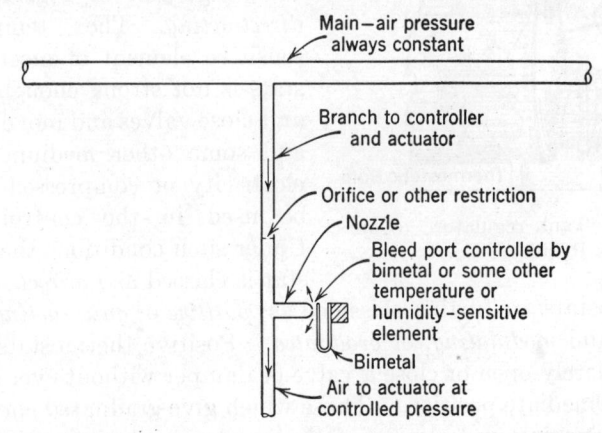

Fig. 21.2. Schematic diagram illustrating the principle of one type of pneumatic controller.

Figure 21.2 illustrates schematically a simple means by which a thermostat, humidistat, or other controller in a pneumatic system can control the air pressure that is supplied to the actuator under its direction. Air from a constant pressure main flows through a branch line to a controller which contains both an orifice or other restriction and a bleed port. The air passing through the bleed port is controlled by movement of a surface attached to some temperature-sensitive element such as a bimetal. If this surface is moved away from the bleed port by action of the temperature-sensitive element, air escapes from this part of the system faster than it enters through the restrictor until the pressure approaches a low value which is determined by the relative areas of the two openings. If the surface is moved tightly against the bleed port, the pressure of the air supplied to the actuator increases until it approaches that in the main.

The arrangement of Fig. 21.2 would bleed air from the system at a very slow rate at all times except when the controller is calling for a maximum pressure in the actuator. Some commercial pneumatic controllers contain features which accomplish the same results with less wasting of compressed air. With some designs it is necessary to exhaust air from the system only when the controller is calling for a reduction of the pressure in the actuator.

21.3. Thermostats. Thermostats are either direct or indirect in their action. When a thermostatic element is powerful enough to operate a valve or damper without another agent, as in the case of the tank regulator of Fig. 21.3, it is *direct-acting*. The temperature-sensitive element of most thermostats is not strong enough to open and close valves and move dampers and some other medium such as electricity or compressed air must be used in the control process. Under such conditions the thermostat is classed as *indirect*.

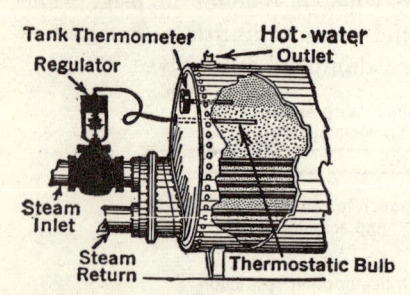

Fig. 21.3. Tank regulator. (Courtesy Powers Regulator Co.)

Thermostats are further classified as: *positive* or *quick-acting*, or *two-position;* and *modulating*, or *graduated*. Positive thermostats quickly and completely open or close a valve or damper without ever stopping at an intermediate position. Those which give graduated control permit slow opening and closing of the regulating device with the possibility of stopping at any intermediate position between fully open and fully closed.

21.4. Indirect, Two-Position Thermostats. Indirect, two-position thermostats are used extensively in controlling heating and summer air-conditioning systems, particularly as applied to homes and small commercial buildings. The majority of thermostats of this class use a bimetal strip in some form as the temperature-sensitive element. The contruction of one form of electric bimetal thermostat is shown in Fig. 21.4 in which the temperature-sensitive element C is mounted upon an arm pivoted at a point near to the center of the element coil. The temperature at which the thermostat functions is determined by the setting of the adjustor which in turn fixes the position of the arm on which C is mounted. The anticipating element is a resistance through which the electric current flows and which serves as a heater unit to warm the bimetal. The prewarming of the bimetal causes the electric circuit to be opened before the temperature of the air has reached that desired. Such action prevents the overrunning of the desired air temperature because of what amounts to thermal lag in the system.

AUTOMATIC CONTROLS

Electric thermostats of the form just described are constructed with two bimetal elements, one of which is used with a heater for control during the daytime and the other a low-temperature unit for night or other periods of reduced temperatures. Such an arrangement can be used in connection with a timing switch (electric clock device) which may be set to shift from one bimetal element to the other at predetermined times.

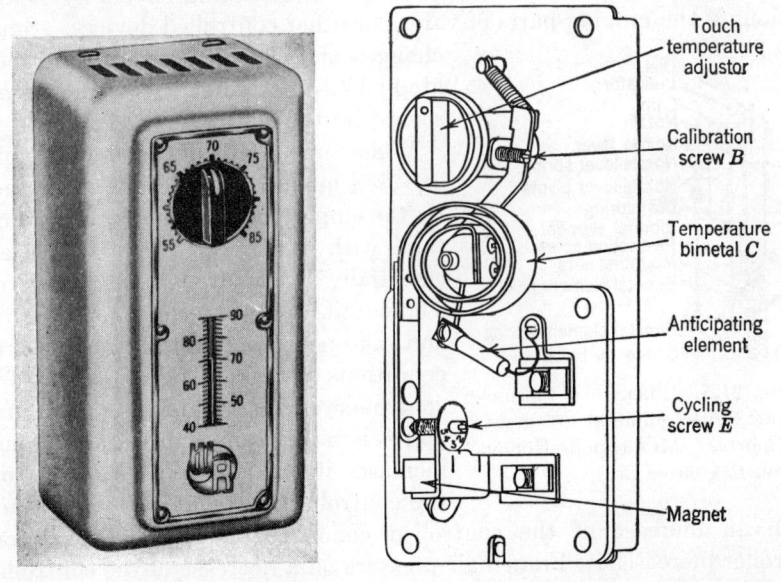

Fig. 21.4. Bimetal electric thermostat. (Courtesy White-Rodgers Electric Co.)

The thermostat of Fig. 21.4 is used with two-wire low-voltage circuits. All actuators used with two-wire thermostats must be spring-returned to assure closing of the valves or dampers they operate whenever the power is cut off either by the action of the thermostat or by the interruption of electric service to the building.

Indirect two-position thermostats are also available which employ three-wire circuits. In this arrangement the free end of the bimetal element which is electrically connected to a *common* contact touches a *high* contact on a rise in room temperature and a *low* contact when the air in the space cools. In this case the valve or damper involved is usually operated by a low-voltage two-position motor of the unidirectional type. The motor through a gear train rotates a shaft through 180 deg when the *high* contact is made and it then completes the cycle to the original position when cooling of the room air causes the bimetal element to make the *low* contact. The motor contains a *maintaining circuit* which assures that the shaft carrying the valve or damper-operating mechanism will rotate a full 180 deg, and that amount only, when either the high or low contact is made in the thermostat.

21.5. Modulating Controls. The thermostat shown in Fig. 21.4 involves on-and-off operation of the equipment which it serves. Proportioning controllers are used to give modulating effects by repositioning the moving parts of valves or other controlled devices. Small changes of the total travel of the actuated parts occur when slight changes are sensed by the controllers.

When pneumatically operated units are used modulating effects are obtained by the employment of devices in connection with the thermostats which are essentially pressure-reducing valves. These units actuate the controlled valves and dampers to maintain the desired conditions of the controlled media. The controllers are designated as *direct-acting* and *reverse-acting*. A direct-acting unit increases its branch (pipe leading from the controller to the motor) air pressure on an increase of the controlled condition; a reverse-acting controller increases its branch air pressure on a decrease of the controlled condition.

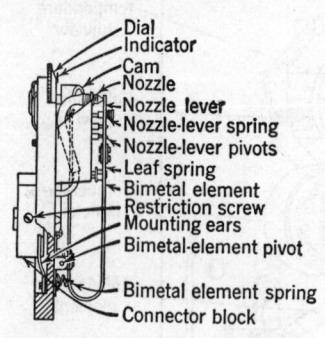

Fig. 21.5. Pneumatic thermostat with bimetal element. (Courtesy Minneapolis-Honeywell Regulator Co.)

The thermostat of Fig. 21.5 is built to operate with compressed air as the moving force. Air-temperature changes about the instrument are detected by the bimetal element which is bent into a U shape. The element is pivoted at a location near the bottom of the bend, and it is loaded by a spring to keep it in contact at all times with a cam whose position is changed as the temperature-adjustment dial is rotated. The motion of the free end of the bimetal element is transmitted through a leaf spring of the nozzle lever which rotates about pivots to either cover or uncover the bleed port (Fig. 21.2). (The bleed port of the thermostat shown in Fig. 21.5 is located at the end of the nozzle. The spring action is sufficient to keep the opening closed against an air pressure of 15 psig in the branch line. When the port is uncovered the compressed air escapes, and the pressure in the line is below that necessary to move the secondary-control parts. When the bimetal element moves to close the port the air pressure in the secondary line increases, and the proper movement of valves and dampers is

AUTOMATIC CONTROLS

made to reduce the air temperature. These thermostats are also built with the bimetal construction reversed so that an increase of air temperature causes the element to open the air port. The restriction-screw adjustment varies the amount of compressed air delivered to the thermostat and the branch line by the main-supply connection. The adjustment is made after installation to give the desired speed of

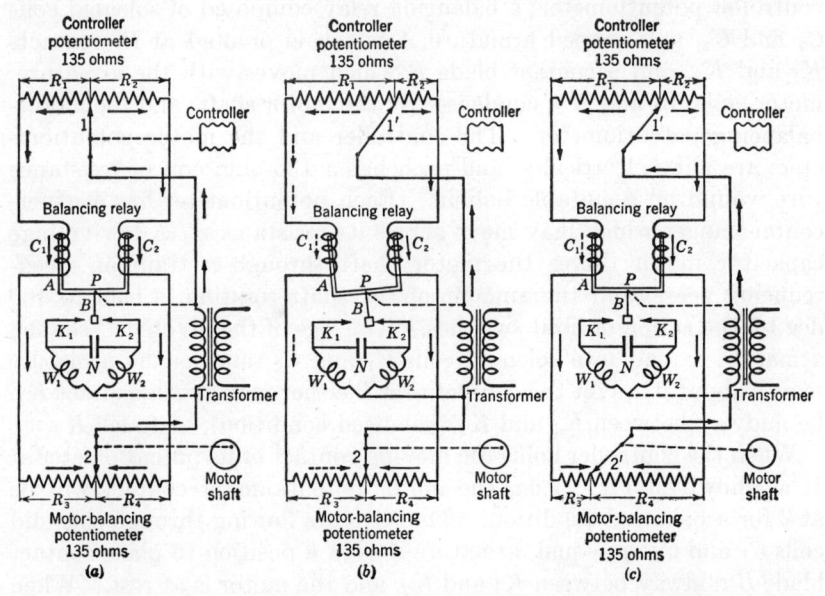

Fig. 21.6. Minneapolis-Honeywell Series-90 control circuit.

operation for the devices controlled. The setting is such that the port can always bleed the branch-line pressure down to 1 psig or less.

Pneumatic thermostats are also contructed with dual arrangements within a single case or housing. One of the elements, with a proper setting, functions to give daytime temperature control, and the other operates at night. Dual thermostats may be automatically changed from day service to night service, and vice versa, by varying the main air pressure from 13 to 17 psig.

Modulating effects may be obtained with electric thermostats by use in the circuits of what amounts to a Wheatstone bridge which when unbalanced actuates the repositioning of an element that moves a valve or a damper. Some of the details of such a device are shown in Fig. 21.6, which indicates the construction and operation of a Minneapolis-Honeywell Regulator Company Series-90 control unit used to provide modulating or proportioning effects when either motorized valves or dampers are used. These circuits operate to position the controlled device (damper or motor valve) between fully open and completely closed conditions and thereby proportion the delivery to the need indicated by the controller mechanism. The Series-90 controllers may be either (1) room thermostats, (2) insertion thermostats, (3) humidity controllers, or (4) pressure controllers.

The essential parts of the mechanism include a transformer; a controller potentiometer; a balancing relay composed of solenoid coils C_1 and C_2, a U-shaped armature A which is pivoted at P, contacts K_1 and K_2, and a contact blade B which moves with the armature; motor coils W_1 and W_2; condenser N; the motor shaft; and the motor-balancing potentiometer. The controller and the motor potentiometers are alike electrically, and each has a 135-ohm coil of resistance wire wound on a suitable bobbin. Each potentiometer has a wiper-contact finger which may move across its resistances. A low-voltage capacitor motor drives the motor shaft through a train of speed-reducing gears, and the amount of the shaft rotation is held to 160 deg by the action of limit switches. The legs of the pivoted U-shaped armature extend into solenoid coils C_1 and C_2, and as the armature moves about its pivot the contact arm B either may touch contact K_1, be midway between K_1 and K_2 (balanced condition), or touch K_2.

When the controller holds the moving contact of its potentiometer at 1, as shown in Fig. 21.6a, the motor potentiometer contact will be at 2 for a balanced condition. The currents flowing through solenoid coils C_1 and C_2 are equal, armature A is in a position to place contact blade B midway between K_1 and K_2, and the motor is at rest. When the controller unit (thermostat or other part) causes its wiper contact to move to position 1', Fig. 21.6b, resistance R_1 is not equal to R_2, and more current flows from the transformer through coil C_2 than flows through C_1; armature A is shifted so that its contact blade B makes an electric circuit at K_2. The motor now receives current through coil W_2 and it rotates in the direction corresponding to this condition. The wiper connection attached to the motor shaft then shifts the motor potentiometer contact to position 2' as shown in Fig. 21.6c. Resistance R_3 then becomes equal to R_2 and R_1 equals R_4. When such a condition exists, the amounts of current flowing through coils C_1 and C_2 are equal and contact blade B is located midway between K_1 and K_2. The motor under these conditions is at rest, and the shaft has positioned the movable part of the damper or valve to meet the demands indicated necessary by the controller unit. The same operating actions take place when the controller wiper contact moves over that portion of the resistance designated as R_1 and the motor wiper contact moves over R_4, except that the motor shaft rotates in an opposite direction.

AUTOMATIC CONTROLS

One fundamental advantage in the proportioning type of control action is that corrective measures are started as soon as the temperature at the sensing element has changed enough to cause any motion of its free end. In the case of off-on or two-position controls the temperature at the sensing element must change the full amount of its differential before any corrective action can occur.

21.6. Insertion Thermostats. An insertion thermostat has its heat-sensitive element placed either within a conduit or a plenum chamber with the remainder of the unit mounted outside it. The external appearance of one form of an insertion thermostat is shown

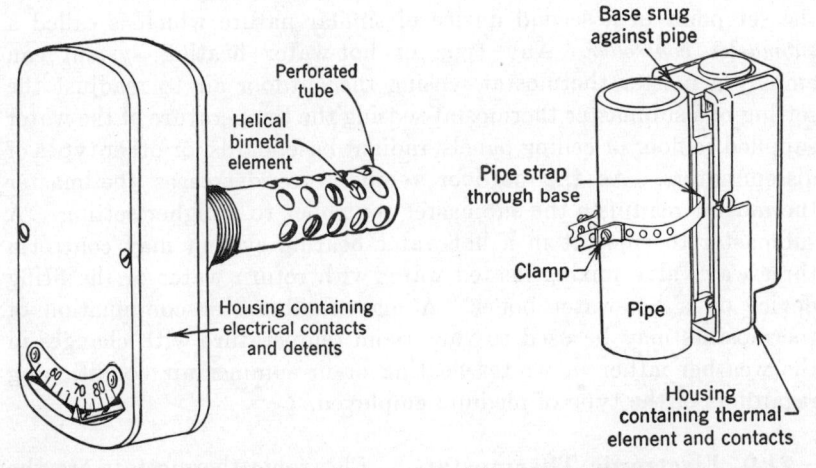

Fig. 21.7. Insertion thermostat for a duct. (Courtesy Barber-Colman Co.)

Fig. 21.8. Surface thermostat. (Courtesy Perfex Corp.)

in Fig. 21.7. This particular device has a spirally wound strip of bimetal as a thermal element which is located within a protecting shield of perforated metal. Other temperature-sensitive materials used in insertion thermostats are volatile fluids and brass bulbs enclosing Invar bars. Insertion thermostats are used for the control of the temperatures of fluids such as water, steam, and brine. The mounting of the mechanism, exclusive of the heat-sensitive part, is external to the fluid whose temperature is to be regulated. The connection of the thermostatic bulb with the remainder of the mechanism is by means of a flexible tube in the case of the direct-acting tank regulator of Fig. 21.3.

21.7. Surface Thermostats. An illustration of a snap-action electric surface control is shown in Fig. 21.8 where the unit is rigidly strapped to a pipe surface so that heat may flow through its base to its thermal element. Applications of the instrument shown in Fig. 21.8

are (1) in hot-water heating systems as a high-limit temperature control placed near to the boiler-water outlets; (2) as a device to prevent operation of a circulator in a hot-water heating system when the boiler-water temperature is low; and (3) when attached to the return pipe of a unit heater ahead of any valve or trap, as a control to prevent the operation of the heater fan when the heater is receiving neither steam nor hot water.

21.8. Master and Submaster Controllers. Many control applications employ the use of remote readjustment of a controller's set point. A *master controller* is one which automatically raises or lowers the set point of a second device of similar nature which is called a *submaster controller*. Any type of hot-water heating system can employ a master thermostat sensing the outdoor air to readjust the setting of a submaster thermostat sensing the temperature of the water supplied to floor or ceiling panels, radiant baseboards, or other types of disseminators. As the outdoor temperature decreases the master thermostat readjusts the submaster controller to a higher setting. A submaster thermostat in a hot-water heating system may control a three-way valve mixing heated water with return water or the firing device on a hot-water boiler. A master-submaster combination of thermostats may be used to vary room temperature with changes in the weather either in winter heating or in summer air conditioning regardless of the type of medium employed.

21.9. Electronic Thermostats.[1] Electronic thermostats are the result of a relatively new concept in the design of temperature regulators. These controllers utilize an electronic amplifier which is a device for increasing the magnitude of a weak signal. The increased signal may then be used through suitable electric or electric and pneumatic apparatus to operate valves or dampers. Because of this magnification a very small change in temperature may cause a change in the position of a valve or a damper and start corrective measures. The over-all effect of this arrangement is to create a two-position thermostat having a very small differential. This feature tends to cause shorter cycling, less severe overshooting, and generally more satisfactory results.

With the addition of a feedback potentiometer and certain other features an electronic thermostat may produce proportional operation of valves or dampers. However, the advantage of proportional operation is not as great in the case of electronic controllers because of the small differential which can be achieved with two-position control.

AUTOMATIC CONTROLS

21.10. Humidistats. The humidistat is a device which is sensitive to moisture changes and which will cause equipment to act to maintain a desired vapor content in the air. Often some form of a thermostat and a humidistat mechanism are placed side by side within a single housing to form a device used for the control of air conditions within a space.

The moisture-sensitive member of a humidistat can be any one of the following: a block of special wood exposing the greatest amount of end grain to the air, human hair, fiber, paper, and membranes. Other

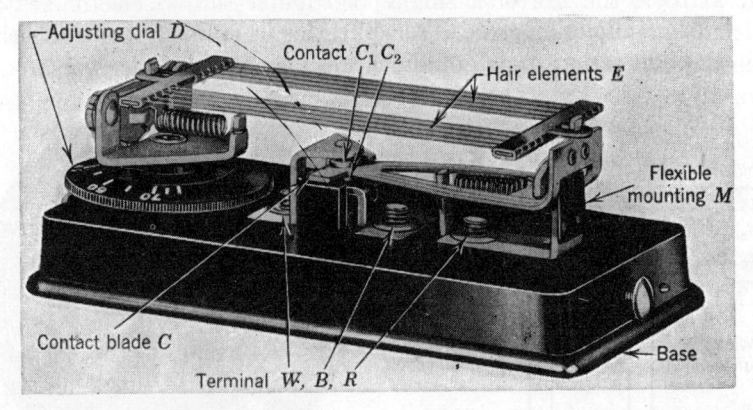

Fig. 21.9. Mechanism of a humidistat. (Courtesy Penn Electric Switch Co.)

humidistats comprise a combination of wet- and dry-bulb thermostats differentially connected.

Humidistats operate either humidifying or dehumidifying equipment. Humidity control may be either two-position or modulating.

The construction of and the application of a wall-mounted human-air element three-wire snap-acting permanent-magnet humidistat are illustrated in Figs. 21.9 and 21.10. The humidistat mechanism, Fig. 21.9, is shown with its cover removed and the unit in a horizontal position. Actual installations are made with vertical placements in which the adjusting dial is at the top of the arrangement. The device includes human-hair elements E; a contact blade C; an adjusting dial D; a flexible mounting of the contact blade M; contacts C_1 and C_2; and terminal connections W, B, and R to which wires with white-, blue-, and red-colored insulations, respectively, are attached at the rear of the instrument.

Movement of the adjusting dial D changes the tension of the hair elements. When the relative humidity of the surrounding air is above that for which the dial is set, the hair elements lengthen and cause blade

C to move against contact C_1 and make an electric circuit through W and R. When the moisture of the air is below that corresponding to the dial setting, the hair elements shorten in length, and blade contact is made at C_2 to complete the circuit through wires B and R. The making of the electric circuit through wires W and R causes the equipment to function to decrease the amount of moisture being held in air; the closing of the circuit through B and R causes air humidification. Wire R is common in either circuit and Fig. 21.10 shows the points of wire attachments at the back of the instrument base. Also shown in Fig. 21.10 is the use of a single-pole double-throw selector switch, which is manually operated, to shift the instrument control either from humidification to dehumidification apparatus or vice versa.

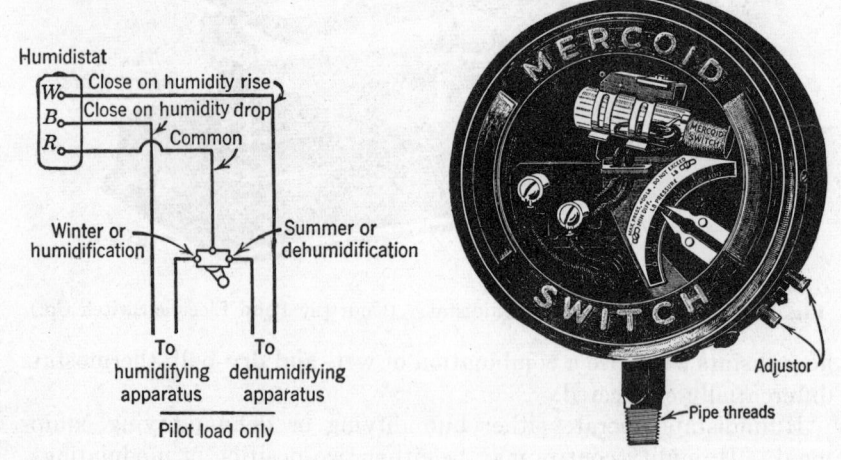

Fig. 21.10. Wiring diagram for humidistat and selector switch. (Courtesy Penn Electric Switch Co.)

Fig. 21.11. Pressure control. (Courtesy The Mercoid Corp.)

21.11. Pressure Controls. Figure 21.11 illustrates one form of pressure control which functions with a tilting electric switch actuated by a Bourdon pressure tube similar to that of the gage shown in Fig. 9.6. The electric device in Fig. 21.11 is called a *mercury switch*. The pressure control of Fig. 21.11, when the pressure drops, moves the mercury switch to close an electric circuit which in turn energizes an electric starter of an electric motor driving a refrigerant compressor or other machine to be regulated. The pressure control illustrated has two adjustments, which may be made from without the case, that allow it to function with a certain differential pressure.

The mercury switch in Fig. 21.11 typifies a type which may be

AUTOMATIC CONTROLS

applied in many kinds of control devices. Switches of this type can be made with either two or more sealed-in electrodes and are adaptable to almost any type of an electric control circuit. The sealed glass container which includes the electrodes is filled with an inert gas which will not corrode them.

The type of pressure control shown in Fig. 21.11 is suitable for regulating pressures of considerable magnitude such as may be found in boilers and in the evaporators and condensers of refrigerating systems. Regulators for controlling the static pressure in a duct of an air-conditioning machine must be equipped with a pressure-sensing element that is much more sensitive than that of the unit of Fig. 21.11. Haines[2] describes a static-pressure regulator which employs two gas-tight bells which when moved vertically operate a switch in either a low-voltage or line-voltage electric circuit. The bells are surrounded with oil to provide a frictionless seal and the arrangement may be used to maintain a constant difference between a controlled pressure and a reference pressure. The reference pressure is usually atomospheric and the device is sensitive to pressure changes amounting to only a fraction of an inch of water.

21.12. Sensitivity of Controls.

In any temperature, humidity, or pressure-control system a change in the state of the conditioned material

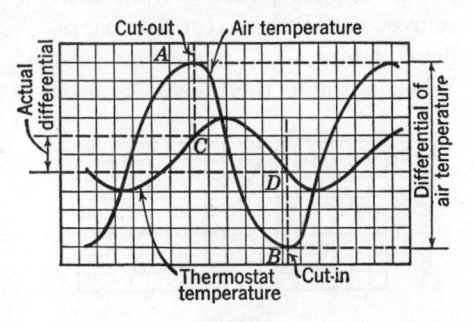

Fig. 21.12. Lag of air temperature.

produces functioning of the apparatus. The desired corrected conditions may come from either two-position or modulating actions. Two-position controls operate well with radiator valves installed in one-pipe steam-heating systems, on-and-off operation of stokers and oil and gas burners, pressure regulators, and various safety devices. Where throttling actions in steam, air, gas, water, and refrigerant valves are necessary to give closer regulation, some form of equipment is required that will give modulating effects. Under some conditions the sensitivity of the control must not be too great, otherwise the unit may be unstable.

The actions of a two-position thermostat of a type that was commonly used in former years are indicated by the test data[3] of Fig. 21.12. The improvement in results which can be effected by an anticipating element such as is incorporated in the thermostat of Fig. 21.4 is shown in Fig. 21.13. Figure 21.13 indicates that without the heat-anticipating device the maximum air-temperature differential was 4 F and with it in service the differential was reduced to 0.5 F as the unit cycled more frequently.

21.13. Locations of Thermostats. The preferable location of a room thermostat from the standpoint of comfort in the space served is on an inside wall about 30 in. above the floor. However, other considerations such as making it inaccessible to small children may cause one to decide to locate it at the 5-ft. level. The location chosen should avoid the possibilities of direct solar radiation or cold drafts from opening doors and it should be as far as possible from heat sources such as radiators, registers, floor lamps, or television sets. When a single thermostat is used to control the temperature of a structure such as a home, careful attention must be given to the selection of the room in which the controller is to be located.

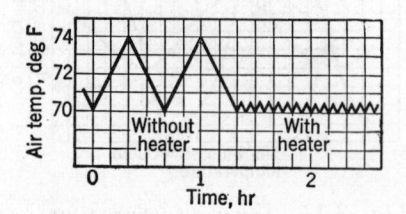

Fig. 21.13. Effect of compensating heater element in thermostat operation.

Fig. 21.14. Motor-operated valve for proportioning control. (Courtesy Barber Colman Co.)

21.14. Actuators. Actuators are sometimes called *primary controls* because they are most directly in command of the heating or cooling medium. The design of any particular unit depends primarily on the device it is to operate and the type of energy to be utilized.

Electric control motors may be of a spring-return type, unidirectional or reversing. The spring-return type is used in two-position, two-wire systems. A unidirectional damper motor is suitable for inclusion in a three-wire circuit. The motor-operated valve of Fig.

AUTOMATIC CONTROLS

21.14 is intended for proportioning control and contains the motor-balancing potentiometer that is indicated in the lower portions of the diagrams of Fig. 21.6.

Another type of electric valve actuator is shown in Fig. 21.15. It is available in line-voltage and low-voltage models and is suitable for two-position control of liquids or gases. The valve may be operated under the action of either a room thermostat, limit control, or safety device. Automatic valves of this type are commonly used for starting

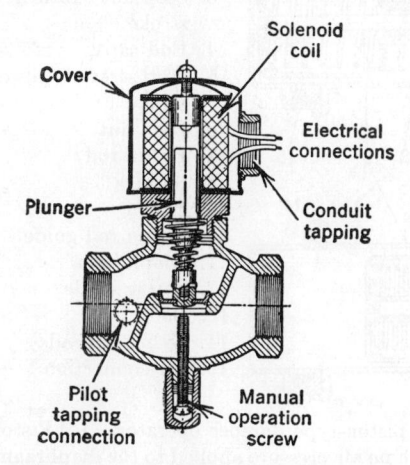

Fig. 21.15. Solenoid globe-type control valve. (Courtesy General Controls Co.)

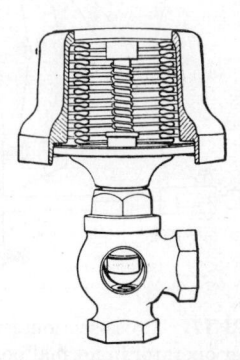

Fig. 21.16. Sylphon-bellows radiator valve. (Courtesy Johnson Service Co.)

or stopping the flow of gas to a burner as directed by a room thermostat. They are also commonly used to shut off the flow of gas to a main burner in case of failure of the pilot flame. In case of current failure in any application the valve automatically closes.

Pneumatic actuators usually employ air under a pressure that is controlled by a suitable thermostat or other controller to balance the force of a spring at the position needed to give the desired result. The radiator valve of Fig. 21.16 is called a *normally open* type because it assumes that position when the gage pressure of the air in the chamber is zero. A pneumatic actuator in which the spring returns the valve or damper to a closed position when the air pressure falls to the critical pressure is known as a *normally closed* type. A pneumatic mixing valve is shown in Fig. 12.17.

Figure 21.17 is a cross-sectional view of a piston-type damper operator. Actuators of this type are well adapted to proportioning control of air flow as they can be made to position the damper leaves between fully open and completely closed at any desired point by maintaining the proper air pressure in the cylinder.

21.15. Use of Dampers in Air-Conditioning Systems. Although dampers have been essential items in every central air-conditioning system since the beginning of the art, very little has been published

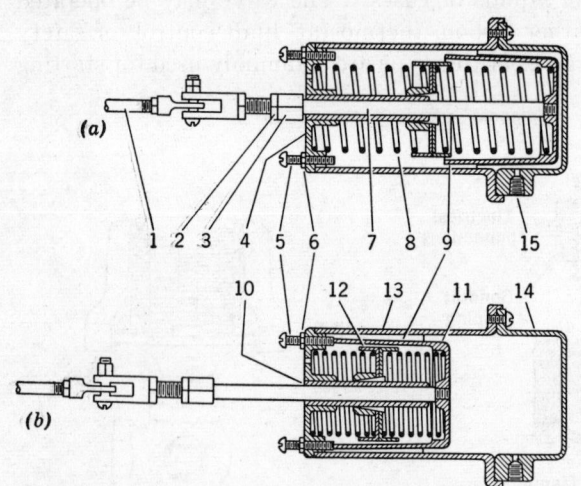

1. Connecting rod
2. Lock nut
3. Stop nut—return stroke
4. End cap
5. Stop screws—power stroke
6. Lock nut
7. Piston rod
8. Springs
9. Piston
10. Piston rod guide
11. Diaphragm
12. Spring guide
13. Body
14. Cylinder head
15. Air connection

Fig. 21.17. Cross-sectional view of a piston-type damper operator. (*a*) Piston damper operator in normal position, with no air pressure applied to the diaphragm. (*b*) Operator in the extreme position with air pressure applied. (*Courtesy Johnson Service Co.*)

about their performance characteristics. *Louver dampers* such as are commonly used in low-pressure systems are made in two styles. The *parallel* type is used in the apparatus depicted in Fig. 20.2 and the *proportioning* type (sometimes called opposed type) is shown in Fig. 21.18. Both forms with variations with respect to details are made by several different companies.

21.16. Damper Applications. The use of two co-ordinating dampers to control the proportions of outdoor and recirculated air is illustrated in Fig. 20.2, The schematic diagram of Fig. 20.6 indicates a third co-ordinating damper which is located in the exhaust duct. The diagram of Fig. 20.6 further indicates that the outdoor-air damper is installed as two completely separate parts. One part is called the *minimum outdoor-air damper* and the other is called the *maximum outdoor-air damper*. In heating and ventilating systems having this arrangement at the outdoor-air intake, the controls are designed to keep the *minimum* damper wide open at all times that the fan is in

AUTOMATIC CONTROLS

operation. The *maximum* damper is usually controlled by a thermostat located in the outdoor air and it remains closed at all times during extremely cold weather. In a rising trend the controls usually begin to open this damper when the outdoor-air temperature reaches 40 F

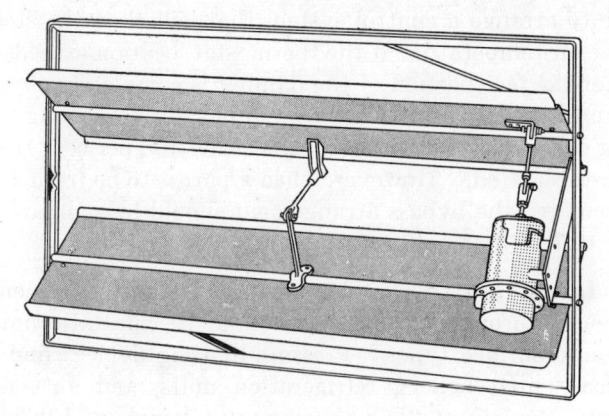

Fig. 21.18. Proportioning damper. (Courtesy Johnson Service Co.)

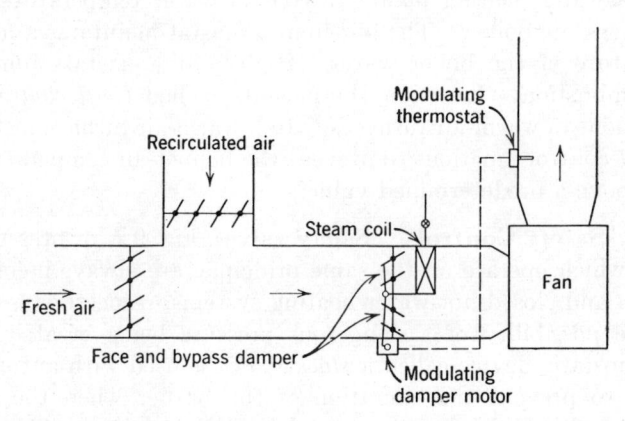

Fig. 21.19. The use of a face and bypass damper with a steam coil.

and the degree of opening is gradually increased with further moderation of the weather. A similar arrangement is frequently used with a summer air conditioner except that the amount of fresh air taken into the system is decreased in a rising trend and the minimum amount only is used when the outdoor-air temperature exceeds a fixed value.

In many cases the temperature of either heated or cooled air can be most effectively controlled through the use of a device known as a *face and bypass damper*, the principle of which is illustrated in Fig. 21.19.

In the arrangement shown it is used to control the final temperature of a mixture of fresh and recirculated air by regulating the proportion that is heated by the steam coil. This control arrangement is usually more satisfactory than controlling the flow of steam to the coil with all of the air passing through it. Because of damper leakage it is advisable to arrange a control system that will close the steam valve in case the air temperature at the thermostat location exceeds a certain level (after the *face* portion of the damper is completely closed).

Face and bypass dampers are also applicable to cooling and dehumidifying coils when the air handled is either 100 per cent fresh or 100 per cent recirculated. However, when a part is to be fresh air and the rest recirculated the bypass arrangement should be as in arrangement 3 of Fig. 19.7 so that only return air will be bypassed.

21.17. Limit Controls. These devices include equipment to control pressures through various ranges and thermostatic units of the insertion and surface types. Pressure-limiting devices find use with mechanically fired boilers, refrigeration units, and air-conditioning equipment. An *aquastat* is a thermostatic device used in connection with hot-water heating plants to control water temperatures as the fluid leaves the boiler. The low-limit aquastat maintains a minimum temperature of the boiler water. High-limit aquastats function to stop combustion. Insertion thermostats, called *limit controls*, have applications in warm-air furnaces. In a warm-air furnace installation the limit control functions to prevent the bonnet-air temperature from rising above a predetermined value.

21.18. Safety Controls. Safety valves, Fig. 9.7, or pressure relief valves, which operate on the same principle, are always incorporated in steam and closed hot-water heating systems, respectively, to eliminate the possibility of a dangerous pressure being created therein. A thermostatic device called a *stack switch* is used with automatic oil burners to prevent the operation of the burner when the flame is extinguished. If the temperature at its location does *not* build up to a certain minimum value in a certain fixed period after the burner is started, the device functions to shut it down.

An automatic safety device of some sort to shut off the flow of fuel to the main burner in case of pilot failure is an essential part of every central heating plant using gas. A common arrangement for this purpose consists of a solenoid valve, Fig. 21.15, in conjunction with a thermocouple circuit which has its hot junction close to the pilot flame. When the pilot flame is burning, sufficient voltage is gener-

AUTOMATIC CONTROLS

ated in the thermocouple circuit to energize the solenoid of the valve and thus cause it to remain open. The valve closes automatically in case of pilot failure. Relighting of the pilot is facilitated if the tube carrying gas to the pilot (Fig. 7.7) is connected into the gas line upstream from the safety shutoff valve. This arrangement is permissible with natural or manufacturered gas but *not* with either propane or butane because they both have a density greater than that of air and tend to accumulate in the vicinity of the burner. In some arrangements the thermocouple performs its function through a magnetic switch in the electric circuit of the thermostat. In this case the gas control valve (Fig. 7.7) serves also as the automatic safety device.

The limit switch of a warm-air heating system and the high-limit aquastat of a hot-water system serve as safety controls as far as danger from overheating the furnace or boiler is concerned.

21.19. Relays. Relays in automatic control systems are devices for converting one kind of control action to another. A simple example of the use of an *electric relay* is in the starting and stopping of the line-voltage motor of a gun-type oil burner as directed by a low-voltage thermostat. In this case closing of the contacts in the thermostat causes low-voltage current to pass through the electromagnet of a switch in the relay which is a part of the line-voltage circuit through the motor. An *electronic relay* is in reality a combination of an amplifier and a relay. A very weak signal is amplified to sufficient strength to operate an electromagnet which in turn operates a switch in a circuit carrying line voltage.

A *pneumatic relay* as used in connection with a pneumatic thermostat or humidistat is a diaphragm-operated reducing valve which supplies air directly to the actuator at the same pressure that is created by the action of the orifice, bleed-port combination (Fig. 21.2). The control pressure from the relay operating on a diaphragm within it is always balanced against the pressure that is controlled by the temperature- or humidity-sensing element. The use of the relay results in faster response at the actuator in providing the corrective action called for by the sensing element. One temperature-sensing element may be arranged to control several relays which in turn may operate several valve or damper actuators. A *pneumatic-electric relay* is an electric switch operated by a pneumatic actuator to permit the starting of electric equipment by means of compressed air. An *electric-pneumatic* relay is an electrically operated diverting valve suitable for transferring air from one control line to another.

REFERENCES

1. "Automatic Temperature and Humidity Control," *Eng. Report No. 429*, Johnson Service Company, Milwaukee, Wis. (1954).
2. *Automatic Control of Heating and Air Conditioning*, J. E. Haines, McGraw-Hill Book Co., New York (1953).
3. "Development of Testing Apparatus for Thermostats," D. D. Wile, *ASHVE Trans.*, **42**, 349 (1936).

BIBLIOGRAPHY

1. "Heating Ventilating Air-Conditioning Guide," ASHAE, New York (1956).
2. *Industrial Automatic Controls*, M. H. LaJoy, Prentice-Hall, New York (1954).
3. "Instrumentation and Control of Boilers," W. G. Holzbock, *Air Conditioning, Heating, Ventilating*, **53**, No. 8, 81 (Aug. 1956).
4. "Room Thermostats and Their Applications," W. G. Holzbock, *Heating and Ventilating*, **50**, No. 5, 95 (May 1953).
5. "Electronic Control of Home Heating," J. M. Wilson, *Electronics*, **23**, No. 12, 84 (Dec. 1950).
6. "Response and Lag in the Control of Panel Heating Systems," F. W. Hutchinson, *ASHVE Trans.*, **53**, 157 (1947).
7. *Automatic Control Terminology*, American Society of Mechanical Engineers, New York (1954).
8. *An Introduction to the Theory of Control in Mechanical Engineering*, R. H. MacMillan, Cambridge University Press, Cambridge (1951).
9. "Control of High-Velocity Single-Duct Systems," W. G. Young, *Heating, Piping, Air Conditioning*, **29**, No. 8, 126 (Aug. 1957).
10. "Control of High-Velocity, Double-Duct Systems," H. W. Alyea, *Heating, Piping, Air Conditioning*, **29**, No. 8, 130 (Aug. 1957).
11. "Pressure Losses and Flow Characteristics of Multiple-Leaf Dampers," E. J. Brown and J. R. Fellows, *Heating, Piping, Air Conditioning*, **29**, No. 8, 119 (Aug. 1957).
12. "Control of Fan Systems in Large Buildings," J. E. Haines, *Air Conditioning, Heating, Ventilating*, **54**, No. 9, 70 (Sept. 1957).

Appendix

TABLE A.1. Areas and Circumferences of Circles

Diameter	Area	Circumference	Diameter	Area	Circumference	Diameter	Area	Circumference
⅛	0.0123	.3927	16	201.06	50.265	54	2290.2	169.646
¼	0.0491	.7854	½	213.82	51.836	55	2375.8	172.788
⅜	0.1104	1.1781	17	226.98	53.407	56	2463.0	175.929
½	0.1963	1.5708	½	240.52	54.978	57	2551.7	179.071
⅝	0.3067	1.9635	18	254.46	56.549	58	2642.0	182.212
¾	0.4417	2.3562	½	268.80	58.119	59	2733.9	135.354
⅞	0.6013	2.7489	19	283.52	59.690	60	2827.4	188.496
1	0.7854	3.1416	½	298.64	61.261	61	2922.4	191.637
⅛	0.9940	3.5343	20	314.16	62.832	62	3019.0	194.779
¼	1.227	3.9270	½	330.06	64.403	63	3117.2	197.920
⅜	1.484	4.3197	21	346.36	65.973	64	3216.9	201.062
½	1.767	4.7124	½	363.05	67.544	65	3318.3	204.204
⅝	2.073	5.1051	22	380.13	69.115	66	3421.2	207.345
¾	2.405	5.4978	½	397.60	70.686	67	3525.6	210.487
⅞	2.761	5.8905	23	415.47	72.257	68	3631.6	213.628
2	3.141	6.2832	½	433.73	73.827	69	3739.2	216.770
¼	3.976	7.0686	24	452.39	75.398	70	3848.4	219.911
½	4.908	7.8540	½	471.43	76.969	71	3959.2	223.053
¾	5.939	8.6394	25	490.87	78.540	72	4071.5	226.195
3	7.068	9.4248	26	530.93	81.681	73	4185.3	229.336
¼	8.295	10.210	27	572.55	84.823	74	4300.8	232.478
½	9.621	10.996	28	615.75	87.965	75	4417.8	235.619
¾	11.044	11.781	29	660.52	91.106	76	4536.4	238.761
4	12.566	12.566	30	706.86	94.248	77	4656.0	241.903
½	15.904	14.137	31	754.76	97.389	78	4778.3	245.044
5	19.635	15.708	32	804.24	100.531	79	4901.6	248.186
½	23.758	17.279	33	855.30	103.673	80	5026.5	251.327
6	28.274	18.850	34	907.92	106.814	81	5153.0	254.469
½	33.183	20.420	35	962.11	109.956	82	5281.0	257.611
7	38.484	21.991	36	1017.8	113.097	83	5410.6	260.752
½	44.178	23.562	37	1075.2	116.239	84	5541.7	263.894
8	50.265	25.133	38	1134.1	119.381	85	5674.5	267.035
½	56.745	26.704	39	1194.5	122.522	86	5808.8	270.177
9	63.617	28.274	40	1256.6	125.664	87	5944.6	273.319
½	70.882	29.845	41	1320.2	128.805	88	6082.1	276.460
10	78.54	31.416	42	1385.4	131.947	89	6221.1	279.602
½	86.59	32.987	43	1452.2	135.088	90	6361.7	282.743
11	95.03	34.558	44	1520.5	138.230	91	6503.8	285.885
½	103.86	36.128	45	1590.4	141.372	92	6647.6	289.027
12	113.09	37.699	46	1661.9	144.513	93	6792.9	292.168
½	122.71	39.270	47	1734.9	147.655	94	6939.7	295.310
13	132.73	40.841	48	1809.5	150.796	95	7088.2	298.451
½	143.13	42.412	49	1885.7	153.938	96	7238.2	301.593
14	153.93	43.982	50	1963.5	157.080	97	7389.8	304.734
½	165.13	45.553	51	2042.8	160.221	98	7542.9	307.876
15	176.71	47.124	52	2123.7	163.363	99	7697.7	311.018
½	188.69	48.695	53	2206.1	166.504	100	7854.0	314.160

TABLE A.2. Graphical Symbols for Drawings*

Piping

HEATING

High Pressure Steam	—//— —//— —//—
Medium Pressure Steam	—/— —/— —/—
Low Pressure Steam	——— ——— ———
High Pressure Return	——//—— ——//—— ——//—
Medium Pressure Return	——/—— ——/—— ——/—
Low Pressure Return	— — — — — — — — —
Boiler Blow Off	
Condensate or Vacuum Pump Discharge	—o— —o— —o—
Feedwater Pump Discharge	—oo— —oo— —oo—
Make-Up Water	
Air Relief Line	
Fuel Oil Flow	———FOF———
Fuel Oil Return	———FOR———
Fuel Oil Tank Vent	———FOV———
Compressed Air	———A———
Hot Water Heating Supply	
Hot Water Heating Return	

AIR CONDITIONING

Refrigerant Discharge	———RD———
Refrigerant Suction	— — —RS— — —
Condenser Water Flow	———C———
Condenser Water Return	— — —CR— — —
Circulating Chilled or Hot Water Flow	———CH———
Circulating Chilled or Hot Water Return	— — —CHR— — —
Make-Up Water	
Humidification Line	—·—H—·—
Drain	———D———
Brine Supply	———B———
Brine Return	— — —BR— — —

PLUMBING

Soil, Waste or Leader (Above Grade)	
Soil, Waste or Leader (Below Grade)	— — — — —
Vent	– – – – –
Cold Water	
Hot Water	———·———
Hot Water Return	———··———
Fire Line	—F——F—
Gas	—G——G—
Acid Waste	ACID
Drinking Water Flow	
Drinking Water Return	
Vacuum Cleaning	—V———V—
Compressed Air	———A———

SPRINKLERS

Main Supplies	———S———
Branch and Head	—o———o—
Drain	—S— — —S—

*Extracted from American Standard Graphical Symbols for Pipe, Fittings, Valves and Piping (*ASA Z32.2.3-1949*) and American Standard Graphical Symbols for Heating, Ventilating and Air-Conditioning (*ASA Z32.2.4-1949*) with permission of the publisher, The American Society of Mechanical Engineers, 29 West 39th St., New York 18, N. Y.

APPENDIX

TABLE A.2 (*Continued*)

Heating

Air Eliminator	⊸△⊸
Anchor	⊸✕ PA
Expansion Joint	⊸▭⊸
Hanger or Support	⊸✕ H
Heat Exchanger	⊸▭─▭⊸
Heat Transfer Surface, Plan (Indicate Type Such as Convector)	▭
Pump (Indicate Type Such as Vacuum)	M─▭
Strainer	⊸▷⊸
Tank (Designate Type)	REC
Thermometer	⊸▯⊸
Thermostat	Ⓣ

Traps

Boiler Return	
Blast Thermostatic	⊸⊗⊸
Float	⊸F⊸
Float and Thermostatic	
Thermostatic	⊸⊗⊸
Unit Heater (Centrifugal Fan), Plan	
Unit Heater (Propeller), Plan	
Unit Ventilator, Plan	

TABLE A.2 (Continued)

Valves

Check	
Diaphragm	
Gate	
Globe	
Lock and Shield	
Motor Operated	
Reducing Pressure	
Relief (Either Pressure or Vacuum)	
Vent Point	

Ventilating

Access Door	
Adjustable Blank Off	
Adjustable Plaque	
Automatic Dampers	
Canvas Connections	

TABLE A.2 (*Continued*)

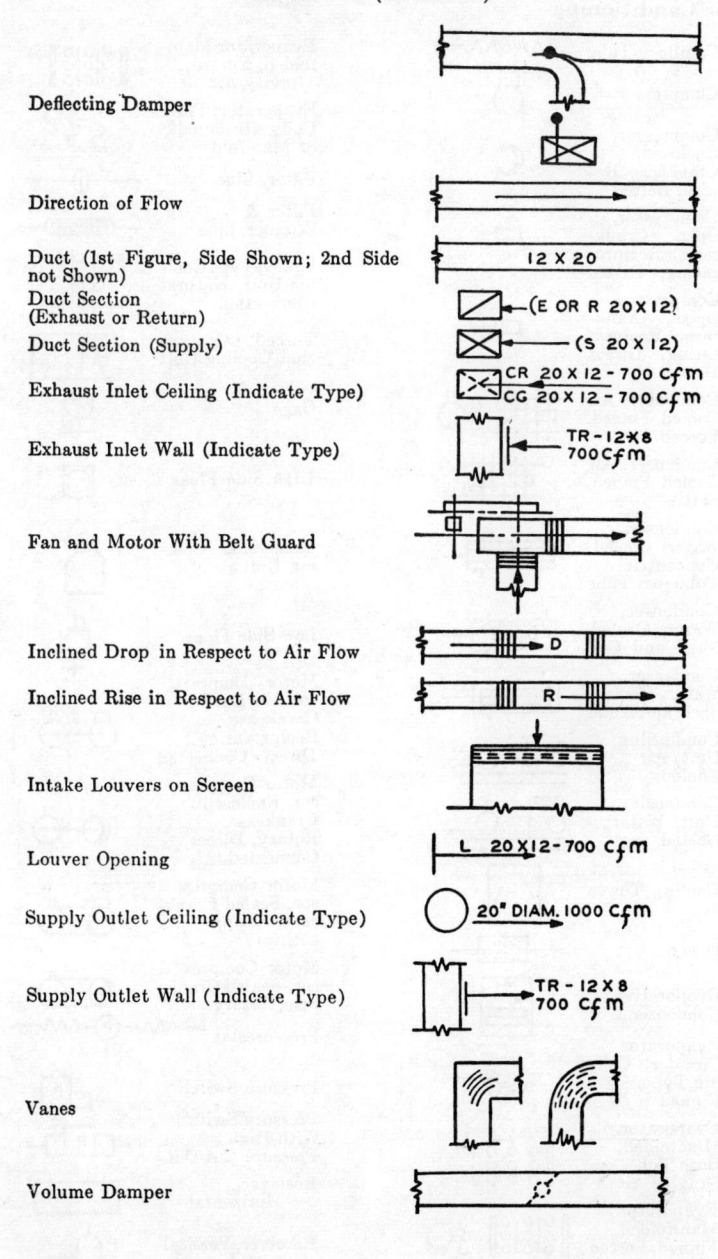

Deflecting Damper	
Direction of Flow	
Duct (1st Figure, Side Shown; 2nd Side not Shown)	12 X 20
Duct Section (Exhaust or Return)	(E OR R 20X12)
Duct Section (Supply)	(S 20 X 12)
Exhaust Inlet Ceiling (Indicate Type)	CR 20 X 12 - 700 Cfm CG 20X12 - 700 Cfm
Exhaust Inlet Wall (Indicate Type)	TR-12X8 700 Cfm
Fan and Motor With Belt Guard	
Inclined Drop in Respect to Air Flow	D
Inclined Rise in Respect to Air Flow	R
Intake Louvers on Screen	
Louver Opening	L 20X12 - 700 Cfm
Supply Outlet Ceiling (Indicate Type)	20" DIAM. 1000 Cfm
Supply Outlet Wall (Indicate Type)	TR - 12 X 8 700 Cfm
Vanes	
Volume Damper	

TABLE A.2 (Continued)

Air Conditioning

Capillary Tube	Evaporator Manifolded, Finned, Gravity Air
Compressor	Evaporator, Plate Coils, Headered or Manifold
Compressor, Enclosed, Crankcase, Rotary, Belted	Filter, Line
Compressor, Open Crankcase, Reciprocating, Belted	Filter & Strainer, Line
	Finned Type Cooling Unit, Natural Convection
Compressor, Open Crankcase, Reciprocating, Direct Drive	Forced Convection Cooling Unit
Condenser, Air Cooled, Finned, Forced Air	Gage
Condenser, Air Cooled, Finned, Static	High Side Float
Condenser, Water Cooled, Concentric Tube in a Tube	Immersion Cooling Unit
Condenser, Water Cooled, Shell and Coil	Low Side Float
Condenser, Water Cooled, Shell and Tube	Motor-Compressor, Enclosed Crankcase, Reciprocating, Direct Connected
Condensing Unit, Air Cooled	
Condensing Unit, Water Cooled	Motor-Compressor, Enclosed Crankcase, Rotary, Direct Connected
Cooling Tower	Motor-Compressor, Sealed Crankcase, Reciprocating
Dryer	Motor-Compressor, Sealed Crankcase, Rotary
Evaporative Condenser	Pressurestat
Evaporator, Circular, Ceiling Type, Finned	Pressure Switch
Evaporator, Manifolded, Bare Tube, Gravity Air	Pressure Switch With High Pressure Cut-Out
	Receiver, Horizontal
Evaporator Manifolded, Finned, Forced Air	Receiver, Vertical

APPENDIX

TABLE A.2 *(Continued)*

Symbol name		Symbol name	
Scale Trap		Evaporator Pressure Regulating, Throttling Type (Evaporator Side)	
Spray Pond		Hand Expansion	
Thermal Bulb		Magnetic Stop	
Thermostat (Remote Bulb)			

Valves

Automatic Expansion		Snap Action	
Compressor Suction Pressure Limiting, Throttling Type (Compressor Side)		Suction Vapor Regulating	
Constant Pressure, Suction		Thermo Suction	
Evaporator Pressure Regulating, Snap Action		Thermostatic Expansion	
Evaporator Pressure Regulating, Thermostatic Throttling Type		Water	
		Vibration Absorber, Line	

Index

Absolute humidity, 19
Absolute pressure, 2
Absolute roughness of pipes, 123
Absolute temperature, 2
Absorbents, 36
  use of, in air conditioning, 502–506
Absorption, dehumidification, 498
  refrigeration, 407, 465, 502
Activated carbon, 326
Actual washer, 338
  design of, for dehumidification, 476–480
Actuators, 530
Adiabatic saturation of air, 27
Adsorption, dehumidification, 36, 465, 468
Aerosols, 327
Air, analysis, 16
  behavior of jet, 315
  changes, infiltration, 74, 75
  circulation, 314
  composition, 16
  compressor, 518
  cooling, 36–39, 460–486

Air, dehumidification, 36–39, 460–486
  density, 8
  diffusion devices, 315–317
  distribution, 314
  ducts, construction of, 311
    design, 307
    insulation, 312
  enthalpy, 19, 24
  exfiltration, 73
  exhaust, 345
  friction in ducts, 297–299
  grilles, 316
  humidification, 30
  humidifier, 35, 338, 496
  infiltration, 73, 74
  ionization, 336
  jet behavior, 315
  measurement of velocity, 279
  mixtures with water vapor, 16
  moisture content, 16–20
  motion, 51
  odors, 324 326
  outlets, 318
  pressure in ducts, 308–310, 499, 503

545

INDEX

Air, pressure losses, 297–310
  properties (table), 22
  recirculation, 467, 497
  removal, 317
  requirements for ventilation, 325
  saturated, 17
  space, 59
  specific heat, 4
  standard air, 290, 326
  sterilization, 327
  still, 45
  stratification, 52
  valve for radiators, 214, 236
  velocities, in ducts, 280, 306
    in heaters, 352, 353
    in occupied zone, 45–47, 314
  vent, 212, 214, 215, 235, 236
  ventilation requirements, 325
  vitiation, 324
  volume and weight, 7, 18, 22
  washers, 337–339
  weight of, as heat carrier, 40
  weight of water vapor to saturate, 17, 22
Air cleaners, 329
  classification, 329
  cyclone type, 334, 335
  schemes employed, 329
  washer type, 337
Air conditioning, 1
  advantages, 365
  all-year residence air conditioners, 513
  applications of all-year, 490
  central systems, all-year, 495
    for summer, 467
  combinations of processes, 491
  dampers, use of, 532
  design calculations, 506
  remote room-cooling units, 461
  requirements, summer and winter, 1, 490
  requirements of machines, 490
  self-contained room-cooling units, 463, 493
  unit, all-year, 493
Air-cooled condenser, 381
Air eliminator, 215, 216
Air filters, types of, bag, 333
  change of direction, 334
  dry, 332

Air filters, electrostatic, 333, 335
  roll, 332
  viscous-impingement, 330
Air horsepower, 289
Air temperature, 2
  design inside, 81, 430
  design outside, 79, 428
  in unheated spaces, 71
Air washers, 337, 496
Air and water vapor mixture, 16, 20, 24
  tabular data, 22
Allowances, expansion, 7, 207
  for pipe fittings, 223
  heat and moisture from applicances, 447
  intermittent heating, 83
All-year air conditioning, 490
  applications, 490
  possible combinations of processes, 491–493
  requirements, 490, 491
Alternating receiver, 216, 217
Altitude angle of sun, 432, 434
Altitude gage, 183
Aluminum heating and cooling panels, 274
Ammonia, 371, 373
  absorption system, 407, 408
Analysis, of air, 16
  of flue gases, 106, 107
  of fuels, 95
Anchor, pipe, 207, 208
Anemometer, hot-wire, 280
  vane, 280
  wheel type, 279
Angle valve, 192
Angstrom units, 327
Anthracite coal, 97
Anticipating (or compensating) element, 521, 529, 530
API gravity, 102
Apparatus dew point, 473, 476
Appraisal of comfort conditions, 46, 52
"Approach," 414
Aquastat, 534
Areas, of circles, 537
  of pipe fittings, 205
Arithmetic mean temperature difference, 355

INDEX

Artificial drying, 29
Ash, 98
ASHAE comfort chart, 45
Aspect ratio, 302, 316
Atmospheric air cleaner, 329
Atmospheric gas burner, 105
Atmospheric pressure, 1
Atomization of oil, 100
Attenuator-diffuser, 502
Attics, fans for, 460
   temperature in, 71, 446
Automatic air filters, 331, 332
Automatic air valves, 214, 216
Automatic bypass, 470
Automatic control, of humidity, 527
   of pressure, 528
   of temperature, 517-526
Automatic stokers, 97
Awnings, effect of, 441
Axial-flow fans, 281-283, 291

Bacteria, 328
Barometer, 2
Barriers, vapor, 84, 85
Baseboard convectors, 165
Baseboard diffuser, 149
Baseboard heating units, 164
   performance, 166
   ratings, 166, 167
Baseboard registers, 137, 143, 147, 148
Basement plan, hot-water heating, 246, 249
Bimetal, 519
Bituminous coal, 97
Blast heaters, 350
   air-friction losses, 354
   performances, with steam, 353
   selection, 356
Bleed port, 519
Blow-through heater, 349
Body heat allowances, 48, 449
Boilers, 169
   capacity, 174, 179, 180
   cast-iron, 171, 172
   classifications, 170
   connections, 256
   efficiency, 180
   gas-fired, 104
   heating load, 180
   heating surface, 170

Boilers, output, 178
   ratings, 172, 174, 178, 179
   selection, 181
   square sectional cast-iron, 171
   steam, 169
   steel, 175, 176, 177
   tests, 180
   types, 169, 170
   warming-up allowances, 181
   water capacity, 238
   water column, 183
   water heating, 169
   water line, 212
   wet-bottom, 172
Boiling temperature of water, 8
Bourdon gage, 182
Bourdon tube, 182
Boyle's law, 7
Brackets, pipe, 207
   radiator, 156
Brake horsepower, fan, 290
   heat equivalent, 448
Breathing-line temperature, 81, 156
British thermal unit (Btu), 3
   definition, 4
Building insulation, 78
Building materials, heat transmission, 62
   thermal storage, 78
Building papers, 89
   classification, 89
Buildings, fuel requirements, 109
   intermittent heating, 83
Burners, adjustment, 106
   coal, 97, 98
   conversion, gas, 105
   gas, 104, 138, 141
   oil, 100
Butane, 103
Bypass factor, 29
Bypass method in summer cooling, 470

Calculation and estimation, of heat gains, 427, 452
   of heat losses, 82
Calculation of vapor pressures and relative humidities, 21
Calorific value of fuels by formula, 96
Capacity of steam return pipes, 228
Capillary cells, 339

INDEX

Capillary tube, 399
Carbon dioxide, 95, 372, 377
  concentration in air, 324
  theoretical percentage from burners, 106
Carbon monoxide, 325
Carburetted water gas, 103
Carrene 7, 372, 377
Carrier equation, 21
Cascade systems, 406
Casings, furnace, 137, 138
Cast-iron boilers, 171
Cast-iron pipe fittings, 190, 191, 193
Cast-iron radiators, 154
Ceilings, heat transmission of, 71
Central all-year air-conditioning plants, 495
  possible arrangements of apparatus, 497
Central-fan heating systems, 137, 139, 140, 273
Centrifugal compressors, 382, 383
Centrifugal fans, 281, 284–288
Centrifugal separation of dust, 334–336
Charging refrigeration systems, 400
Charles' law, 7
Check valves, 193
Chimneys, 108
Circles, areas of, 537
Circular equivalent, 299
Circulation of night air, 461
Circulator, hot-water, selection of, 255
Cleaners, air, 329
  cleaning efficiency, 328
Clearance, window sash, 75
Clearance volume of compressor, 379
  clearance volumetric efficiency, 380
Climatic conditions in United States, 80, 429
Closed expansion tank, 236
  sizing, 238
Closed-tank hot-water heating systems, 236, 238
Coal, analysis, 96, 98
  anthracite, 97
  bituminous, 97
  calorific value, 96
  sizes, 97, 98
Cocks, air, 236
Codes, boiler rating and testing, 172, 178

Coefficient of heat transmission, 57
  by calculation, 69, 70
  by test, 57
  combination, 71
  of air space, 64
  of surface, 61
Coefficient of performance (COP), 369, 373
Coils, cooling, 38
    selection, 480
  direct expansion, 394
  heating, 350–356
  preheater, 497
  reheater, 467–497
Coke, 98
Coke-oven gas, 103
Cold, effects on human body, 48, 52
Cold-air (or return-air) ducts, 137, 139, 140
Combination burners, 105
Combination coefficients, 71
Combination oil and gas burners, 105
Combination panel and convection systems, 275
Combustion, 95
  air required, 106
Comfort, conditions, appraisal of, 52
  effective temperature, 44
  factors affecting, 46
Comfort chart, 44, 45
  discussion, 46
  effective temperature, 44
  psychrometer, 20
Commercial pipe, 188
Common contact (thermostat), 521
Comparison of fuel costs, 111
Components of cooling load, 427
Composition, air, 16
  water, 8
Compression air cock, 236
Compression refrigeration system, 366
  charging, 400
  control, 402
  elimination of water from, 400
  leakage (tests for), 402
  pressure-enthalpy diagram, 368
  system diagram, 367
  temperature-entropy diagram, 369
Compressor displacement, 373
Compressors, air, 518

INDEX

Compressors, centrifugal, 382
   conditions of rating, 383, 384
   piston, 378
   reciprocating, 378
   refrigerant, 378
   rotary, 381
Condensate return pumps, 218, 220, 222
Condensers, refrigeration, 386
   air-cooled, 378, 381, 387
      ratings, 392
   effective surface (table), 391
   evaporative, 388
   prediction of capacity, 389
   selection, 389
Condensers, water-cooled, 383, 388
Conductance, air space, 59, 64
   building materials, 62
   insulation, 62
   surface or film, 61, 62, 64
Conduction of heat, 5
Conductivity, thermal, 5, 61
   of various substances, 5, 62
Conduit, pipe, 231, 232
Connections, for boilers, 212, 216, 218, 222, 257
   Hartford (safety) loop, 218
   radiators, 212, 215, 235, 236
Constants, gas law, 8
Control of compression refrigeration systems, 402
Controls, automatic, 517
   electric, 517, 521
   electronic, 526
   essential parts, 517–519
   hot-water heating systems, 256
   humidity, 527
   pneumatic, 517
   pressure, 528
   refrigeration systems, 402
   temperature, 520
   unit heaters, 362
   warm-air furnaces, 150
Convection, heat transfer, 6
Convectors, 159
   control, 161
   enclosures, 160
   heat emission, 162
   rating corrections, 162
   ratings, 161
   types, 159

Conversion burners, 99–101, 105
Cooling, evaporative, 29
   for comfort, 471
Cooling coils, 38
   direct expansion type, 394, 482
   selection, 480
   water type, 469
Cooling load, 427
   calculations, example of, 452–456
   components, 427
   design temperatures, inside and outside, 429–431
   estimation, 427, 456
   survey, 452
Cooling towers, 412–414
   performance, 414
Cooling water problem, 412
Copper tubing, 196
   dimensions, 196
   ferrule, 198
   fittings, 197
   flared joint, 198
   heat losses, 206
   soldered joint, 198
   types, 196
   weight, 196
Coverings, pipe, 200–205
Crack, door, 75
   window, 75
Crawl space, 140
   prevention of condensation, 84
Critical velocity in steam pipes, 226

Dalton's law, 17
Dampers, 496, 532, 533
   use of, 532
Decalorator, 411
Decibel, 312
Declination of sun, 435
Defrosting, 403
   methods, 404
Degree day, 111
Dehumidification of air, 36
   apparatus for, by absorption, 465
      by adsorption, 465, 468
   by cooling, 36
   methods, 36
   with coils, 38
Density, of air, 8
   of mercury, 2

INDEX

Density, of saturated water vapor, 9
   of water (1/specific volume), 12, 13
Density effect on fan performance, 296
Design procedure, for all-year air conditioning, 506
   for summer air conditioning, 473
     design calculations, 506
     design of actual washer, 476
Design temperatures, inside dry bulb, 81, 430
   outside dry bulb, 80, 428
   outside wet bulb, 429
Dew-point temperature, 24
   apparatus, 476
Diaphragm and motor valves, 201, 530, 531
Dichlorodifluoromethane, 373
Dichloromonofluoromethane, 377
Dichlorotetrafluoroethane (Freon-114), 377
Differential vacuum heating, 219
Diffuse radiation, 432
   table of values, 435
Diffusers, air, for double duct system, 502
   for high-velocity system, 500
   types, baseboard, 149
     ceiling, 317
     floor, 148
     wall, 137, 147
Direct acting (pneumatic controls), 522
Direct-acting thermostat, 520
Direct-expansion coils, 394
   performance, 482
Direct radiators, 154
Direct solar radiation, 432, 433
   table of values, 435
Dirt, 328
Discomfort from cold or hot surfaces, 52
Disk fans, 281
Dissemination of heat, 5
District heating, 229
Double duct system, 500–502
Downfeed heating systems, hot-water, 235
   steam, 211
Downward system of ventilation, 344
Draft, 108
   divertor (or hood), 104
   requirements, 108
Drawing symbols, 538

Drawings for plans, 229, 249
Draw-through blast heater, 304
Draw-through heater, 349
Dry-bulb temperature, 20
   optimum temperature and relative humidity for industrial processes, 491
Dry return, 215
Drying of materials, 29
Ducts, branch take-off losses, 304–306
   circular equivalent, 299
   design, 307
   friction, 297
   individual duct system, 303
   insulation, 312
   joining of edges, 311
   metal gage, 311
   recirculating, 137, 139, 140
   traverse, 118
   trunk line, 304
   velocities, 306
   velocity measurement, 279, 280
Dust, 327
   counts, 329
   electrical precipitation, 335

Edge loss factors, 72
Edr (equivalent direct radiation), 156
Effect of system resistance, fans, 293
Effective temperature, 44
   charts, 45, 47
Ejector system of ventilation, 344
Elbow equivalents, air ducts, 300, 302
   hot-water pipes, 243
   steam pipes, 225
   tees, 129, 130, 242
Elbows, cast-iron, 190
   copper, 197
   sheet metal, 302
   welding, 195
Electric heating panels, 271
   using conductive rubber, 271
   using glass, 272
Electric relay, 535
   electric-pneumatic, 535
Electrical precipitation of dust, 336, 337
Electronic devices, relays, 535
   thermostat, 526
Eliminator, air, 338

INDEX

Emissivity factors for heat from surfaces, 65, 66
   air spaces, 65
Energy, 3
Enthalpy, 5
   of air, any condition, 24
   of liquid, steam, 9
   of saturated air, 19
   of vaporization, steam, 9
   total, steam, 9
Enthalpy potential, 414
Entrance losses, 301
Equivalent, diameter, 124, 125
   direct radiation, 156
   elbows, for copper and iron fittings, 243
   length of pipe, 225, 300, 302
   temperature differentials, 442
      values for roofs, table, 443
      values for walls, table, 444
   wind velocities, 74
Estimating fuel consumption, 109
Estimating heat gains, 427
   from appliances, 447
   from people, 449
   through glass areas, 438-441
   through roofs, 443
   through walls, 444
Eupatheoscope, 52
Evaporation of water from human body, 48, 449
Evaporative cooling, 29, 471, 472
Evaporators (refrigerant), 394
   direct expansion coils, 394
   selection of, 395, 396
   specifications for, 397
   types, dry, 393
      flooded, 394
      spray, 383, 394
Exfiltration, 73, 74
Expansion, 6
   coefficients, 7
   formula, 6
   of oil, 7
   of water, 7
Expansion devices, 398
   automatic expansion valve, 400
   capillary tube, 399
   high side float valve, 400
   low side float valve, 400
   thermostatic valve, 398, 399

Expansion joints, 207, 208
Expansion tanks, 235, 236, 237-240
   advantages of closed tanks, 240
   sizing closed tanks, 238
   sizing open tanks, 237
Expansion valves, 398-400
Extended surface, 159, 160
   baseboard heater, 165
   coils, 350
   condensers, 390

Face area, of heater, 352
Face and bypass damper, 533
Fan, air horsepower, 289
   arrangements of drive, 288
   attic, 460, 461
   axial-flow, 281-283
   blades, 284-286
   brake horsepower, 290
   centrifugal, 281, 284-286
   characteristics, 291, 292
   classification, 282
   designation, 288
   effect of blade shapes, 286
   effects of system resistance, 293
   efficiency, 289
   laws, 296
   mechanical efficiency, 289
   multiblade, 284
   nomenclature, 287
   performance, 290-292
   power characteristic, 292
   propeller or disk, 281, 282
   ratings, 294
   selection, 297, 307
   static efficiency, 290
   steel plate, 284
   tables, 294, 295
   test code, 290
   total efficiency, 290
   tubeaxial, 281, 282
   types, 281
   vaneaxial, 281, 283
Fan-coil heating systems, 349
Feeder, boiler water, 184
Filters, air, 329-333
Fire-tube boilers, 170, 175, 176
   spiral ribbon spinners, 176
Fittings (pipe), areas of, 205
   cast-iron, 190

INDEX

Fittings (pipe), copper, 197
  copper, compression, 198
    flare, 198
    solder, 198
  equivalents, 243
  flanged, 192
  lift, 218
  screwed, 190
  transition, 301
  welding, 195
Flange union, 191
Flanges, pipe, 192
Flash gas, 368
Floor furnaces, 149
Flow of air, in pipes and ducts, measurement of, 279, 280
Flow-control valve, 258
Flow fittings, hot-water, 250, 251, 253
  resistance, 253
Flow of steam through pipes, 223
Flue gases, analysis, 107
  composition, 106–108
Flue size and chimney height, 174
Foot-pound, 3
Force, 2
  pound force, 2
Forced-circulation air coolers, 463, 466, 467
Forced convectors, 163
Freon-12, 372
  pressure-enthalpy diagram, 374
  properties (table), 376
  viscosity, of liquid, 375
    of vapor, 375
Freons, 371, 372
Fresh air requirements, 325
Friction charts, 125, 224, 241, 242, 298
Friction factor, 119, 120
Friction losses, in air ducts, 297
  in heater coils, 354
  in hot-water piping, 240
  in pipe and duct fittings, 126–132
  in steam piping, 223
  in tees, 129
Fuel, classification, 95
  costs, 109
  gaseous, 102
  heating value, 96
  oils, 99, 101
  solid, 97, 98

Fumes, 327
Furnace capacity, warm-air, 144, 146

Gages, 1
  altitude, 183
  compound (pressure and vacuum), 182
Gas analyzer, Orsat, 107
Gas boilers, 104
Gas burners, 103
  conversion, 105
Gas constants, 8
Gas equations, 7
Gas laws, 7
Gases, fuel, 102
  perfect, 7
Gas-fired heating units, 103
Gate valves, 192, 193
Glass heating panels, 272
Globe thermometer, 52, 53
Globe valves, 192, 193
Gradients, temperature, 60, 85
Graphical symbols for drawings, 538
  for air-conditioning apparatus, 542, 543
  for heating equipment, 539, 540
  for piping, 538
  for refrigerating equipment, 542, 543
  for ventilating equipment, 540, 541
Grate area, boiler, 174
Grilles, register, 147
Gun-type oil burner, 101

Hangers, pipe, 207
Hartford (safety) loop, 218
Heat, 3
  conductance, 59
  conduction, 5
  conductivity, 5
  convection, 6
  definition, 3
  dissemination, 5, 156–167
    from radiators, 156–158, 162
  emissivity factors, 66
  emitted by occupants, 48, 449
  exchangers, chilled water, 383
  from motors, 448
  latent, 3
  measurement, 57
  meter, Nicholls, 57

INDEX

Heat, of evaporation, 3
  of fusion, 3
  production and regulation in man, 48
  pump, 415
  sensible, 3
  solar, 431
  sources, 415
  specific, 4
  transfer, blast heaters, 353
  transmission, of building materials, 62
    of radiators, 156–159
  utilization and losses in combustion, 109
Heat gains, from appliances, 447
  from fan work, 451
  from occupants, 48–51, 448, 449
  from outside air, 449
  from solar radiation, 438–442
  of residence, 456
    through duct walls, 311, 451
    through interior walls, floors, ceilings, 446
    through multiple glass and blocks, 440, 441
    through outside walls and roofs, 442
Heat losses, by infiltration, 77
  calculation, 82
  from air ducts, 311
  from copper tubes, 205, 206
  from human body, 48, 49, 50
  from pipes, 201–204
Heat pump performance ratio (HPPR), 418, 419
Heat pumps, 415–419
  applications, 419
  performance, 418, 419
  possible arrangements, 416
    with ground coils, 417
Heat transfer, by conduction, 5, 66
  by convection, 6, 66, 67
  by radiation, 6, 65
  in refrigerant condensers, 390
Heat transmission, from basements, 72
  from crawl spaces, 72
  in heat exchangers, 356, 361, 390, 396
  through basement floors, 73
  through concrete slabs, 72
Heaters, air, 31, 32
  blast, calculations for, 356
  friction in, 355

Heaters, unit, 357
Heating, apparatus arrangements, 497 504
  effect of radiators, 158
  fan-coil systems, 349
  hot-water, 234
  intermittent, 83
  reversed-refrigeration, 415
  steam, 211
  vapor, 215
  warm-air systems, 135
Heating boilers, 169
  boiler-burner units, 104, 176, 177
  cast-iron sectional, 171, 172, 174
  classification, 170
  efficiency, 180, 181
  gas burners, 104, 105
  heating surface, 170
  oil burners, 100
  output, 180
  ratings, 172, 174, 178, 179
  safety rules, 182
  safety valve, 182
  steel, 175
  stokers, 97
  tests, 180
  water column, 183
  wet-bottom, 172
Heating coils, 349–356, 469
  selection, 356
Heating plans, 229, 249
Heating value, by calculation, 96
  higher, 96
  lower, 96
Hermetic compressors, 381
High contact (thermostat), 521
High side float valve, 367
High-temperature media other than water, 259
High-temperature water heating systems, 258
High-velocity systems, 499
Horizontal return-tube boilers, 175–176
Horsepower, air, 289
  brake, 290
  pump, 256
  requirements of refrigerant compressors, 384
    maximum, 386
    theoretical, 385

Hot-blast heaters, 349
  finned tube, 350
  performance, 353
Hot-water heating systems, applications, 234
  classification, 234
  closed, 236
  design, 244, 250
  forced-circulation, 236, 246, 249
  friction loss, 240–249, 252–254
  gravity-flow, 235
  manifold system, 255
  open, 235
  piping details, 236, 244
  pressure systems without expansion tanks, 240
  series loop system, 253
  sizing of pipes, 244
Hot-water radiators, 156
Human comfort, 44
Humidification, 30–35
  by preheating, 31
  moisture required, 144
  necessity for, 142
  with heated water, 34
  with steam, 34–36
Humidifiers, 35, 138, 496
Humidifying efficiency, 28
  of commercial washers, 28
Humidistat, 527
  wiring diagram, 528
Humidity, absolute, 19
  control, 527
  effect on comfort, 45
  ratio, 19
  relative, 21
  specific, 19
Hydrogen, 95
  in fuels, 95
Hygrometers, 21

I = B = R ratings of cast-iron boilers, 172–174
Ice storage air conditioning, 486
Ideal washer, 32, 473–476
Incident, angle, 432
  radiation, 433
Induction units, 498, 499
Industrial air and gas cleaners, 329
Infiltration of air, 73

Infiltration of air, based on, air changes, 75
  cracks, 77
  tabular data, 76
Injection-type air grille, 316
Inside air temperatures, 44–47, 81, 430, 431
  at different levels, 81
Insulating materials, for buildings, 78
  heat transmission, 62
  types, 78
  weights, 62
Insulation, pipe, 200
Intermediate media, 415
Intermittent heating, 83
Invisible condensation, 84
Ionization of air, 336

Jet behavior, air, 315
Joints, expansion, 207

Kata thermometer, 280
Kinematic viscosity, 122
  of air, 121
  of water, 122

Laminar flow, 122
Latent heat, 3
  load, 428, 447, 449
  of water vapor, 9–13
Limit controls, 534
Liquified petroleum gases, 103
Lithium chloride, 504
  equilibria with water (data), 505
  equipment for employing in air conditioning, 504
  properties, 504
Load, boiler heating, 180
  heating, 82
  seasonal and monthly, 109
Location of thermostats, 530
Logarithmic mean temperature difference, 355
Losses, entrance, 301
Low contact (thermostat), 521
Low-temperature refrigeration, 405
  compression ratios, 405
Low-water alarm, 184
  and cutoff, 184

INDEX

Machinery as heat source, 448
Mains, dry-return, 215
  hot-water heating, 235, 236, 246, 249
  steam, 212
  wet-return, 212
Maintaining circuit (thermostat), 522
Manifold hot-water heating system, 255
Manufactured gases, 103
Mass, 3
  pound mass, 3
Master thermostat, 526
Maximum outdoor-air damper, 501, 532
Mean effective temperature difference (METD), 395
Mechanical refrigeration, 365
  applications, 365
  fundamentals of compression system, 366
  types of systems, 366
Mechanical ventilation, 340, 341
Mechanical warm-air furnace, 136, 137, 138, 141
Mercoid (mercury) switch, 528
Methyl chloride, 372, 377
Micron, 328
Milinch, 245
  conversion to feet, 248
Minimum outdoor-air damper, 501, 532
Mists, 328
Mixed gases, 103
Mixing calculations, air, 39, 40
Mixing unit, 501, 502
  selection data for, 503
Mixing valve, 258
Mixture, air and water vapor, 16
Modulating controls, 522
Moisture, 8
  content of air, 50
  eliminator, 338
  gain through walls, 450
  loss by human body, 48, 51
  permeability of various materials, 86
  weight from appliances, 447
Moisture emitted, by appliances, 447
  by humans, 51, 449
Moisture transmission, 84, 450
Mole, 7
Monochlorodifluoromethane (Freon-22), 376

Monochlorotrifluoromethane (Freon-13), 377
Motion of air, 51
Motor-balancing potentiometer, 523
Multiblade fans, 284
Multistage compression, 405

Natural gas, 102
Natural ventilation, 340
  calculations, 341
  roof ventilators, 341
Net refrigerating effect, 371
Nicholls heat meter, 57
Night air cooling, 461
Nipples, pipe, 190
  radiator, 155
Nitrogen, 8, 16
Noise, control, 314
  measurement, 312
Noise levels, 312–314
Noncondensable gases in refrigerati systems, 401
  removal, 401
Nozzles, 478
Nut union, 191

Odors, 326
Oils, 99
  atomization, 100
  burning point, 102
  classification, 99
  flash point, 102
  furnace requirements, 100
  pour point, 102
  specific gravity, 102
  specifications, 101
  viscosity, 101
Oil burners, 100
Oil gas, 103
Oil-logged, 372
One-pipe heating systems, downfeed, 211, 235
  hot-water, 235, 236, 249, 250–255
  steam, 211
  upfeed, 212
Open-tank, hot-water systems, 235
  sizing of tanks, 237
Orifice, hot-water heating systems, 244
  sizes, 245
  in pneumatic control system, 519

556 INDEX

Orifice, in steam radiators and convectors, 229
Orsat test apparatus, 107
Outside air temperatures, 79, 428
  daily variation, 430
  range, 430
  table, 80, 429
Over-all coefficient, 57, 59
  combination coefficient, 71
  in refrigerant condensers, 390
Overhead downfeed heating systems,
    hot-water, 235
  steam, 211
Oxygen, 8, 16
Ozone, 327

Packed tower, 339, 340
Panel cooling, 274
Panel heating, 262
  advantages, 264
  coil design, 267
  control, 269
  copper vs. steel pipes, 266
  definition, 262
  design of, hot-water panels, 270
  disadvantages, 265
  electricity as heating medium, 271
  floor vs. wall vs. ceiling panels, 265
  fundamentals, 262
  heat outputs of panels, 266
  pipe coils under wood floors, 268
  warm-air panels, 272–274
Partial pressures, Dalton's law, 17
Per cent saturation of air, 21
Perfect gases, 7
  laws of, 7
Performance factor (washers), 477
Performance tests of boilers, 172–174, 180
Permeability of materials to water
    vapor, 86, 87
  calculation, 88
  perm, 88
  permeance, 87
Petroleum oils, 99
  composition, 99
Piezometer ring, 115, 116
Pilot flame, 104, 141
  failure, 534
Pipe (piping), 188
  bare, 203

Pipe (piping), conduits, 231, 232
  copper, 196
  couplings, 190
  details, 230
  dimensions, 189, 196
  drawings, 538
  expansion, 207, 208
  fittings, 190–198
  hangers, and supports, 207
  heat losses from, 201–206
  hot-water systems, 235, 236, 246, 249
  installation, 205
  insulation, 200, 202
  materials, 188
  nominal diameters, 188
  pitch of, 212, 216
  plastic, 208
  sizes, 189, 196
    for hot-water systems, 243–254
    for refrigeration systems, 403
    for steam systems, 223–228
  steam, 230
  steel, 188, 189
  table, 189, 196
  threads, 190
  tunnels, 231
  unions, 191
  vapor heating, 215
  weights, 189, 196
  welds, 194
  wrought iron, 188, 189
Pipe traverse with Pitot tube, 118
Pipeless warm-air furnace, 149
Pitot tubes, 116, 279
Pitot-static tube, 116, 117
Plans, heating, 229
Plate fan, steel, 284
Plenum chamber, 139, 140
Pneumatic controller, 519
Pneumatic-electric, 535
Pneumatic relay, 535
Pollens, 328
Pollution of air, 324, 327, 328
Portable steel boilers, 175, 176, 177
Positive-action thermostat, 520
Positive-return vapor systems, 216
Pot-type oil burner, 99
Pour point of oil, 102

INDEX

Power, 3
  consumption of fans, 289, 291, 294
  for pumping in hot-water systems, 256
Pressure, barometric, 2
  controls, 528
  drop in mains, 223–228, 249, 254, 307
  fan, 291–293
  gage, 1
Pressure, heating boilers, 170
  loss due to area changes, 301
    in hot-water systems, 240–255
    in noncircular ducts, 124
    in steam piping, 223
  loss in air ducts, 297–307
  loss in general, 119
  reducing valve, 199, 201
  static, 114
  total, 114
  velocity, 115
  water vapor, 21
Pressure and altitude gage, 183
Pressure systems of hot-water heating, 234
  no tank, 240
  tank in basement, 236
Pressure-volume relationship in a cylinder, 379
Primary controls, 530
Procedure for designing, actual washer, 476
  all-year air conditioning, 506
  summer air conditioning, 473
Producer gas, 103
Products of combustion, 95, 107
Propane, 103
Propeller fans, 281
Propylene glycol, 327
Proximate analyses of coals, 98
Psychrometers, 20
Psychrometric chart, 25
  ASHAE chart, 28
  use of, 26
Psychrometric data, 20, 22
  calculated data vs. standard chart, 26
Pumps, circulating, 249, 255, 257
  condensation, 213, 222
  vacuum, 218, 220
Purging, 401
Push nipples, 155

Quality of steam, 9
Quantity of air, for ventilation, 325
  heat carrier, 40

Radiant baseboards, 164, 165, 272
Radiant heating, 262
Radiation, 6
  from body, 50, 52, 263
Radiation, of heat, 6
  solar, 431
  Stefan-Boltzmann law, 6
Radiators, air valves, 214
  condensation rates, 158
  connections, sizes of, 155
  correction of radiation factors, 157
  heat transmission, 156, 157
  heating effect, 158
  hot-water type, 156
  leg sections, 155
  loop sections, 155
  nipples, 155
  number of sections required, 157
  performance, 156
  ratings, 155–157
  selection, 157
  standard dimensions, 155
  steam type, 156
  tappings, 155
  testing, 156
  traps, 221
  tubular, 154
  valves, 221
  volume, 238
Range, 414
Ratings, of heating boilers, 172, 178
  of refrigeration machines, 383
  of warm-air furnaces, 144
Receivers, alternating, 216, 217
Reciprocating compressors, 378
Recirculating duct, 137, 139, 140
Recirculating register, 137, 147
Recirculation of air, 137, 140, 467, 497, 501
Rectangular ducts, circular equivalent of, 299
Reducing fittings, 191, 195, 197
Refrigerant compressors, 378
  reciprocating, 378–380
  types, 378

INDEX

Refrigerants, 370
  comparative characteristics, 372, 373
  desired properties, 370
Refrigeration, 365
  absorption, 407
  compression, 366
  condensers, 386–394
  media, 415
  reversed cycle, 415
Refrigeration, steam jet, 411
  water vapor, 409
Regain, static pressure, 116, 309
Registers, 145, 147–149
  recirculating, 137, 139, 147
  temperatures, 135
  warm-air, 137, 139, 140, 147, 148
Regulators, hot-water tank, 520
Reheating of air, 31, 32, 483
Relative humidity, 21
  control, 30, 505, 527
  effect on bacteria, 327
Relays, 535
Remote room-cooling units, 461
Resistance, of building materials, 62
  of pipe fittings, 126–132
  to flow in hot-water systems, 240
  total for fans, 287
Respiration, effect of, 324
Return, dry, 215, 216
  piping, 212, 215, 216, 218, 228
  traps, 217
  tubular boiler, 175, 176
  wet, 212
Reverse acting (pneumatic controls), 522
Reversed Carnot cycle, 370
Reynolds number, 121
Roof ventilators, 341
Room conditions, 46, 81, 430, 485
Rotary compressors, 381
Roto-Clone air cleaner, 336
Rubber heating panels, 271

Safety controls, 534
Safety valve, 182
Saturated air, 17, 19, 27
Saturated steam or vapor, 9
  properties, 10–13
Saturation of air, adiabatic, 27
  per cent, 21

SBI ratings of steel boilers, 178
Screw nipple, 155, 190
Screwed fittings, 190
Scrubber plates, 338
Seasonal heating efficiency, 110
Sectional cast-iron boilers, 171
Selection of packaged all-year air conditioners, 512
Selective temperature steam heating, 222
Self-contained, all-year, room air conditioners, 493
  advantages, 495
  rating data (table), 494
Self-contained room cooling units, 463
  water-cooled type, 465
  window type, 464
Semihermetic compressors, 381
Sensible heat, 3, 427
Sensible-heat factor (SHF), 37
Sensitivity of controls, 529
Separation factor, 334
Separator, steam, 199
Series-loop, one-pipe hot-water heating system, 253
Series-90 control circuit, 523
Setting of boilers, 170
Shading of glass areas, 441
Sheet metal, gage for ducts, 311
Shock losses, 299
Short cycling, 526, 530
Silica gel, 36, 468
Sink (heat), 415
Sky radiation, 432, 435
Sling psychrometer, 20
Smoke, 328
Smoke pipe design, 108
Snow melting, 271
Soft coal, 97
Sol-air temperature, 442
Solar angles, 432
  altitude angle, table of values, 434
  azimuth angle, table of values, 434
  definitions, 432
Solar radiation, 431–438
  angles involved in, 432
  lag in transmission through walls and roofs, 442
  through bare and shaded windows, 438–441

INDEX

Sorbents, 36
Sound, 312
  control, 314
  intensity table, 313
  typical levels, 313
Source (heat), 415
Spaces, unheated, 71
Specific gravity of oil, 102
Specific heat, 4
  mean, of air, 4
  of superheated steam, 13
  of refrigerants, 373
  of various substances, 4
Specific humidity, 19
Specific volume, air, 8
  refrigerants, 373
  saturated steam, 9, 10–13
Spinners, 176
Split system, 341
Spontaneous combustion, 98
Spray chamber, air washer, 338, 496
Spray nozzles, 338, 478
Spray water, cooling, 473
  heating, 34
Stack switch, 534
Stacks, warm-air, 137
Standard codes, boilers, 172, 178
  warm-air heating, 152
Standard cycle theoretical horsepower, 372
Standard pipe, 189, 196
  fittings, 190, 197
Static pressure, 114, 116
  control in ducts, 317
  regain, 116, 309, 310
Steam, 9
  conduits, 231
  density, 9
  enthalpy, 9
  flow through pipes, 223
  generation, 169
  properties of saturated, 10
  quality, 9
  specific volume, 9
  superheated, 13
  tunnels, 231
  viscosity, 123
Steam consumption of buildings, 110
Steam flow, chart, 224
  equation, 121

Steam heater for water, 185
Steam heating, 211
  central, 211–229
  district, 229
Steam-heating systems, 211
  air-vent, 212
  capacity of return piping, 228
  classification, 211
  condensate return pump, 213
  design, 225, 227
  equivalent length of pipe run, 225
  gravity systems, 211, 212, 215
    one-pipe, 211–214
    two-pipe, 214–220
  mechanical, 213
  orifice control, 229
  pipe sizes, 224
  piping details, 230
  pressure drops, 224
  selective temperature, 222
  subatmospheric, 219
  types, 211, 222
  vacuum, 217
  vapor, 215
    one-pipe, 211
    two-pipe, 214
  zone control, 222
Steam-jet refrigeration, 411
Steam main relay, 213
Steam separator, 199
Steam tables, 10–13
Steam traps, 199, 200
Steel Boiler Institute (SBI) ratings of steel boilers, 178
Steel boilers, 175
  data, 179
Steel-sash window infiltration, 76
Stefan-Boltzmann law, 6
Sterilization of air, 327
Still air, 45
Stokers, underfeed, 97
Stop valves, 192
Storage-type water heater, 184, 185, 257
Storm sash, 75, 77
Stratification of air, 52
Streamline flow, 122
Subatmospheric steam systems, 219
Submaster thermostat, 526
Submerged water heater, 184
Sulfur, 95

Sulfur dioxide, 95, 372, 377
Summer, air conditioning, 29, 36
   air-conditioning apparatus, 460
      central systems, 464
         remote units, 461
         self-contained units, 463
   air temperatures, 46, 430
Summer, comfort zone, 45–47
Sun effect on roofs, 442
Superheated steam, 13
   specific heat, 13–14
   total enthalpy, 13
Superheated vapor, 368, 369
Surface, boiler heating, 170
   cooling, 480
   extended, 163, 165, 394
   radiator heating, 155, 156
Surface conductance, 58, 61
   for building materials, 58, 61
Surfaces, cold and hot, effects of, 52
Switches, mercury, 528
Systems of air distribution, 343, 344, 467, 497

Tank heaters, 184, 185
Tank-in-basement pressure system, 236
Tank regulator, 185
Tanks, expansion, 237–240
Tappings, radiator, 155
Tees, cast-iron, 129, 190, 242, 243
   copper, 197, 242, 243
Temperature, 2
   absolute, 2
   apparatus dew-point, 473
   base for degree day, 111
   body, 48, 263
   control, 517
   dew-point, 24
   dry-bulb, 20
   effective, 44
   inside surfaces, 60, 70
   mean radiant, 263
   of occupied spaces, 81, 431
   records of cities, 80, 429
   register, 135
   surface of man, 263
   systems for control, 517
   TAC design, 80
   unheated spaces, 71
   wet-bulb, 20

Temperature control, automatic, 517
   bypass method, 470
Temperature differential, hot-water heating, 243
Temperature gradients, in rooms, 158
   of frame walls, 60
Temperature range of water in hot-water heating, 243
Temperature regulation, automatic, 517
Temperatures, air, of unheated spaces, 71
   final in blast heaters, 353
   inside air, 81, 431
   mean, in blast heaters, 355
   outside air, 79, 428
   proper level of air, 81
Terms, 1
Tests, of boilers, 180
   of fans, 290
   of radiators, 156
Therm, 111
Thermal conductivities, 5, 61
Thermal storage of building materials, 78
Thermocouple, 2, 534
Thermo-integrator, 53
Thermometers, globe, 53
   hot-water, 183
   kata, 280
   mercurial, 20, 53, 183
Thermostatic control of cooling and heating, 526
Thermostats, 517, 518, 520
   compound (two elements), 521
   direct-acting, 520
   electrical, 521, 523
   gradual movement, 522, 523
   immersion, 520
   indirect-acting, 520
   insertion, 525
   location, 530
   master, 526
   motor control, 523
   pneumatic, 519
   positive action, 520
   submaster, 526
   surface, 525
   two-position, 520
   two-wire, 521
Throttling process, 368

INDEX

Throw of air, 315
Time relationship, heat flow, 442
Ton of refrigeration, 370
Total pressure, 114, 116
  developed by fan, 287
Tower, cooling, 412
Transition fittings, 301
Transmission coefficients, 59
Transmitted direct and diffuse solar radiation, 439
Traps, 199
  bucket, 200
  float, 200
  float and thermostatic, 200
  return, 217
  thermostatic, 222
Traverse, 118, 281
Trichlorotrifluoroethane (Freon-113), 377
Triethylene glycol, 327
Trunk line ducts, 303
Tube, Bourdon, 182
Tubing, copper, 196
  dimensions, 196
  steel, 188
Tubular radiators, 154
Tunnels (for steam pipes), 231
Turning vanes, in duct elbows, 300
Two-pipe heating systems, overhead downfeed hot-water, 235
  steam, 214
  upfeed hot-water, 235, 246
  upfeed steam, 215, 216, 218

U values (for heat loss calculations), calculation, 58, 68–71
  numerical examples, 69
Ultimate analysis, 96
Underground tunnel, 231
Underwriter's loop, 213
Unheated spaces, air temperatures in, 71
Unidirectional damper motor, 521, 530
Unions, 191
Unit air conditioners, 461, 463, 493
Unit coolers, 461, 463
Unit heaters, 357
  air discharge from, 360
  classification, 357
  control, 362
  location, 360

Unit heaters, rating, 361
  types, 357
    direct-fired, 359
    floor, 359, 360
    gas-fired, 359
    revolving-outlet, 358
    suspended, 357
    vertical-discharge, 358
  typical performance data, 360
Unit ventilators, 343
Upfeed heating systems, hot-water, 235, 236, 246, 249
  steam, 212, 215, 218
Upward system of ventilation, 343

Vacuum heating systems, 217
  differential, 219
  return-line, 219
Vacuum pumps, 220
Vacuum steam systems, 217
Valves, 193
  air, 214
  angle, 192
  check, 193
  diaphragm, 201, 531
  equivalent length, 131, 225, 243
  flanged, 192
  flow-control, 249, 257, 258
  gate, 192
  globe, 192
  graduated control, 214
  materials, 193
  motor-operated, 530
  pneumatically operated, 258, 531
  pressure-reducing, 199, 201
  radiator, 214, 221
  safety, 182
  water relief, 240, 257
Vanes, turning, 300
Vapor barrier, 84
Vapor heating systems, one-pipe, 214
  positive return, 217
  two-pipe, 215
Vapor (water) transmission, 84–89, 450
Velocity, air, 114, 279
  allowable, in ducts, 306
    in hot-water heating systems, 244
    in pipes handling refrigerants, 403
    in pipes of one-pipe steam systems, 226

INDEX

Velocity, calculation from velocity pressure, 118, 281
  measurement of air, 279
  method of duct design, 306
  of flow in hot-water systems, 241, 242
  pressure, 114, 115, 116, 281
  wind, 80, 429
Velocity diagrams, fans, 286
Velometer, 280
Ventilating, fans, 281, 461
  systems, 340–345
Ventilation, air recirculation, 137, 139, 467
  central systems, 341
  downward flow systems, 344
  necessity for, 324
  upward flow systems, 343
Ventilators, roof, 341
  unit, 343
Vents, air, 341, 344
Viscosity, absolute, 122
  conversion of units, 122
  of steam, 123
Viscous air filters, 330
Viscous flow, 122
Visible condensation, 83
Volume, of air and saturated vapor, 18
  of direct radiators, 238
  of expansion tanks, 237, 238
  of saturated steam, 10–13
  of water, 12, 13
Volumetric efficiency of compressor, 380
Volumetric flue-gas analysis, 107

Wall, air leakage through, 74
  solar radiation on, 435, 442
  time lag of heat transmission, 442
Wall-solar azimuth angle, 432, 434
Wall stacks, 137
Warm-air furnace systems, 135
  advantages, 135
  applications and classifications, 135
  control, 150
  design, 149
  gravity flow vs. forced circulation, 136
  installation, 137
  perimeter-loop system, 137
  perimeter-radial system, 139
  types, attic stowaway, 142

Warm-air furnace systems, type, blend-air, 142
  conventional forced-air, 136
  crawl-space plenum, 141
Warm-air furnaces, 144
  casings, 137, 138
  floor furnaces, 149
  gas-fired unit, 138
  oil-fired unit, 100
  selection, 145
  sizes and ratings, 144, 146
  wall furnaces, 150, 151
Warm-air registers, 147–149
Washers, air, 337
  cellular type, 339
Water, 8
  boiling point, 8
  composition, 8
  density, 9
  evaporation, 9
  friction losses in pipes, 119, 240
  properties, 10–13
  required for hot-water heating, 238
  specific volume, 10–13
  specific weight (1/specific volume), 10–13
  thermal properties, 10–13
  vapor refrigeration, 409
Water circulator, 255
Water cooling by spray, 388, 411, 412, 413
Water feeder, boiler, 184
Water gas, 103
Water heaters, 172, 184, 185, 257
Water line, boiler, 212
Water tube boiler, 177
Water vapor, 8, 95
  latent heat, 9, 10–13
  mixture with air, 16
    calculation of pressure, 21
  pressure in walls, 85
  transmission, 85
Water vapor absorption system, 409
Weather Bureau records, 79, 428
  air temperatures, 79, 428
  wind, 79, 428
Weather stripping, 75
Weight, 3
  of air (density), 8, 18
  of insulation, 62

INDEX

Weight, of steam (density = 1/specific volume), 10–13
  of water (density = 1/specific volume), 10–13
  of water vapor to saturate air, 17, 22, 23
Wet-bulb temperature, 20
Wet return, 212
Wetted cyclones, 334
Wheatstone bridge, 523
Wind, directions and velocities, 80, 429
  effect on heat transmission, 64
  equivalent velocities, 74
  heating season, 80
  velocity and air infiltration, 76
Window, air leakage through, 76
  clearance of sash, 76
  comparison of shades, 441
  cracks, 77
  solar radiation, 438

Window, storm sash, 75
  ventilation, 340
  weather stripping, 75
Window air conditioners, 464
Winter, air-conditioning apparatus, 135, 490
  air humidification, 27, 30, 35, 138, 493, 496, 504
  air temperatures, 80, 81
    comfort zone, 45, 47
    relative humidity, 45, 46, 83
    wind directions and velocities, 74, 80
Wood frame wall construction, 60
Wood stave, pipe conduit, 123, 232
Work, 3

Zero, absolute temperature, 2
Zoning in summer air-conditioning systems, 470

MODERN BOOKSTALL
POONA 1: